Advances in Mechanics and Mathematics

Advances in Continuum Mechanics

Volume 43

More information about this subseries at http://www.springer.com/series/16097

Reuven Segev • Marcelo Epstein
Editors

Geometric Continuum Mechanics

Editors
Reuven Segev
Department of Mechanical Engineering
Ben-Gurion University of the Negev
Beer-Sheva, Israel

Marcelo Epstein
Department of Mechanical and Manufacturing Engineering
University of Calgary
Calgary, AB, Canada

ISSN 1571-8689 ISSN 1876-9896 (electronic)
Advances in Mechanics and Mathematics
ISSN 2524-4639 ISSN 2524-4647 (electronic)
Advances in Continuum Mechanics
ISBN 978-3-030-42682-8 ISBN 978-3-030-42683-5 (eBook)
https://doi.org/10.1007/978-3-030-42683-5

Mathematics Subject Classification (2020): 74A05, 74A10, 74A99, 58D15, 22A22, 58A25, 58A30

This book is published under the imprint Birkhäuser, www.birkhauser-science.com by the registered company Springer Nature Switzerland AG.
The registered company address is: Gewerbestrasse 11, 6330 Cham, Switzerland

Geometric Continuum Mechanics

This volume contains a compilation of extended articles on the applications of various topics in modern differential geometry to the foundations of continuum mechanics.

The application of differential geometry to the mechanics of systems having a finite number of degrees of freedom, as appeared initially in the works of Abraham, Marsden, Souriau, Arnold, Smale, and others, and led to the vast literature that followed, needs no introduction. Quoting from A.D. Lewis,[1] one learns that:

> What has resulted from the merging of mechanics and differential geometry has been a deep understanding of the structures that contribute to the mathematical foundations of mechanics. …… Moreover, mechanics has given life and breadth to some areas of differential geometry, just as one might hope with the interplay of mathematics and its applications.

Continuum mechanics, as a field theory, poses an even greater challenge. To quote C. Truesdell:[2]

> …the mechanics of deformable bodies, …is inherently not only subtler, more beautiful, and grander but also far closer to nature than is the rather arid special case called "analytical mechanics"….

Similarly, W. Noll writes:[3]

> It is true that the mechanics of systems of a finite number of mass points has been on a sufficiently rigorous basis since Newton. Many textbooks on theoretical mechanics dismiss

[1]Andrew D. Lewis, The bountiful intersection of differential geometry, mechanics, and control theory, *Annual Review of Control, Robotics, and Autonomous Systems*, 1, 1.135–1.158, 2018, https://doi.org/10.1146/annurev-control-060117-105033.

[2]Clifford Truesdell, *A First Course in Rational Continuum Mechanics*, Vol. 1, Academic Press, 1991.

[3]Walter Noll, The foundations of classical mechanics in the light of recent advances in continuum mechanics, in *The Axiomatic Method, with Special Reference to Geometry and Physics*, Henkin, L.; Suppes, P. & Tarski, A. (Eds.), North-Holland, 1959, 266–281.

continuous bodies with the remark that they can be regarded as the limiting case of a particle system with an increasing number of particles. They cannot.

Thus, taking into account this inherent subtlety of continuum mechanics, it is not surprising that the applications of modern differential geometry to the foundations of continuum mechanics did not progress on par with that of the mechanics of systems having a finite number of degrees of freedom.

It is hoped that the collection of articles compiled in this volume will serve as an introduction to the current work on the interaction of differential geometry and continuum mechanics. The continuum mechanical aspects of the articles are mainly: (1) kinematics of continuous bodies, and force and stress theories, (2) the geometry of dislocations as described from various points of view including Noll's uniformity and homogeneity theory. With the assumption that the reader is familiar with the basics of modern differential geometry, the articles contain applications of global analysis, algebraic geometry, geometric measure theory, and the theory of groupoids and algebroids. A chapter containing a comprehensive introduction to manifolds of mappings, which also includes some new results, is also contained in the volume for the benefit of readers interested in the global geometric approach to continuum mechanics.

Special attempts have been made by the authors to make their presentations self-contained and accessible to a wide readership.

The subjects discussed in the book are situated in the intersection between mathematical analysis and geometry, physics, and engineering. We believe that readers from these disciplines should find the book to be of relevance and interest. Hopefully, this volume will stimulate young scientists and make current literature on the subject accessible to them.

It is a common belief among the authors of the various articles that the geometric formulation of continuum mechanics is an elegant and fascinating subject, which leads to better understanding of both disciplines.

The original version of this book was revised: Volume number has been changed from 42 to 43. The correction to this book is available at https://doi.org/10.1007/978-3-030-42683-5_10

Contents

Part I
Kinematics, Forces and Stress Theory

Manifolds of Mappings for Continuum Mechanics

Peter W. Michor

Contents

Abstract After an introduction to convenient calculus in infinite dimensions, the foundational material for manifolds of mappings is presented. The central character is the smooth convenient manifold $C^\infty(M, N)$ of all smooth mappings from a finite dimensional Whitney manifold germ M into a smooth manifold N. A Whitney manifold germ is a smooth (in the interior) manifold with a very general boundary, but still admitting a continuous Whitney extension operator. This notion is developed here for the needs of geometric continuum mechanics.

2010 Mathematics Subject Classification 58B20, 58D15, 35Q31

P. W. Michor (✉)
Fakultät für Mathematik, Universität Wien, Wien, Austria
e-mail: peter.michor@univie.ac.at.

R. Segev, M. Epstein (eds.), *Geometric Continuum Mechanics*, Advances in Mechanics and Mathematics 43, https://doi.org/10.1007/978-3-030-42683-5_1

1 Introduction

At the birthplace of the notion of manifolds, in the Habilitationsschrift [93, end of section I], Riemann mentioned infinite dimensional manifolds explicitly. The translation into English in [94] reads as follows:

> There are manifoldnesses in which the determination of position requires not a finite number, but either an endless series or a continuous manifoldness of determinations of quantity. Such manifoldnesses are, for example, the possible determinations of a function for a given region, the possible shapes of a solid figure, etc.

Reading this with a lot of good will one can interpret it as follows: When Riemann sketched the general notion of a manifold, he also had in mind the notion of an infinite dimensional manifold of mappings between manifolds. He then went on to describe the notion of Riemannian metric and to talk about curvature.

The dramatis personae of this foundational chapter are named in the following diagram:

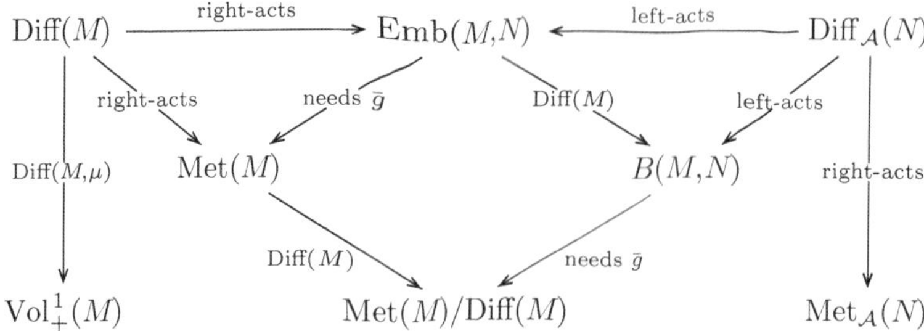

In this diagram:

- M is a finite dimensional compact smooth manifold.
- N is a finite dimensional smooth manifolds without boundary, and $\bar{g}$ is one fixed background Riemannian metric on N which we always assume to be of bounded geometry; see Sect. 5.
- $\operatorname{Met}(N) = \Gamma(S^2_+ T^*N)$ is the space of all Riemannian metrics on N.
- $\operatorname{Diff}(M)$ is the regular Fréchet Lie group of all diffeomorphisms on the compact manifold M with corners.
- $\operatorname{Diff}_{\mathcal{A}}(N)$, $\mathcal{A} \in \{H^\infty, \mathcal{S}, c\}$ the regular Lie group of all smooth diffeomorphisms of decay $\mathcal{A}$ towards Id_N.
- $\operatorname{Emb}(M, N)$ is the infinite dimensional smooth manifold of all embeddings $M \to N$, which is the total space of a smooth principal fiber bundle $\operatorname{Emb}(M, N) \to B(M, N) = \operatorname{Imm}(M, N)/\operatorname{Diff}(M)$ with structure group $\operatorname{Diff}(M)$ and base manifold $B(M, N)$, the space of all smooth submanifolds of N of type M. It is possible to extend $\operatorname{Emb}(M, N)$ to the manifold of $\operatorname{Imm}(M, N)$ and $B(M, N)$ to the infinite dimensional orbifold $B_i(M, N)$.
- $\operatorname{Vol}^1_+(M) \subset \Gamma(\operatorname{vol}(M))$ is the space of all positive smooth probability densities on the manifold M with corners.

Since it will be of importance for geometric continuum mechanics, I will allow the source manifold M to be quite general: M can be a manifold with corners; see Sect. 3. This setting is worked out in detail in [69]. Or M can be a Whitney manifold germ, a notion originating in this paper; see Sect. 4.

In this foundational chapter I will describe the theory of manifolds of mappings, of groups of diffeomorphisms, of manifolds of submanifolds (with corners), and of some striking results about weak Riemannian geometry on these spaces. See [10] for an overview article which is much more comprehensive for the aspect of shape spaces.

An explicit construction of manifolds of smooth mappings modeled on Fréchet spaces was described by Eells [28]. Differential calculus beyond the realm of Banach spaces has some inherent difficulties even in its definition; see Sect. 2. Smoothness of composition and inversion was first treated on the group of all smooth diffeomorphisms of a compact manifold in [63]; however, there was a gap in the proof, which was first filled by Gutknecht [48]. Manifolds of C^k-mappings and/or mappings of Sobolev classes were treated by Eliasson [31] and Eells [27], Smale–Abraham [1], and [92]. Since these are modeled on Banach spaces, they allow solution methods for equations and have found a lot of applications. See in particular [26].

In preparation of this chapter I noticed that the canonical chart construction for the manifold $C^\infty(M, N)$ even works if we allow M to be a *Whitney manifold germ*. These are modeled on open subsets of closed subsets of $\mathbb{R}^m$ which (1) admit a continuous Whitney extension operator and (2) are the closure of their interior. See Sect. 4 for a thorough discussion. Many results for them described below are preliminary, e.g., Theorem 6.4, Sect. 7.2. I expect that they can be strengthened considerably, but I had not enough time to pursue them during the preparation of this chapter.

I thank Reuven Segev and Marcelo Epstein for asking me for a contribution to this volume, and I thank them and Leonhard Frerick, Andreas Kriegl, Jochen Wengenroth, and Armin Rainer for helpful discussions.

2 A Short Review of Convenient Calculus in Infinite Dimensions

Traditional differential calculus works well for finite dimensional vector spaces and for Banach spaces. Beyond Banach spaces, the main difficulty is that composition of linear mappings stops to be jointly continuous at the level of Banach spaces, for any compatible topology. Namely, if for a locally convex vector space E and its dual E' the evaluation mapping $\text{ev} : E \times E' \to \mathbb{R}$ is jointly continuous, then there are open neighborhoods of zero $U \subset E$ and $U' \subset E'$ with $\text{ev}(U \times U') \subset [-1, 1]$. But then U' is contained in the polar of the open set U, and thus is bounded. So E' is normable, and a fortiori E is normable.

For locally convex spaces which are more general than Banach spaces, we sketch here the convenient approach as explained in [44] and [55].

The name *convenient calculus* mimics the paper [98] whose results (but not the name "convenient") was predated by Brown [17–19]. They discussed compactly generated spaces as a cartesian closed category for algebraic topology. Historical remarks on only those developments of calculus beyond Banach spaces that led to convenient calculus are given in [55, end of chapter I, p. 73ff].

2.1 The c^∞-Topology

Let E be a locally convex vector space. A curve $c : \mathbb{R} \to E$ is called *smooth* or C^∞ if all derivatives exist and are continuous. Let $C^\infty(\mathbb{R}, E)$ be the space of smooth curves. It can be shown that the set $C^\infty(\mathbb{R}, E)$ does not entirely depend on the locally convex topology of E, only on its associated bornology (system of bounded sets); see [55, 2.11]. The final topologies with respect to the following sets of mappings into E (i.e., the finest topology on E such that each map is continuous) coincide; see [55, 2.13]:

1. $C^\infty(\mathbb{R}, E)$.
2. The set of all Lipschitz curves (so that $\{\frac{c(t)-c(s)}{t-s} : t \neq s, |t|, |s| \leq C\}$ is bounded in E, for each C).
3. The set of injections $E_B \to E$ where B runs through all bounded absolutely convex subsets in E, and where E_B is the linear span of B equipped with the Minkowski functional $\|x\|_B := \inf\{\lambda > 0 : x \in \lambda B\}$.
4. The set of all Mackey-convergent sequences $x_n \to x$ (i.e., those for which there exists a sequence $0 < \lambda_n \nearrow \infty$ with $\lambda_n(x_n - x)$ bounded).

The resulting unique topology is called the c^∞-topology on E and we write $c^\infty E$ for the resulting topological space.

In general (on the space $\mathcal{D}$ of test functions, for example) it is finer than the given locally convex topology, it is not a vector space topology, since addition is no longer jointly continuous. Namely, even $c^\infty(\mathcal{D} \times \mathcal{D}) \neq c^\infty\mathcal{D} \times c^\infty\mathcal{D}$.

The finest among all locally convex topologies on E which are coarser than $c^\infty E$ is the bornologification of the given locally convex topology. If E is a Fréchet space, then $c^\infty E = E$.

2.2 Convenient Vector Spaces

A locally convex vector space E is said to be a *convenient vector space* if one of the following equivalent conditions holds (called c^∞-completeness); see [55, 2.14]:

1. For any $c \in C^\infty(\mathbb{R}, E)$ the (Riemann-) integral $\int_0^1 c(t)dt$ exists in E.
2. Any Lipschitz curve in E is locally Riemann integrable.
3. A curve $c : \mathbb{R} \to E$ is C^∞ if and only if $\lambda \circ c$ is C^∞ for all $\lambda \in E^*$, where E^* is the dual of all continuous linear functionals on E.

- Equivalently, for all $\lambda \in E'$, the dual of all bounded linear functionals.
- Equivalently, for all $\lambda \in \mathcal{V}$, where $\mathcal{V}$ is a subset of E' which recognizes bounded subsets in E; see [55, 5.22]

We call this *scalarwise* C^∞.

4. Any Mackey-Cauchy sequence (i.e., $t_{nm}(x_n - x_m) \to 0$ for some $t_{nm} \to \infty$ in $\mathbb{R}$) converges in E. This is visibly a mild completeness requirement.
5. If B is bounded closed absolutely convex, then E_B is a Banach space.
6. If $f : \mathbb{R} \to E$ is scalarwise Lip^k, then f is Lip^k, for $k > 1$.
7. If $f : \mathbb{R} \to E$ is scalarwise C^∞, then f is differentiable at 0.

Here a mapping $f : \mathbb{R} \to E$ is called Lip^k if all derivatives up to order k exist and are Lipschitz, locally on $\mathbb{R}$. That f is scalarwise C^∞ (resp., Lip^k) means $\lambda \circ f$ is C^∞ (resp., Lip^k) for all continuous (equiv., bounded) linear functionals on E.

2.3 *Smooth Mappings*

Let E and F be convenient vector spaces, and let $U \subset E$ be c^∞-open. A mapping $f : U \to F$ is called *smooth* or C^∞, if $f \circ c \in C^\infty(\mathbb{R}, F)$ for all $c \in C^\infty(\mathbb{R}, U)$. See [55, 3.11].

If E is a Fréchet space, then this notion coincides with all other reasonable notions of C^∞-mappings; see below. Beyond Fréchet spaces, as a rule, there are more smooth mappings in the convenient setting than in other settings, e.g., C_c^∞. Moreover, any smooth mapping is continuous for the c^∞-topologies, but in general not for the locally convex topologies: As shown in the beginning of Sect. 2, the evaluation mapping $\text{ev} : E \times E' \to \mathbb{R}$ is continuous only if E is normable. On Fréchet spaces each smooth mapping is continuous; see the end of Sect. 2.1.

2.4 *Main Properties of Smooth Calculus*

In the following all locally convex spaces are assumed to be convenient:

1. For maps on Fréchet spaces the notion of smooth mapping from Sect. 2.3 coincides with all other reasonable definitions. On $\mathbb{R}^2$ this is a nontrivial statement; see [16] or [55, 3.4].
2. Multilinear mappings are smooth if and only if they are bounded; see [55, 5.5].
3. If $E \supseteq U \xrightarrow{f} F$ is smooth, then the derivative $df : U \times E \to F$ is smooth, and also $df : U \to L(E, F)$ is smooth where $L(E, F)$ denotes the convenient space of all bounded linear mappings with the topology of uniform convergence on bounded subsets; see [55, 3.18].
4. The chain rule holds; see [55, 3.18].
5. The space $C^\infty(U, F)$ is again a convenient vector space where the structure is given by the injection

$$C^\infty(U,F) \xrightarrow{C^\infty(c,\lambda)} \prod_{c\in C^\infty(\mathbb{R},U),\lambda\in F^*} C^\infty(\mathbb{R},\mathbb{R}), \quad f\mapsto (\lambda\circ f\circ c)_{c,\lambda},$$

and where $C^\infty(\mathbb{R},\mathbb{R})$ carries the topology of compact convergence in each derivative separately; see [55, 3.11 and 3.7].

6. The exponential law holds; see [55, 3.12].: For c^∞-open $V \subset F$,

$$C^\infty(U, C^\infty(V,G)) \cong C^\infty(U\times V, G)$$

is a linear diffeomorphism of convenient vector spaces.
Note that this result (for $U = \mathbb{R}$) is the main assumption of variational calculus. Here it is a theorem.

7. A linear mapping $f : E \to C^\infty(V,G)$ is smooth (by (2) equivalent to bounded) if and only if $E \xrightarrow{f} C^\infty(V,G) \xrightarrow{\mathrm{ev}_v} G$ is smooth for each $v\in V$. (*Smooth uniform boundedness theorem*; see [55, theorem 5.26].)
8. A mapping $f : U \to L(F,G)$ is smooth if and only if

$$U \xrightarrow{f} L(F,G) \xrightarrow{\mathrm{ev}_v} G$$

is smooth for each $v \in F$, because then it is scalarwise smooth by the classical uniform boundedness theorem.

9. The following canonical mappings are smooth. This follows from the exponential law by simple categorical reasoning; see [55, 3.13]:

$$\begin{aligned}
&\mathrm{ev} : C^\infty(E,F)\times E \to F, \quad \mathrm{ev}(f,x) = f(x)\\
&\mathrm{ins} : E \to C^\infty(F, E\times F), \quad \mathrm{ins}(x)(y) = (x,y)\\
&(\quad)^\wedge : C^\infty(E, C^\infty(F,G)) \to C^\infty(E\times F, G)\\
&(\quad)^\vee : C^\infty(E\times F, G) \to C^\infty(E, C^\infty(F,G))\\
&\mathrm{comp} : C^\infty(F,G)\times C^\infty(E,F) \to C^\infty(E,G)\\
&C^\infty(\ \ ,\ \ ) : C^\infty(F,F_1)\times C^\infty(E_1,E) \to\\
&\qquad\qquad\qquad \to C^\infty(C^\infty(E,F), C^\infty(E_1,F_1))\\
&\qquad (f,g) \mapsto (h\mapsto f\circ h\circ g)\\
&\prod : \prod C^\infty(E_i,F_i) \to C^\infty(\prod E_i, \prod F_i).
\end{aligned}$$

This ends our review of the standard results of convenient calculus. Just for the curious reader and to give a flavor of the arguments, we enclose a lemma that is used many times in the proofs of the results above.

Lemma (Special Curve Lemma, [55, 2.8]) *Let E be a locally convex vector space. Let x_n be a sequence which converges fast to x in E; i.e., for each $k\in\mathbb{N}$ the* *sequence* $n^k(x_n - x)$ *is bounded. Then the infinite polygon through the* x_n *can be parameterized as a smooth curve* $c : \mathbb{R} \to E$ *such that* $c(\frac{1}{n}) = x_n$ *and* $c(0) = x$.

2.5 Remark Convenient calculus (i.e., having properties 6 and 7) exists also for:

- Real analytic mappings; see [54] or [55, Chapter II].
- Holomorphic mappings; see [62] or [55, Chapter II] (using the notion of [35, 36]).
- Many classes of Denjoy–Carleman ultradifferentiable functions, both of Beurling type and of Roumieu type, see [57–59, 61].
- With some adaptations, Lip^k; see [44]. One has to adapt the exponential law Sect. 2.4(9) in the obvious way.
- With more adaptations, even $C^{k,\alpha}$ (the k-th derivative is Hölder-continuous with index $0 < \alpha \leq 1$); see [37, 38]. Namely, if f is $C^{k,\alpha}$ and g is $C^{k,\beta}$, then $f \circ g$ is $C^{k,\alpha\beta}$.

Differentiability C^n cannot be described by a convenient approach (i.e., allowing result like Sect. 2.4). Only such differentiability notions allow this, which can be described by boundedness conditions only.

We shall treat C^n mapping spaces using the following result.

2.6 Recognizing Smooth Curves

The following result is very useful if one wants to apply convenient calculus to spaces which are not tied to its categorical origin, like the Schwartz spaces $\mathcal{S}$, $\mathcal{D}$, or Sobolev spaces; for its uses see [77] and [60]. In what follows $\sigma(E, \mathcal{V})$ denotes the initial (also called weak) topology on E with respect to a set $\mathcal{V} \subset E'$.

Theorem ([44, Theorem 4.1.19]) *Let* $c : \mathbb{R} \to E$ *be a curve in a convenient vector space* E. *Let* $\mathcal{V} \subset E'$ *be a subset of bounded linear functionals such that the bornology of* E *has a basis of* $\sigma(E, \mathcal{V})$*-closed sets. Then the following are equivalent:*

(1) c *is smooth*
(2) *There exist locally bounded curves* $c^k : \mathbb{R} \to E$ *such that* $\lambda \circ c$ *is smooth* $\mathbb{R} \to \mathbb{R}$ *with* $(\lambda \circ c)^{(k)} = \lambda \circ c^k$, *for each* $\lambda \in \mathcal{V}$.

If $E = F'$ *is the dual of a convenient vector space* F, *then for any point separating subset* $\mathcal{V} \subseteq F$ *the bornology of* E *has a basis of* $\sigma(E, \mathcal{V})$*-closed subsets, by* [44, 4.1.22].

This theorem is surprisingly strong: note that $\mathcal{V}$ does not need to recognize bounded sets. We shall use the theorem in situations where $\mathcal{V}$ is just the set of all point evaluations on suitable Sobolev spaces.

2.7 Frölicher Spaces

Following [55, Section 23] we describe here the following simple concept: A *Frölicher space* or a space with *smooth structure* is a triple $(X, \mathcal{C}_X, \mathcal{F}_X)$ consisting of a set X, a subset $\mathcal{C}_X$ of the set of all mappings $\mathbb{R} \to X$, and a subset $\mathcal{F}_X$ of the set of all functions $X \to \mathbb{R}$, with the following two properties:

1. A function $f : X \to \mathbb{R}$ belongs to $\mathcal{F}_X$ if and only if $f \circ c \in C^\infty(\mathbb{R}, \mathbb{R})$ for all $c \in \mathcal{C}_X$.
2. A curve $c : \mathbb{R} \to X$ belongs to $\mathcal{C}_X$ if and only if $f \circ c \in C^\infty(\mathbb{R}, \mathbb{R})$ for all $f \in \mathcal{F}_X$.

Note that a set X together with any subset $\mathcal{F}$ of the set of functions $X \to \mathbb{R}$ generates a unique Frölicher space $(X, \mathcal{C}_X, \mathcal{F}_X)$, where we put in turn:

$$\mathcal{C}_X := \{c : \mathbb{R} \to X : f \circ c \in C^\infty(\mathbb{R}, \mathbb{R}) \text{ for all } f \in \mathcal{F}\},$$
$$\mathcal{F}_X := \{f : X \to \mathbb{R} : f \circ c \in C^\infty(\mathbb{R}, \mathbb{R}) \text{ for all } c \in \mathcal{C}_X\},$$

so that $\mathcal{F} \subseteq \mathcal{F}_X$. The set $\mathcal{F}$ will be called a *generating set of functions for the Frölicher space*. Similarly a set X together with any subset $\mathcal{C}$ of the set of curves $\mathbb{R} \to X$ generates a Frölicher space; $\mathcal{C}$ is then called a *generating set of curves* for this Frölicher structure. Note that *a locally convex space E is convenient if and only if it is a Frölicher space with the structure whose space $\mathcal{C}_E$ of smooth curves is the one described in Sect. 2.1, or whose space $\mathcal{F}_E$ of smooth functions is described in Sect. 2.3*. This follows directly from Sect. 2.2.

On each Frölicher space we shall consider the final topology with respect to all smooth curves $c : \mathbb{R} \to X$ in $\mathcal{C}_X$; i.e., the coarsest topology such that each such c is continuous. This is in general finer that the initial topology with respect to all functions in $\mathcal{F}_X$.

A mapping $\varphi : X \to Y$ between two Frölicher spaces is called *smooth* if one of the following three equivalent conditions hold

3. For each $c \in \mathcal{C}_X$ the composite $\varphi \circ c$ is in $\mathcal{C}_Y$. Note that here $\mathcal{C}_X$ can be replaced by a generating set $\mathcal{C}$ of curves in X.
4. For each $f \in \mathcal{F}_Y$ the composite $f \circ \varphi$ is in $\mathcal{F}_X$. Note that $\mathcal{F}_Y$ can be replaced by a generating set of functions.
5. For each $c \in \mathcal{C}_X$ and for each $f \in \mathcal{F}_Y$ the composite $f \circ \varphi \circ c$ is in $C^\infty(\mathbb{R}, \mathbb{R})$.

The set of all smooth mappings from X to Y will be denoted by $C^\infty(X, Y)$. Then we have $C^\infty(\mathbb{R}, X) = \mathcal{C}_X$ and $C^\infty(X, \mathbb{R}) = \mathcal{F}_X$.

Frölicher spaces and smooth mappings form a category which is complete, cocomplete, and cartesian closed, by Kriegl and Michor [55, 23.2].

Note that there is the finer notion of diffeological spaces X introduced by Souriau: These come equipped with a set of mappings from open subsets of $\mathbb{R}^n$'s into X subject to some obvious properties concerning reparameterizations by C^∞-

mappings; see [51]. The obvious functor associating the generated Frölicher space to a diffeological space is both left and right adjoint to the embedding of the category of Frölicher spaces into the category of diffeological spaces. A characterization of those diffeological spaces which are Frölicher spaces is in [106, Section 2.3].

3 Manifolds with Corners

In this section we collect some results which are essential for the extension of the convenient setting for manifolds of mappings to a source manifold which has corners and which need not be compact.

3.1 Manifolds with Corners

For more information we refer to [25, 66, 69]. Let $Q = Q^m = \mathbb{R}^m_{\geq 0}$ be the positive orthant or quadrant. By Whitney's extension theorem or Seeley's theorem (see also the discussion in Sects. 4.1–4.3), the restriction $C^\infty(\mathbb{R}^m) \to C^\infty(Q)$ is a surjective continuous linear mapping which admits a continuous linear section (extension mapping); so $C^\infty(Q)$ is a direct summand in $C^\infty(\mathbb{R}^m)$. A point $x \in Q$ is called a *corner of codimension (or index)* $q > 0$ if x lies in the intersection of q distinct coordinate hyperplanes. Let $\partial^q Q$ denote the set of all corners of codimension q.

A manifold with corners (recently also called a quadrantic manifold) M is a smooth manifold modeled on open subsets of Q^m. We assume that it is connected and second countable; then it is paracompact and each open cover admits a subordinated smooth partition of unity.

We do not assume that M is oriented, but for Moser's theorem we will eventually assume that M is compact. Let $\partial^q M$ denote the set of all corners of codimension q. Then $\partial^q M$ is a submanifold without boundary of codimension q in M; it has finitely many connected components if M is compact. We shall consider ∂M as stratified into the connected components of all $\partial^q M$ for $q > 0$. Abusing notation we will call $\partial^q M$ the boundary stratum of codimension q; this will lead to no confusion. Note that ∂M itself is not a manifold with corners. We shall denote by $j_{\partial^q M} : \partial^q M \to M$ the embedding of the boundary stratum of codimension q into M, and by $j_{\partial M} : \partial M \to M$ the whole complex of embeddings of all strata.

Each diffeomorphism of M restricts to a diffeomorphism of ∂M and to a diffeomorphism of each stratum $\partial^q M$. The Lie algebra of Diff(M) consists of all vector fields X on M such that $X|_{\partial^q M}$ is tangent to $\partial^q M$. We shall denote this Lie algebra by $\mathfrak{X}(M, \partial M)$.

3.2 Lemma *Any manifold with corners M is a submanifold with corners of an open manifold $\tilde{M}$ of the same dimension, and each smooth function on M extends to a smooth function on $\tilde{M}$. Each smooth vector bundle over M extends to a smooth* *vector bundle over $\tilde{M}$. Each immersion (embedding) of M into a smooth manifold N without boundary is the restriction of an immersion (embedding) of a (possibly smaller) $\tilde{M} \supset M$ into N.*

Proof Choose a vector field X on M which is complete, and along ∂M is nowhere 0 and pointing into the interior. Then for $\varepsilon > 0$ we can replace M by the flow image $\mathrm{Fl}^X_\varepsilon(M)$ which is contained in the interior $\tilde{M} = M \setminus \partial M$. The extension properties follow from the Whitney extension theorem. An immersion extends, since its rank cannot fall locally. An embedding f extends since $\{(f(x), f(y)) : (x, y) \in M \times M \setminus \mathrm{Diag}_M\}$ has positive distance to the closed Diag_N in $N \times N$, locally in M, and we can keep it that way; see [69, 5.3] for too many details. □

3.3 Differential Forms on Manifolds with Corners

There are several differential complexes on a manifold with corners. If M is not compact there are also the versions with compact support.

- Differential forms that vanish near ∂M. If M is compact, this is the same as the differential complex $\Omega_c(M \setminus \partial M)$ of differential forms with compact support in the open interior $M \setminus \partial M$.
- $\Omega(M, \partial M) = \{\alpha \in \Omega(M) : j^*_{\partial^q M}\alpha = 0 \text{ for all } q \geq 1\}$, the complex of differential forms that pull back to 0 on each boundary stratum.
- $\Omega(M)$, the complex of all differential forms. Its cohomology equals singular cohomology with real coefficients of M, since $\mathbb{R} \to \Omega^0 \to \Omega^1 \to \dots$ is a fine resolution of the constant sheaf on M; for that one needs existence of smooth partitions of unity and the Poincaré lemma which hold on manifolds with corners. The Poincaré lemma can be proved as in [73, 9.10] in each quadrant.

If M is an oriented manifold with corners of dimension m and if $\mu \in \Omega^m(M)$ is a nowhere vanishing form of top degree, then $\mathfrak{X}(M) \ni X \mapsto i_X\mu \in \Omega^{m-1}(M)$ is a linear isomorphism. Moreover, $X \in \mathfrak{X}(M, \partial M)$ (tangent to the boundary) if and only if $i_X\mu \in \Omega^{m-1}(M, \partial M)$.

3.4 Towards the Long Exact Sequence of the Pair $(M, \partial M)$

Let us consider the short exact sequence of differential graded algebras

$$0 \to \Omega(M, \partial M) \to \Omega(M) \to \Omega(M)/\Omega(M, \partial M) \to 0\,.$$

The complex $\Omega(M)/\Omega(M, \partial M)$ is a subcomplex of the product of $\Omega(N)$ for all connected components N of all $\partial^q M$. The quotient consists of forms which extend continuously over boundaries to ∂M with its induced topology in such a way that one can extend them to smooth forms on M; this is contained in the space of "stratified forms" as used in [104]. There Stokes' formula is proved for stratified forms.

3.5 Proposition (Stokes' Theorem) *For a connected oriented manifold M with corners of dimension* $\dim(M) = m$ *and for any* $\omega \in \Omega_c^{m-1}(M)$ *we have*

$$\int_M d\omega = \int_{\partial^1 M} j_{\partial^1 M}^* \omega \,.$$

Note that $\partial^1 M$ may have several components. Some of these might be non-compact.

We shall deduce this result from Stokes' formula for a manifold with boundary by making precise the fact that $\partial^{\geq 2} M$ has codimension 2 in M and has codimension 1 with respect to $\partial^1 M$. The proof also works for manifolds with more general boundary strata, like manifolds with cone-like singularities. A lengthy full proof can be found in [24].

Proof We first choose a smooth decreasing function f on $\mathbb{R}_{\geq 0}$ such that $f = 1$ near 0 and $f(r) = 0$ for $r \geq \varepsilon$. Then $\int_0^\infty f(r)dr < \varepsilon$ and for $Q^m = \mathbb{R}^m_{\geq 0}$ with $m \geq 2$,

$$\begin{aligned}\Big|\int_{Q^m} f'(|x|)\,dx\Big| &= C_m\Big|\int_0^\infty f'(r)r^{m-1}\,dr\Big| = C_m\Big|\int_0^\infty f(r)(r^{m-1})'\,dr\Big| \\ &= C_m\int_0^\varepsilon f(r)(r^{m-1})'\,dr \leq C_m\varepsilon^{m-1}\,,\end{aligned}$$

where C_m denotes the surface area of $S^{m-1} \cap Q^m$. Given $\omega \in \Omega_c^{m-1}(M)$ we use the function f on quadrant charts on M to construct a function g on M that is 1 near $\partial^{\geq 2} M = \bigcup_{q \geq 2} \partial^q M$, has support close to $\partial^{\geq 2} M$ and satisfies $\left|\int_M dg \wedge \omega\right| < \varepsilon$. Then $(1-g)\omega$ is an $(m-1)$-form with compact support in the manifold with boundary $M \setminus \partial^{\geq 2} M$, and Stokes' formula (cf. [73, 10.11]) now says

$$\int_{M\setminus\partial^{\geq 2}M} d((1-g)\omega) = \int_{\partial^1 M} j_{\partial^1 M}^*((1-g)\omega)\,.$$

But $\partial^{\geq 2} M$ is a null set in M and the quantities

$$\Big|\int_M d((1-g)\omega) - \int_M d\omega\Big| \quad \text{and} \quad \Big|\int_{\partial^1 M} j_{\partial^1 M}^*((1-g)\omega) - \int_{\partial^1 M} j_{\partial^1 M}^*\omega\Big|$$

are small if ε is small enough. □

3.6 Riemannian Manifolds with Bounded Geometry

If M is not necessarily compact without boundary we equip M with a Riemannian metric g of bounded geometry which exists by [47, Theorem 2']. This means that

- (I) The injectivity radius of (M, g) is positive.
- (B_∞) Each iterated covariant derivative of the curvature is uniformly g-bounded: $\|\nabla^i R\|_g < C_i$ for $i = 0, 1, 2, \dots$.

The following is a compilation of special cases of results collected in [30, chapter 1].

Proposition ([29, 53]) *If (M, g) satisfies (I) and (B_∞), then the following holds*

(1) *(M, g) is complete.*
(2) *There exists $\varepsilon_0 > 0$ such that for each $\varepsilon \in (0, \varepsilon_0)$ there is a countable cover of M by geodesic balls $B_\varepsilon(x_\alpha)$ such that the cover of M by the balls $B_{2\varepsilon}(x_\alpha)$ is still uniformly locally finite.*
(3) *Moreover there exists a partition of unity $1 = \sum_\alpha \rho_\alpha$ on M such that $\rho_\alpha \geq 0$, $\rho_\alpha \in C_c^\infty(M)$, $\operatorname{supp}(\rho_\alpha) \subset B_{2\varepsilon}(x_\alpha)$, and $|D_u^\beta \rho_\alpha| < C_\beta$ where u are normal (Riemannian exponential) coordinates in $B_{2\varepsilon}(x_\alpha)$.*
(4) *In each $B_{2\varepsilon}(x_\alpha)$, in normal coordinates, we have $|D_u^\beta g_{ij}| < C'_\beta$, $|D_u^\beta g^{ij}| < C''_\beta$, and $|D_u^\beta \Gamma_{ij}^m| < C'''_\beta$, where all constants are independent of α.*

3.7 Riemannian Manifolds with Bounded Geometry Allowing Corners

If M has corners, we choose an open manifold $\tilde{M}$ of the same dimension which contains M as a submanifold with corners; see 3.1. It is very desirable to prove that then there exists a Riemannian metric $\tilde{g}$ on $\tilde{M}$ with bounded geometry such that each boundary component of each $\partial^q M$ is totally geodesic.

For a compact manifold with boundary (no corners of codimension ≥ 2), existence of such a Riemannian metric was proven in [45, 2.2.3] in a more complicated context. A simple proof goes as follows: Choose a tubular neighborhood U of ∂M in $\tilde{M}$ and use the symmetry $\varphi(u) = -u$ in the vector bundle structure on U. Given a metric $\tilde{g}$ on $\tilde{M}$, then ∂M is totally geodesic for the metric $\frac{1}{2}(\tilde{g} + \varphi^* \tilde{g})$ on U, since ∂M (the zero section) is the fixed point set of the isometry φ. Now glue this metric to the $\tilde{g}$ using a partition of unity for the cover U and $\tilde{M} \setminus V$ for a closed neighborhood V of ∂M in U.

Existence of a geodesic spray on a manifold with corners which is tangential to each boundary component $\partial^q M$ was proved in [69, 2.8, see also 10.3]. A direct proof of this fact can be distilled from the proof of lemma in Sect. 5.9 below. This is sufficient for constructing charts on the diffeomorphism group $\operatorname{Diff}(M)$ in Sect. 6.1 below.

4 Whitney Manifold Germs

More general than manifolds with corners, Whitney manifold germs allow for quite singular boundaries but still controlled enough so that a continuous Whitney extension operator to an open neighborhood manifold exists.

4.1 Compact Whitney Subsets

Let $\tilde{M}$ be an open smooth connected m-dimensional manifold. A closed connected subset $M \subset \tilde{M}$ is called a *Whitney subset*, or $\tilde{M} \supset M$ is called a *Whitney pair*, if

(1) M is the closure of its open interior in $\tilde{M}$, and
(2) There exists a continuous linear extension operator

$$\mathcal{E} : \mathcal{W}(M) \to C^\infty(\tilde{M}, \mathbb{R})$$

from the linear space $\mathcal{W}(M)$ of all Whitney jets of infinite order with its natural Fréchet topology (see below) into the space $C^\infty(\tilde{M}, \mathbb{R})$ of smooth functions on $\tilde{M}$ with the Fréchet topology of uniform convergence on compact subsets in all derivatives separately.

We speak of a *compact Whitney subset* or *compact Whitney pair* if M is compact. In this case, in (2), we may equivalently require that $\mathcal{E}$ is linear continuous into the Fréchet space $C^\infty_L(\tilde{M}, \mathbb{R}) \subset C^\infty_c(\tilde{M}, \mathbb{R})$ of smooth functions with support in a compact subset L which contains M in its interior, by using a suitable bump function.

The property of being a Whitney pair is obviously invariant under diffeomorphisms (of $\tilde{M}$) which act linearly and continuously both on $\mathcal{W}(M)$ and $C^\infty(\tilde{M}, \mathbb{R})$ in a natural way.

This property of being a Whiney pair is local in the following sense: If $\tilde{M}_i \supset M_i$ covers $\tilde{M} \supset M$, then $\tilde{M} \supset M$ is a Whitney pair if and only if each $\tilde{M}_i \supset M_i$ is a Whitney pair, see Theorem 4.4 below.

More Details For $\mathbb{R}^m \supset M$, by an extension operator $\mathcal{E} : \mathcal{W}(M) \to C^\infty(\tilde{M}, \mathbb{R})$ we mean that $\partial_\alpha \mathcal{E}(F)|_M = F^{(\alpha)}$ for each multi-index $\alpha \in \mathbb{N}^m_{\geq 0}$ and each Whitney jet $F \in \mathcal{W}(M)$. We recall the definition of a Whitney jet F. If $M \subset \mathbb{R}^m$ is compact, then

$$F = (F^{(\alpha)})_{\alpha \in \mathbb{N}^m_{\geq 0}} \in \prod_\alpha C^0(M) \qquad \text{such that for}$$

$$T^n_y(F)(x) = \sum_{|\alpha| \leq n} \frac{F^{(\alpha)}(y)}{\alpha!}(x-y)^\alpha \qquad \text{the remainder seminorm}$$

$$q_{n,\varepsilon}(F) := \sup\left\{\frac{|F^{(\alpha)}(x) - \partial^\alpha T^n_y(F)(x)|}{|x-y|^{n-|\alpha|}} : {}^{|\alpha|\le n, x,y\in M}_{0<|x-y|\le\varepsilon}\right\} = o(\varepsilon);$$

so $q_{n,\varepsilon}(F)$ goes to 0 for $\varepsilon \to 0$, for each n separately. The n-th Whitney seminorm is then

$$\|F\|_n = \sup\{|F^{(\alpha)}(x)| : x \in M, |\alpha| \le n\} + \sup\{q_{n,\varepsilon}(F) : \varepsilon > 0\}.$$

For closed but non-compact M one uses the projective limit over a countable compact exhaustion of M. This describes the natural Fréchet topology on the space of Whitney jets for closed subsets of $\mathbb{R}^m$. The extension to manifolds is obvious.

Whitney proved in [107] that a linear extension operator always exists for a closed subset $M \subset \mathbb{R}^m$, but not always a continuous one, for example, for M a point. For a finite differentiability class C^n there exists always a continuous extension operator.

4.2 Proposition *For a Whitney pair $\tilde M \supset M$, the space of $\mathcal{W}(M)$ of Whitney jets on M is linearly isomorphic to the space*

$$C^\infty(M,\mathbb{R}) := \{f|_M : f \in C^\infty(\tilde M,\mathbb{R})\}.$$

Proof This follows from [40, 3.11], where the following is proved: If $f \in C^\infty(\mathbb{R}^m,\mathbb{R})$ vanishes on a Whitney subset $M \subset \mathbb{R}^m$, then $\partial^\alpha f|_M = 0$ for each multi-index α. Thus any continuous extension operator is injective. □

4.3 Examples and Counterexamples of Whitney Pairs

We collect here results about closed subsets of $\mathbb{R}^m$ which are or are not Whitney subsets.

(a) If M is a manifold with corners, then $\tilde M \supset M$ is a Whitney pair. This follows from Mityagin [79] or Seeley [97].
(b) If M is closed in $\mathbb{R}^m$ with dense interior and with Lipschitz boundary, then $\mathbb{R}^m \supset M$ is a Whitney pair; by Stein [99, p 181]. In [15, Theorem I] Bierstone proved that a closed subset $M \subset \mathbb{R}^n$ with dense interior is a Whitney pair, if it has Hölder $C^{0,\alpha}$-boundary for $0 < \alpha \le 1$ which may be chosen on each $M \cap \{x : N \le |x| \le N+2\}$ separately. A fortiori, each subanalytic subset in $\mathbb{R}^n$ gives a Whitney pair, [15, Theorem II].
(c) If $f \in C^\infty(\mathbb{R}_{\ge 0},\mathbb{R})$ which is flat at 0 (all derivatives vanish at 0), and if M is a closed subset containing 0 of $\{(x,y) : x \ge 0, |y| \le |f(x)|\} \subset \mathbb{R}^2$, then $\mathbb{R}^2 \supset M$ is not a Whitney pair; see [101, Beispiel 2].

(d) For $r \geq 1$, the set $\{x \in \mathbb{R}^m : 0 \leq x_1 \leq 1, x_2^2 + \cdots + x_m^2 \leq x_1^{2r}\}$ is called the parabolic cone of order r. Then the following result [101, Satz 4.6] holds:
A closed subset $M \subset \mathbb{R}^m$ is a Whitney subset, if the following condition holds: For each compact $K \subset \mathbb{R}^m$ there exists a parabolic cone S and a family $\varphi_i : S \to \phi_i(S) \subset M \subset \mathbb{R}^m$ of diffeomorphisms such that $K \cap M \subseteq \bigcup_i \overline{\varphi_i(S)}$ and $\sup_i |\varphi_i|_k < \infty$, $\sup_i |\varphi_i^{-1}|_k < \infty$ for each k separately.

(e) A characterization of closed subsets admitting continuous Whitney extension operators has been found by Frerick [40, 4.11] in terms of local Markov inequalities, which, however, is very difficult to check directly.
Let $M \subset \mathbb{R}^m$ be closed. Then the following are equivalent:

(e1) *M admits a continuous linear Whitney extension operator*

$$\mathcal{E} : \mathcal{W}(M) \to C^\infty(\mathbb{R}^m, \mathbb{R}) .$$

(e2) *For each compact $K \subset M$ and $\theta \in (0, 1)$ there is $r \geq 0$ and $\varepsilon_0 > 0$ such that for all $k \in \mathbb{N}_{\geq 1}$ there is $C \geq 1$ such that*

$$|dp(x_0)| \leq \frac{C}{\varepsilon^r} \sup_{\substack{|y-x_0| \leq \varepsilon \\ y \in \mathbb{R}^m}} |p(y)|^\theta \sup_{\substack{|x-x_0| \leq \varepsilon \\ x \in M}} |p(x)|^{1-\theta}$$

for all $p \in \mathbb{C}[x_1, \ldots, x_m]$ of degree $\leq k$, for all $x_0 \in K$, and for all $\varepsilon_0 > \varepsilon > 0$.

(e3) *For each compact $K \subset M$ there exists a compact L in $\mathbb{R}^m$ containing K in its interior, such that for all $\theta \in (0, 1)$ there is $r \geq 1$ and $C \geq 1$ such that*

$$\sup_{x \in K} |dp(x)| \leq C \deg(p)^r \sup_{y \in L} |p(y)|^\theta \sup_{z \in L \cap M} |p(z)|^{1-\theta}$$

for all $p \in \mathbb{C}[x_1, \ldots, x_m]$.

(f) Characterization (e) has been generalized to a characterization of compact subsets of $\mathbb{R}^m$ which admit a continuous Whitney extension operator with linear (or even affine) loss of derivatives, in [41]. In the paper [42] a similar characterization is given for an extension operator without loss of derivative, and a sufficient geometric condition is formulated [42, Corollary 2] which even implies that there are closed sets with empty interior admitting continuous Whitney extension operators, like the Sierpiński triangle or Cantor subsets. Thus we cannot omit assumption (Sect. 4.1.1) that M is the closure of its open interior in $\tilde{M}$ in our definition of Whitney pairs.

(g) The following result by Frerick [40, Theorem 3.15] gives an easily verifiable sufficient condition:

Let $K \subset \mathbb{R}^m$ be compact and assume that there exist $\varepsilon_0 > 0$, $\rho > 0$, $r \geq 1$ such that for all $z \in \partial K$ and $0 < \varepsilon < \varepsilon_0$ there is $x \in K$ with $B_{\rho\varepsilon^r}(x) \subset K \cap B_\varepsilon(z)$. Then K admits a continuous linear Whitney extension operator $\mathcal{W}(F) \to C^\infty(\mathbb{R}^m, \mathbb{R})$.

This implies (a), (b), and (d).

4.4 Theorem *Let $\tilde{M}$ be an open manifold and let $M \subset \tilde{M}$ be a compact subset that is the closure of its open interior. $M \subset \tilde{M}$ is a Whitney pair if and only if for every smooth atlas $(\tilde{M} \supset U_\alpha, u_\alpha : U_\alpha \to u_\alpha(U_\alpha) \subset \mathbb{R}^m)_{\alpha \in A}$ of the open manifold $\tilde{M}$, each $u_\alpha(M \cap U_\alpha) \subset u_\alpha(U_\alpha)$ is a Whitney pair.*

Consequently, for a Whitney pair $M \subset \tilde{M}$ and $U \subset \tilde{M}$ open, $M \cap U \subset \tilde{M} \cap U$ is also a Whitney pair.

Proof

(1) We consider a locally finite countable smooth atlas $(\tilde{M} \supset U_\alpha, u_\alpha : U_\alpha \to u_\alpha(U_\alpha) \subset \mathbb{R}^m)_{\alpha \in \mathbb{N}}$ of $\tilde{M}$ such that each $u_\alpha(U_\alpha) \supset u_\alpha(M \cap U_\alpha)$ is a Whitney pair.

We use a smooth "partition of unity" $f_\alpha \in C_c^\infty(U_\alpha, \mathbb{R}_{\geq 0})$ on $\tilde{M}$ with $\sum_\alpha f_\alpha^2 = 1$. The following mappings induce linear embeddings onto closed direct summands of the Fréchet spaces:

$$\begin{array}{ccc} C^\infty(\tilde{M},\mathbb{R}) & \underset{\Sigma_\alpha f_\alpha \cdot g_\alpha \leftarrow (g_\alpha)_\alpha}{\overset{f \mapsto (f_\alpha \cdot f)_\alpha}{\rightleftarrows}} & \Pi_\alpha C^\infty(U_\alpha,\mathbb{R}) \\ \mathcal{W}(M) & \rightleftarrows & \Pi_\alpha \mathcal{W}(U_\alpha \cap M) \end{array}$$

If each $u_\alpha(U_\alpha) \supset u_\alpha(U_\alpha \cap M)$ is a Whitney pair, then so is $U_\alpha \supset U_\alpha \cap M$, via the isomorphisms induced by u_α, and

$$\begin{array}{ccc} \mathcal{W}(M) & \xrightarrow[f \mapsto (f_\alpha \cdot f)_\alpha]{} & \Pi_\alpha \mathcal{W}(U_\alpha \cap M) \\ & & \Big\downarrow \Pi_\alpha \mathcal{E}_\alpha \\ C^\infty(\tilde{M},\mathbb{R}) & \xleftarrow{\Sigma_\alpha f_\alpha \cdot g_\alpha \leftarrow (g_\alpha)_\alpha} & \Pi_\alpha C^\infty(U_\alpha,\mathbb{R}) \end{array}$$

is a continuous Whitney extension operator, so that $\tilde{M} \supset M$ is a Whitney pair. This proves the easy direction of the theorem.

The following argument for the converse direction is inspired by Frerick and Wengenroth [43].

(2) (See [40, Def. 3.1], [65, Section 29–31]) A Fréchet space E is said to have *property* (DN) if for one (equivalently, any) increasing system $(\| \cdot \|_n)_{n \in \mathbb{N}}$ of seminorms generating the topology the following holds:

- There exists a continuous seminorm $\| \ \|$ on E (called a *dominating norm*, hence the name (DN)) such that for all (equivalently, one) $0 < \theta < 1$ and all $m \in \mathbb{N}$ there exist $k \in \mathbb{N}$ and $C > 0$ with

$$\| \ \|_m \leq C \| \ \|_k^{\theta} \cdot \| \ \|^{1-\theta} .$$

The property (DN) is an isomorphism invariant and is inherited by closed linear subspaces. The Fréchet space $\mathfrak{s}$ of rapidly decreasing sequences has property (DN).

(3) ([101, Satz 2.6], see also [40, Theorem 3.3]) *A closed subset M in $\mathbb{R}^m$ admits a continuous linear extension operator $\mathcal{W}(M) \to C^\infty(\mathbb{R}^m, \mathbb{R})$ if and only if for each $x \in \partial M$ there exists a compact neighborhood K of x in $\mathbb{R}^m$ such that*

$$\mathcal{W}_K(M) := \left\{ f \in \mathcal{W}(M) : \operatorname{supp}(f^{(\alpha)}) \subset K \text{ for all } \alpha \in \mathbb{N}^m_{\geq 0} \right\}$$

has property (DN).

We assume from now on that $\tilde{M} \supset M$ is a Whitney pair.

(4) Given a compact set $K \subset \tilde{M}$, let $L \subset \tilde{M}$ be a compact smooth manifold with smooth boundary which contains K in its interior. Let $\tilde{L}$ be the double of L, i.e., L smoothly glued to another copy of L along the boundary; $\tilde{L}$ is a compact smooth manifold containing L as a submanifold with boundary.

Then $C^\infty(\tilde{L}, \mathbb{R})$ is isomorphic to the space $\mathfrak{s}$ of rapidly decreasing sequences: This is due to [105]. In fact, using a Riemannian metric g on $\tilde{L}$, the expansion in an orthonormal basis of eigenvectors of $1 + \Delta^g$ of a function $h \in L^2$ has coefficients in $\mathfrak{s}$ if and only if $h \in C^\infty(\tilde{L}, \mathbb{R})$, because $1 + \Delta^g : H^{k+2}(\tilde{L}) \to H^k(\tilde{L})$ is an isomorphism for Sobolev spaces H^k with $k \geq 0$, and since the eigenvalues μ_n of Δ^g satisfy $\mu_n \sim C_{\tilde{L}} \cdot n^{2/\dim(\tilde{L})}$ for $n \to \infty$, by Weyl's asymptotic formula. Thus $C^\infty(\tilde{L}, \mathbb{R})$ has property (DN).

Moreover, $C^\infty_L(\tilde{M}, \mathbb{R}) = \{ f \in C^\infty(\tilde{M}, \mathbb{R}) : \operatorname{supp}(f) \subset L \}$ is a closed linear subspace of $C^\infty(\tilde{L}, \mathbb{R})$, by extending each function by 0. Thus also $C^\infty_L(\tilde{M}, \mathbb{R})$ has property (DN).

We choose now a function $g \in C^\infty_L(\tilde{M}, \mathbb{R}_{\geq 0})$ with $g|_K = 1$ and consider

$$\begin{array}{ccc} \mathcal{W}_K(M) & \xrightarrow{\ \mathcal{E}_K\ } & C^\infty_L(\tilde{M},\mathbb{R}) \\ \big\downarrow & & \big\uparrow {\scriptstyle f \mapsto g.f} \\ \mathcal{W}(M) & \xrightarrow{\ \mathcal{E}_M\ } & C^\infty(\tilde{M},\mathbb{R}) \end{array}$$

The resulting composition $\mathcal{E}_K$ is a continuous linear embedding onto a closed linear subspace of the space $C^\infty_L(\tilde{M}, \mathbb{R})$ which has (DN). Thus we proved:

(5) **Claim** *If $\tilde{M} \supset M$ is a Whitney pair and K is compact in $\tilde{M}$, the Fréchet space $\mathcal{W}_K(M)$ has property* (DN).

(6) We consider now a smooth chart $\tilde{M} \supset U \xrightarrow{u} u(U) = \mathbb{R}^m$. For $x \in \partial u(M)$ let K be a compact neighborhood of x in $\mathbb{R}^m$. The chart u induces a linear isomorphism

$$\mathcal{W}_K(u(M \cap U)) \xrightarrow{u^*} \mathcal{W}_{u^{-1}(K)}(U \cap M) \cong \mathcal{W}_{u^{-1}(K)}(M),$$

where the right-hand side mapping is given by extending each $f^{(\alpha)}$ by 0. By claim (5) the Fréchet space $W_{u^{-1}(K)}(M)$ has property (DN); consequently also the isomorphic space $\mathcal{W}_K(u(M \cap U))$ has property (DN). By (3) we conclude that $\mathbb{R}^m = u(U) \supset u(M \cap U)$ is a Whitney pair.

(7) If we are given a general chart $\tilde{M} \supset U \xrightarrow{u} u(U) \subset \mathbb{R}^m$, we cover U by a locally finite atlas $(U \supset U_\alpha, u_\alpha : U_\alpha \to u_\alpha(U_\alpha) = \mathbb{R}^m)_{\alpha \in \mathbb{N}}$. By (6) each $\mathbb{R}^m = u_\alpha(U_\alpha) \supset u_\alpha(M \cap U_\alpha)$ is a Whitney pair, and by the argument in (1) the pair $U \supset M \cap U$ is a Whitney pair, and thus the diffeomorphic $u(U) \supset u(U \cap M)$ is also a Whitney pair.

□

4.5 Our Use of Whitney Pairs

We consider an equivalence class of Whitney pairs $\tilde{M}_i \supset M_i$ for $i = 0, 1$ where $\tilde{M}_0 \supset M_0$ is equivalent to $\tilde{M}_1 \supset M_1$ if there exist an open submanifolds $\tilde{M}_i \supset \hat{M}_i \supset M_i$ and a diffeomorphism $\varphi : \hat{M}_0 \to \hat{M}_1$ with $\varphi(M_0) = M_1$. By a *germ of a Whitney manifold* we mean an equivalence class of Whitney pairs as above. Given a Whitney pair $\tilde{M} \supset M$ and its corresponding germ, we may keep M fixed and equip it with all open connected neighborhoods of M in $\tilde{M}$; each neighborhood is then a representative of this germ; called an *open neighborhood manifold* of M. In the following we shall speak of a *Whitney manifold germ M* and understand that it comes with open manifold neighborhoods $\tilde{M}$. If we want to stress a particular neighborhood we will write $\tilde{M} \supset M$.

The *boundary* ∂M of a Whitney manifold germ is the topological boundary of M in $\tilde{M}$. It can be a quite general set as seen from the examples in Sect. 4.3 and the discussion in Sect. 4.9. But infinitely flat cusps cannot appear.

4.6 Other Approaches

We remark that there are other settings, like the concept of a *manifold with rough boundary*; see [95] and literature cited there. The main idea there is to start with closed subsets $C \subset \mathbb{R}^m$ with dense interior, to use the space of functions which are C^n in the interior of C such that all partial derivatives extend continuously to C. Then one looks for sufficient conditions (in particular for $n = \infty$) on C such that there exists a continuous Whitney extension operator on the space of these functions, and builds manifolds from that. The condition in [95] are in the spirit of Sect. 4.3(d). By extending these functions and restricting their jets to C we see that the manifolds with rough boundary are Whitney manifold germs.

Another possibility is to consider closed subsets $C \subset \mathbb{R}^m$ with dense interior such there exists a continuous linear extension operator on the space $C^\infty(C) = \{f|_C : f \in C^\infty(\mathbb{R}^m)\}$ with the quotient locally convex topology. These are exactly the Whitney manifold pairs $\mathbb{R}^m \supset M$, by Proposition 4.2. In this setting, for C^n with $n < \infty$ there exist continuous extension operators $C^n_b(C) \to C^n_b(\mathbb{R}^m)$ (where the subscript b means bounded for all derivatives separately) for arbitrary subsets $C \subset \mathbb{R}^m$; see [39].

We believe that our use of Whitney manifold germs is quite general, simple, and avoids many technicalities. But it is aimed at the case C^∞; for C^k or $W^{k,p}$ other approaches, like the one in [95], might be more appropriate.

4.7 Tangent Vectors and Vector Fields on Whitney Manifold Germs

In line with the more general convention for vector bundles in Sect. 4.8 below, we define the tangent bundle TM of a Whitney manifold germ M as the restriction $TM = T\tilde{M}|_M$. For $x \in \partial M$, a tangent vector $X_x \in T_xM$ is said to be *interior pointing* if there exist a curve $c : [0, 1) \to M$ which is smooth into $\tilde{M}$ with $c'(0) = X_x$. And $X_x \in T_xM$ is called *tangent to the boundary* if there exists a curve $c : (-1, 1) \to \partial M$ which is smooth into $\tilde{M}$ with $c'(0) = X_x$. The *space of vector fields on* M is given as

$$\mathfrak{X}(M) = \{X|_M : X \in \mathfrak{X}(\tilde{M})\}.$$

Using a continuous linear extension operator, $\mathfrak{X}(M)$ is isomorphic to a locally convex direct summand in $\mathfrak{X}(\tilde{M})$. If M is a compact Whitney manifold germ, $\mathfrak{X}(M)$ is a direct summand even in $\mathfrak{X}_L(\tilde{M}) = \{X \in \mathfrak{X}(\tilde{M}) : \operatorname{supp}(X) \subseteq L\}$ where $L \subset \tilde{M}$ is a compact set containing M in its interior. We define the *space of vector fields on* M *which are tangent to the boundary* as

$$\mathfrak{X}_\partial(M) = \Big\{X|_M : X \in \mathfrak{X}(\tilde{M}), x \in \partial M \implies \mathrm{Fl}^X_t(x) \in \partial M \\ \text{for all } t \text{ for which } \mathrm{Fl}^X_t(x) \text{ exists in } \tilde{M}\Big\},$$

where Fl^X_t denotes the flow mapping of the vector field X up to time t which is locally defined on $\tilde{M}$. Obviously, for $X \in \mathfrak{X}_\partial(M)$ and $x \in \partial M$ the tangent vector $X(x)$ is tangent to the boundary in the sense defined above. I have no proof that the converse is true:

Question Suppose that $X \in \mathfrak{X}(\tilde{M})$ has the property that for each $x \in \partial M$ the tangent vector $X(x)$ is tangent to the boundary. Is it true that then $X|_M \in \mathfrak{X}_\partial(M)$?

A related question for which I have no answer is:

Question For each $x \in \partial M$ and tangent vector $X_x \in T_x M$ which is tangent to the boundary, is there a smooth vector field $X \in \mathfrak{X}_{c,\partial}(M)$ with $X(x) = X_x$?

Lemma *For a Whitney manifold germ M, the space $\mathfrak{X}_\partial(M)$ of vector field tangent to the boundary is a closed linear sub Lie algebra of $\mathfrak{X}(M)$. The space $\mathfrak{X}_{c,\partial}(M)$ of vector fields with compact support tangent to the boundary is a closed linear sub Lie algebra of $\mathfrak{X}_c(M)$.*

Proof By definition, for $X \in \mathfrak{X}(\tilde{M})$ the restriction $X|_M$ is in $\mathfrak{X}_\partial(M)$ if and only if $x \in \partial M$ implies that $\mathrm{Fl}^X_t(x) \in \partial M$ for all t for which $\mathrm{Fl}^X_t(x)$ exists in $\tilde{M}$. These conditions describe a set of continuous equations, since $(X, t, x) \mapsto \mathrm{Fl}^X_t(x)$ is smooth; see the proof of Sect. 6.1 for a simple argument. Thus $X \in \mathfrak{X}(\tilde{M})$ is closed.

The formulas (see, e.g., [81, pp. 56,58])

$$\lim_{n\to\infty} (Fl^X_{t/n} \circ \mathrm{Fl}^Y_{t/n})^n(x) = \mathrm{Fl}^{X+Y}_t(x)$$

$$\lim_{n\to\infty} \left(\mathrm{Fl}^Y_{-(t/n)^{1/2}} \circ \mathrm{Fl}^X_{-(t/n)^{1/2}} \circ \mathrm{Fl}^Y_{(t/n)^{1/2}} \circ \mathrm{Fl}^X_{(t/n)^{1/2}} \right)^n (x) = \mathrm{Fl}^{[X,Y]}_t(x)$$

shows that $\mathfrak{X}_\partial(M)$ is a Lie subalgebra. □

The Smooth Partial Stratifications of the Boundary of a Whitney Manifold Germ Given a Whitney manifold germ $\tilde{M} \supset M$ of dimension m, for each $x \in \partial M$ we denote by $\mathcal{L}^\infty(x)$ the family consisting of each maximal connected open smooth submanifold L of $\tilde{M}$ which contains x and is contained in ∂M. Note that for $L \in \mathcal{L}^\infty(x)$ and $y \in L$ we have $L \in \mathcal{L}^\infty(y)$. $\{T_x L : L \in \mathcal{L}^\infty(x)\}$ is a set of linear subspaces of $T_x\tilde{M}$. The collective of these for all $x \in \partial M$ is something like a "field of quivers of vector spaces" over ∂M. It might be the key to eventually construct charts for the regular Frölicher Lie group Diff(M) treated in Sect. 6.3 below, and for constructing charts for the Frölicher space Emb(M, N) in Sect. 7.2 below.

4.8 Mappings, Bundles, and Sections

Let M be Whitney manifold germ and let N be a manifold without boundary. By a smooth mapping $f : M \to N$ we mean $f = \tilde{f}|_M$ for a smooth mapping $\tilde{f} : \tilde{M} \to N$ for an open manifold neighborhood $\tilde{M} \supset M$. Whitney jet on M naively make sense only if they take values in a vector space or, more generally, in a vector bundle. One could develop the notion of Whitney jets of infinite order with values in a manifold as sections of $J^\infty(M, N) \to M$ with Whitney conditions of each order. We do not know whether this has been written down formally. But we can circumvent this easily by considering a closed embedding $i : N \to \mathbb{R}^p$ and a tubular neighborhood $p : U \to i(N)$; i.e., U is an open neighborhood and is (diffeomorphic to) the total space of a smooth vector bundle which projection p.

Then we can consider a Whitney jet on M with values in $\mathbb{R}^p$ (in other words, a p-tuple of Whitney jets) such that the 0-order part lies in $i(N)$. Using a continuous Whitney extension operator, we can extend the Whitney jet to a smooth mapping $\tilde{f} : \tilde{M} \to \mathbb{R}^p$. Then consider the open set $\tilde{f}^{-1}(U) \subset \tilde{M}$ instead of $\tilde{M}$, and replace $\tilde{f}$ by $p \circ \tilde{f}$. So we just extended the given Whitney jet to a smooth mapping $\tilde{M} \to N$, and also showed, that the space of Whitney jets is isomorphic to the space

$$C^\infty(M, N) = \{f|_M : f \in C^\infty(\tilde{M}, N), \tilde{M} \supset M\}.$$

Note that the neighborhood $\tilde{M}$ can be chosen independently of the mapping f, but dependent on N. This describes a nonlinear extension operator $C^\infty(M, N) \to C^\infty(\tilde{M}, N)$; we shall see in Sect. 5 that this extension operator is continuous and even smooth in the manifold structures.

For finite n we shall need the space $C^{\infty,n}(\mathbb{R} \times M, \mathbb{R}^p)$ of restrictions to M of mappings $\mathbb{R} \times \tilde{M} \ni (t, x) \mapsto f(t, x) \in \mathbb{R}^p$ which are C^∞ in t and C^n in x. If $\tilde{M}$ is open in $\mathbb{R}^m$ we mean by this that any partial derivative $\partial_t^k \partial_x^\alpha f$ of any order $k \in \mathbb{N}_{\geq 0}$ in t and of order $|\alpha| \leq n$ in x exists and is continuous on $\mathbb{R} \times \tilde{M}$. This carries over to an open manifold $\tilde{M}$, and finally, using again a tubular neighborhood $p : U \to i(N)$ as above, to the space $C^{\infty,n}(\mathbb{R} \times M, N)$, for any open manifold N. For a treatment of $C^{m,n}$-maps leading to an exponential law see [2]; since C^n is not accessible to a convenient approach, a more traditional calculus has to be used there.

By a (vector or fiber) bundle $E \to M$ over a germ of a Whitney manifold M we mean the restriction to M of a (vector or fiber) bundle $\tilde{E} \to \tilde{M}$, i.e., of a (vector or fiber) bundle over an open manifold neighborhood. By a smooth section of $E \to M$ we mean the restriction of a smooth section of $\tilde{E} \to \tilde{M}$ for a neighborhood $\tilde{M}$. Using classifying smooth mappings into a suitable Grassmannian for vector bundles over M and using the discussion above one could talk about Whitney jets of vector bundles and extend them to a manifold neighborhood of M.

We shall use the following spaces of sections of a vector bundle $E \to M$ over a Whitney manifold germ M. This is more general than [55, Section 30], since we allow Whitney manifold germs as base.

- $\Gamma(E) = \Gamma(M \leftarrow E)$, the space of smooth sections, i.e., restrictions of smooth sections of $\tilde{E} \to \tilde{M}$ for a fixed neighborhood $\tilde{M}$, with the Fréchet space topology of compact convergence on the isomorphic space of Whitney jets of sections.
- $\Gamma_c(E)$, the space of smooth sections with compact support, with the inductive limit (LF)-topology.
- $\Gamma_{C^n}(E)$, the space of C^n-section, with the Fréchet space topology of compact convergence on the space of Whitney n-jets. If M is compact and n finite, $\Gamma_{C^n}(E)$ is a Banach space.
- $\Gamma_{H^s}(E)$, the space of Sobolev H^s-sections, for $0 \leq s \in \mathbb{R}$. Here M should be a compact Whitney manifold germ. The measure on M is the restriction of the volume density with respect to a Riemannian metric on $\tilde{M}$. One also needs a

fiber metric on E. The space $\Gamma_{H^k}(E)$ is independent of all choices, but the inner product depends on the choices. One way to define $\Gamma_{H^k}(E)$ is to use a finite atlas which trivializes $\tilde{E}|_L$ over a compact manifold with smooth boundary L which is a neighborhood of M in $\tilde{M}$ and a partition of unity, and then use the Fourier transform description of the Sobolev space. For a careful description see [7]. For $0 \leq k < s - \dim(M)/2$ we have $\Gamma_{H^s}(E) \subset \Gamma_{C^k}(E)$ continuously.

- More generally, for $0 \geq s \in \mathbb{R}$ and $1 < p < \infty$ we also consider $\Gamma_{W^{s,p}}(E)$, the space of $W^{s,p}$-sections: For integral s, all (covariant) derivatives up to order s are in L^p. For $0 \leq k < s - \dim(M)/p$ we have $\Gamma_{H^s}(E) \subset \Gamma_{C^k}(E)$ continuously.

4.9 Is Stokes' Theorem Valid for Whitney Manifold Germs?

This seems far from obvious. Here is an example, due to [43]:

By the first answer to the MathOverflow question [50] there is a set K in $[0, 1] \subset \mathbb{R}$ which is the closure of its open interior such that the boundary is a Cantor set with positive Lebesgue measure. Moreover, $\mathbb{R} \supset K$ is a Whitney pair by Tidten [102], or by the local Markov inequalities [40, Proposition 4.8], or by Frerick et al. [41]. To make this connected, consider $K_2 := (K \times [0, 2]) \cup ([0, 1] \times [1, 2])$ in $\mathbb{R}^2$. Then $\mathbb{R}^2 \supset K_2$ is again a Whitney pair, but ∂K_2 has positive 2-dimensional Lebesgue measure.

As an aside we remark that Cantor-like closed sets in $\mathbb{R}$ might or might not admit continuous extension operators; see [101, Beispiel 1], [102], and the final result in [5], where a complete characterization is given in terms of the logarithmic dimension of the Cantor-like set.

4.10 Theorem ([52, Theorem 4]) *Let M be a connected compact oriented Whitney manifold germ. Let $\omega_0, \omega \in \Omega^m(M)$ be two volume forms (both > 0) with $\int_M \omega_0 = \int_M \omega$. Suppose that there is a diffeomorphism $f : M \to M$ such that $f^*\omega|_U = \omega_0|_U$ for an open neighborhood of ∂M in M.*

Then there exists a diffeomorphism $\tilde{f} : M \to M$ with $\tilde{f}^*\omega = \omega_0$ such that $\tilde{f}$ equals f on an open neighborhood of ∂M.

This relative Moser theorem for Whitney manifold germs is modeled on the standard proof of Moser's theorem in [73, Theorem 31.13]. The proof of [52, Theorem 4] is for manifolds with corners, but it works without change for Whitney manifold germs.

5 Manifolds of Mappings

In this section we demonstrate how convenient calculus allows for very short and transparent proofs of the core results in the theory of manifolds of smooth mappings. We follow [55] but we allow the source manifold to be a Whitney manifold germ. In

[69] M was allowed to have corners. We will treat manifolds of smooth mappings, and of C^n-mappings, and we will also mention the case of Sobolev mappings.

5.1 Lemma (Smooth Curves into Spaces of Sections of Vector Bundles) *Let $p\colon E \to M$ be a vector bundle over a compact smooth manifold M, possibly with corners.*

(1) *Then the space $C^\infty(\mathbb{R}, \Gamma(E))$ of all smooth curves in $\Gamma(E)$ consists of all $c \in C^\infty(\mathbb{R} \times M, E)$ with $p \circ c = \mathrm{pr}_2 : \mathbb{R} \times M \to M$.*
(2) *Then the space $C^\infty(\mathbb{R}, \Gamma_{C^n}(E))$ of all smooth curves in $\Gamma_{C^n}(E)$ consists of all $c \in C^{\infty,n}(\mathbb{R} \times M, E)$ (see Sect. 4.8) with $p \circ c = \mathrm{pr}_2 : \mathbb{R} \times M \to M$.*
(3) *If M is a compact manifold or a compact Whitney manifold germ, then for each $1 < p < \infty$ and $s \in (\dim(M)/p, \infty)$ the space $C^\infty(\mathbb{R}, \Gamma_{W^{s,p}}(E))$ of smooth curves in $\Gamma_{W^{s,p}}(E)$ consists of all continuous mappings $c : \mathbb{R} \times M \to E$ with $p \circ c = \mathrm{pr}_2 : \mathbb{R} \times M \to M$ such that the following two conditions hold:*

- *For each $x \in M$ the curve $t \mapsto c(t, x) \in E_x$ is smooth; let $(\partial_t^k c)(t, x) = \partial_t^k(c(\ , x))(t)$.*
- *For each $k \in \mathbb{N}_{\geq 0}$, the curve $\partial_t^k c$ has values in $\Gamma_{W^{s,p}}(E)$ so that $\partial_t^k c : \mathbb{R} \to \Gamma_{W^{s,p}}(E)$, and $t \mapsto \|\partial_t^k c(t, \cdot)\|_{\Gamma_{W^{s,p}(E)}}$ is bounded, locally in t.*

(4) *If M is an open manifold, then the space $C^\infty(\mathbb{R}, \Gamma_c(E))$ of all smooth curves in the space $\Gamma_c(E)$ of smooth sections with compact support consists of all $c \in C^\infty(\mathbb{R} \times M, E)$ with $p \circ c = \mathrm{pr}_2 : \mathbb{R} \times M \to M$ such that*

- *for each compact interval $[a, b] \subset \mathbb{R}$ there is a compact subset $K \subset M$ such that $c(t, x) = 0$ for $(t, x) \in [a, b] \times (M \setminus K)$.*

Likewise for the space $C^\infty(\mathbb{R}, \Gamma_{C^n,c}(E))$ of smooth curves in the space of C^n-sections with compact support.
(5) *Let $p\colon E \to M$ be a vector bundle over a compact Whitney manifold germ. Then the space $C^\infty(\mathbb{R}, \Gamma(E))$ of smooth curves in $\Gamma(E)$ consists of all smooth mappings $c : \mathbb{R} \times \tilde{M} \to \tilde{E}$ with $p \circ c = \mathrm{pr}_2 : \mathbb{R} \times \tilde{M} \to \tilde{M}$ for some open neighborhood manifold $\tilde{M}$ and extended vector bundle $\tilde{E}$. We may even assume that there is a compact submanifold with smooth boundary $L \subset \tilde{M}$ containing M in its interior such that $c(t, x) = 0$ for $(t, x) \in \mathbb{R} \times (\tilde{M} \setminus L)$. Using the last statement of Sect. 4.1, this is equivalent to the space of all smooth mappings $c : \mathbb{R} \times M \to E \subset \tilde{E}$ with $p \circ c = \mathrm{pr}_2 : \mathbb{R} \times M \to M$.*
(6) *Let $p\colon E \to M$ be a vector bundle over a non-compact Whitney manifold germ $M \subset \tilde{M}$, then the space $C^\infty(\mathbb{R}, \Gamma_c(E))$ of all smooth curves in the space*

$$\Gamma_c(E) = \{s|_M : s \in \Gamma_c(\tilde{M} \leftarrow \tilde{E})\}$$

of smooth sections with compact support (see Sect. 4.8) consists of all smooth mappings $c : \mathbb{R} \times \tilde{M} \to \tilde{E}$ with $p \circ c = \mathrm{pr}_2 : \mathbb{R} \times \tilde{M} \to \tilde{M}$ such that

- *for each compact interval $[a, b] \subset \mathbb{R}$ there is a compact subset $K \subset \tilde{M}$ such that $c(t, x) = 0$ for $(t, x) \in [a, b] \times (M \setminus K)$.*

Proof

(1) This follows from the exponential law in Sect. 2.4.6 after trivializing the bundle.
(2) We trivialize the bundle, assume that M is open in $\mathbb{R}^m$, and then prove this directly. In [55, 3.1 and 3.2] one finds a very explicit proof of the case $n = \infty$, which one can restrict to our case here.
(3) To see this we first choose a second vector bundle $F \to M$ such that $E \oplus_M F$ is a trivial bundle, i.e., isomorphic to $M \times \mathbb{R}^n$ for some $n \in \mathbb{N}$. Then $\Gamma_{W^{s,p}}(E)$ is a direct summand in $W^{s,p}(M, \mathbb{R}^n)$, so that we may assume without loss that E is a trivial bundle, and then, that it is 1-dimensional. So we have to identify $C^\infty(\mathbb{R}, W^{s,p}(M, \mathbb{R}))$. But in this situation we can just apply Theorem 2.6 for the set $\mathcal{V} \subset W^{s,p}(M, \mathbb{R})'$ consisting of all point evaluations $\mathrm{ev}_x : H^s(M, \mathbb{R}) \to \mathbb{R}$ and use that $W^{s,p}(M, \mathbb{R})$ is a reflexive Banach space.
(4) This is like (1) or (2) where we have to assure that the curve c takes values in the space of sections with compact support which translates to the condition.
(5) and (6) follow from (4) after extending to $\tilde{E} \to \tilde{M}$.

□

5.2 Lemma *Let E_1, E_2 be vector bundles over smooth manifold or a Whitney manifold germ M, let $U \subset E_1$ be an open neighborhood of the image of a smooth section, let $F : U \to E_2$ be a fiber preserving smooth mapping. Then the following statements hold:*

(1) *If M is compact, the set $\Gamma(U) := \{h \in \Gamma(E_1) : h(M) \subset U\}$ is open in $\Gamma(E_1)$, and the mapping $F_* : \Gamma(U) \to \Gamma(E_2)$ given by $h \mapsto F \circ h$ is smooth. Likewise for spaces $\Gamma_c(E_i)$, if M is not compact.*
(2) *If M is compact, for $n \in \mathbb{N}_{\geq 0}$ the set*

$$\Gamma_{C^n}(U) := \{h \in \Gamma_{C^n}(E_1) : h(M) \subset U\}$$

is open in $\Gamma_{C^n}(E_1)$, and the mapping $F_* : \Gamma_{C^n}(U) \to \Gamma_{C^n}(E_2)$ given by $h \mapsto F \circ h$ is smooth.
(3) *If M is compact and $s > \dim(M)/p$, the set*

$$\Gamma_{W^{s,p}}(U) := \{h \in \Gamma_{W^{s,p}}(E_1) : h(M) \subset U\}$$

is open in $\Gamma_{W^{s,p}}(E_1)$, and the mapping $F_* : \Gamma_{W^{s,p}}(U) \to \Gamma_{W^{s,p}}(E_2)$ given by $h \mapsto F \circ h$ is smooth.

If the restriction of F to each fiber of E_1 is real analytic, then F_* is real analytic; but in this paper we concentrate on C^∞ only. This lemma is a variant of the so-called Omega-lemma; e.g., see [69]. Note how simple the proof is using convenient calculus.

Proof Without loss suppose that $U = E_1$.

(1) and (2) follow easily since F_* maps smooth curves to smooth curves; see their description in Lemma 5.1(1) and (2).

(3) Let $c : \mathbb{R} \ni t \mapsto c(t,\ ) \in \Gamma_{W^{s,p}}(E_1)$ be a smooth curve. As $s > \dim(M)/2$, it holds for each $x \in M$ that the mapping $\mathbb{R} \ni t \mapsto F_x(c(t,x)) \in (E_2)_x$ is smooth. By the Faà di Bruno formula (see [34] for the 1-dimensional version, preceded in [3] by 55 years), we have for each $p \in \mathbb{N}_{>0}$, $t \in \mathbb{R}$, and $x \in M$ that

$$\partial_t^p F_x(c(t,x)) =$$

$$= \sum_{j \in \mathbb{N}_{>0}} \sum_{\substack{\alpha \in \mathbb{N}_{>0}^j \\ \alpha_1 + \cdots + \alpha_j = p}} \frac{1}{j!} d^j(F_x)(c(t,x)) \Big(\frac{\partial_t^{(\alpha_1)} c(t,x)}{\alpha_1!}, \ldots, \frac{\partial_t^{(\alpha_j)} c(t,x)}{\alpha_j!} \Big).$$

For each $x \in M$ and $\alpha_x \in (E_2)_x^*$ the mapping $s \mapsto \langle s(x), \alpha_x \rangle$ is a continuous linear functional on the Hilbert space $\Gamma_{W^{s,p}}(E_2)$. The set $\mathcal{V}_2$ of all of these functionals separates points and therefore satisfies the condition of Theorem 2.6. We also have for each $p \in \mathbb{N}_{>0}$, $t \in \mathbb{R}$, and $x \in M$ that

$$\partial_t^p \langle F_x(c(t,x)), \alpha_x \rangle = \langle \partial_t^p F_x(c(t,x)), \alpha_x \rangle = \langle \partial_t^p F_x(c(t,x)), \alpha_x \rangle.$$

Using the explicit expressions for $\partial_t^p F_x(c(t,x))$ from above we may apply Lemma (5.1.3) to conclude that $t \mapsto F(c(t,\ ))$ is a smooth curve $\mathbb{R} \to \Gamma_{H^s}(E_1)$. Thus, F_* is a smooth mapping. □

5.3 The Manifold Structure on $C^\infty(M,N)$ and $C^k(M,N)$

Let M be a compact or open finite dimensional smooth manifold or even a compact Whitney manifold germ, and let N be a smooth manifold. We use an auxiliary Riemannian metric $\bar{g}$ on N and its exponential mapping $\exp^{\bar{g}}$; some of its properties are summarized in the following diagram:

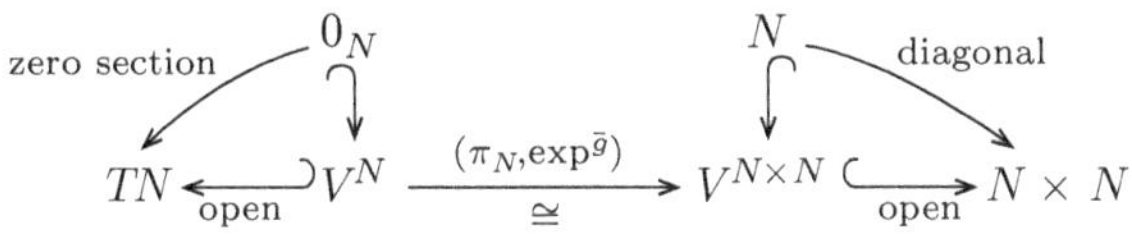

$$\begin{array}{ccccccc} & \overset{\text{zero section}}{\swarrow} & 0_N & & N & \overset{\text{diagonal}}{\searrow} & \\ & & \downarrow & & \downarrow & & \\ TN & \underset{\text{open}}{\hookleftarrow} & V^N & \xrightarrow[\cong]{(\pi_N, \exp^{\bar{g}})} & V^{N\times N} & \underset{\text{open}}{\hookrightarrow} & N \times N \end{array}$$

Without loss we may assume that $V^{N\times N}$ is symmetric:

$$(y_1, y_2) \in V^{N\times N} \iff (y_2, y_1) \in V^{N\times N}.$$

- If M is compact, then $C^\infty(M,N)$, the space of smooth mappings $M \to N$ has the following manifold structure. A chart, centered at $f \in C^\infty(M,N)$, is

$$C^\infty(M,N) \supset U_f = \{g : (f,g)(M) \subset V^{N\times N}\} \xrightarrow{u_f} \tilde U_f \subset \Gamma(M \leftarrow f^*TN)$$
$$u_f(g) = (\pi_N, \exp^{\bar g})^{-1} \circ (f,g), \quad u_f(g)(x) = (\exp^{\bar g}_{f(x)})^{-1}(g(x))$$
$$(u_f)^{-1}(s) = \exp^{\bar g}_f \circ s, \qquad (u_f)^{-1}(s)(x) = \exp^{\bar g}_{f(x)}(s(x)).$$

Note that $\tilde U_f$ is open in $\Gamma(M \leftarrow f^*TN)$ if M is compact.

- If M is open, then the compact C^∞-topology on $\Gamma(f^*TN)$ is not suitable since $\tilde U_f$ is in general not open. We have to control the behavior of sections near infinity on M. One solution is to use the space $\Gamma_c(f^*TN)$ of sections with compact support as modeling spaces and to adapt the topology on $C^\infty(M,N)$ accordingly. This has been worked out in [69] and [55].
- If M is compact Whitney manifold germ with neighborhood manifold $\tilde M \supset M$ we use the Fréchet space $\Gamma(M \leftarrow f^*TN) = \{s|_M : s \in \Gamma_L(\tilde M \leftarrow \tilde f^*TN)\}$ where $L \subset \tilde M$ is a compact set containing M in its interior and $\tilde f : \tilde M \to N$ is an extension of f to a suitable manifold neighborhood of M. Via an extension operator the Fréchet space $\Gamma(M \leftarrow f^*TN)$ is a direct summand in the Fréchet space $\Gamma_L(\tilde M \leftarrow \tilde f^*TN)$ of smooth sections with support in L.
- Likewise, for a non-compact Whitney manifold germ we use the convenient (LF)-space
$$\Gamma_c(M \leftarrow f^*TN) = \{s|_M : s \in \Gamma_c(\tilde M \leftarrow \tilde f^*TN)\}$$
of sections with compact support.
- On the space $C^k(M,N,)$ for $k \in \mathbb{N}_{\geq 0}$ we use only charts as described above with the center $f \in C^\infty(M,N)$, namely
$$C^k(M,N) \supset U_f = \{g : (f,g)(M) \subset V^{N\times N}\} \xrightarrow{u_f} \tilde U_f \subset \Gamma_{C^k}(M \leftarrow f^*TN)\,.$$
We claim that these charts cover $C^k(M,N)$: Since $C^\infty(M,N)$ is dense in $C^k(M,N)$ in the Whitney C^k-topology, for any $g \in C^k(M,N)$ there exists $f \in C^\infty(M,N,) \cap U_g$. But then $g \in U_f$ since $V^{N\times N}$ is symmetric. This is true for compact M. For a compact Whitney manifold germ we can apply this argument in a compact neighborhood L of M in $\tilde M$, replacing $\tilde M$ by the interior of L after the fact.
- On the space $W^{s,p}(M,N)$ for $\dim(M)/p < s \in \mathbb{R}$ we use only charts as described above with the center $f \in C^\infty(M,N)$, namely
$$W^{s,p}(M,N) \supset U_f = \{g : (f,g)(M) \subset V^{N\times N}\} \xrightarrow{u_f}$$
$$\xrightarrow{u_f} \tilde U_f \subset \Gamma_{W^{s,p}}(M \leftarrow f^*TN)\,.$$
These charts cover $W^{s,p}(M,N)$, by the following argument: Since $C^\infty(M,N)$ is dense in $W^{s,p}(M,N)$ and since $W^{s,p}(M,N) \subset C^k(M,N)$ via a continuous injection for $0 \leq k < s - \dim(M)/p$, a suitable C^0 – sup-norm neighborhood

of $g \in W^{s,p}(M,N)$ contains a smooth $f \in C^\infty(M,N)$, thus $f \in U_g$ and by symmetry of $V^{N\times N}$ we have $g \in U_f$. This is true for compact M. For a compact Whitney manifold germ we can apply this argument in a compact neighborhood which is a manifold with smooth boundary L of M in $\tilde{M}$ and apply the argument there.

In each case, we equip $C^\infty(M,N)$ or $C^k(M,N)$ or $W^{s,p}(M,N)$ with the initial topology with respect to all chart mappings described above: The coarsest topology, so that all chart mappings u_f are continuous.

For non-compact M the direct limit $\Gamma_c(f^*TN) = \varinjlim_L \Gamma_L(f^*TN)$ over a compact exhaustion L of M in the category of locally convex vector spaces is strictly coarser that the direct limit in the category of Hausdorff topological spaces. It is more *convenient* to use the latter topology which is called c^∞ topology; compare with Sect. 2.1.

5.4 Lemma

(1) *If M is a compact smooth manifold or is a compact Whitney manifold germ,*

$$C^\infty(\mathbb{R}, \Gamma(M \leftarrow f^*TN)) = \Gamma(\mathbb{R}\times M \leftarrow {\rm pr_2}^* f^*TN)\,.$$

For smooth $f \in C^\infty(M,N)$,

$$C^\infty(\mathbb{R}, \Gamma_{C^n}(M \leftarrow f^*TN)) = \Gamma_{C^{\infty,n}}(\mathbb{R}\times M \leftarrow {\rm pr_2}^* f^*TN)\,.$$

(2) *If M is a non-compact smooth manifold of Whitney manifold germ, the sections on the right-hand side have to satisfy the corresponding conditions of Lemma 5.1(4).*

For a compact Whitney manifold germ M the space $\Gamma(\mathbb{R}\times M \leftarrow {\rm pr_2}^* f^*TN)$ is a direct summand in the space $\Gamma_{\mathbb{R}\times L}(\mathbb{R}\times \tilde{M} \leftarrow {\rm pr_2}^* f^*TN)$ of sections with support in $\mathbb{R}\times L$ for a fixed compact set $L \subset \tilde{M}$ containing M in its interior. Likewise $\Gamma_{C^{\infty,n}}(\mathbb{R}\times M \leftarrow {\rm pr_2}^* f^*TN)$ is a direct summand in the space $\Gamma_{C^{\infty,n},\mathbb{R}\times L}(\mathbb{R}\times \tilde{M} \leftarrow {\rm pr_2}^* f^*TN)$ of $C^{\infty,n}$-sections. One could introduce similar notation for $C^\infty(\mathbb{R}, \Gamma_{W^{s,p}}(M \leftarrow f^*TN))$.

Proof This follows from Lemma 5.1. □

5.5 Lemma *Let M be a smooth manifold or Whitney manifold germ, compact or not, and let N be a manifold. Then the chart changes for charts centered on smooth mappings are smooth (C^∞) on the space $C^\infty(M,N)$, also on $C^k(M,N)$ for $k \in \mathbb{N}_{\geq 0}$, and on $W^{s,p}(M,N)$ for $1 < p < \infty$ and $s >$ $\dim(M)/p$:*

$$\tilde{U}_{f_1} \ni s \mapsto (u_{f_2,f_1})_*(s) := (\exp^{\bar{g}}_{f_2})^{-1} \circ \exp^{\bar{g}}_{f_1} \circ s \in \tilde{U}_{f_2}\,.$$

Proof This follows from Lemma 5.2, since any chart change is just compositions from the left by a smooth fiber respecting locally defined diffeomorphism. □

5.6 Lemma

(1) *If M is a compact manifold or a compact Whitney manifold germ, then*

$$C^\infty(\mathbb{R}, C^\infty(M, N)) \cong C^\infty(\mathbb{R} \times M, N).$$

(2) *If M is not compact, $C^\infty(\mathbb{R}, C^\infty(M, N))$ consists of all smooth $c : \mathbb{R}\times M \to N$ such that*

- *for each compact interval $[a, b] \subset \mathbb{R}$ there is a compact subset $K \subset M$ such that $c(t, x)$ is constant in $t \in [a, b]$ for each $x \in M \setminus K$.*

Proof By Lemma 5.4. □

5.7 Lemma *Composition $(f, g) \mapsto g \circ f$ is smooth as a mapping*

$$C^\infty(P, M) \times C^\infty(M, N) \to C^\infty(P, N)$$
$$C^k(P, M) \times C^\infty(M, N) \to C^k(P, N)$$
$$W^{s,p}(P, M) \times C^\infty(M, N) \to W^{s,p}(P, N)$$

for P a manifold or a Whitney manifold germ, compact or not, and for M and N manifolds.

For more general M the description becomes more complicated. See the special case of the diffeomorphism group of a Whitney manifold germ M in Sect. 6.3 below.

Proof Since it maps smooth curves to smooth curves. □

5.8 Corollary *For M a manifold or a Whitney manifold germ and a manifold N, the tangent bundle of the manifold $C^\infty(M, N)$ of mappings is given by*

$$TC^\infty(M, N) = C^\infty(M, TN) \xrightarrow{C^\infty(M,\pi_N)=(\pi_N)_*} C^\infty(M, N),$$
$$TC^k(M, N) = C^k(M, TN) \xrightarrow{C^k(M,\pi_N)=(\pi_N)_*} C^k(M, N),$$
$$TW^{s,p}(M, N) = W^{s,p}(M, TN) \xrightarrow{W^{s,p}(M,\pi_N)=(\pi_N)_*} W^{s,p}(M, N).$$

Proof This follows from the chart structure and the fact that sections of $f^*TN \to M$ correspond to mappings $s : M \to TN$ with $\pi_N \circ s = f$. □

5.9 Sprays Respecting Fibers of Submersions

Sprays are versions of Christoffel symbols and lead to exponential mappings. They are easier to adapt to fibered manifolds than Riemannian metrics. Recall that a spray S on a manifold N without boundary is a smooth mapping $S : TN \to T^2N$ with the following properties:

- $\pi_{TN} \circ S = \mathrm{Id}_{TN}$; S is a vector field.
- $T(\pi_N) \circ S = \mathrm{Id}_{TN}$; S is a "differential equation of second order."
- Let $m_t^N : TN \to TN$ and $m_t^{TN} : T^2N \to T^2N$ be the scalar multiplications. Then $S \circ m_t^N = T(m_t^N).m_t^{TN}.S$.

Locally, in charts of TN and T^2N induced by a chart of N, a spray looks like $S(x, v) = (x, v; v; \Gamma(x, v))$ where Γ is quadratic in v. For a spray $S \in \mathfrak{X}(TN)$ on a manifold N, we let $\exp(X) := \pi_N(\mathrm{Fl}_1^S(X))$, then the mapping $\exp : TN \supset V \to N$ is smooth, defined on an open neighborhood V of the zero section in TN, which is called the *exponential mapping* of the spray S. Since $T_{0_x}(\exp|_{T_xN}) = \mathrm{Id}_{T_xN}$ (via $T_{0_x}(T_xN) = T_xN$), by the inverse function theorem $\exp_x := \exp|_{T_xN}$ is a diffeomorphism near 0_x in TN onto an open neighborhood of x in N. Moreover the mapping $(\pi_N, \exp) : TN \supset \tilde{V} \to N \times N$ is a diffeomorphism from an open neighborhood $\tilde{V}$ of the zero section in TN onto an open neighborhood of the diagonal in $N \times N$.

Lemma *Let $q : N \to M$ be a smooth surjective submersion between connected manifolds without boundary. Then there exists a spray S on N which is tangential to the fibers of q, i.e., $S(T(q^{-1}(x))) \subset T^2(q^{-1}(x))$ for each $x \in M$.*

This is a simplified version of [69, 10.9].

Proof In suitable charts on N and M the submersion q looks like a linear projection $(y_1, y_2) \mapsto y_1$. The local expression $T(\text{chart}) \to T^2(\text{chart})$ of a spray is

$$
\begin{aligned}
&S\big((y_1, y_2), (v_1, v_2)\big) = \\
&\qquad = \big((y_1, y_2), (v_1, v_2); (v_1, v_2), (\Gamma^1(y_1, y_2; v_1, v_2), \Gamma^2(y_1, y_2, v_1, v_2))\big),
\end{aligned}
$$

where $\Gamma^i(y_1, y_2, v_1, v_2)$ is quadratic in (v_1, v_2). The spray is tangential to the fibers of q if and only if $\Gamma^1(y_1, y_2, 0, v_2) = 0$. This clearly exists locally (e.g., choose $\Gamma^1 = 0$). Now we use a partition of unity (φ_α) subordinated to a cover $N = \bigcup_\alpha U_\alpha$ with such charts and glue local sprays with the induced partition of unity $(\varphi_\alpha \circ \pi_N)$ subordinated to the cover $TN = \bigcup_\alpha TU\alpha$ for the vector bundle $\pi_{TN} : T^2N \to TN$. Locally this looks like (where $y = (y_1, y_2)$, etc.)

$$
\begin{aligned}
\Big(\sum_\alpha (\varphi_\alpha \circ \pi_N).S_\alpha\Big)(y, v) &= \Big(y, v; \sum_\alpha \varphi_\alpha(y)v, \sum_\alpha \varphi_\alpha(y)\big(\Gamma^1_\alpha(y, v), (\Gamma^2_\alpha(y, v)\big)\Big) \\
&= \Big(y, v; v, \big(\sum_\alpha \varphi_\alpha(y)\Gamma^1_\alpha(y, v), \sum_\alpha \varphi_\alpha(y)\Gamma^2_\alpha(y, v)\big)\Big)
\end{aligned}
$$

and is therefore a spray which is tangential to the fibers of q. □

5.10 Proposition ([69, 10.10]) *Let $q : N \to M$ be a smooth surjective submersion between connected manifolds without boundary. The space $S^q(M, N)$ of all smooth sections of q is a splitting smooth submanifold of $C^\infty(M, N)$. Similarly, the spaces*

$S^q_{C^N}(M,N)$ *and* $S^q_{W^{s,p}}(M,N)$ *of* C^N*-sections and* $W^{s,p}$*-sections are smooth splitting submanifolds of* $C^N(M,N)$ *or* $W^{s,p}(M,N)$ *(for* $s > \dim(M)/p$*), respectively.*

The proof given here is simpler than the one in [69, 10.10].

Proof Let us first assume that M is compact. Given a smooth section $f \in S^q(M,N)$, consider the chart centered at f from Sect. 5.3

$$C^\infty(M,N) \supset U_f = \{g : (f,g)(M) \subset V^{N\times N}\} \xrightarrow{u_f} \tilde{U}_f \subset \Gamma(M \leftarrow f^*TN)$$
$$u_f(g) = (\pi_N, \exp^S)^{-1} \circ (f,g), \quad u_f(g)(x) = (\exp^S_{f(x)})^{-1}(g(x))$$
$$(u_f)^{-1}(s) = \exp^S_f \circ s, \qquad (u_f)^{-1}(s)(x) = \exp^S_{f(x)}(s(x)),$$

where we use the exponential mapping with respect to a spray S on N which is tangential to the fibers of q. Using an unrelated auxiliary Riemannian metric $\bar{g}$ on N we can smoothly split the tangent bundle $TN = V^q(N) \oplus H^q(N)$ into the vertical bundle of all vectors tangent to the fibers of q, and into its orthogonal complement with respect to $\bar{g}$. The orthonormal projections $P^{\bar{g}} : TN \to V^q(N)$ and $\mathrm{Id}_{TN} - P^{\bar{g}} : TN \to H^q(N)$ induce the direct sum decomposition

$$\Gamma(M \leftarrow f^*TN) = \Gamma(M \leftarrow f^*V^q(N)) \oplus \Gamma(M \leftarrow f^*TN) \quad s \mapsto (P^{\bar{g}}.s, s - P^{\bar{g}}.s).$$

Now $g \in U_f$ is in $S^q(M,N)$ if and only if $u_f(g) \in \Gamma(f^*V^q(N))$.

If M is not compact we may use the spaces of sections with compact support as described in Sect. 5.3. Similarly for the cases of C^N-sections or $W^{s,p}$-sections. □

5.11 Corollary *Let* $p : E \to M$ *be a fiber bundle over a compact Whitney manifold germ* M*. Then the space* $\Gamma(E)$ *of smooth sections is a splitting smooth submanifold of* $C^\infty(M,E)$*. Likewise for the spaces* $\Gamma_{C^N}(E)$ *and* $\Gamma_{W^{s,p}}(E)$ *of* C^N*-sections and* $W^{s,p}$*-sections.*

Proof Recall from Sect. 4.8 that $E = \tilde{E}|_M$ for a smooth fiber bundle $\tilde{E} \to \tilde{M}$. There the result follows from 5.10. Using (fixed) extension operators

$$\Gamma(M \leftarrow f^*TE) \to \Gamma_L(\tilde{M} \leftarrow \tilde{f}^*T\tilde{E}),$$

etc. we can extend this the case of Whitney manifold germs. □

6 Regular Lie Groups

6.1 Regular Lie Groups

We consider a smooth Lie group G with Lie algebra $\mathfrak{g} = T_eG$ modeled on convenient vector spaces. The notion of a regular Lie group is originally due to [86–

91] for Fréchet Lie groups, was weakened and made more transparent by Milnor [78], and then carried over to convenient Lie groups in [56], see also [55, 38.4]. We shall write $\mu : G \times G \to G$ for the multiplication with $x.y = \mu(x, y) = \mu_x(y) = \mu^y(x)$ for left and right translation.

A Lie group G is called *regular* if the following holds:

- For each smooth curve $X \in C^\infty(\mathbb{R}, \mathfrak{g})$ there exists a curve $g \in C^\infty(\mathbb{R}, G)$ whose right logarithmic derivative is X, i.e.,

$$\begin{cases} g(0) & = e \\ \partial_t g(t) & = T_e(\mu^{g(t)})X(t) = X(t).g(t). \end{cases}$$

  The curve g is uniquely determined by its initial value $g(0)$, if it exists.
- Put $\mathrm{evol}^r_G(X) = g(1)$ where g is the unique solution required above. Then $\mathrm{evol}^r_G : C^\infty(\mathbb{R}, \mathfrak{g}) \to G$ is required to be C^∞ also. We have $\mathrm{Evol}^X_t := g(t) = \mathrm{evol}^r_G(tX)$.

Of course we could equivalently use the left logarithmic derivative and the corresponding left evolution operator. Group inversion maps the two concepts into each other. See [55, Section 38] for more information. Up to now, every Lie group modeled on convenient vector spaces is regular.

There are other notions of regularity for infinite dimensional Lie groups: For example, one may require that each curve $X \in L^1_{\mathrm{loc}}(\mathbb{R}, \mathfrak{g})$ admits an absolutely continuous curve $\mathrm{Evol}^X : \mathbb{R} \to G$ whose right logarithmic derivative is X. See [46] or [49] and references therein. It might be that all these notions of regularity are equivalent for Lie groups modeled on convenient vector spaces.

6.2 Theorem *For each manifold M with or without corners, the diffeomorphism group* $\mathrm{Diff}(M)$ *is a regular Lie group. Its Lie algebra is the space $\mathfrak{X}(M)$ of all vector fields with the negative of the usual bracket as Lie bracket, if M is compact without boundary. It is the space $\mathfrak{X}_c(M)$ of fields with compact support, if M is an open manifold. It is the space $\mathfrak{X}_\partial(M)$ of Sect. 4.7 of vector fields tangent to the boundary, if M is a compact manifold with corners. If M is not compact with corners, then the Lie algebra is the space $\mathfrak{X}_{c,\partial}(M)$ of boundary respecting vector fields with compact support.*

Proof If M is a manifold without boundary then $\mathrm{Diff}(M) \xrightarrow{open} C^\infty(M, M)$. If M is open, then the group of diffeomorphisms differing from the identity only on a compact set is open in $\mathrm{Diff}(M)$.

If M has corners we use an open manifold $\tilde{M}$ containing M as a submanifold with corners as in Lemma 3.2. In the description of the chart structure in Sect. 5.3 for $\mathrm{Diff}(\tilde{M})$ we have to use the exponential mapping for a geodesic spray on $\tilde{M}$ such that each component of each $\partial^q M$ is totally geodesic. This spray exists; see Sect. 3.7 or Sect. 5.9. Restricting all sections to M then yields a smooth chart centered at the identity for $\mathrm{Diff}(M)$. Then we use right translations of this chart. The explicit chart structure on $\mathrm{Diff}(M)$ is described in [69, 10.16]. Extending all sections to $\tilde{M}$ via a fixed continuous linear Whitney extension operator respecting compact support identifies $\mathrm{Diff}(M)$ as a splitting smooth closed submanifold of $\mathrm{Diff}(\tilde{M})$, but not as a subgroup.

Composition is smooth by restricting it from $C^\infty(M, M) \times C^\infty(M, M)$, using Lemma 5.7 and its extension to the situation with corners.

Inversion is smooth: If $t \mapsto f(t, \quad)$ is a smooth curve in $\mathrm{Diff}(M)$, then $f(t, \quad)^{-1}$ satisfies the implicit equation $f(t, f(t, \quad)^{-1}(x)) = x$, so by the finite dimensional implicit function theorem, $(t, x) \mapsto f(t, \quad)^{-1}(x)$ is smooth. So inversion maps smooth curves to smooth curves, and is smooth.

Let $X(t, x)$ be a time-dependent vector field on M (in $C^\infty(\mathbb{R}, \mathfrak{X}(M))$). Then $\mathrm{Fl}_s^{\partial_t \times X}(t, x) = (t + s, \mathrm{Evol}^X(t, x))$ satisfies the ordinary differential equation

$$\partial_t \,\mathrm{Evol}(t, x) = X(t, \mathrm{Evol}(t, x)).$$

If $X(s, t, x) \in C^\infty(\mathbb{R}^2, \mathfrak{X}(M))$ is a smooth curve of smooth curves in $\mathfrak{X}(M)$, then obviously the solution of the equation depends smoothly also on the further variable s, thus evol maps smooth curves of time dependent vector fields to smooth curves of diffeomorphism. □

6.3 *The Diffeomorphism Group of a Whitney Manifold Germ*

For a Whitney manifold germ $\tilde{M} \supset M$, we consider the diffeomorphism group

$$\begin{aligned}\mathrm{Diff}(M) = \{\varphi|_M : \varphi \in C^\infty(\tilde{M}, \tilde{M}),\ \varphi(M) = M,\\ \varphi \text{ is a diffeomorphism on an open neighborhood of } M\}.\end{aligned}$$

We also consider the following set $\mathcal{C}$ of smooth curves: Those $c : \mathbb{R} \to \mathrm{Diff}(M)$ which are of the form $c = \tilde{c}|_{\mathbb{R}\times M}$ for a smooth

$$\tilde{c} : \mathbb{R} \times \tilde{M} \to \tilde{M} \quad \text{ with } \tilde{c}(t, \quad)|_M \in \mathrm{Diff}(M) \text{ for each } t \in \mathbb{R}.$$

Note that for t in a compact interval $\tilde{c}(t, \quad)$ is a diffeomorphism on a fixed open neighborhood of M in $\tilde{M}$.

6.4 Theorem *For a Whitney manifold germ M the group* $\mathrm{Diff}(M)$ *is a Frölicher space and a group with smooth composition and inversion. It has a convenient Lie algebra $\mathfrak{X}_{c,\partial}(M)$ with the negative of the usual bracket as Lie bracket, and it is regular: There exists an evolution operator and it is smooth.*

Proof The Frölicher space structure is the one induced by the set $\mathcal{C}$ of smooth curves described above. I do not know whether this set of smooth curves is saturated, i.e., $\mathcal{C} = \mathcal{C}_{\mathrm{Diff}(M)}$ in the notation of Sect. 2.7; this might depend on the structure of the boundary.

The proof is now quite similar to the one of Sect. 6.1. We claim that composition maps $\mathcal{C} \times \mathcal{C}$ to $\mathcal{C} \subseteq \mathcal{C}_{\mathrm{Diff}(M)}$, and that inversion maps $\mathcal{C}$ to $\mathcal{C} \subseteq \mathcal{C}_{\mathrm{Diff}(M)}$. Since by definition each curve $c \in \mathcal{C}$ extend to a smooth mapping $\tilde{c} : \mathbb{R} \times \tilde{M} \to \tilde{M}$ we can actually use a slight adaption of the proof of Sect. 6.1 for open manifolds. □

6.5 *The Connected Component of* Diff(M) *for a Whitney Manifold Germ M*

We consider a Whitney manifold germ $M \subset \tilde{M}$. As usual for Frölicher space, we equip Diff(M) with the final topology with respect to all smooth curves in in the generating set $\mathcal{C}$ as described in Sect. 6.3. Diff(M) is actually a topological group, with the refined topology (i.e., the c^∞-topology) on Diff$(M) \times$ Diff(M). Let Diff$_0(M)$ be the connected component of the identity in Diff(M) with respect to this topology.

Theorem *For a Whitney manifold germ $M \subset \tilde{M}$ we actually have*

$$\mathrm{Diff}_0(M) = \{\tilde{\varphi}|_M : \tilde{\varphi} \in \mathrm{Diff}_0(\tilde{M}),\ \tilde{\varphi}(M) = M\}.$$

Consequently, the subgroup

$$\mathrm{Diff}^{\tilde{}}(M) = \{\tilde{\varphi}|_M : \tilde{\varphi} \in \mathrm{Diff}(\tilde{M}),\ \tilde{\varphi}(M) = M\}$$

is an open subgroup in Diff(M) *and thus a normal subgroup, and the corresponding generating set $\mathcal{C}$ of smooth curves in* Diff$^{\tilde{}}(M)$ *is saturated.*

Proof Let $\varphi \in$ Diff$_0(M)$. Then there exists a smooth curve $\varphi : \mathbb{R} \to$ Diff(M) with $\varphi(0) =$ Id and $\varphi(1) = \varphi$ of the form $\varphi = \tilde{c}|_{\mathbb{R}\times M}$ where $\tilde{c} : \mathbb{R} \times \tilde{M} \to \tilde{M}$ is a smooth mapping with $\tilde{c}(t,\ )|_M \in$ Diff(M) for each $t \in \mathbb{R}$. Then $X(t,x) = (\partial_t\varphi(t))(\varphi(t)^{-1}(x))$ gives us a time-dependent vector field which is defined on $[0,1] \times U$ for some open neighborhood U of M in $\tilde{M}$, by the definition of Diff(M) in Sect. 6.3. Using a continuous extension operator on $X|_{[0,1]\times M}$ and a smooth bump function gives us a smooth time-dependent vector field $\tilde{X} : [0,1] \times \tilde{M} \to T\tilde{M}$ with support in a fixed open neighborhood, say, such that $\tilde{X}|_{[0,1]\times M} = X|_{[0,1]\times M}$. Solving the ODE $\partial_t\tilde{\varphi}(t.x) = \tilde{X}(t, \tilde{\varphi}(t,x))$ on $\tilde{M}$ gives us for $t = 1$ a diffeomorphism $\tilde{\varphi} \in$ Diff$(\tilde{M})$ which extends φ.

Given any $\varphi \in$ Diff$^{\tilde{}}(M)$, the coset φ.Diff$_0(M) \subset$ Diff(M) is the connected component of φ in Diff(M). This shows that Diff$^{\tilde{}}(M)$ is open in Diff(M). □

The construction in the proof above actually describes a smooth mapping

$$\mathcal{E} : \{c \in C^\infty(\mathbb{R}, \mathrm{Diff}_0(M)) : c(0) = \mathrm{Id}\} \to \{\tilde{\varphi} \in \mathrm{Diff}_0(\tilde{M}) : \tilde{\varphi}(M) = M\}$$

such that $\mathcal{E}(c)|_M = c(1)$, since another smooth real parameter s goes smoothly through solving the ODE.

6.6 Remark

In this paper I refrain from trying to give a general definition of a regular Frölicher group, which would be an abstract concept that catches the essential properties of $\mathrm{Diff}(M)$ for a Whitney manifold germ $M \subset \tilde{M}$. Let me just remark, that it probably would fit into the concept of manifolds based on smooth curves instead of charts as developed in [72]; those among them whose tangent spaces are Banach spaces turn out to be Banach manifolds. Some Lie theoretic tools are developed in the beginning of Sect. 8.5 below.

6.7 Regular (Right) Half Lie Groups

A smooth manifold G modeled on convenient vector spaces is called a *(right) half Lie group*, if it is a group such that multiplication $\mu : G \times G \to G$ and inversion $\nu : G \to G$ are continuous (note that here we have to take the induced c^∞-topology on the product $G \times G$ if the model spaces are not Fréchet), but each right translation $\mu^x : G \to G$, $\mu^x(y) = y.x$ is smooth. The notion of a half Lie group was coined in [60]. See [64] for a study of half Lie groups in general, concentrating on semidirect products with representation spaces.

Not every tangent vector in T_eG can be extended to a left invariant vector field on the whole group, but they can be extended to right invariant vector fields, which are only continuous and not differentiable in general. The same holds for right invariant Riemannian metrics. The tangent space at the identity is not a Lie algebra in general; thus we refrain from calling it $\mathfrak{g}$. Have a look at the examples in Theorem 6.8 to get a feeling for this.

Let us discuss regularity on a (right) half Lie group G: For a smooth curve $g : \mathbb{R} \to G$ the velocity curve $g' : \mathbb{R} \to TG$ is still smooth, and for fixed t the right logarithmic derivative $X(t) := g'(t).g(t)^{-1} = T(\mu^{g(t)^{-1}}).g'(t)$ lies in T_eG, but $t \mapsto X(t)$ is only continuous $\mathbb{R} \to T_eG$. A (right) half Lie group G is called *C^0-regular* if for every C^0-curve $X : \mathbb{R} \to T_eG$ there exists a C^1-curve $\mathrm{Evol}^X = g : \mathbb{R} \to G$ with $g(0) = e$ and $g'(t) = X(t).g(t) = T(\mu^{g(t)}).X(t)$. We also require that $X \mapsto \mathrm{Evol}^X$ is smooth $C^0(\mathbb{R}, T_eG) \to C^1(\mathbb{R}, G)$.

6.8 Theorem (Diffeomorphism Groups of Finite Degrees of Differentiability)

(1) *For a compact smooth manifold M, possibly with corners, and for any $n \in \mathbb{N}_{\geq 1}$ the group $\mathrm{Diff}_{C^n}(M)$ of C^n-diffeomorphism of M is a C^0-regular half Lie group.*
(2) *For a compact smooth manifold M, possibly with corners, and for any $s \geq \dim M/p + 1$, the group $\mathrm{Diff}_{W^{s,p}}(M)$ of Sobolev $W^{s,p}$-diffeomorphism of M is a C^0-regular half Lie group.*

Note that the group of homeomorphisms of M is **not** open in $C^0_{\text{nice}}(M, M)$; see the proof below for C^∞_{nice}. Also note that $T_{\text{Id}}\operatorname{Diff}_{C^n}(M) = \mathfrak{X}_{\partial,C^n}(M)$ is the space of C^n-vector fields which are tangent to the boundary. This is not a Lie algebra, since the Lie bracket of two C^n fields is a C^{n-1} field in general.

Proof

(1) Following [69, 10.16], we construct the smooth manifold structure by using the exponential mapping of a spray on M which is tangential to the boundary; for existence see Sects. 3.7 and 5.9. Let $C^n_{\text{nice}}(M, M,)$ be the set of all C^n-mappings $f : M \to M$ with $f^{-1}(\partial^q M) = \partial^q M$ for each q. Then we use the (restriction of the) chart structure described in Sect. 5.3, using this exponential mappings, and using only charts centered at smooth mappings $f \in C^\infty_{\text{nice}}(M, M)$, as follows:

$$
\begin{aligned}
C^n_{\text{nice}}(M, N) \supset U_f &= \{g : (f, g)(M) \subset V^{M\times M}\} \xrightarrow{u_f} \tilde{U}_f \subset \\
&\subset \{s \in C^n(M, TM) : \pi_M \circ s = f, s(\partial^q M) \subset T(\partial^q M)\} \subset \Gamma_{C^n}(f^* T\tilde{M}), \\
u_f(g) &= (\pi_N, \exp^{\bar{g}})^{-1} \circ (f, g), \quad u_f(g)(x) = (\exp^{\bar{g}}_{f(x)})^{-1}(g(x)), \\
(u_f)^{-1}(s) &= \exp^{\bar{g}}_f \circ s, \qquad (u_f)^{-1}(s)(x) = \exp^{\bar{g}}_{f(x)}(s(x)).
\end{aligned}
$$

By the symmetry of $V^{M\times M}$ (see Sect. 5.3) these charts cover $C^n_{\text{nice}}(M, M)$, and the chart changes are smooth since they map smooth curves (as described in Lemma 5.1(2)) to smooth curves; compare to Lemma 5.7. The group $\operatorname{Diff}_{C^n}(M)$ is open in $C^n_{\text{nice}}(M, M)$, by the implicit function theorem and some easy arguments.

Continuity of composition and inversion are easy to check. Right translations are smooth since they map smooth curves.

C^1-regularity follows easily: Given $X \in C^0(\mathbb{R}, T_{\text{Id}}\operatorname{Diff}_{C^n}(M))$, view it as a time-dependent C^n-vector field on M which is tangential to the boundary, a continuous curve in $\mathfrak{X}_\partial(M)$ and solve the corresponding ODE. The evolution operator Evol is smooth, since it maps smooth curves to smooth curves by standard ODE-arguments.

(2) This follows easily by adapting the proof of (1) above, using that $\operatorname{Diff}_{W^{s,p}} M \subset \operatorname{Diff}_{C^1}(M)$ by the Sobolev embedding lemma.

□

6.9 *Groups of Smooth Diffeomorphisms on* $\mathbb{R}^n$

If we consider the group of all orientation preserving diffeomorphisms $\operatorname{Diff}(\mathbb{R}^n)$ of $\mathbb{R}^n$, it is not an open subset of $C^\infty(\mathbb{R}^n, \mathbb{R}^n)$ with the compact C^∞-topology. So it is not a smooth manifold in the usual sense, but we may consider it as a Lie group in the cartesian closed category of Frölicher spaces, see [55, Section 23], with the structure induced by the injection $f \mapsto (f, f^{-1}) \in C^\infty(\mathbb{R}^n, \mathbb{R}^n) \times C^\infty(\mathbb{R}^n, \mathbb{R}^n)$. Or one can use the setting of "manifolds" based on smooth curves instead of charts, with lots of extra structure (tangent bundle, parallel transport, geodesic structure), described in [72]; this gives a category of smooth "manifolds" where those which have Banach spaces as tangent fibers are exactly the usual smooth manifolds modeled on Banach spaces, which is cartesian closed: $C^\infty(M, N)$ and $\mathrm{Diff}(M)$ are always "manifolds" for "manifolds" M and N, and the exponential law holds.

We shall now describe regular Lie groups in $\mathrm{Diff}(\mathbb{R}^n)$ which are given by diffeomorphisms of the form $f = \mathrm{Id}_{\mathbb{R}} + g$ where g is in some specific convenient vector space of bounded functions in $C^\infty(\mathbb{R}^n, \mathbb{R}^n)$. Now we discuss these spaces on $\mathbb{R}^n$, we describe the smooth curves in them, and we describe the corresponding groups. These results are from [77] and from [60, 61] for the more exotic groups.

The Group $\mathrm{Diff}_{\mathcal{B}}(\mathbb{R}^n)$ The space $\mathcal{B}(\mathbb{R}^n)$ (called $\mathcal{D}_{L^\infty}(\mathbb{R}^n)$ by Schwartz [96]) consists of all smooth functions which have all derivatives (separately) bounded. It is a Fréchet space. By Vogt [105], the space $\mathcal{B}(\mathbb{R}^n)$ is linearly isomorphic to $\ell^\infty \hat{\otimes} \mathfrak{s}$ for any completed tensor-product between the projective one and the injective one, where $\mathfrak{s}$ is the nuclear Fréchet space of rapidly decreasing real sequences. Thus $\mathcal{B}(\mathbb{R}^n)$ is not reflexive, not nuclear, not smoothly paracompact.

The space $C^\infty(\mathbb{R}, \mathcal{B}(\mathbb{R}^n))$ of smooth curves in $\mathcal{B}(\mathbb{R}^n)$ consists of all functions $c \in C^\infty(\mathbb{R}^{n+1}, \mathbb{R})$ satisfying the following property:

- *For all $k \in \mathbb{N}_{\geq 0}$, $\alpha \in \mathbb{N}^n_{\geq 0}$ and each $t \in \mathbb{R}$ the expression $\partial_t^k \partial_x^\alpha c(t, x)$ is uniformly bounded in $x \in \mathbb{R}^n$, locally in t.*

To see this use Theorem 2.6 for the set $\{\mathrm{ev}_x : x \in \mathbb{R}\}$ of point evaluations in $\mathcal{B}(\mathbb{R}^n)$. Here $\partial_x^\alpha = \frac{\partial^{|\alpha|}}{\partial x^\alpha}$ and $c^k(t) = \partial_t^k f(t, \quad)$.

$\mathrm{Diff}^+_{\mathcal{B}}(\mathbb{R}^n) = \big\{f = \mathrm{Id} + g : g \in \mathcal{B}(\mathbb{R}^n)^n, \det(\mathbb{I}_n + dg) \geq \varepsilon > 0\big\}$ *denotes the corresponding group*, see below.

The Group $\mathrm{Diff}_{W^{\infty,p}}(\mathbb{R}^n)$ For $1 \leq p < \infty$, the space

$$W^{\infty,p}(\mathbb{R}^n) = \bigcap_{k \geq 1} L^p_k(\mathbb{R}^n)$$

is the intersection of all L^p-Sobolev spaces, the space of all smooth functions such that each partial derivative is in L^p. It is a reflexive Fréchet space. It is called $\mathcal{D}_{L^p}(\mathbb{R}^n)$ in [96]. By Vogt [105], the space $W^{\infty,p}(\mathbb{R}^n)$ is linearly isomorphic to $\ell^p \hat{\otimes} \mathfrak{s}$. Thus it is not nuclear, not Schwartz, not Montel, and smoothly paracompact only if p is an even integer.

The space $C^\infty(\mathbb{R}, H^\infty(\mathbb{R}^n))$ of smooth curves in $W^{\infty,p}(\mathbb{R}^n)$ consists of all functions $c \in C^\infty(\mathbb{R}^{n+1}, \mathbb{R})$ satisfying the following property:

- *For all $k \in \mathbb{N}_{\geq 0}$, $\alpha \in \mathbb{N}^n_{\geq 0}$ the expression $\|\partial_t^k \partial_x^\alpha f(t, \quad)\|_{L^p(\mathbb{R}^n)}$ is locally bounded near each $t \in \mathbb{R}$.*

The proof is literally the same as for $\mathcal{B}(\mathbb{R}^n)$, noting that the point evaluations are continuous on each Sobolev space L^p_k with $k > \frac{n}{p}$.
$\mathrm{Diff}^+_{W^{\infty,p}}(\mathbb{R}^n) = \{f = \mathrm{Id} + g : g \in W^{\infty,p}(\mathbb{R}^n)^n, \det(\mathbb{I}_n + dg) > 0\}$ *denotes the corresponding group.*

The Group $\mathrm{Diff}_{\mathcal{S}}(\mathbb{R}^n)$ The algebra $\mathcal{S}(\mathbb{R}^n)$ of rapidly decreasing functions is a reflexive nuclear Fréchet space.
The space $C^\infty(\mathbb{R}, \mathcal{S}(\mathbb{R}^n))$ *of smooth curves in* $\mathcal{S}(\mathbb{R}^n)$ *consists of all functions* $c \in C^\infty(\mathbb{R}^{n+1}, \mathbb{R})$ *satisfying the following property:*

- *For all* $k, m \in \mathbb{N}_{\geq 0}$ *and* $\alpha \in \mathbb{N}^n_{\geq 0}$, *the expression* $(1 + |x|^2)^m \partial_t^k \partial_x^\alpha c(t, x)$ *is uniformly bounded in* $x \in \mathbb{R}^n$, *locally uniformly bounded in* $t \in \mathbb{R}$.

$\mathrm{Diff}^+_{\mathcal{S}}(\mathbb{R}^n) = \{f = \mathrm{Id} + g : g \in \mathcal{S}(\mathbb{R}^n)^n, \det(\mathbb{I}_n + dg) > 0\}$ *is the corresponding group.*

The Group $\mathrm{Diff}_c(\mathbb{R}^n)$ The algebra $C^\infty_c(\mathbb{R}^n)$ of all smooth functions with compact support is a nuclear (LF)-space.
The space $C^\infty(\mathbb{R}, C^\infty_c(\mathbb{R}^n))$ *of smooth curves in* $C^\infty_c(\mathbb{R}^n)$ *consists of all functions* $f \in C^\infty(\mathbb{R}^{n+1}, \mathbb{R})$ *satisfying the following property:*

- *For each compact interval* $[a, b]$ *in* $\mathbb{R}$ *there exists a compact subset* $K \subset \mathbb{R}^n$ *such that* $f(t, x) = 0$ *for* $(t, x) \in [a, b] \times (\mathbb{R}^n \setminus K)$.

$\mathrm{Diff}_c(\mathbb{R}^n) = \{f = \mathrm{Id} + g : g \in C^\infty_c(\mathbb{R}^n)^n, \det(\mathbb{I}_n + dg) > 0\}$ *is the corresponding group.* The case $\mathrm{Diff}_c(\mathbb{R}^n)$ is well-known since 1980.

Ideal Properties of Function Spaces The function spaces discussed are boundedly mapped into each other as follows:

$$C^\infty_c(\mathbb{R}^n) \longrightarrow \mathcal{S}(\mathbb{R}^n) \longrightarrow W^{\infty,p}(\mathbb{R}^n) \xrightarrow{p\leq q} W^{\infty,q}(\mathbb{R}^n) \longrightarrow \mathcal{B}(\mathbb{R}^n)$$

and each space is a bounded locally convex algebra and a bounded $\mathcal{B}(\mathbb{R}^n)$-module. Thus each space is an ideal in each larger space.

6.10 Theorem ([77] and [60]) *The sets of diffeomorphisms*

$$\mathrm{Diff}_c(\mathbb{R}^n), \quad \mathrm{Diff}_{\mathcal{S}}(\mathbb{R}^n), \quad \mathrm{Diff}_{H^\infty}(\mathbb{R}^n), \textit{ and } \mathrm{Diff}_{\mathcal{B}}(\mathbb{R}^n)$$

are all smooth regular Lie groups. We have the following smooth injective group homomorphisms:

$$\mathrm{Diff}_c(\mathbb{R}^n) \longrightarrow \mathrm{Diff}_{\mathcal{S}}(\mathbb{R}^n) \longrightarrow \mathrm{Diff}_{W^{\infty,p}}(\mathbb{R}^n) \longrightarrow \mathrm{Diff}_{\mathcal{B}}(\mathbb{R}^n).$$

Each group is a normal subgroup in any other in which it is contained, in particular in $\mathrm{Diff}_{\mathcal{B}}(\mathbb{R}^n)$.

The proof of this theorem relies on repeated use of the Faà di Bruno formula for higher derivatives of composed functions. This offers difficulties on non-compact manifolds, where one would need a non-commutative Faà di Bruno formula for iterated covariant derivatives. In the paper [60] many more similar groups are discussed, modeled on spaces of Denjoy–Carleman ultradifferentiable functions. It is also shown that for $p > 1$ the group $\mathrm{Diff}_{W^{\infty,p}\cap L^1}(\mathbb{R}^n)$ is only a topological group with smooth right translations—a property which is similar to the one of finite order Sobolev groups $\mathrm{Diff}_{W^{k,p}}(\mathbb{R}^n)$. Some of these groups were used extensively in [80].

6.11 Corollary *$\mathrm{Diff}_{\mathcal{B}}(\mathbb{R}^n)$ acts on Γ_c, $\Gamma_{\mathcal{S}}$ and Γ_{H^∞} of any tensor bundle over $\mathbb{R}^n$ by pullback. The infinitesimal action of the Lie algebra $\mathfrak{X}_{\mathcal{B}}(\mathbb{R}^n)$ on these spaces by the Lie derivative maps each of these spaces into itself. A fortiori, $\mathrm{Diff}_{H^\infty}(\mathbb{R}^n)$ acts on $\Gamma_{\mathcal{S}}$ of any tensor bundle by pullback.*

6.12 Trouvé Groups

For the following see [85, 103, 108]. Trouvé groups are useful for introducing topological metrics on certain groups of diffeomorphism on $\mathbb{R}^d$ starting from a suitable reproducing kernel Hilbert space of vector fields without using any Lie algebra structure; see Sect. 8.12 below.

Consider a time-dependent vector field $X : [0, 1] \times \mathbb{R}^d \to \mathbb{R}$ of sufficient regularity (e.g., continuous in $t \in [0, 1]$ and Lipschitz continuous in $x \in \mathbb{R}^d$ with t-integrable global Lipschitz constant) so that

$$x(t) = x_0 + \int_0^t X(s, x(s))\, ds$$

is uniquely solvable for all $t \in [0, 1]$ and $x_0 \in \mathbb{R}^d$. Then we consider the *evolution* $\mathrm{evol}^X(x_0) = x(1)$. For $X \in L^1([0,1], C_b^1(\mathbb{R}^d, \mathbb{R})^d)$ (where $f \in C_b^k$ if all iterated partial derivatives of order between 0 and k are continuous and globally bounded) we have $\mathrm{evol}^X \in \mathrm{Id} + C_b^1(\mathbb{R}^d, \mathbb{R}^d)$ and is a diffeomorphism with $(\mathrm{evol})^{-1} \in \mathrm{Id} + C_b^1(\mathbb{R}^d, \mathbb{R}^d)$. Given a convenient locally convex vector space $\mathcal{A}(\mathbb{R}^d, \mathbb{R}^d)$ of mappings $\mathbb{R}^d \to \mathbb{R}^d$ which continuously embeds into $C_b^1(\mathbb{R}^d, \mathbb{R}^d)$ and a suitable family of mappings $[0, 1] \to \mathcal{A}(\mathbb{R}^d, \mathbb{R}^d)$, the associated *Trouvé group* is given by

$$\mathcal{G}_{\mathcal{A}} := \{\mathrm{evol}^X : X \in \mathcal{F}_{\mathcal{A}}\},$$

where $\mathcal{F}_{\mathcal{A}} = \mathcal{F}([0,1], \mathcal{A}(\mathbb{R}^d, \mathbb{R}^d))$ is a suitable vector space of time-dependent vector fields. It seems that for a wide class of spaces $\mathcal{A}$ the Trouvé group $\mathcal{G}_{\mathcal{A}}$ is independent of the choice of $\mathcal{F}_{\mathcal{A}}$ if the latter contains the piecewise smooth curves and is contained in the curves which are integrable by seminorms; a precise statement is still lacking, but see [82, 84, 85], and citations therein. The space $\mathcal{A}$ is called *$\mathcal{F}_{\mathcal{A}}$-ODE-closed* if $\mathrm{evol}^X \in \mathrm{Id} + \mathcal{A}(\mathbb{R}^d, \mathbb{R}^d)$ for each $X \in \mathcal{F}_{\mathcal{A}}$. For ODE-closed $\mathcal{A}$ the Trouvé group $\mathcal{G}_{\mathcal{A}}$ is contained in $\mathrm{Id} + \mathcal{A}(\mathbb{R}^d, \mathbb{R}^d)$.

For some spaces $\mathcal{A}$ it has been proved that $\mathcal{F}_{\mathcal{A}}$ is equal to the connected component of the identity of

$$\{\mathrm{Id}+f : f \in \mathcal{A}(\mathbb{R}^d, \mathbb{R}^d), \inf_{x\in\mathbb{R}^d} \det df(x) > -1\},$$

namely

- For Sobolev spaces $W^{k,2}$ with $k > d/2$ by Bruveris and Vialard [22]; $\mathcal{G}_{\mathcal{A}}$ is a half Lie group.
- For Hölder spaces by Nenning and Rainer [84].
- For Besov spaces by Nenning [83].
- For $\mathcal{B}$, $W^{\infty,p}$, Schwartz functions $\mathcal{S}$, C_c^∞, and many classes of Denjoy–Carleman functions, where $\mathcal{G}_{\mathcal{A}}$ is always a regular Lie group; see [85].

7 Spaces of Embeddings or Immersions, and Shape Spaces

This is the main section in this chapter.

7.1 The Principal Bundle of Embeddings

For finite dimensional manifolds M, N with M compact, $\mathrm{Emb}(M, N)$, the space of embeddings of M into N, is open in $C^\infty(M, N)$, so it is a smooth manifold. $\mathrm{Diff}(M)$ acts freely and smoothly from the right on $\mathrm{Emb}(M, N)$.

Theorem $\mathrm{Emb}(M, N) \to \mathrm{Emb}(M, N)/\mathrm{Diff}(M) = B(M, N)$ *is a smooth principal fiber bundle with structure group* $\mathrm{Diff}(M)$. *Its base is a smooth manifold.*

This result was proved in [70] for M an open manifold without boundary; see also [69]. Note that $B(M, N)$ is the smooth manifold of all submanifolds of N which are of diffeomorphism type M. Therefore it is also called the *nonlinear Grassmannian* in [45], where this theorem is extended to the case when M has boundary. From another point of view, $B(M, N)$ is called the *differentiable Chow variety* in [68]. It is an example of a *shape space*.

Proof We use an auxiliary Riemannian metric $\bar{g}$ on N. Given an embedding $f \in \mathrm{Emb}(M, N)$, we view $f(M)$ as a submanifold of N and we split the tangent bundle of N along $f(M)$ as $TN|_{f(M)} = \mathrm{Nor}(f(M)) \oplus Tf(M)$. The exponential mapping describes a tubular neighborhood of $f(M)$ via

$$\mathrm{Nor}(f(M)) \xrightarrow[\cong]{\exp^{\bar{g}}} W_{f(M)} \xrightarrow{p_{f(M)}} f(M).$$

If $g : M \to N$ is C^1-near to f, then $\varphi(g) := f^{-1} \circ p_{f(M)} \circ g \in \mathrm{Diff}(M)$ and we may consider $g \circ \varphi(g)^{-1} \in \Gamma(f^* W_{f(M)}) \subset \Gamma(f^* \mathrm{Nor}(f(M)))$. This is the required local splitting. □

7.2 The Space of Immersions and the Space of Embeddings of a Compact Whitney Manifold Germ

Let $\tilde{M} \supset M$ be a compact Whitney manifold germ, and let N be a smooth manifold with $\dim(M) \leq \dim(N)$. We define the space of immersions as

$$\mathrm{Imm}(M, N) = \{f = \tilde{f}|_M, f \in C^\infty(\tilde{M}, N), T_x\tilde{f} \text{ is injective for } x \in M\}$$

which is open in the smooth manifold $C^\infty(M, N)$ and is thus itself a smooth manifold. Note that any extension of an immersion $f \in \mathrm{Imm}(M, N)$ to $\tilde{f} \in C^\infty(\tilde{M}, N)$ is still an immersion on an open neighborhood of M in $\tilde{M}$.

Likewise we let

$$\begin{aligned}\mathrm{Emb}(M, N) = \{f|_M, f \in C^\infty(\tilde{M}, N), T_x f \text{ is injective for } x \in M,\\ f : M \to N \text{ is a topological embedding}\}.\end{aligned}$$

Since M is compact, any extension of an embedding $f \in \mathrm{Emb}(M, N)$ to $\tilde{f} \in C^\infty(\tilde{M}, N)$ is an embedding on some open neighborhood of M in $\tilde{M}$; see [69, 5.3] for a proof a related result.

Theorem *For a compact Whitney germ M and a smooth manifold N with* $\dim(M) < \dim(N)$ *the projection*

$$\pi : \mathrm{Emb}(M, N) \to \mathrm{Emb}(M, N)/\,\mathrm{Diff}(M) = B(M, N)$$

is a smooth principal fiber bundle of Frölicher spaces with structure group the Frölicher group Diff(M) *from Theorem* 6.4. *Its base is the quotient Frölicher space.*

Proof Since I do not know that Diff(M) is a smooth manifold, we treat all spaces here as Frölicher spaces. By definition, the right action of Diff(M) on Emb(M, N) is free, and smooth between the Frölicher spaces. The quotient $B(M, N)$ carries the quotient Frölicher structure with generating set of curves $\{\pi \circ c : c \in C^\infty(\mathbb{R}, \mathrm{Emb}(M, N))\}$, i.e., those which lift to a smooth curve. □

7.3 The Orbifold Bundle of Immersions

Let M be a (not necessarily compact) manifold without boundary. Let N be an open manifold with $\dim(M) \leq \dim(N)$. Then Imm(M, N), the space of immersions $M \to N$, is open in $C^\infty(M, N)$, and is thus a smooth manifold. The regular Lie group (or Frölicher group if M is a Whitney manifold germ) Diff(M) acts smoothly from the right, but no longer freely.

An immersion $i : M \to N$ is called *free* if $\mathrm{Diff}(M)$ acts freely on it: $i \circ f = i$ for $f \in \mathrm{Diff}(M)$ implies $f = \mathrm{Id}_M$.

The space $B_i(M, N) = \mathrm{Imm}(M, N)/\mathrm{Diff}(M)$ is an example of a *shape space*. It appeared in the form of $B_i(S^1, \mathbb{R}^2)$, the shape space of plane immersed curves, in [75] and [76]. The following theorem is essentially due to [23]; since this paper contains some annoying misprints and is difficult to understand, we give here an extended version with a more detailed proof. The reader may skip this proof and jump directly to Sect. 7.2 below.

Theorem ([23]) *Let M be a finite dimensional smooth manifold. Let N be smooth finite dimensional manifolds with* $\dim(M) \leq \dim(N)$. *Then the following holds:*

(1) *The diffeomorphism group* $\mathrm{Diff}(M)$ *acts smoothly from the right on the manifold* $\mathrm{Imm}_{\mathrm{prop}}(M, N)$ *of all smooth proper immersions $M \to N$, which is an open subset of $C^\infty(M, N)$.*
(2) *The space of orbits* $\mathrm{Imm}_{prop}(M, N)/\mathrm{Diff}(M)$ *is Hausdorff in the quotient topology.*
(3) *The set* $\mathrm{Imm}_{\mathrm{free,prop}}(M, N)$ *of all proper free immersions is open in $C^\infty(M, N)$ and is the total space of a smooth principal fiber bundle* $\mathrm{Imm}_{\mathrm{free,prop}}(M, N) \to \mathrm{Imm}_{\mathrm{free,prop}}(M, N)/\mathrm{Diff}(M)$.
(4) *Let $i \in$* $\mathrm{Imm}(M, N)$ *be an immersion which is not free. So we have a nontrivial isotropy subgroup* $\mathrm{Diff}(M)_i \subset \mathrm{Diff}(M)$ *consisting of all $f \in$* $\mathrm{Diff}(M)$ *with $i \circ f = i$. Then the isotropy group* $\mathrm{Diff}(M)_i$ *acts properly discontinuously on M. Thus the projection $q_1 : M \to M_1 := M/$*$\mathrm{Diff}(M)_i$ *is a covering mapping onto a smooth manifold M_1. There exists an immersion $i_1 : M_1 \to N$ with $i = i_1 \circ q_1$. In particular,* $\mathrm{Diff}(M)_i$ *is countable, and is finite if M is compact. There exists a further covering $q_2 : M \to M_1 \to M_2$ and a* free *immersion $i_2 : M_2 \to N$ with $i = i_2 \circ q_2$.*
(5) *Let M have the property that for any covering $M \to M_1$ of smooth manifolds, any diffeomorphism $M_1 \to M_1$ admits a lift $M \to M$; e.g., M simply connected, or $M = S^1$. Let $i \in Imm(M, N)$ be an immersion which is not free, i.e., has nontrivial isotropy group* $\mathrm{Diff}(M)_i$, *and let $q_1 : M \to M_1 := M/$*$\mathrm{Diff}(M)_i$ *be the corresponding covering map. Then in the following commutative diagram the bottom mapping*

$$\begin{array}{ccc}
\mathrm{Imm}_{\mathrm{free}}(M_1,N) & \xrightarrow{(q_1)^*} & \mathrm{Imm}(M,N) \\
\downarrow \pi & & \pi \downarrow \\
\mathrm{Imm}_{\mathrm{free}}(M_1,N)/\mathrm{Diff}(M_1) & \longrightarrow & \mathrm{Imm}(M,N)/\mathrm{Diff}(M)
\end{array}$$

is the inclusion of a (possibly non-Hausdorff) manifold, the stratum of $\pi(i)$ in the stratification of the orbit space. This stratum consists of the orbits of all immersions which have $\mathrm{Diff}(M)_i$ *as isotropy group. See (23) and (24) below for a more complete description of the orbit structure near i.*

(6) [100] *We have a right action of* $\operatorname{Diff}(M)$ *on* $\operatorname{Imm}(M, N) \times M$ *which is given by* $(i, x).f = (i \circ f, f^{-1}(x))$. *This action is free.*

$$(\operatorname{Imm}(M, N) \times M, \pi, (\operatorname{Imm}(M, N) \times M)/\operatorname{Diff}(M), \operatorname{Diff}(M))$$

is a smooth principal fiber bundle with structure group $\operatorname{Diff}(M)$ *and a smooth base manifold* $S(M, N) := (\operatorname{Imm}(M, N) \times M)/\operatorname{Diff}(M)$ *which might possibly be non-Hausdorff. If we restrict to the open subset* $\operatorname{Imm}_{\text{prop}}(M, N) \times M$ *of proper immersions times* M *then the base space is Hausdorff.*

Proof Without loss, let M be connected. Fix an immersion $i : M \to N$. We will now describe some data for i which we will use throughout the proof. If we need these data for several immersions, we will distinguish them by appropriate superscripts.

(7) Setup There exist sets $W_\alpha \subset \overline{W}_\alpha \subset U_\alpha \subset \overline{U}_\alpha \subset V_\alpha \subset M$ such that (W_α) is an open cover of M, $\overline{W}_\alpha$ is compact, and V_α is an open locally finite cover of M, each W_α, U_α, and V_α is connected, and such that $i|V_\alpha : V_\alpha \to N$ is an embedding for each α.

Let g be a fixed Riemannian metric on N and let $\exp^N$ be the induced geodesic exponential mapping. Then let $p : \mathcal{N}(i) \to M$ be the *normal bundle* of i, defined in the following way: For $x \in M$ let $\mathcal{N}(i)_x := (T_x i(T_x M))^\perp \subset T_{i(x)}N$ be the g-orthogonal complement in $T_{i(x)}N$. Then

$$\begin{array}{ccc} \mathcal{N}(i) & \xrightarrow{\;\bar{i}\;} & TN \\ {\scriptstyle p}\downarrow & & \downarrow{\scriptstyle \pi_N} \\ M & \xrightarrow{\;i\;} & N \end{array}$$

is a vector bundle homomorphism over i, which is fiberwise injective.

Now let $U^i = U$ be an open neighborhood of the zero section of $\mathcal{N}(i)$ which is so small that $(\exp^N \circ \bar{i})|(U|_{V_\alpha}) : U|_{V_\alpha} \to N$ is a diffeomorphism onto its image which describes a tubular neighborhood of the submanifold $i(V_\alpha)$ for each α. Let

$$\tau = \tau^i := (\exp^N \circ \bar{i})|_U : \mathcal{N}(i) \supset U \to N.$$

It will serve us as a substitute for a tubular neighborhood of $i(M)$.

For any $f \in \operatorname{Diff}(M)_i = \{f \in \operatorname{Diff}(M) : i \circ f = i\}$ we have an induced vector bundle homomorphism $\bar{f}$ over f:

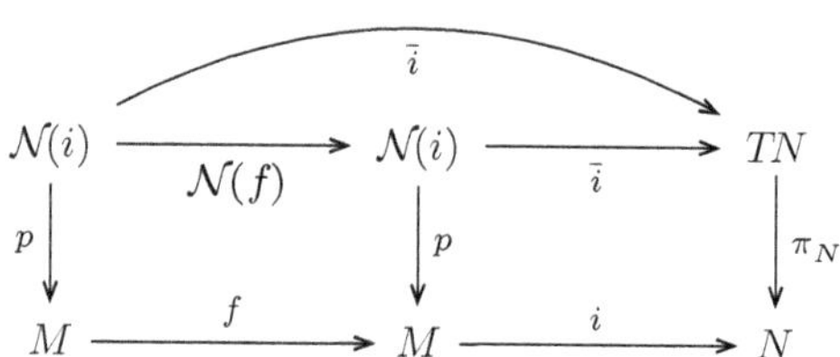

(8) Claim *Let $i \in \operatorname{Imm}(M, N)$ and let $f \in \operatorname{Diff}(M)$ have a fixed point $x_0 \in M$ and satisfy $i \circ f = i$. Then $f = Id_M$.*

Namely, we consider the sets (U_α) for the immersion i of (7). Let us investigate $f(U_\alpha) \cap U_\alpha$. If there is an $x \in U_\alpha$ with $y = f(x) \in U_\alpha$, we have $(i|_{U_\alpha})(x) = ((i \circ f)|U_\alpha)(x) = (i|_{U_\alpha})(f(x)) = (i|_{U_\alpha})(y)$. Since $i|_{U_\alpha}$ is injective we have $x = y$, and

$$f(U_\alpha) \cap U_\alpha = \{x \in U_\alpha : f(x) = x\}.$$

Thus $f(U_\alpha) \cap U_\alpha$ is closed in U_α. Since it is also open and since U_α is connected, we have $f(U_\alpha) \cap U_\alpha = \emptyset$ or $= U_\alpha$.

Now we consider the set $\{x \in M : f(x) = x\}$. We have just shown that it is open in M. Since it is also closed and contains the fixed point x_0, it coincides with M. Claim (7) follows.

(9) Claim *If for an immersion $i \in \operatorname{Imm}(M, N)$ there is a point in $i(M)$ with only one preimage, then i is a free immersion.*

Let $x_0 \in M$ be such that $i(x_0)$ has only one preimage. If $i \circ f = i$ for $f \in \operatorname{Diff}(M)$ then $f(x_0) = x_0$ and $f = Id_M$ by claim (8).

Note that there are free immersions without a point in $i(M)$ with only one preimage: Consider a figure eight which consists of two touching circles. Now we may map the circle to the figure eight by going first n times around the upper circle, then m around the lower one with $n, m \geq 2$.

(10) Claim *Let i be a free immersion $M \to N$. Then there is an open neighborhood $\mathcal{W}(i)$ in $\operatorname{Imm}(M, N)$ which is saturated for the $\operatorname{Diff}(M)$-action and which splits smoothly as*

$$\mathcal{W}(i) = \mathcal{Q}(i) \times \operatorname{Diff}(M).$$

Here $\mathcal{Q}(i)$ is a smooth splitting submanifold of $\operatorname{Imm}(M, N)$, diffeomorphic to an open neighborhood of the zero section in $\Gamma_c(M \leftarrow \mathcal{N}(i))$. In particular the space $\operatorname{Imm}_{\text{free}}(M, N)$ is open in $C^\infty(M, N)$.

Let $\pi : \operatorname{Imm}(M, N) \to \operatorname{Imm}(M, N)/\operatorname{Diff}(M) = B_i(M, N)$ be the projection onto the orbit space, which is equipped with the quotient topology. Then the mapping $\pi|_{\mathcal{Q}(i)} : \mathcal{Q}(i) \to \pi(\mathcal{Q}(i))$ is bijective onto an open subset of the quotient. If i runs through $\operatorname{Imm}_{\text{free,prop}}(M, N)$ of all free and proper immersions these mappings define a smooth atlas for the quotient space, so that

$$(\operatorname{Imm}_{\text{free,prop}}(M, N), \pi, \operatorname{Imm}_{\text{free,prop}}(M, N)/\operatorname{Diff}(M), \operatorname{Diff}(M))$$

is a smooth principal fiber bundle with structure group $\operatorname{Diff}(M)$.

The restriction to proper immersions is necessary because we are only able to show that $\operatorname{Imm}_{\text{prop}}(M, N)/\operatorname{Diff}(M)$ is Hausdorff in (11) below.

For the proof of claim (10), we consider the setup (7) for the free immersion i. Let

$$\tilde{\mathcal{U}}(i) := \{j \in \mathrm{Imm}(M,N) : j(\overline{W}^i_\alpha) \subseteq \tau^i(U^i|_{U^i_\alpha}) \text{ for all } \alpha,\ j \sim i\},$$

where $j \sim i$ means that $j = i$ off some compact set in M. Then by Sect. 5.3 (for open M) the set $\tilde{\mathcal{U}}(i)$ is an open neighborhood of i in $\mathrm{Imm}(M,N)$. For each $j \in \tilde{\mathcal{U}}(i)$ we define

$$\begin{aligned}
&\varphi_i(j) : M \to U^i \subseteq \mathcal{N}(i),\\
&\varphi_i(j)(x) := (\tau^i|_{(U^i|_{U^i_\alpha})})^{-1}(j(x)) \text{ if } x \in W^i_\alpha.
\end{aligned}$$

Note that $\varphi_i(j)$ is defined piecewise on M, but the pieces coincide when they overlap. Therefore a smooth curve through j is mapped to a smooth curve and so $\varphi_i : \tilde{\mathcal{U}}(i) \to C^\infty(M, \mathcal{N}(i))$ is a smooth mapping which is bijective onto the open set

$$\tilde{\mathcal{V}}(i) := \{h \in C^\infty(M, \mathcal{N}(i)) : h(\overline{W}^i_\alpha) \subseteq U^i|_{U^i_\alpha} \text{ for all } \alpha,\ h \sim 0\}$$

in $C^\infty(M, \mathcal{N}(i))$. Its inverse is given by the smooth mapping $\tau^i_* : h \mapsto \tau^i \circ h$. Now we consider the open subsets

$$\begin{aligned}
\mathcal{V}(i) &:= \{h \in \tilde{\mathcal{V}}(i) : p \circ h \in \mathrm{Diff}_c(M)\} \subset \tilde{\mathcal{V}}(i)\\
\mathcal{U}(i) &:= \tau^i_*(\mathcal{V}(i)) \subset \tilde{\mathcal{U}}(i)
\end{aligned}$$

and the diffeomorphism $\varphi_i : \mathcal{U}(i) \to \mathcal{V}(i)$. For $h \in \mathcal{V}(i)$ we have $\tau^i_*(h \circ f) = \tau^i_*(h) \circ f$ for those $f \in \mathrm{Diff}(M)$ which are near enough to the identity so that $h \circ f \in \mathcal{V}(i)$. And if $\tau^i \circ h \circ f = \tau^i \circ h$ then $h \circ f = h$ by the construction of $\mathcal{N}(i)$ in (7), and then $f = \mathrm{Id}_M$ since i is a free immersion; see the second diagram in (7).

We consider now the open set

$$\{h \circ f : h \in \mathcal{V}(i),\ f \in \mathrm{Diff}(M)\} \subseteq C^\infty(M, U^i).$$

Consider the smooth mapping from it into $\Gamma_c(M \leftarrow U^i) \times \mathrm{Diff}(M)$ given by $h \mapsto (h \circ (p \circ h)^{-1}, p \circ h)$, where $\Gamma_c(M \leftarrow U^i)$ is the space of sections with compact support of $U^i \to M$. So if we let $\mathcal{Q}(i) := \tau^i_*(\Gamma_c(M \leftarrow U^i) \cap \mathcal{V}(i)) \subset \mathrm{Imm}(M,N)$ we have

$$\mathcal{W}(i) := \mathcal{U}(i) \circ \mathrm{Diff}_c(M) \cong \mathcal{Q}(i) \times \mathrm{Diff}(M) \cong (\Gamma_c(M \leftarrow U^i) \cap \mathcal{V}(i)) \times \mathrm{Diff}(M),$$

since the action of $\mathrm{Diff}(M)$ on i is free and by the argument above. Consequently $\mathrm{Diff}(M)$ acts freely on each immersion in $\mathcal{W}(i)$, so $\mathrm{Imm}_{\mathrm{free}}(M,N)$ is open in $C^\infty(M,N)$. Furthermore

$$\pi|_{\mathcal{Q}(i)} : \mathcal{Q}(i) \to \mathrm{Imm}_{\mathrm{free}}(M, N)/\operatorname{Diff}(M)$$

is bijective onto an open set in the quotient.

We consider

$$\varphi_i \circ (\pi|_{\mathcal{Q}(i)})^{-1} : \pi(\mathcal{Q}(i)) \to \Gamma_c(M \leftarrow U^i) \subset C_c^\infty(N, \mathcal{N}(i))$$

as a chart for the quotient space.

In order to investigate the chart change let $j \in \mathrm{Imm}_{\mathrm{free}}(M, N)$ be such that $\pi(\mathcal{Q}(i)) \cap \pi(\mathcal{Q}(j)) \neq \emptyset$. Then there is an immersion $h \in \mathcal{W}(i) \cap \mathcal{Q}(j)$, so there exists a unique $f_0 \in \operatorname{Diff}(M)$ (given by $f_0 = p \circ \varphi_i(h)$) such that $h \circ f_0^{-1} \in \mathcal{Q}(i)$. If we consider $j \circ f_0^{-1}$ instead of j and call it again j, we have $\mathcal{Q}(i) \cap \mathcal{Q}(j) \neq \emptyset$ and consequently $\mathcal{U}(i) \cap \mathcal{U}(j) \neq \emptyset$. Then the chart change is given as follows:

$$\begin{aligned}\varphi_i \circ (\pi|_{\mathcal{Q}(i)})^{-1} \circ \pi \circ (\tau^j)_* &: \Gamma_c(M \leftarrow U^j) \to \Gamma_c(M \leftarrow U^i)\\ s \mapsto \tau^j \circ s &\mapsto \varphi_i(\tau^j \circ s) \circ (p^i \circ \varphi_i(\tau^j \circ s))^{-1}.\end{aligned}$$

This is of the form $s \mapsto \beta \circ s$ for a locally defined diffeomorphism $\beta : \mathcal{N}(j) \to \mathcal{N}(i)$ which is not fiber respecting, followed by $h \mapsto h \circ (p^i \circ h)^{-1}$. Both composants are smooth by the general properties of manifolds of mappings. So the chart change is smooth.

We have to show that the quotient space $\mathrm{Imm}_{\mathrm{prop,free}}(M, N)/\operatorname{Diff}(M)$ is Hausdorff.

(11) Claim *The orbit space* $\mathrm{Imm}_{\mathrm{prop}}(M, N)/\operatorname{Diff}(M)$ *of the space of all proper immersions under the action of the diffeomorphism group is Hausdorff in the quotient topology.*

This follows from (18) below. I am convinced that the whole orbit space $\mathrm{Imm}(M, N)/\operatorname{Diff}(M)$ is Hausdorff, but I was unable to prove this.

(12) Claim *Let i and $j \in$* $\mathrm{Imm}_{\mathrm{prop}}(M, N)$ *with $i(M) \neq j(M)$ in N. Then their projections $\pi(i)$ and $\pi(j)$ are different and can be separated by open subsets in* $\mathrm{Imm}_{\mathrm{prop}}(M, N)/\operatorname{Diff}(M)$.

We suppose that $i(M) \not\subseteq \overline{j(M)} = j(M)$ (since proper immersions have closed images). Let $y_0 \in i(M) \setminus \overline{j(M)}$, then we choose open neighborhoods V of y_0 in N and W of $j(M)$ in N such that $V \cap W = \emptyset$. We consider the sets

$$\begin{aligned}\mathcal{V} &:= \{k \in \mathrm{Imm}_{\mathrm{prop}}(M, N) : k(M) \cap V \neq \emptyset\} \quad \text{and}\\ \mathcal{W} &:= \{k \in \mathrm{Imm}_{\mathrm{prop}}(M, N) : k(M) \subseteq W\}.\end{aligned}$$

Then $\mathcal{V}$ and $\mathcal{W}$ are $\operatorname{Diff}(M)$-saturated disjoint open neighborhoods of i and j, respectively, so $\pi(\mathcal{V})$ and $\pi(\mathcal{W})$ separate $\pi(i)$ and $\pi(j)$ in the quotient space $\mathrm{Imm}_{\mathrm{prop}}(M, N)/\operatorname{Diff}(M)$.

(13) Claim *For a proper immersion $i : M \to N$ and $x \in i(M)$ let $\delta(x) \in \mathbb{N}$ be the number of points in $i^{-1}(x)$. Then $\delta : i(M) \to \mathbb{N}$ is upper semicontinuous, i.e., the set $\{x \in i(M) : \delta(x) \le k\}$ is open in $i(M)$ for each k.*

Let $x \in i(M)$ with $\delta(x) = k$ and let $i^{-1}(x) = \{y_1, \dots, y_k\}$. Then there are pairwise disjoint open neighborhoods W_n of y_n in M such that $i|_{W_n}$ is an embedding for each n. The set $M \setminus (\bigcup_n W_n)$ is closed in M, and since i is proper the set $i(M \setminus (\bigcup_n W_n))$ is also closed in $i(M)$ and does not contain x. So there is an open neighborhood U of x in $i(M)$ which does not meet $i(M \setminus (\bigcup_n W_n))$. Obviously $\delta(z) \le k$ for all $z \in U$.

(14) Claim *Consider two proper immersions i_1 and $i_2 \in \mathrm{Imm}_{\mathrm{prop}}(M, N)$ such that $i_1(M) = i_2(M) =: L \subseteq N$. Then we have mappings $\delta_1, \delta_2 : L \to \mathbb{N}$ as in (13). If $\delta_1 \ne \delta_2$ then the projections $\pi(i_1)$ and $\pi(i_2)$ are different and can be separated by disjoint open neighborhoods in $\mathrm{Imm}_{\mathrm{prop}}(M, N)/\mathrm{Diff}(M)$.*

Let us suppose that $m_1 = \delta_1(y_0) \ne \delta_2(y_0) = m_2$. There is a small connected open neighborhood V of y_0 in N such that $i_1^{-1}(V)$ has m_1 connected components and $i_2^{-1}(V)$ has m_2 connected components. These assertions describe Whitney C^0-open neighborhoods in $\mathrm{Imm}_{\mathrm{prop}}(M, N)$ of i_1 and i_2 which are closed under the action of $\mathrm{Diff}(M)$, respectively. Obviously these two neighborhoods are disjoint.

(15) Assumption We assume that we are given two immersions i_1 and $i_2 \in \mathrm{Imm}_{\mathrm{prop}}(M, N)$ with $i_1(M) = i_2(M) =: L$ such that the functions from (14) are equal: $\delta_1 = \delta_2 =: \delta$.

Let $(L_\beta)_{\beta \in B}$ be the partition of L consisting of all pathwise connected components of level sets $\{x \in L : \delta(x) = c\}$, c some constant.

Let B_0 denote the set of all $\beta \in B$ such that the interior of L_β in L is not empty. Since M is second countable, B_0 is countable.

(16) Claim *$\bigcup_{\beta \in B_0} L_\beta$ is dense in L.*

Let k_1 be the smallest number in $\delta(L)$ and let B_1 be the set of all $\beta \in B$ such that $\delta(L_\beta) = k_1$. Then by claim (13) each L_β for $\beta \in B_1$ is open. Let L^1 be the closure of $\bigcup_{\beta \in B_1} L_\beta$. Let k_2 be the smallest number in $\delta(L \setminus L^1)$ and let B_2 be the set of all $\beta \in B$ with $\beta(L_\beta) = k_2$ and $L_\beta \cap (L \setminus L^1) \ne \emptyset$. Then by claim (13) again $L_\beta \cap (L \setminus L^1) \ne \emptyset$ is open in L so L_β has non-empty interior for each $\beta \in B_2$. Then let L^2 denote the closure of $\bigcup_{\beta \in B_1 \cup B_2} L_\beta$ and continue the process. If $\delta(L)$ is bounded, the process stops. If $\delta(L)$ is unbounded, by claim (13) we always find new L_β with non-empty interior, we finally exhaust L and claim (16) follows.

Let $(M^1_\lambda)_{\lambda \in C^1}$ be a suitably chosen cover of M by subsets of the sets $i_1^{-1}(L_\beta)$ such that:

(i) Each $i_1|_{\mathrm{int}\, M^1_\lambda}$ is an embedding for each λ.

(ii) The set C^1_0 of all λ with M^1_λ having non empty interior is at most countable.

Let $(M^2_\mu)_{\mu \in C^2}$ be a cover chosen in a similar way for i_2.

(iii) For each pair $(\mu, \lambda) \in C_0^2 \times C_0^1$ the two open sets $i_2(\text{int}(M_\mu^2))$ and $i_1(\text{int}(M_\lambda^1))$ in L are either equal or disjoint.

Note that the union $\bigcup_{\lambda \in C_0^1} \text{int}\, M_\lambda^1$ is dense in M and thus $\bigcup_{\lambda \in C_0^1} \overline{M_\lambda^1} = M$; similarly for the M_μ^2.

(17) Procedure Given immersions i_1 and i_2 as in (15) we will try to construct a diffeomorphism $f : M \to M$ with $i_2 \circ f = i_1$. If we meet obstacles to the construction this will give enough control on the situation to separate i_1 from i_2.

Choose $\lambda_0 \in C_0^1$; so $\text{int}\, M_{\lambda_0}^1 \neq \emptyset$. Then $i_1 : \text{int}\, M_{\lambda_0}^1 \to L_{\beta_1(\lambda_0)}$ is an embedding, where $\beta_1 : C^1 \to B$ is the mapping satisfying $i_1(M_\lambda^1) \subseteq L_{\beta_1(\lambda)}$ for all $\lambda \in C^1$.

We choose $\mu_0 \in \beta_2^{-1}\beta_1(\lambda_0) \subset C_0^2$ such that $f := (i_2|_{\text{int}\, M_{\mu_0}^2})^{-1} \circ i_1|_{\text{int}\, M_{\lambda_0}^1}$ is a diffeomorphism $\text{int}\, M_{\lambda_0}^1 \to \text{int}\, M_{\mu_0}^2$; this follows from (iii). Note that f is uniquely determined by the choice of μ_0, if it exists, by claim (8). So we will repeat the following construction for every $\mu_0 \in \beta_2^{-1}\beta_1(\lambda_0) \subset C_0^2$.

Now we try to extend f. We choose $\lambda_1 \in C_0^1$ such that $\overline{M}_{\lambda_0}^1 \cap \overline{M}_{\lambda_1}^1 \neq \emptyset$.

Case a Only $\lambda_1 = \lambda_0$ is possible. So $M_{\lambda_0}^1$ is dense in M since M is connected and we may extend f by continuity to a diffeomorphism $f : M \to M$ with $i_2 \circ f = i_1$.

Case b We can find $\lambda_1 \neq \lambda_0$. We choose $x \in \overline{M}_{\lambda_0}^1 \cap \overline{M}_{\lambda_1}^1$ and a sequence (x_n) in $M_{\lambda_0}^1$ with $x_n \to x$. Then we have a sequence $(f(x_n))$ in M.

Case ba $y := \lim f(x_n)$ exists in M. Then there is $\mu_1 \in C_0^2$ such that $y \in \overline{M}_{\mu_0}^2 \cap \overline{M}_{\mu_1}^2$.

Let $U_{\alpha_1}^1$ be an open neighborhood of x in M such that $i_1|U_{\alpha_1}^1$ is an embedding and let similarly $U_{\alpha_2}^2$ be an open neighborhood of y in M such that $i_2|U_{\alpha_2}^2$ is an embedding. We consider now the set $i_2^{-1}i_1(U_{\alpha_1}^1)$. There are two cases possible.

Case baa The set $i_2^{-1}i_1(U_{\alpha_1}^1)$ is a neighborhood of y. Then we extend f to $i_1^{-1}(i_1(U_{\alpha_1}^1) \cap i_2(U_{\alpha_2}^2))$ by $i_2^{-1} \circ i_1$. Then f is defined on some open subset of $\text{int}\, M_{\lambda_1}^1$ and by the situation chosen in (15) and by (iii), the diffeomorphism f extends to the whole of $\text{int}\, M_{\lambda_1}^1$.

Case bab The set $i_2^{-1}i_1(U_{\alpha_1}^1)$ is not a neighborhood of y. This is a definite obstruction to the extension of f.

Case bb The sequence (x_n) has no limit in M. This is a definite obstruction to the extension of f.

If we meet an obstruction we stop and try another μ_0. If for all admissible μ_0 we meet obstructions we stop and remember the data. If we do not meet an obstruction we repeat the construction with some obvious changes.

(18) Claim *The construction of (17) in the setting of (15) either produces a diffeomorphism $f : M \to M$ with $i_2 \circ f = i_1$ or we may separate i_1 and i_2 by open sets in* $\operatorname{Imm}_{\text{prop}}(M, N)$ *which are saturated with respect to the action of* $\operatorname{Diff}(M)$

If for some μ_0 we do not meet any obstruction in the construction (17), the resulting f is defined on the whole of M and it is a continuous mapping $M \to M$ with $i_2 \circ f = i_1$. Since i_1 and i_2 are locally embeddings, f is smooth and of maximal rank. Since i_1 and i_2 are proper, f is proper. So the image of f is open and closed and since M is connected, f is a surjective local diffeomorphism, thus a covering mapping $M \to M$. But since $\delta_1 = \delta_2$ the mapping f must be a 1-fold covering, i.e., a diffeomorphism.

If for all $\mu_0 \in \beta_2^{-1}\beta_1(\lambda_0) \subset C_0^2$ we meet obstructions we choose small mutually distinct open neighborhoods V_λ^1 of the sets $i_1(M_\lambda^1)$. We consider the Whitney C^0-open neighborhood $\mathcal{V}_1$ of i_1 consisting of all immersions j_1 with $j_1(M_\lambda^1) \subset V_\lambda^1$ for all λ. Let $\mathcal{V}_2$ be a similar neighborhood of i_2.

We claim that $\mathcal{V}_1 \circ \operatorname{Diff}(M)$ and $\mathcal{V}_2 \circ \operatorname{Diff}(M)$ are disjoint. For that it suffices to show that for any $j_1 \in \mathcal{V}_1$ and $j_2 \in \mathcal{V}_2$ there does not exist a diffeomorphism $f \in \operatorname{Diff}(M)$ with $j_2 \circ f = j_1$. For that to be possible the immersions j_1 and j_2 must have the same image L and the same functions $\delta(j_1), \delta(j_2) : L \to \mathbb{N}$. But now the combinatorial relations of the slightly distinct new sets M_λ^1, L_β, and M_μ^2 are contained in the old ones, so any try to construct such a diffeomorphism f starting from the same λ_0 meets the same obstructions.

Statements (2) and (3) of the theorem are now proved.

(19) Claim *For a non-free immersion $i \in \operatorname{Imm}(M, N)$, the nontrivial isotropy subgroup $\operatorname{Diff}(M)_i = \{f \in \operatorname{Diff}(M) : i \circ f = i\}$ acts properly discontinuously on M, so the projection $q_1 : M \to M_1 := M/\operatorname{Diff}(M)_i$ is a covering map onto a smooth manifold on M_1. There is an immersion $i_1 : M_1 \to N$ with $i = i_1 \circ q_1$. In particular $\operatorname{Diff}(M)_i$ is countable, and is finite if M is compact.*

We have to show that for each $x \in M$ there is an open neighborhood U such that $f(U) \cap U = \emptyset$ for $f \in \operatorname{Diff}(M)_i \setminus \{Id\}$. We consider the setup (7) for i. By the proof of (8) we have $f(U_\alpha^i) \cap U_\alpha^i = \{x \in U_\alpha^i : f(x) = x\}$ for any $f \in \operatorname{Diff}(M)_i$. If f has a fixed point then $f = \text{Id}$, by (8), so $f(U_\alpha^i) \cap U_\alpha^i = \emptyset$ for all $f \in \operatorname{Diff}(M)_i \setminus \{Id\}$. The rest is clear.

The factorized immersion i_1 is in general not a free immersion. The following is an example for that: Let $M_0 \xrightarrow{\alpha} M_1 \xrightarrow{\beta} M_2 \xrightarrow{\gamma} M_3$ be a sequence of covering maps with fundamental groups $1 \to G_1 \to G_2 \to G_3$. Then the group of deck transformations of γ is given by $\mathcal{N}_{G_3}(G_2)/G_2$, the normalizer of G_2 in G_3, and the group of deck transformations of $\gamma \circ \beta$ is $\mathcal{N}_{G_3}(G_1)/G_1$. We can easily arrange that $\mathcal{N}_{G_3}(G_2) \not\subseteq \mathcal{N}_{G_3}(G_1)$, then γ admits deck transformations which do not lift to M_1. Then we thicken all spaces to manifolds, so that $\gamma \circ \beta$ plays the role of the immersion i.

(20) Claim *Let $i \in \operatorname{Imm}(M, N)$ be an immersion which is not free. Then there is a submersive covering map $q_2 : M \to M_2$ such that i factors to an immersion $i_2 : M_2 \to N$ which is free.*

Let $q_0 : M_0 \to M$ be the universal covering of M and consider the immersion $i_0 = i \circ q_0 : M_0 \to N$ and its isotropy group $\mathrm{Diff}(M_0)_{i_0}$. By (19) it acts properly discontinuously on M_0 and we have a submersive covering $q_{02} : M_0 \to M_2$ and an immersion $i_2 : M_2 \to N$ with $i_2 \circ q_{02} = i_0 = i \circ q_0$. By comparing the respective groups of deck transformations it is easily seen that $q_{02} : M_0 \to M_2$ factors over $q_1 \circ q_0 : M_0 \to M \to M_1$ to a covering $q_{12} : M_1 \to M_2$. The mapping $q_2 := q_{12} \circ q_1 : M \to M_2$ is the looked for covering: If $f \in \mathrm{Diff}(M_2)$ fixes i_2, it lifts to a diffeomorphism $f_0 \in \mathrm{Diff}(M_0)$ which fixes i_0, so $f_0 \in \mathrm{Diff}(M_0)_{i_0}$, so $f = \mathrm{Id}$.

Statement (4) of the theorem follows from (19) and (20).

(21) Convention In order to avoid complications we assume now that M is such a manifold that

- For any covering $M \to M_1$, any diffeomorphism $M_1 \to M_1$ admits a lift $M \to M$.

If M is simply connected, this condition is satisfied. Also for $M = S^1$ the condition is easily seen to be valid. So what follows is applicable to loop spaces.

Condition (21) implies that in the proof of claim (20) we have $M_1 = M_2$.

(22) Description of a Neighborhood of a Singular Orbit Let M be a manifold satisfying (21). In the situation of (19) we consider the normal bundles $p_i : \mathcal{N}(i) \to M$ and $p_{i_1} : \mathcal{N}(i_1) \to M_1$. Then the covering map $q_1 : M \to M_1$ lifts uniquely to a vector bundle homomorphism $\mathcal{N}(q_1) : \mathcal{N}(i) \to \mathcal{N}(i_1)$ which is also a covering map, such that $\tau^{i_1} \circ \mathcal{N}(q_1) = \tau^i$.

We have $M_1 = M/\mathrm{Diff}(M)_i$ and the group $\mathrm{Diff}(M)_i$ acts also as the group of deck transformations of the covering $\mathcal{N}(q_1) : \mathcal{N}(i) \to \mathcal{N}(i_1)$ by $\mathrm{Diff}(M)_i \ni f \mapsto \mathcal{N}(f)$, where

$$\begin{array}{ccc} \mathcal{N}(i) & \xrightarrow{\mathcal{N}(f)} & \mathcal{N}(i) \\ \downarrow & & \downarrow \\ M & \xrightarrow{f} & M \end{array}$$

is a vector bundle isomorphism for each $f \in \mathrm{Diff}(M)_i$; see the end of (7). If we equip $\mathcal{N}(i)$ and $\mathcal{N}(i_1)$ with the fiber Riemann metrics induced from the fixed Riemannian metric g on N, the mappings $\mathcal{N}(q_1)$ and all $\mathcal{N}(f)$ are fiberwise linear isometries.

Let us now consider the right action of $\mathrm{Diff}(M)_i$ on the space of sections $\Gamma_c(M \leftarrow \mathcal{N}(i))$ given by $f^*s := \mathcal{N}(f)^{-1} \circ s \circ f$.

From the proof of claim (10) we recall now the sets

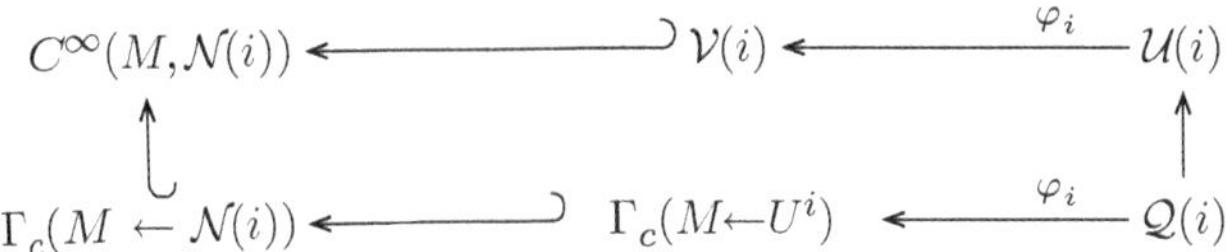

Both mappings φ_i are diffeomorphisms. But since the action of $\text{Diff}(M)$ on i is not free we cannot extend the splitting submanifold $\mathcal{Q}(i)$ to an orbit cylinder as we did in the proof of claim (10). $\mathcal{Q}(i)$ is a smooth transversal for the orbit though i.

For any $f \in \text{Diff}(M)$ and $s \in \Gamma_c(M \leftarrow U^i) \subset \Gamma_c(M \leftarrow \mathcal{N}(i))$ we have

$$\varphi_i^{-1}(f^* s) = \tau_*^i(f^* s) = \tau_*^i(s) \circ f.$$

So the space $q_1^* \Gamma_c(M \leftarrow \mathcal{N}(i_1))$ of all sections of $\mathcal{N}(i) \to M$ which factor to sections of $\mathcal{N}(i_1) \to M_1$, is exactly the space of all fixed points of the action of $\text{Diff}(M)_i$ on $\Gamma_c(M \leftarrow \mathcal{N}(i))$; and they are mapped by $\tau_*^i = \varphi_i^{-1}$ to such immersions in $\mathcal{Q}(i)$ which have again $\text{Diff}(M)_i$ as isotropy group.

If $s \in \Gamma_c(M \leftarrow U^i) \subset \Gamma_c(M \leftarrow \mathcal{N}(i))$ is an arbitrary section, the orbit through $\tau_*^i(s) \in \mathcal{Q}(i)$ hits the transversal $\mathcal{Q}(i)$ again in the points $\tau_*^i(f^* s)$ for $f \in \text{Diff}(M)_i$.

Statement (5) of the theorem is now proved.

(23) The Orbit Structure *We have the following description of the orbit structure near i in* $\text{Imm}(M, N)$*: For fixed $f \in \text{Diff}(M)_i$ the set of fixed points* $\text{Fix}(f) := \{j \in \mathcal{Q}(i) : j \circ f = j\}$ *is called a* generalized wall. *The union of all generalized walls is called the* diagram $\mathcal{D}(i)$ *of i. A connected component of the complement* $\mathcal{Q}(i) \setminus \mathcal{D}(i)$ *is called a* generalized Weyl chamber. *The group* $\text{Diff}(M)_i$ *maps walls to walls and chambers to chambers. The immersion i lies in every wall. We shall see shortly that there is only one chamber and that the situation is rather distinct from that of reflection groups.*

If we view the diagram in the space $\Gamma_c(M \leftarrow U^i) \subset \Gamma_c(M \leftarrow \mathcal{N}(i))$ which is diffeomorphic to $\mathcal{Q}(i)$, then it consists of traces of closed linear subspaces, because the action of $\text{Diff}(M)_i$ on $\Gamma_c(M \leftarrow \mathcal{N}(i))$ consists of linear isometries in the following way. Let us tensor the vector bundle $\mathcal{N}(i) \to M$ with the natural line bundle of half densities on M, and let us remember one positive half density to fix an isomorphism with the original bundle. Then $\text{Diff}(M)_i$ still acts on this new bundle $\mathcal{N}_{1/2}(i) \to M$ and the pullback action on sections with compact support is isometric for the inner product

$$\langle s_1, s_2 \rangle := \int_M g(s_1, s_2).$$

We now extend the walls and chambers from

$$\mathcal{Q}(i) = \Gamma_c(M \leftarrow U^i) \subset \Gamma_c(M \leftarrow \mathcal{N}(i))$$

to the whole space $\Gamma_c(M \leftarrow \mathcal{N}(i)) = \Gamma_c(M \leftarrow \mathcal{N}_{1/2}(i))$; recall from (22) that $\text{Diff}(M)_i$ acts on the whole space.

(24) Claim *Each wall in $\Gamma_c(M \leftarrow \mathcal{N}_{1/2}(i))$ is a closed linear subspace of infinite codimension. Since there are at most countably many walls, there is only one chamber.*

From the proof of claim (19) we know that $f(U_\alpha^i) \cap U_\alpha^i = \emptyset$ for all $f \in \text{Diff}(M)_i$ and all sets U_α^i from the setup (7). Take a section s in the wall of fixed points of f.

Choose a section s_α with support in some U^i_α and let the section s be defined by $s|_{U^i_\alpha} = s_\alpha|_{U^i_\alpha}$, $s|_{f^{-1}(U^i_\alpha)} = -f^*s_\alpha$, 0 elsewhere. Then obviously $\langle s, s'\rangle = 0$ for all s' in the wall of f. But this construction furnishes an infinite dimensional space contained in the orthogonal complement of the wall of f.

(25) The Action of $\operatorname{Diff}(M)$ **on** $\operatorname{Imm}(M, N) \times M$ **Proof of (6)**
Here we will consider the right action $(i, x).f = (i \circ f, f^{-1}(x))$ of $\operatorname{Diff}(M)$ on $\operatorname{Imm}(M, N) \times M$. This action is free: If $(i \circ f, f^{-1}(x)) = (i, x)$ then from claim (8) we get $f = Id_M$.

Claim *Let* $(i, x) \in \operatorname{Imm}(M, N) \times M$. *Then there is an open neighborhood* $\mathcal{W}(i, x)$ *in* $\operatorname{Imm}(M, N) \times M$ *which is saturated for the* $\operatorname{Diff}(M)$*-action and which splits smoothly as*

$$\mathcal{W}(i, x) = \mathcal{Q}(i, x) \times \operatorname{Diff}(M).$$

Here $\mathcal{Q}(i, x)$ *is a smooth splitting submanifold of* $\operatorname{Imm}(M, N) \times M$, *diffeomorphic to an open neighborhood of* $(0, x)$ *in* $C^\infty(\mathcal{N}(i))$.

Let $\pi : \operatorname{Imm}(M, N) \times M \to (\operatorname{Imm}(M, N) \times M)/\operatorname{Diff}(M) = S(M, N)$ *be the projection onto the orbit space, which we equip with the quotient topology. Then* $\pi|_{\mathcal{Q}(i,x)} : \mathcal{Q}(i, x) \to \pi(\mathcal{Q}(i, x))$ *is bijective onto an open subset of the quotient. If* (i, x) *runs through* $\operatorname{Imm}(M, N) \times M$ *these mappings define a smooth atlas for the quotient space, so that*

$$(\operatorname{Imm}(M, N) \times M, \pi, (\operatorname{Imm}(M, N) \times M)/\operatorname{Diff}(M), \operatorname{Diff}(M))$$

is a smooth principal fiber bundle with structure group $\operatorname{Diff}(M)$.

If we restrict to the open subset $\operatorname{Imm}_{\text{prop}}(M, N) \times M$ *of proper immersions times* M *then the base space is Hausdorff.*

By claim (19), the isotropy subgroup $\operatorname{Diff}(M)_i = \{f \in \operatorname{Diff}(M) : i \circ f = i\}$ acts properly discontinuously on M, so $q_1 : M \to M/\operatorname{Diff}(M)_i =: M_1$ is a covering. We choose an open neighborhood W_x of x in M such that $q_1 : W_x \to M_1$ is injective.

Now we adapt the second half of the proof of claim (10) and use freely all the notation from there. We consider the open set

$$\{(h \circ f, f^{-1}(y)) : h \in \mathcal{V}(i), y \in W_x, f \in \operatorname{Diff}(M)\} \subset \\ \subset C^\infty(M, U^i) \times M \subset C^\infty(M, \mathcal{N}(i)) \times M.$$

We have a smooth mapping from it into $\Gamma_c(M \leftarrow U^i) \times W_x \times \operatorname{Diff}(M)$ which is given by $(h, y) \mapsto (h \circ (p \circ h)^{-1}, (p \circ h)(y), p \circ h)$, where $\Gamma_c(M \leftarrow U^i)$ is the space of sections with compact support of $U^i \to M$. We now put

$$\mathcal{Q}(i, x) := \tau^i_*(\Gamma_c(M \leftarrow U^i) \cap \mathcal{V}(i)) \times W_x \subset \operatorname{Imm}(M, N) \times M.$$

Then we have

$$\begin{aligned}\mathcal{W}(i,x) :&= \{(h\circ f, f(y)) : h\in\mathcal{U}(i), y\in W_x, f\in \operatorname{Diff}(M)\}\\ &\cong \mathcal{Q}(i,x)\times \operatorname{Diff}(M)\cong (\Gamma_c(M\leftarrow U^i)\cap\mathcal{V}(i))\times W_x\times\operatorname{Diff}(M),\end{aligned}$$

since the action of $\operatorname{Diff}(M)$ is free. The quotient mapping $\pi|_{\mathcal{Q}(i)} : \mathcal{Q}(i) \to \operatorname{Imm}_{\text{free}}(M,N)/\operatorname{Diff}(M)$ is bijective onto an open set in the quotient. We now use $(\varphi_i\times Id_{W_x})\circ(\pi|_{\mathcal{Q}(i,x)})^{-1} : \pi(\mathcal{Q}(i,x)) \to \Gamma_c(M\leftarrow U^i)\times W_x$ as a chart for the quotient space. In order to investigate the chart change let $(j,y)\in\operatorname{Imm}(M,N)\times M$ be such that $\pi(\mathcal{Q}(i,x))\cap\pi(\mathcal{Q}(j,y))\neq\emptyset$. Then there exists $(h,z)\in\mathcal{W}(i,x)\cap\mathcal{Q}(j,y)$, so there exists a unique $f\in\operatorname{Diff}(M)$ (given by $f=p\circ\varphi_i(h)$) such that $(h\circ f^{-1}, f(z))\in\mathcal{Q}(i,x)$. If we consider $(j\circ f^{-1}, f(y))$ instead of (j,y) and call it again (j,y), we have $\mathcal{Q}(i,x)\cap\mathcal{Q}(j,y)\neq\emptyset$ and consequently $\mathcal{U}(i)\cap\mathcal{U}(j)\neq\emptyset$. Now the first component of the chart change is smooth by the argument in the end of the proof of claim (10), and the second component is just $Id_{W_x\cap W_y}$.

The result about Hausdorff follows from claim (11). The fibers over $\operatorname{Imm}(M,N)/\operatorname{Diff}(M)$ can be read off the following diagram:

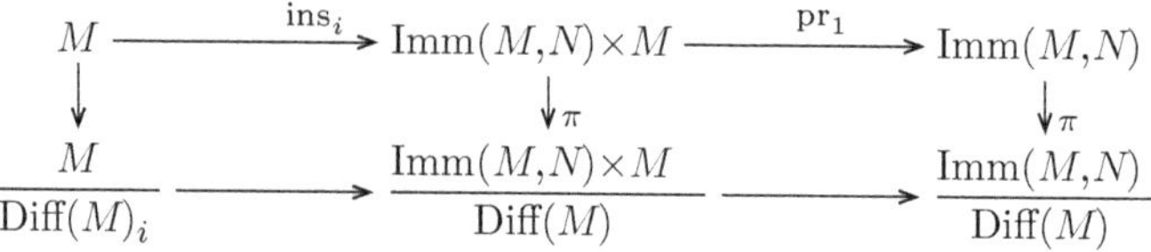

This finishes the proof of Theorem 7.3. □

8 Weak Riemannian Manifolds

If an infinite dimensional manifold is not modeled on a Hilbert space, then a Riemannian metric cannot describe the topology on each tangent space. We have to deal with more complicated situations.

8.1 Manifolds, Vector Fields, Differential Forms

Let M be a smooth manifold modeled on convenient vector spaces. Tangent vectors to M are kinematic ones.

The reason for this is that eventually we want flows of vector fields, and that there are too many derivations in infinite dimensions, even on a Hilbert space H: Let $\alpha\in L(H,H)$ be a continuous linear functional which vanishes on the subspace of compact operators, thus also on $H \otimes H$. Then the linear functional $f \mapsto \alpha(d^2 f(0))$ is a derivation at 0 on $C^\infty(H)$, since

$$\alpha(d^2(f.g)(0)) = \alpha\big(d^2 f(0).g(0) + df(0) \otimes dg(0) + dg(0) \otimes df(0) + f(0).d^2 g(0)\big)$$

and α vanishes on the two middle terms. There are even non-zero derivations which differentiate 3 times, see [55, 28.4].

The (kinematic) tangent bundle TM is then a smooth vector bundle as usual. Differential forms of degree k are then smooth sections of the bundle $L^k_{\text{skew}}(TM; \mathbb{R})$ of skew symmetric k-linear functionals on the tangent bundle, since this is the only version which admits exterior derivative, Lie derivatives along vector field, and pullbacks along arbitrary smooth mappings; see [55, 33.21]. The de Rham cohomology equals singular cohomology with real coefficients if the manifold is smoothly paracompact; see [71] and [55, Section 34]. If a vector field admits a flow, then each integral curve is uniquely given as a flow line; see [55, 32.14].

8.2 Weak Riemannian Manifolds

Let M be a smooth manifold modeled on convenient locally convex vector spaces. A smooth Riemannian metric g on M is called weak if $g_x : T_x M \to T_x^* M$ is only injective for each $x \in M$. The image $g(TM) \subset T^*M$ is called the *smooth cotangent bundle* associated to g. Then g^{-1} is the metric on the smooth cotangent bundle as well as the morphism $g(TM) \to TM$. We have a special class of 1-forms $\Omega^1_g(M) := \Gamma(g(TM))$ for which the musical mappings make sense: $\alpha^\sharp = g^{-1}\alpha \in \mathfrak{X}(M)$ and $X^\flat = gX$. These 1-forms separate points on TM. The exterior derivative is defined by $d : \Omega^1_g(M) \to \Omega^2(M) = \Gamma(L^2_{\text{skew}}(TM; \mathbb{R}))$ since the embedding $g(TM) \subset T^*M$ is a smooth fiber linear mapping.

Existence of the Levi-Civita covariant derivative is equivalent to: *The metric itself admits* symmetric *gradients with respect to itself.* Locally this means: If M is c^∞-open in a convenient vector space V_M. Then

$$D_{x,X} g_x(X, Y) = g_x(X, \text{grad}_1 g(x)(X, Y)) = g_x(\text{grad}_2 g(x)(X, X), Y),$$

where $D_{x,X}$ denote the directional derivative at x in the direction X, and where the mappings $\text{grad}_1 g$ and $\text{sym}\,\text{grad}_2 g : M \times V_M \times V_M \to V_M$, given by $(x, X) \mapsto \text{grad}_{1,2}\, g(x)(X, X)$, are smooth and quadratic in $X \in V_M$. The geodesic equation then is (again locally) given by

$$c_{tt} = \tfrac{1}{2} \text{grad}_1 g(c)(c_t, c_t) - \text{grad}_2 g(c)(c_t, c_t)\,.$$

This formula corresponds to the usual formula for the geodesic flow using Christoffel symbols, expanded out using the first derivatives of the metric tensor. For the existence of the covariant derivative see [68, 2.4], and for the geodesic equation see

[76, 2.1 and 2.4]; there this is done in a special case, but the method works in the general case without changes. See also [12, 4.2, 4.3, and 4.4] for a derivation in another special case.

8.3 Weak Riemannian Metrics on Spaces of Immersions

For a compact manifold M and a finite dimensional Riemannian manifold $(N, \bar{g})$ we can consider the following weak Riemannian metrics on the manifold $\operatorname{Imm}(M, N)$ of smooth immersions $M \to N$:

$$G^0_f(h,k) = \int_M \bar{g}(h,k)\operatorname{vol}(f^*\bar{g}) \qquad \text{the } L^2\text{-metric,}$$

$$G^s_f(h,k) = \int_M \bar{g}((1+\Delta^{f^*\bar{g}})^s h, k)\operatorname{vol}(f^*\bar{g}) \qquad \text{a Sobolev metric of order } s,$$

$$G^\Phi_f(h,g) = \int_M \Phi(f)\bar{g}(h,k)\operatorname{vol}(f^*\bar{g}) \qquad \text{an almost local metric.}$$

Here $\operatorname{vol}(f^*\bar{g})$ is the volume density on M of the pullback metric $g = f^*\bar{g}$, and $\Delta^{f^*\bar{g}}$ is the (Bochner) Laplacian with respect to g and $\bar{g}$ acting on sections of f^*TN, and $\Phi(f)$ is a positive function of the total volume $\operatorname{Vol}(f^*g) = \int_M \operatorname{vol}(f^*g)$, of the scalar curvature $\operatorname{Scal}(f^*\bar{g})$, and of the mean curvature $\operatorname{Tr}(S^f)$, S^f being the second fundamental form. See [12, 13] for more information. All these metrics are invariant for the right action of the reparameterization group $\operatorname{Diff}(M)$, so they descend to metrics on shape space $B_i(M, N)$ (off the singularities) such that the projection $\operatorname{Imm}(M, N) \to B_i(M, N)$ is a Riemannian submersion of a benign type: the G-orthogonal component to the tangent space to the $\operatorname{Diff}(M)$-orbit consists always of smooth vector fields. So there is no need to use the notion of robust weak Riemannian metrics discussed below.

8.4 Theorem *The Riemannian metrics on* $\operatorname{Imm}(M, N)$ *defined in Sect. 8.3 have the following properties:*

(1) *Geodesic distance on* $\operatorname{Imm}(M, N)$*, defined as the infimum of path-lengths of smooth isotopies between two immersions, vanishes for the* L^2*-metric* G^0.
(2) *Geodesic distance is positive on* $B_i(M, N)$ *for the almost local metric* G^Φ *if* $\Phi(f) \geq 1 + A\operatorname{Tr}(S^F)$*, or if* $\Phi(f) \geq A\operatorname{Vol}(f^*\bar{g})$*, for some* $A > 0$.
(3) *Geodesic distance is positive on* $B_i(M, N)$ *for the Sobolev metric* G^s *if* $s \geq 1$.
(4) *The geodesic equation is locally well-posed on* $\operatorname{Imm}(M, N)$ *for the Sobolev metric* G^s *if* $s \geq 1$*, and globally well-posed (and thus geodesically complete) on* $\operatorname{Imm}(S^1, \mathbb{R}^n)$*, if* $s \geq 2$.

(1) is due to [75] for $B_i(S^1, \mathbb{R}^2)$, to [74] for $B_i(M, N)$ and for $\operatorname{Diff}(M)$, which combines to the result for $\operatorname{Imm}(M, N)$ as noted in [6]. (2) is proved in [13]. For (3) see [12]. (4) is due to [21] and [20].

8.5 *Analysis Tools on Regular Lie Groups and on* Diff(M) *for a Whitney Manifold Germ*

Let G be a regular convenient Lie group, with Lie algebra $\mathfrak{g}$. We also consider a Frölicher group $G = \text{Diff}(M)$ for a Whitney manifold germ $M \subset \tilde{M}$ with Lie algebra $\mathfrak{g} = \mathfrak{X}_{c,\partial}(M)$, with the negative of the usual Lie bracket, as described in Sects. 6.3–6.6.

Let $\mu : G \times G \to G$ be the group multiplication, μ_x the left translation and μ^y the right translation, $\mu_x(y) = \mu^y(x) = xy = \mu(x, y)$. The adjoint action $\text{Ad} : G \to GL(\mathfrak{g})$ is given by $\text{Ad}(g)X = T(\mu^{g^{-1}}).T(\mu_g)X$. Let $L, R : \mathfrak{g} \to \mathfrak{X}(G)$ be the left and right invariant vector field mappings, given by $L_X(g) = T_e(\mu_g).X$ and $R_X = T_e(\mu^g).X$, respectively. They are related by $L_X(g) = R_{\text{Ad}(g)X}(g)$. Their flows are given by

$$\text{Fl}_t^{L_X}(g) = g.\exp(tX) = \mu^{\exp(tX)}(g),$$
$$\text{Fl}_t^{R_X}(g) = \exp(tX).g = \mu_{\exp(tX)}(g).$$

The right Maurer–Cartan form $\kappa = \kappa^r \in \Omega^1(G, \mathfrak{g})$ is given by $\kappa_x(\xi) := T_x(\mu^{x^{-1}}) \cdot \xi$. *It satisfies the left Maurer–Cartan equation* $d\kappa^r - \frac{1}{2}[\kappa^r, \kappa^r]^\wedge_{\mathfrak{g}} = 0$, where $[\ ,\ ]^\wedge$ denotes the wedge product of $\mathfrak{g}$-valued forms on G induced by the Lie bracket. Note that $\frac{1}{2}[\kappa^r, \kappa^r]^\wedge(\xi, \eta) = [\kappa^r(\xi), \kappa^r(\eta)]$.
Namely, evaluate $d\kappa^r$ on right invariant vector fields R_X, R_Y for $X, Y \in \mathfrak{g}$.

$$\begin{aligned}(d\kappa^r)(R_X, R_Y) &= R_X(\kappa^r(R_Y)) - R_Y(\kappa^r(R_X)) - \kappa^r([R_X, R_Y]) \\ &= R_X(Y) - R_Y(X) + [X, Y] = 0 - 0 + [\kappa^r(R_X), \kappa^r(R_Y)].\end{aligned}$$

The left Maurer–Cartan form $\kappa^l \in \Omega^1(G, \mathfrak{g})$ is given by $\kappa^l_x(\xi) := T_x(\mu_{x^{-1}}) \cdot \xi$. *The left Maurer–Cartan form* κ^l *satisfies the right Maurer–Cartan equation* $d\kappa^l + \frac{1}{2}[\kappa^l, \kappa^l]^\wedge_{\mathfrak{g}} = 0$.

The (exterior) derivative of the function $\text{Ad} : G \to GL(\mathfrak{g})$ satisfies

$$d\,\text{Ad} = (\text{ad} \circ \kappa^r).\text{Ad} = \text{Ad}.(\text{ad} \circ \kappa^l)$$

since we have

$$\begin{aligned}d\,\text{Ad}(T\mu^g.X) &= \partial_t|_0 \text{Ad}(\exp(tX).g) = \partial_t|_0 \text{Ad}(\exp(tX)).\text{Ad}(g) \\ &= \text{ad}(\kappa^r(T\mu^g.X)).\text{Ad}(g)\,,\end{aligned}$$
$$d\,\text{Ad}(T\mu_g.X) = \partial_t|_0 \text{Ad}(g.\exp(tX)) = \text{Ad}(g).\text{ad}(\kappa^l(T\mu_g.X))\,.$$

8.6 Right Invariant Weak Riemannian Metrics on Regular Lie Groups and on Diff(M) for a Whitney Manifold Germ

We continue under the assumptions of Sect. 8.5, Let $\gamma = \mathfrak{g} \times \mathfrak{g} \to \mathbb{R}$ be a positive definite bounded (weak) inner product. Then

$$\gamma_x(\xi, \eta) = \gamma\big(T(\mu^{x^{-1}}) \cdot \xi,\ T(\mu^{x^{-1}}) \cdot \eta\big) = \gamma\big(\kappa(\xi),\ \kappa(\eta)\big)$$

is a right invariant (weak) Riemannian metric on G and any (weak) right invariant bounded Riemannian metric is of this form, for suitable γ. Denote by $\check{\gamma} : \mathfrak{g} \to \mathfrak{g}^*$ the mapping induced by γ, from the Lie algebra into its dual (of bounded linear functionals) and by $\langle \alpha, X\rangle_{\mathfrak{g}}$ the duality evaluation between $\alpha \in \mathfrak{g}^*$ and $X \in \mathfrak{g}$.

Let $g : [a, b] \to G$ be a smooth curve. The velocity field of g, viewed in the right trivializations, coincides with the right logarithmic derivative

$$\delta^r(g) := T(\mu^{g^{-1}}) \cdot \partial_t g = \kappa(\partial_t g) = (g^*\kappa)(\partial_t).$$

The energy of the curve $g(t)$ is given by

$$E(g) = \frac{1}{2}\int_a^b \gamma_g(g', g')dt = \frac{1}{2}\int_a^b \gamma\big((g^*\kappa)(\partial_t), (g^*\kappa)(\partial_t)\big)\, dt.$$

For a variation $g(s, t)$ with fixed endpoints we then use that

$$d(g^*\kappa)(\partial_t, \partial_s) = \partial_t(g^*\kappa(\partial_s)) - \partial_s(g^*\kappa(\partial_t)) - 0,$$

partial integration, and the left Maurer–Cartan equation to obtain

$$\begin{aligned}
\partial_s E(g) &= \frac{1}{2}\int_a^b 2\gamma\big(\partial_s(g^*\kappa)(\partial_t),\ (g^*\kappa)(\partial_t)\big)\, dt \\
&= \int_a^b \gamma\big(\partial_t(g^*\kappa)(\partial_s) - d(g^*\kappa)(\partial_t, \partial_s),\ (g^*\kappa)(\partial_t)\big)\, dt \\
&= -\int_a^b \gamma\big((g^*\kappa)(\partial_s),\ \partial_t(g^*\kappa)(\partial_t)\big)\, dt \\
&\qquad - \int_a^b \gamma\big([(g^*\kappa)(\partial_t), (g^*\kappa)(\partial_s)],\ (g^*\kappa)(\partial_t)\big)\, dt \\
&= -\int_a^b \big\langle \check{\gamma}(\partial_t(g^*\kappa)(\partial_t)),\ (g^*\kappa)(\partial_s)\big\rangle_{\mathfrak{g}}\, dt
\end{aligned}$$

$$- \int_a^b \langle \check{\gamma}((g^*\kappa)(\partial_t)), \operatorname{ad}_{(g^*\kappa)(\partial_t)}(g^*\kappa)(\partial_s)\rangle_{\mathfrak{g}}\, dt$$

$$= - \int_a^b \langle \check{\gamma}(\partial_t(g^*\kappa)(\partial_t)) + (\operatorname{ad}_{(g^*\kappa)(\partial_t)})^* \check{\gamma}((g^*\kappa)(\partial_t)), (g^*\kappa)(\partial_s)\rangle_{\mathfrak{g}}\, dt.$$

Thus the curve $g(0, t)$ is critical for the energy if and only if

$$\check{\gamma}(\partial_t(g^*\kappa)(\partial_t)) + (\operatorname{ad}_{(g^*\kappa)(\partial_t)})^* \check{\gamma}((g^*\kappa)(\partial_t)) = 0.$$

In terms of the right logarithmic derivative $u : [a, b] \to \mathfrak{g}$ of $g : [a, b] \to G$, given by $u(t) := g^*\kappa(\partial_t) = T_{g(t)}(\mu^{g(t)^{-1}}) \cdot g'(t)$, the *geodesic equation* has the expression

$$\boxed{\partial_t u = -\check{\gamma}^{-1} \operatorname{ad}(u)^* \check{\gamma}(u)}\,.$$

Thus the geodesic equation exists in general if and only if $\operatorname{ad}(X)^*\check{\gamma}(X)$ is in the image of $\check{\gamma} : \mathfrak{g} \to \mathfrak{g}^*$, i.e.,

$$\operatorname{ad}(X)^*\check{\gamma}(X) \in \check{\gamma}(\mathfrak{g})$$

for every $X \in \mathfrak{X}$; this leads to the existence of the Christoffel symbols. Arnold [4] asked for the more restrictive condition $\operatorname{ad}(X)^*\check{\gamma}(Y) \in \check{\gamma}(\mathfrak{g})$ for all $X, Y \in \mathfrak{g}$. The geodesic equation for the *momentum* $p := \gamma(u)$ is

$$p_t = -\operatorname{ad}(\check{\gamma}^{-1}(p))^* p.$$

There are situations, see Theorem 8.11 or [9], where only the more general condition is satisfied, but where the usual transpose $\operatorname{ad}^\top(X)$ of $\operatorname{ad}(X)$,

$$\operatorname{ad}^\top(X) := \check{\gamma}^{-1} \circ \operatorname{ad}_X^* \circ \check{\gamma}$$

does not exist for all X.

We describe now the *covariant derivative* and the *curvature*. The right trivialization $(\pi_G, \kappa^r) : TG \to G \times \mathfrak{g}$ induces the isomorphism $R : C^\infty(G, \mathfrak{g}) \to \mathfrak{X}(G)$, given by $R(X)(x) := R_X(x) := T_e(\mu^x) \cdot X(x)$, for $X \in C^\infty(G, \mathfrak{g})$ and $x \in G$. Here $\mathfrak{X}(G) := \Gamma(TG)$ denotes the Lie algebra of all vector fields. For the Lie bracket and the Riemannian metric we have

$$\begin{aligned}
[R_X, R_Y] &= R(-[X, Y]_{\mathfrak{g}} + dY \cdot R_X - dX \cdot R_Y),\\
R^{-1}[R_X, R_Y] &= -[X, Y]_{\mathfrak{g}} + R_X(Y) - R_Y(X),\\
\gamma_x(R_X(x), R_Y(x)) &= \gamma(X(x), Y(x))\,,\ x \in G.
\end{aligned}$$

In what follows, we shall perform all computations in $C^\infty(G, \mathfrak{g})$ instead of $\mathfrak{X}(G)$. In particular, we shall use the convention

$$\nabla_X Y := R^{-1}(\nabla_{R_X} R_Y) \quad \text{for } X, Y \in C^\infty(G, \mathfrak{g})$$

to express the Levi-Civita covariant derivative.

8.7 Lemma ([9, 3.3]) *Assume that for all* $\xi \in \mathfrak{g}$ *the element* $\operatorname{ad}(\xi)^* \check{\gamma}(\xi) \in \mathfrak{g}^*$ *is in the image of* $\check{\gamma} : \mathfrak{g} \to \mathfrak{g}^*$ *and that* $\xi \mapsto \check{\gamma}^{-1} \operatorname{ad}(\xi)^* \check{\gamma}(\xi)$ *is bounded quadratic (or, equivalently, smooth). Then the Levi-Civita covariant derivative of the metric* γ *exists and is given for any* $X, Y \in C^\infty(G, \mathfrak{g})$ *in terms of the isomorphism* R *by*

$$\nabla_X Y = dY.R_X + \rho(X)Y - \frac{1}{2} \operatorname{ad}(X)Y,$$

where

$$\rho(\xi)\eta = \tfrac{1}{4}\check{\gamma}^{-1}\big(\operatorname{ad}^*_{\xi+\eta} \check{\gamma}(\xi+\eta) - \operatorname{ad}^*_{\xi-\eta} \check{\gamma}(\xi-\eta)\big) = \tfrac{1}{2}\check{\gamma}^{-1}\big(\operatorname{ad}^*_\xi \check{\gamma}(\eta) + \operatorname{ad}^*_\eta \check{\gamma}(\xi)\big)$$

is the polarized version. The mapping $\rho : \mathfrak{g} \to L(\mathfrak{g}, \mathfrak{g})$ *is bounded, and we have* $\rho(\xi)\eta = \rho(\eta)\xi$. *We also have*

$$\gamma\big(\rho(\xi)\eta, \zeta\big) = \frac{1}{2}\gamma(\xi, \operatorname{ad}(\eta)\zeta) + \frac{1}{2}\gamma(\eta, \operatorname{ad}(\xi)\zeta),$$

$$\gamma(\rho(\xi)\eta, \zeta) + \gamma(\rho(\eta)\zeta, \xi) + \gamma(\rho(\zeta)\xi, \xi) = 0.$$

For $X, Y \in C^\infty(G, \mathfrak{g})$ we have

$$[R_X, \operatorname{ad}(Y)] = \operatorname{ad}(R_X(Y)) \quad \text{and} \quad [R_X, \rho(Y)] = \rho(R_X(Y)).$$

The *Riemannian curvature* is then computed as follows:

$$\begin{aligned}
\mathcal{R}(X, Y) &= [\nabla_X, \nabla_Y] - \nabla_{-[X,Y]_\mathfrak{g} + R_X(Y) - R_Y(X)} \\
&= [R_X + \rho_X - \tfrac{1}{2} \operatorname{ad}_X, R_Y + \rho_Y - \tfrac{1}{2} \operatorname{ad}_Y] \\
&\quad - R(-[X, Y]_\mathfrak{g} + R_X(Y) - R_Y(X)) - \rho(-[X, Y]_\mathfrak{g} + R_X(Y) - R_Y(X)) \\
&\quad + \frac{1}{2} \operatorname{ad}(-[X, Y]_\mathfrak{g} + R_X(Y) - R_Y(X)) \\
&= [\rho_X, \rho_Y] + \rho_{[X,Y]_\mathfrak{g}} - \frac{1}{2}[\rho_X, \operatorname{ad}_Y] + \frac{1}{2}[\rho_Y, \operatorname{ad}_X] - \frac{1}{4} \operatorname{ad}_{[X,Y]_\mathfrak{g}}
\end{aligned}$$

which is visibly a tensor field.

For the numerator of the sectional curvature we obtain

$$\begin{aligned}\gamma\big(\mathcal{R}(X,Y)X,Y\big) &= \gamma(\rho_X\rho_Y X,Y) - \gamma(\rho_Y\rho_X X,Y) + \gamma(\rho_{[X,Y]}X,Y)\\ &\quad - \frac{1}{2}\gamma(\rho_X[Y,X],Y) + \frac{1}{2}\gamma([Y,\rho_X X],Y)\\ &\quad + 0 - \frac{1}{2}\gamma([X,\rho_Y X],Y) - \frac{1}{4}\gamma([[X,Y],X],Y)\\ &= \gamma(\rho_X X,\rho_Y Y) - \|\rho_X Y\|_\gamma^2 + \frac{3}{4}\|[X,Y]\|_\gamma^2\\ &\quad - \frac{1}{2}\gamma(X,[Y,[X,Y]]) + \frac{1}{2}\gamma(Y,[X,[X,Y]])\\ &= \gamma(\rho_X X,\rho_Y Y) - \|\rho_X Y\|_\gamma^2 + \frac{3}{4}\|[X,Y]\|_\gamma^2\\ &\quad - \gamma(\rho_X Y,[X,Y]]) + \gamma(Y,[X,[X,Y]]).\end{aligned}$$

If the adjoint $\operatorname{ad}(X)^\top : \mathfrak{g} \to \mathfrak{g}$ exists, this is easily seen to coincide with Arnold's original formula [4],

$$\begin{aligned}\gamma(\mathcal{R}(X,Y)X,Y) = &-\frac{1}{4}\|\operatorname{ad}(X)^\top Y + \operatorname{ad}(Y)^\top X\|_\gamma^2 + \gamma(\operatorname{ad}(X)^\top X, \operatorname{ad}(Y)^\top Y)\\ &+ \frac{1}{2}\gamma(\operatorname{ad}(X)^\top Y - \operatorname{ad}(Y)^\top X, \operatorname{ad}(X)Y) + \frac{3}{4}\|[X,Y]\|_\gamma^2.\end{aligned}$$

8.8 Examples of Weak Right Invariant Riemannian Metrics on Diffeomorphism Groups

Let M be a finite dimensional manifold. We consider the following regular Lie groups: $\operatorname{Diff}(M)$, the group of all diffeomorphisms of M if M is compact. $\operatorname{Diff}_c(M)$, the group of diffeomorphisms with compact support, if M is not compact. If $M = \mathbb{R}^n$, we also may consider one of the following: $\operatorname{Diff}_{\mathcal{S}}(\mathbb{R}^n)$, the group of all diffeomorphisms which fall rapidly to the identity. $\operatorname{Diff}_{W^{\infty,p}}(\mathbb{R}^n)$, the group of all diffeomorphisms which are modeled on the space $W^{\infty,p}(\mathbb{R}^n)^n$, the intersection of all $W^{k,p}$-Sobolev spaces of vector fields. The last type of groups works also for a *Riemannian manifold of bounded geometry* $(M,\bar g)$; see [30] for Sobolev spaces on them. In the following we write $\operatorname{Diff}_{\mathcal{A}}(M)$ for any of these groups. The Lie algebras are the spaces $\mathfrak{X}_{\mathcal{A}}(M)$ of vector fields, where $\mathcal{A} \in \{C_c^\infty, \mathcal{S}, W^{\infty,p}\}$, with the negative of the usual bracket as Lie bracket.

Most of the following weak Riemannian metrics also make sense on $\operatorname{Diff}(M)$ for a compact Whitney manifold germ $M \subset \tilde M$, but their behavior has not been investigated. In particular, I do not know how the Laplacian $1 + \Delta^g$ behaves on $\mathfrak{X}_\partial(M)$ and its Sobolev completions.

A right invariant weak inner product on $\mathrm{Diff}_{\mathcal{A}}(M)$ is given by a smooth positive definite inner product γ on the Lie algebra $\mathfrak{X}_{\mathcal{A}}(M)$ which is described by the *inertia operator* $L = \check{\gamma} : \mathfrak{X}_{\mathcal{A}}(M) \to \mathfrak{X}_{\mathcal{A}}(M)'$ and we shall denote its inverse by $K = L^{-1} : L(\mathfrak{X}_{\mathcal{A}}(M)) \to \mathfrak{X}_{\mathcal{A}}(M)$. Under suitable conditions on L (like an elliptic coercive (pseudo) differential operator of high enough order) the operator K turns out to be the reproducing kernel of a Hilbert space of vector fields which is contained in the space of either C^1_b (bounded C^1 with respect to $\bar{g}$) or C^2_b vector fields. See [108, Chapter 12], [68], and [80] for uses of the reproducing Hilbert space approach. The right invariant metric is then defined as in Sect. 8.5, where $\langle\ ,\ \rangle_{\mathfrak{X}_{\mathcal{A}}(M)}$ is the duality:

$$G^L_\varphi(X\circ\varphi, Y\circ\varphi) = G^L_{\mathrm{Id}}(X,Y) = \gamma(X,Y) = \langle L(X), Y\rangle_{\mathfrak{X}_{\mathcal{A}}(M)}.$$

For example, the Sobolev metric of order s corresponds to the inertia operator $L(X) = (1+\Delta^{\bar{g}})^s(X).\operatorname{vol}(\bar{g})$. Examples of metrics are

$$G^0_{\mathrm{Id}}(X,Y) = \int_M \bar{g}(X,Y)\operatorname{vol}(\bar{g}) \qquad \text{the } L^2 \text{ metric},$$

$$G^s_{\mathrm{Id}}(X,Y) = \int_M \bar{g}((1+\Delta^{\bar{g}})^s X, Y)\operatorname{vol}(\bar{g}) \qquad \text{a Sobolev metric of order } s,$$

$$G^{\dot{H}^1}_{\mathrm{Id}}(X,Y) = \int_{\mathbb{R}} X'.Y' dx = -\int_{\mathbb{R}} X''Y\,dx \qquad \text{where } X,Y\in\mathfrak{X}_{\mathcal{A}}(\mathbb{R}).$$

As explained in Sect. 8.8, the geodesic equation on $\mathrm{Diff}_{\mathcal{A}}(M)$ is given as follows: Let $\varphi : [a,b] \to \mathrm{Diff}_{\mathcal{A}}(M)$ be a smooth curve. In terms of its right logarithmic derivative

$$u : [a,b] \to \mathfrak{X}_{\mathcal{A}}(M), \quad u(t) := \varphi^*\kappa(\partial_t) = \varphi'(t)\circ\varphi(t)^{-1},$$

the geodesic equation is

$$L(u_t) = L(\partial_t u) = -\operatorname{ad}(u)^* L(u).$$

The *condition for the existence of the geodesic equation* is as follows:

$$X \mapsto K(\operatorname{ad}(X)^* L(X))$$

is bounded quadratic $\mathfrak{X}_{\mathcal{A}}(M) \to \mathfrak{X}_{\mathcal{A}}(M)$. Using *Lie derivatives*, the computation of ad^*_X is especially simple. Namely, for any section ω of $T^*M\otimes \operatorname{vol}$ and vector fields $\xi,\eta\in\mathfrak{X}_{\mathcal{A}}(M)$, we have

$$\int_M (\omega, [\xi,\eta]) = \int_M (\omega, \mathcal{L}_\xi(\eta)) = -\int_M (\mathcal{L}_\xi(\omega), \eta),$$

hence $\operatorname{ad}^*_\xi(\omega) = +\mathcal{L}_\xi(\omega)$. Thus the Hamiltonian version of the geodesic equation on the smooth dual $L(\mathfrak{X}_{\mathcal{A}}(M)) \subset \Gamma_{C^2_b}(T^*M \otimes \text{vol})$ becomes

$$\partial_t \alpha = -\operatorname{ad}^*_{K(\alpha)} \alpha = -\mathcal{L}_{K(\alpha)}\alpha,$$

or, keeping track of everything,

$$\partial_t \varphi = u \circ \varphi, \qquad \partial_t \alpha = -\mathcal{L}_u \alpha \qquad u = K(\alpha) = \alpha^\sharp, \qquad \alpha = L(u) = u^\flat.$$

8.9 Theorem *Geodesic distance vanishes on* $\operatorname{Diff}_{\mathcal{A}}(M)$ *for any Sobolev metric of order* $s < \frac{1}{2}$. *If* $M = S^1 \times C$ *with* C *compact, then geodesic distance vanishes also for* $s = \frac{1}{2}$. *It also vanishes for the* L^2*-metric on the Virasoro group* $\mathbb{R} \rtimes \operatorname{Diff}_{\mathcal{A}}(\mathbb{R})$.

Geodesic distance is positive on $\operatorname{Diff}_{\mathcal{A}}(M)$ *for any Sobolev metric of order* $s \geq 1$. *If* $\dim(M) = 1$ *then geodesic distance is also positive for* $s > \frac{1}{2}$.

This is proved in [8, 14], and [6]. Note that low order Sobolev metrics have geodesic equations corresponding to well-known nonlinear PDEs: On $\operatorname{Diff}(S^1)$ or $\operatorname{Diff}_{\mathcal{A}}(\mathbb{R})$ the L^2-geodesic equation is Burgers' equation, on the Virasoro group it is the KdV equation, and the (standard) H^1-geodesic is (in both cases a variant of) the Camassa–Holm equation; see [10, 7.2] for a more comprehensive overview. All these are completely integrable infinite dimensional Hamiltonian systems.

8.10 Theorem *Let* $(M, \bar{g})$ *be a compact Riemannian manifold. Then the geodesic equation is locally well-posed on* $\operatorname{Diff}_{\mathcal{A}}(M)$ *and the geodesic exponential mapping is a local diffeomorphism for a Sobolev metric of integer order* $s \geq 1$. *For a Sobolev metric of integer order* $s > \frac{\dim(M)+3}{2}$ *the geodesic equation is even globally well-posed, so that* $(\operatorname{Diff}_{\mathcal{A}}(M), G^s)$ *is geodesically complete. This is also true for non-integer order* s *if* $M = \mathbb{R}^n$.

For $M = S^1$, *the geodesic equation is locally well-posed even for* $s \geq \frac{1}{2}$.

For these results see [11, 12, 32, 33].

8.11 Theorem ([9]) *For* $\mathcal{A} \in \{C^\infty_c, \mathcal{S}, W^{\infty,1}\}$ *let*

$$\mathcal{A}_1(\mathbb{R}) = \{f \in C^\infty(R) \,:\, f' \in \mathcal{A}(\mathbb{R}),\ f(-\infty) = 0\}$$

and let $\operatorname{Diff}_{\mathcal{A}_1}(\mathbb{R}) = \{\varphi = \operatorname{Id} + f \,:\, f \in \mathcal{A}_1(\mathbb{R}),\ f' > -1\}$. *These are all regular Lie groups. The right invariant weak Riemannian metric*

$$G^{\dot{H}^1}_{\operatorname{Id}}(X, Y) = \int_{\mathbb{R}} X'Y'\,dx$$

is positive definite both on $\operatorname{Diff}_{\mathcal{A}}(\mathbb{R})$ *where it does not admit a geodesic equation (a non-robust weak Riemannian manifold), and on* $\operatorname{Diff}_{\mathcal{A}_1}(\mathbb{R})$ *where it admits a geodesic equation but not in the stronger sense of Arnold. On* $\operatorname{Diff}_{\mathcal{A}_1}(\mathbb{R})$ *the geodesic equation is the Hunter-Saxton equation*

$$(\varphi_t)\circ\varphi^{-1}=u, \qquad u_t=-uu_x+\frac{1}{2}\int_{-\infty}^{x}(u_x(z))^2\,dz\,,$$

and the induced geodesic distance is positive. We define the R-map by

$$R:\mathrm{Diff}_{\mathcal{A}_1}(\mathbb{R})\to\mathcal{A}\big(\mathbb{R},\mathbb{R}_{>-2}\big)\subset\mathcal{A}(\mathbb{R},\mathbb{R}), \quad R(\varphi)=2\left((\varphi')^{1/2}-1\right).$$

The R-map is invertible with inverse

$$R^{-1}:\mathcal{A}\big(\mathbb{R},\mathbb{R}_{>-2}\big)\to\mathrm{Diff}_{\mathcal{A}_1}(\mathbb{R}), \quad R^{-1}(\gamma)(x)=x+\frac{1}{4}\int_{-\infty}^{x}\gamma^2+4\gamma\,dx\,.$$

The pullback of the flat L^2-metric via R is the $\dot H^1$-metric on $\mathrm{Diff}_{\mathcal{A}}(\mathbb{R})$*, i.e., $R^*\langle\cdot,\cdot\rangle_{L^2(dx)} = G^{\dot H^1}$. Thus the space* $\left(\mathrm{Diff}_{\mathcal{A}_1}(\mathbb{R}),\dot H^1\right)$ *is a flat space in the sense of Riemannian geometry. There are explicit formulas for geodesics, geodesic distance, and geodesic splines, even for more restrictive spaces $\mathcal{A}_1$ like Denjoy–Carleman ultradifferentiable function classes. There are also soliton-like solutions.* $(\mathrm{Diff}_{\mathcal{A}_1}(\mathbb{R}),G^{\dot H^1})$ *is geodesically convex, but not geodesically complete; the geodesic completion is the smooth semigroup*

$$\mathrm{Mon}_{\mathcal{A}_1}=\{\varphi=\mathrm{Id}+f\ :\ f\in\mathcal{A}_1(\mathbb{R})\,,\ f'\geq-1\}\,.$$

Any geodesic can hit the subgroup $\mathrm{Diff}_{\mathcal{A}}(\mathbb{R})\subset\mathrm{Diff}_{\mathcal{A}_1}(\mathbb{R})$ *at most twice.*

8.12 Trouvé Groups for Reproducing Kernel Hilbert Spaces

This is the origin of the notion of a Trouvé group. It puts the approach of Sect. 8.1 to Theorem 8.11 upside down and gets rid of the use of the Lie algebra structure on the space of vector fields. If the generating space $\mathcal{A}$ of vector fields on $\mathbb{R}^d$ for the Trouvé group $\mathcal{G}_{\mathcal{A}}$ (see Sect. 6.12) is a reproducing kernel Hilbert space $(\mathcal{A}(\mathbb{R}^d,\mathbb{R}^d),\langle\ ,\ \rangle_{\mathcal{A}})$ contained in C^1_b, then

$$\mathrm{dist}(\mathrm{Id},\varphi):=\inf\Big\{\int_0^1\|X(t)\|_{\mathcal{A}}\,dt\ :\ X\in\mathcal{F}_{\mathcal{A}},\ \mathrm{evol}^X=\varphi\Big\}$$

defines a metric which makes the Trouvé group $\mathcal{G}_{\mathcal{A}}$ into a topological group; see [103, 108]. This is widely used for the *Large Deformation Diffeomorphic Metric Matching* (LDDMM) method in image analysis and computational anatomy. The most popular reproducing kernel Hilbert space is the one where the kernel is a Gaussian $e^{-|x|^2/\sigma}$. Here the space $\mathcal{A}$ is a certain space of entire real analytic functions, and a direct description of the Trouvé group is severely lacking.

9 Robust Weak Riemannian Manifolds and Riemannian Submersions

9.1 Robust Weak Riemannian Manifolds

Some constructions may lead to vector fields whose values do not lie in T_xM, but in the Hilbert space completion $\overline{T_xM}$ with respect to the weak inner product g_x. We need that $\bigcup_{x\in M}\overline{T_xM}$ forms a smooth vector bundle over M. In a coordinate chart on open $U\subset M$, $TM|_U$ is a trivial bundle $U\times V$ and all the inner products $g_x, x\in U$ define inner products on the same topological vector space V. They all should be bounded with respect to each other, so that the completion $\overline{V}$ of V with respect to g_x does not depend on x and $\bigcup_{x\in U}\overline{T_xM}\cong U\times\overline{V}$. This means that $\bigcup_{x\in M}\overline{T_xM}$ forms a smooth vector bundle over M with trivializations the linear extensions of the trivializations of the tangent bundle $TM\to M$. Chart changes should respect this. This is a compatibility property between the weak Riemannian metric and some smooth atlas of M.

Definition A convenient weak Riemannian manifold (M,g) will be called a *robust* Riemannian manifold if

- The Levi-Civita covariant derivative of the metric g exists: The symmetric gradients should exist and be smooth.
- The completions $\overline{T_xM}$ form a smooth vector bundle as above.

9.2 Theorem *If a right invariant weak Riemannian metric on a regular Lie group admits the Levi-Civita covariant derivative, then it is already robust.*

Proof By right invariance, each right translation $T\mu^g$ extends to an isometric isomorphisms $\overline{T_xG}\to\overline{T_{xg}G}$. By the smooth uniform boundedness theorem these isomorphisms depend smoothly on $g\in G$. □

9.3 Covariant Curvature and O'Neill's Formula

In [68, 2.2] one finds the following formula for the numerator of sectional curvature, which is valid for *closed smooth* 1-forms $\alpha,\beta\in\Omega^1_g(M)$ on a weak Riemannian manifold (M,g). Recall that we view $g: TM\to T^*M$ and so g^{-1} is the dual inner product on $g(TM)$ and $\alpha^\sharp=g^{-1}(\alpha)$.

$$
\begin{aligned}
&g\big(R(\alpha^\sharp,\beta^\sharp)\alpha^\sharp,\beta^\sharp\big) = \\
&-\tfrac12\alpha^\sharp\alpha^\sharp(\|\beta\|^2_{g^{-1}}) - \tfrac12\beta^\sharp\beta^\sharp(\|\alpha\|^2_{g^{-1}}) + \tfrac12(\alpha^\sharp\beta^\sharp+\beta^\sharp\alpha^\sharp)g^{-1}(\alpha,\beta) \\
&\qquad\qquad\big(\text{last line } = -\alpha^\sharp\beta([\alpha^\sharp,\beta^\sharp]) + \beta^\sharp\alpha([\alpha^\sharp,\beta^\sharp]])\big) \\
&-\tfrac14\|d(g^{-1}(\alpha,\beta))\|^2_{g^{-1}} + \tfrac14 g^{-1}\big(d(\|\alpha\|^2_{g^{-1}}), d(\|\beta\|^2_{g^{-1}})\big) \\
&+\tfrac34\big\|[\alpha^\sharp,\beta^\sharp]\big\|^2_g.
\end{aligned}
$$

This is called Mario's formula since Mario Micheli derived the coordinate version in his 2008 thesis. Each term depends only on g^{-1} with the exception of the last term. The role of the last term (which we call the O'Neill term) will become clear in the next result. Let $p : (E, g_E) \to (B, g_B)$ be a Riemannian submersion between infinite dimensional robust Riemannian manifolds; i.e., for each $b \in B$ and $x \in E_b := p^{-1}(b)$ the tangent mapping $T_x p : (T_x E, g_E) \to (T_b B, g_B)$ is a surjective metric quotient map so that

$$
\|\xi_b\|_{g_B} := \inf\big\{\|X_x\|_{g_E} \;:\; X_x \in T_x E, T_x p.X_x = \xi_b\big\}.
$$

The infinimum need not be attained in $T_x E$ but will be in the completion $\overline{T_x E}$. The orthogonal subspace $\{Y_x : g_E(Y_x, T_x(E_b)) = 0\}$ will therefore be taken in $\overline{T_x(E_b)}$ in $T_x E$. If $\alpha_b = g_B(\alpha_b^\sharp, \quad) \in g_B(T_b B) \subset T_b^* B$ is an element in the g_B-smooth dual, then $p^*\alpha_b := (T_x p)^*(\alpha_b) = g_B(\alpha_b^\sharp, T_x p \quad) : T_x E \to \mathbb{R}$ is in $T_x^* E$ but in general it is not an element in the smooth dual $g_E(T_x E)$. It is, however, an element of the Hilbert space completion $\overline{g_E(T_x E)}$ of the g_E-smooth dual $g_E(T_x E)$ with respect to the norm $\| \quad \|_{g_E^{-1}}$, and the element $g_E^{-1}(p^*\alpha_b) =: (p^*\alpha_b)^\sharp$ is in the $\| \quad \|_{g_E}$-completion $\overline{T_x E}$ of $T_x E$. We can call $g_E^{-1}(p^*\alpha_b) =: (p^*\alpha_b)^\sharp$ the *horizontal lift* of $\alpha_b^\sharp = g_B^{-1}(\alpha_b) \in T_b B$.

9.4 Theorem ([68, 2.6]) *Let $p : (E, g_E) \to (B, g_B)$ be a Riemannian submersion between infinite dimensional robust Riemannian manifolds. Then for* ***closed*** *1-forms $\alpha, \beta \in \Omega^1_{g_B}(B)$ O'Neill's formula holds in the form:*

$$
\begin{aligned}
g_B\big(R^B(\alpha^\sharp,\beta^\sharp)\beta^\sharp,\alpha^\sharp\big) &= g_E\big(R^E((p^*\alpha)^\sharp,(p^*\beta)^\sharp)(p^*\beta)^\sharp,(p^*\alpha)^\sharp\big) \\
&\quad + \tfrac34\|[(p^*\alpha)^\sharp,(p^*\beta)^\sharp]^{ver}\|^2_{g_E}.
\end{aligned}
$$

Proof The last (O'Neill) term is the difference between curvature on E and the pullback of the curvature on B. □

9.5 Semilocal Version of Mario's Formula, Force, and Stress

In all interesting examples of orbits of diffeomorphisms groups through a template shape, Mario's covariant curvature formula leads to complicated and impenetrable formulas. Efforts to break this down to comprehensible pieces led to the concepts of symmetrized force and (shape-) stress explained below. Since acceleration sits in the second tangent bundle, one either needs a covariant derivative to map it down to the tangent bundle, or at least rudiments of local charts. In [68] we managed the local version. Interpretations in mechanics or elasticity theory are still lacking.

Let (M, g) be a robust Riemannian manifold, $x \in M$, $\alpha, \beta \in g_x(T_xM)$. Assume we are given local smooth vector fields X_α and X_β such that:

1. $X_\alpha(x) = \alpha^\sharp(x), \quad X_\beta(x) = \beta^\sharp(x)$,
2. Then $\alpha^\sharp - X_\alpha$ is zero at x. Therefore it has a well-defined derivative $D_x(\alpha^\sharp - X_\alpha)$ lying in $\operatorname{Hom}(T_xM, T_xM)$. For a vector field Y we have $D_x(\alpha^\sharp - X_\alpha).Y_x = [Y, \alpha^\sharp - X_\alpha](x) = \mathcal{L}_Y(\alpha^\sharp - X_\alpha)|_x$. The same holds for β.
3. $\mathcal{L}_{X_\alpha}(\alpha) = \mathcal{L}_{X_\alpha}(\beta) = \mathcal{L}_{X_\beta}(\alpha) = \mathcal{L}_{X_\beta}(\beta) = 0$,
4. $[X_\alpha, X_\beta] = 0$.

Locally constant 1-forms and vector fields will do. We then define

$$\begin{aligned}
\mathcal{F}(\alpha, \beta) &:= \tfrac{1}{2} d(g^{-1}(\alpha, \beta)), \qquad \text{a 1-form on } M \text{ called the } \textit{force}, \\
\mathcal{D}(\alpha, \beta)(x) &:= D_x(\beta^\sharp - X_\beta).\alpha^\sharp(x) \\
&= d(\beta^\sharp - X_\beta).\alpha^\sharp(x), \qquad \in T_xM \text{ called the } \textit{stress}. \\
&\Longrightarrow \mathcal{D}(\alpha, \beta)(x) - \mathcal{D}(\beta, \alpha)(x) = [\alpha^\sharp, \beta^\sharp](x).
\end{aligned}$$

Then in terms of force and stress the numerator of sectional curvature looks as follows:

$$\begin{aligned}
& g\big(R(\alpha^\sharp, \beta^\sharp)\beta^\sharp, \alpha^\sharp\big)(x) = R_{11} + R_{12} + R_2 + R_3\,, \qquad \text{where} \\
& R_{11} = \tfrac{1}{2}\big(\mathcal{L}^2_{X_\alpha}(g^{-1})(\beta, \beta) - 2\mathcal{L}_{X_\alpha}\mathcal{L}_{X_\beta}(g^{-1})(\alpha, \beta) + \mathcal{L}^2_{X_\beta}(g^{-1})(\alpha, \alpha)\big)(x)\,, \\
& R_{12} = \langle \mathcal{F}(\alpha, \alpha), \mathcal{D}(\beta, \beta)\rangle + \langle \mathcal{F}(\beta, \beta), \mathcal{D}(\alpha, \alpha)\rangle - \langle \mathcal{F}(\alpha, \beta), \mathcal{D}(\alpha, \beta) + \mathcal{D}(\beta, \alpha)\rangle \\
& R_2 = \big(\|\mathcal{F}(\alpha, \beta)\|^2_{g^{-1}} - \big\langle \mathcal{F}(\alpha, \alpha)), \mathcal{F}(\beta, \beta)\big\rangle_{g^{-1}}\big)(x)\,, \\
& R_3 = -\tfrac{3}{4}\|\mathcal{D}(\alpha, \beta) - \mathcal{D}(\beta, \alpha)\|^2_{g_x}\,.
\end{aligned}$$

9.6 Landmark Space as Homogeneous Space of Solitons

This subsection is based on [67]; the method explained here has many applications in computational anatomy and elsewhere, under the name LDDMM (large diffeomorphic deformation metric matching).

A *landmark* $q = (q_1, \dots, q_N)$ is an N-tuple of distinct points in $\mathbb{R}^n$; landmark space $\mathrm{Land}^N(\mathbb{R}^n) \subset (\mathbb{R}^n)^N$ is open. Let $q^0 = (q_1^0, \dots, q_N^0)$ be a fixed standard template landmark. Then we have the surjective mapping

$$\begin{aligned} &\mathrm{ev}_{q^0} : \mathrm{Diff}_{\mathcal{A}}(\mathbb{R}^n) \to \mathrm{Land}^N(\mathbb{R}^n), \\ &\varphi \mapsto \mathrm{ev}_{q^0}(\varphi) = \varphi(q^0) = (\varphi(q_1^0), \dots, \varphi(q_N^0)). \end{aligned}$$

Given a Sobolev metric of order $s > \frac{n}{2} + 2$ on $\mathrm{Diff}_{\mathcal{A}}(\mathbb{R}^n)$, we want to induce a Riemannian metric on $\mathrm{Land}^N(\mathbb{R}^n)$ such that ev_{q_0} becomes a Riemannian submersion.

The fiber of ev_{q^0} over a landmark $q = \varphi_0(q^0)$ is

$$\begin{aligned} \{\varphi \in \mathrm{Diff}_{\mathcal{A}}(\mathbb{R}^n) : \varphi(q^0) = q\} &= \varphi_0 \circ \{\varphi \in \mathrm{Diff}_{\mathcal{A}}(\mathbb{R}^n) : \varphi(q^0) = q^0\} \\ &= \{\varphi \in \mathrm{Diff}_{\mathcal{A}}(\mathbb{R}^n) : \varphi(q) = q\} \circ \varphi_0 \,. \end{aligned}$$

The tangent space to the fiber is

$$\{X \circ \varphi_0 : X \in \mathfrak{X}_{\mathcal{S}}(\mathbb{R}^n), X(q_i) = 0 \text{ for all } i\}.$$

A tangent vector $Y \circ \varphi_0 \in T_{\varphi_0} \mathrm{Diff}_{\mathcal{S}}(\mathbb{R}^n)$ is $G^L_{\varphi_0}$-perpendicular to the fiber over q if and only if

$$\int_{\mathbb{R}^n} \langle LY, X \rangle \, dx = 0 \quad \forall X \text{ with } X(q) = 0.$$

If we require Y to be smooth then $Y = 0$. So we assume that $LY = \sum_i P_i.\delta_{q_i}$, a distributional vector field with support in q. Here $P_i \in T_{q_i}\mathbb{R}^n$. But then

$$\begin{aligned} Y(x) &= L^{-1}\Big(\sum_i P_i.\delta_{q_i}\Big) = \int_{\mathbb{R}^n} K(x-y) \sum_i P_i.\delta_{q_i}(y)\, dy = \sum_i K(x - q_i).P_i, \\ T_{\varphi_0}(\mathrm{ev}_{q^0}).(Y \circ \varphi_0) &= Y(q_k)_k = \sum_i (K(q_k - q_i).P_i)_k \,. \end{aligned}$$

Now let us consider a tangent vector $P = (P_k) \in T_q \mathrm{Land}^N(\mathbb{R}^n)$. Its horizontal lift with footpoint φ_0 is $P^{\mathrm{hor}} \circ \varphi_0$ where the vector field P^{hor} on $\mathbb{R}^n$ is given as follows: Let $K^{-1}(q)_{ki}$ be the inverse of the $(N \times N)$-matrix $K(q)_{ij} = K(q_i - q_j)$. Then

$$\begin{aligned} P^{\mathrm{hor}}(x) &= \sum_{i,j} K(x - q_i) K^{-1}(q)_{ij} P_j \,, \\ L(P^{\mathrm{hor}}(x)) &= \sum_{i,j} \delta(x - q_i) K^{-1}(q)_{ij} P_j \,. \end{aligned}$$

Note that P^{hor} is a vector field of class H^{2l-1}.

The Riemannian metric on the finite dimensional manifold Land^N induced by the g^L-metric on $\text{Diff}_{\mathcal{S}}(\mathbb{R}^n)$ is given by

$$\begin{aligned}
g^L_q(P,Q) &= G^L_{\varphi_0}(P^{\text{hor}}, Q^{\text{hor}}) = \int_{\mathbb{R}^n} \langle L(P^{\text{hor}}), Q^{\text{hor}}\rangle\, dx \\
&= \int_{\mathbb{R}^n} \Big\langle \sum_{i,j} \delta(x-q_i) K^{-1}(q)_{ij} P_j, \sum_{k,l} K(x-q_k)K^{-1}(q)_{kl} Q_l \Big\rangle\, dx \\
&= \sum_{i,j,k,l} K^{-1}(q)_{ij} K(q_i-q_k) K^{-1}(q)_{kl} \langle P_j, Q_l\rangle \\
g^L_q(P,Q) &= \sum_{k,l} K^{-1}(q)_{kl}\langle P_k, Q_l\rangle.
\end{aligned}$$

The *geodesic equation* in vector form is

$$\begin{aligned}
\ddot{q}_n = &-\frac{1}{2}\sum_{k,i,j,l} K^{-1}(q)_{ki}\, \text{grad}\, K(q_i-q_j)(K(q)_{in}-K(q)_{jn})K^{-1}(q)_{jl}\langle \dot{q}_k, \dot{q}_l\rangle \\
&+ \sum_{k,i} K^{-1}(q)_{ki}\Big\langle \text{grad}\, K(q_i-q_n), \dot{q}_i-\dot{q}_n\Big\rangle \dot{q}_k\,.
\end{aligned}$$

The cotangent bundle $T^*\text{Land}^N(\mathbb{R}^n) = \text{Land}^N(\mathbb{R}^n)\times((\mathbb{R}^n)^N)^* \ni (q,\alpha)$. We treat $\mathbb{R}^n$ like scalars; $\langle\quad,\quad\rangle$ is always the standard inner product on $\mathbb{R}^n$.
The inverse metric is then given by

$$(g^L)^{-1}_q(\alpha,\beta) = \sum_{i,j} K(q)_{ij}\langle \alpha_i, \beta_j\rangle, \qquad K(q)_{ij} = K(q_i-q_j).$$

The energy function is

$$E(q,\alpha) = \tfrac{1}{2}(g^L)^{-1}_q(\alpha,\alpha) = \tfrac{1}{2}\sum_{i,j} K(q)_{ij}\langle \alpha_i, \alpha_j\rangle$$

and its Hamiltonian vector field (using $\mathbb{R}^n$-valued derivatives to save notation) is

$$H_E(q,\alpha) = \sum_{i,k=1}^{N}\Big(K(q_k-q_i)\alpha_i\frac{\partial}{\partial q_k} + \text{grad}\, K(q_i-q_k)\langle\alpha_i,\alpha_k\rangle\frac{\partial}{\partial \alpha_k}\Big).$$

So the *Hamiltonian version of the geodesic equation* is the flow of this vector field:

$$\begin{cases} \dot{q}_k &= \sum_i K(q_i-q_k)\alpha_i \\ \dot{\alpha}_k &= -\sum_i \text{grad}\, K(q_i-q_k)\langle\alpha_i,\alpha_k\rangle. \end{cases}$$

We shall use *stress and force* to express the geodesic equation and curvature:

$$\alpha_k^\sharp = \sum_i K(q_k - q_i)\alpha_i, \quad \alpha^\sharp = \sum_{i,k} K(q_k - q_i)\langle\alpha_i, \tfrac{\partial}{\partial q^k}\rangle$$

$$\mathcal{D}(\alpha,\beta) := \sum_{i,j} dK(q_i - q_j)(\alpha_i^\sharp - \alpha_j^\sharp)\Big\langle\beta_j, \frac{\partial}{\partial q_i}\Big\rangle, \quad \text{the } \textit{stress}.$$

$$\mathcal{D}(\alpha,\beta) - \mathcal{D}(\beta,\alpha) = (D_{\alpha^\sharp}\beta^\sharp) - D_{\beta^\sharp}\alpha^\sharp = [\alpha^\sharp, \beta^\sharp], \quad \textit{Lie bracket}.$$

$$\mathcal{F}_i(\alpha,\beta) = \frac{1}{2}\sum_k \operatorname{grad} K(q_i - q_k)(\langle\alpha_i, \beta_k\rangle + \langle\beta_i, \alpha_k\rangle)$$

$$\mathcal{F}(\alpha,\beta) := \sum_i \langle\mathcal{F}_i(\alpha,\beta), dq_i\rangle = \frac{1}{2} d\, g^{-1}(\alpha,\beta) \quad \text{the } \textit{force}.$$

The geodesic equation on $T^* \operatorname{Land}^N(\mathbb{R}^n)$ then becomes

$$\begin{cases} \dot q &= \alpha^\sharp \\ \dot\alpha &= -\mathcal{F}(\alpha,\alpha)\,. \end{cases}$$

Next we shall compute *curvature via the cotangent bundle*. From the semilocal version of Mario's formula for the numerator of the sectional curvature for constant 1-forms α, β on landmark space, where $\alpha_k^\sharp = \sum_i K(q_k - q_i)\alpha_i$, we get directly:

$$\begin{aligned}
&g^L\big(R(\alpha^\sharp,\beta^\sharp)\alpha^\sharp,\beta^\sharp\big) = \\
&= \big\langle\mathcal{D}(\alpha,\beta) + \mathcal{D}(\beta,\alpha), \mathcal{F}(\alpha,\beta)\big\rangle \\
&\quad - \big\langle\mathcal{D}(\alpha,\alpha), \mathcal{F}(\beta,\beta)\big\rangle - \big\langle\mathcal{D}(\beta,\beta), \mathcal{F}(\alpha,\alpha)\big\rangle \\
&\quad - \tfrac{1}{2}\sum_{i,j}\Big(d^2K(q_i - q_j)(\beta_i^\sharp - \beta_j^\sharp, \beta_i^\sharp - \beta_j^\sharp)\langle\alpha_i, \alpha_j\rangle \\
&\qquad\qquad - 2d^2K(q_i - q_j)(\beta_i^\sharp - \beta_j^\sharp, \alpha_i^\sharp - \alpha_j^\sharp)\langle\beta_i, \alpha_j\rangle \\
&\qquad\qquad + d^2K(q_i - q_j)(\alpha_i^\sharp - \alpha_j^\sharp, \alpha_i^\sharp - \alpha_j^\sharp)\langle\beta_i, \beta_j\rangle\Big) \\
&\quad - \|\mathcal{F}(\alpha,\beta)\|_{g^{-1}}^2 + g^{-1}\big(\mathcal{F}(\alpha,\alpha), \mathcal{F}(\beta,\beta)\big). \\
&\quad + \tfrac{3}{4}\|[\alpha^\sharp,\beta^\sharp]\|_g^2
\end{aligned}$$

9.7 Shape Spaces of Submanifolds as Homogeneous Spaces for the Diffeomorphism Group

Let M be a compact manifold and $(N, \bar{g})$ a Riemannian manifold of bounded geometry as in Sect. 3.6. The diffeomorphism group $\mathrm{Diff}_{\mathcal{A}}(N)$ acts also from the left on the manifold of $\mathrm{Emb}(M, N)$ embeddings and also on the *nonlinear Grassmannian* or *differentiable Chow variety* $B(M, N) = \mathrm{Emb}(M, N)/\mathrm{Diff}(M)$. For a Sobolev metric of order $s > \frac{\dim(N)}{2} + 2$ one can then again induce a Riemannian metric on each $\mathrm{Diff}_{\mathcal{A}}(N)$-orbit, as we did above for landmark spaces. This is done in [68], where the geodesic equation is computed and where curvature is described in terms of stress and force.

References

1. R. Abraham. *Lectures of Smale on differential topology*. Columbia University, New York, 1962. Lecture Notes.
2. H. Alzaareer and A. Schmeding. Differentiable mappings on products with different degrees of differentiability in the two factors. *Expo. Math.*, 33(2):184–222, 2015.
3. L. F. A. Arbogast. *Du calcul des dérivations*. Levrault, Strasbourg, 1800.
4. V. Arnold. Sur la géometrie differentielle des groupes de Lie de dimension infinie et ses applications à l'hydrodynamique des fluides parfaits. *Ann. Inst. Fourier*, 16:319–361, 1966.
5. B. Arslan, A. P. Goncharov, and M. Kocatepe. Spaces of Whitney functions on Cantor-type sets. *Canad. J. Math.*, 54(2):225–238, 2002.
6. M. Bauer, M. Bruveris, P. Harms, and P. W. Michor. Vanishing geodesic distance for the Riemannian metric with geodesic equation the KdV-equation. *Ann. Global Anal. Geom.*, 41(4):461–472, 2012.
7. M. Bauer, M. Bruveris, P. Harms, and P. W. Michor. Smooth perturbations of the functional calculus and applications to Riemannian geometry on spaces of metrics, 2018. arXiv:1810.03169.
8. M. Bauer, M. Bruveris, and P. W. Michor. Geodesic distance for right invariant Sobolev metrics of fractional order on the diffeomorphism group ii. *Ann. Global Anal. Geom.*, 44(4):361–368, 2013.
9. M. Bauer, M. Bruveris, and P. W. Michor. The homogeneous Sobolev metric of order one on diffeomorphism groups on the real line. *Journal of Nonlinear Science*, 24(5):769–808, 2014.
10. M. Bauer, M. Bruveris, and P. W. Michor. Overview of the geometries of shape spaces and diffeomorphism groups. *J. Math. Imaging Vis.*, 50:60–97, 2014.
11. M. Bauer, J. Escher, and B. Kolev. Local and global well-posedness of the fractional order EPDiff equation on $\mathbb{R}^d$. *Journal of Differential Equations*, 2015. arXiv:1411.4081.
12. M. Bauer, P. Harms, and P. W. Michor. Sobolev metrics on shape space of surfaces. *J. Geom. Mech.*, 3(4):389–438, 2011.
13. M. Bauer, P. Harms, and P. W. Michor. Almost local metrics on shape space of hypersurfaces in n-space. *SIAM J. Imaging Sci.*, 5(1):244–310, 2012.
14. M. Bauer, P. Harms, and P. W. Michor. Sobolev metrics on the manifold of all Riemannian metrics. *J. Differential Geom.*, 94(2):187–208, 2013.
15. E. Bierstone. Extension of Whitney fields from subanalytic sets. *Invent. Math.*, 46:277–300, 1978.
16. J. Boman. Differentiability of a function and of its compositions with functions of one variable. *Math. Scand.*, 20:249–268, 1967.

17. R. Brown. *Some problems of algebraic topology*. PhD thesis, 1961. Oxford.
18. R. Brown. Ten topologies for $X \times Y$. *Quart. J. Math.*, 14:303–319, 1963.
19. R. Brown. Function spaces and product topologies. *Quart. J. Math.*, 15:238–250, 1964.
20. M. Bruveris. Completeness properties of Sobolev metrics on the space of curves. *J. Geom. Mech.*, 7(2):125–150, 2015. arXiv:1407.0601.
21. M. Bruveris, P. W. Michor, and D. Mumford. Geodesic completeness for Sobolev metrics on the space of immersed plane curves. *Forum Math. Sigma*, 2:e19 (38 pages), 2014.
22. M. Bruveris and F.-X. Vialard. On completeness of groups of diffeomorphisms. *J. Eur. Math. Soc.*, 19(5):1507–1544, 2017.
23. V. Cervera, F. Mascaro, and P. W. Michor. The orbit structure of the action of the diffeomorphism group on the space of immersions. *Diff. Geom. Appl.*, 1:391–401, 1991.
24. B. Conrad. Stokes' theorem with corners. Math 396 Handout, Stanford University. http://math.stanford.edu/~conrad/diffgeomPage/handouts/stokescorners.pdf.
25. A. Douady and L. Hérault. Arrondissement des variétés à coins. *Comment. Math. Helv.*, 48:484–491, 1973. Appendice à A.Borel and J.-P. Serre: Corners and arithmetic groups.
26. D. Ebin and J. Marsden. Groups of diffeomorphisms and the motion of an incompressible fluid. *Ann. Math.*, 92:102–163, 1970.
27. J. Eells. A setting for global analysis. *Bull AMS*, 72:571–807, 1966.
28. J. Eells, Jr. On the geometry of function spaces. In *Symposium internacional de topologia algebraica International symposium on algebraic topology*, pages 303–308. Universidad Nacional Autónoma de México and UNESCO, Mexico City, 1958.
29. J. Eichhorn. Gauge theory on open manifolds of bounded geometry. *Internat. J. Modern Physics*, 7:3927–3977, 1992.
30. J. Eichhorn. *Global analysis on open manifolds*. Nova Science Publishers Inc., New York, 2007.
31. H. I. Eliasson. Geometry of manifolds of maps. *J. Differential Geometry*, 1:169–194, 1967.
32. J. Escher and B. Kolev. Geodesic completeness for Sobolev H^s-metrics on the diffeomorphisms group of the circle. *Journal of Evolution Equations*, 2014.
33. J. Escher and B. Kolev. Right-invariant Sobolev metrics of fractional order on the diffeomorphism group of the circle. *J. Geom. Mech.*, 6:335–372, 2014.
34. C. F. Faà di Bruno. Note sur une nouvelle formule du calcul différentielle. *Quart. J. Math.*, 1:359–360, 1855.
35. L. Fantappié. I functionali analitici. *Atti Accad. Naz. Lincei*, Mem. 3–11:453–683, 1930.
36. L. Fantappié. Überblick über die Theorie der analytischen Funktionale und ihre Anwendungen. *Jahresber. Deutsch. Mathem. Verein.*, 43:1–25, 1933.
37. C.-A. Faure. *Théorie de la différentiation dans les espaces convenables*. PhD thesis, Université de Genéve, 1991.
38. C.-A. Faure and A. Frölicher. Hölder differentiable maps and their function spaces. In *Categorical topology and its relation to analysis, algebra and combinatorics (Prague, 1988)*, pages 135–142. World Sci. Publ., Teaneck, NJ, 1989.
39. C. Fefferman. C^m extension by linear operators. *Ann. of Math. (2)*, 166(3):779–835, 2007.
40. L. Frerick. Extension operators for spaces of infinite differentiable Whitney jets. *J. Reine Angew. Math.*, 602:123–154, 2007.
41. L. Frerick, E. Jordá, and J. Wengenroth. Tame linear extension operators for smooth Whitney functions. *J. Funct. Anal.*, 261(3):591–603, 2011.
42. L. Frerick, E. Jordá, and J. Wengenroth. Whitney extension operators without loss of derivatives. *Rev. Mat. Iberoam.*, 32(2):377–390, 2016.
43. L. Frerick and J. Wengenroth. Private communication. 2019.
44. A. Frölicher and A. Kriegl. *Linear spaces and differentiation theory*. J. Wiley, Chichester, 1988. Pure and Applied Mathematics.
45. F. Gay-Balmaz and C. Vizman. Principal bundles of embeddings and nonlinear Grassmannians. *Ann. Global Anal. Geom.*, 46(3):293–312, 2014.
46. H. Glöckner. Measurable regularity properties of infinite-dimensional Lie groups, 2015. arXiv:1601.02568.

47. R. E. Greene. Complete metrics of bounded curvature on noncompact manifolds. *Arch. Math. (Basel)*, 31(1):89–95, 1978/79.
48. J. Gutknecht. *Die C^∞_Γ-Struktur auf der Diffeomorphismengruppe einer kompakten Mannigfaltigkeit*. 1977. Doctoral Thesis, ETH Zürich.
49. M. Hanusch. Regularity of Lie groups, 2017. arXiv:1711.03508.
50. K. Hart. Sets with positive Lebesgue measure boundary. MathOverflow question, 2010. https://mathoverflow.net/q/26000.
51. P. Iglesias-Zemmour. *Diffeology*, volume 185 of *Mathematical Surveys and Monographs*. American Mathematical Society, Providence, RI, 2013.
52. E. Klassen and P. W. Michor. Closed surfaces with different shapes that are indistinguishable by the SRNF. *Achivum Mathematicum (Brno)*, 56:107–114, 2020. https://doi.org/10.5817/AM2020-2-107. arxiv:1910.10804.
53. Y. A. Kordyukov. L^p-theory of elliptic differential operators on manifolds of bounded geometry. *Acta Appl. Math.*, 23(3):223–260, 1991.
54. A. Kriegl and P. W. Michor. A convenient setting for real analytic mappings. *Acta Mathematica*, 165:105–159, 1990.
55. A. Kriegl and P. W. Michor. *The Convenient Setting for Global Analysis*. AMS, Providence, 1997. 'Surveys and Monographs 53'.
56. A. Kriegl and P. W. Michor. Regular infinite-dimensional Lie groups. *J. Lie Theory*, 7(1):61–99, 1997.
57. A. Kriegl, P. W. Michor, and A. Rainer. The convenient setting for non-quasianalytic Denjoy–Carleman differentiable mappings. *J. Funct. Anal.*, 256:3510–3544, 2009.
58. A. Kriegl, P. W. Michor, and A. Rainer. The convenient setting for quasianalytic Denjoy–Carleman differentiable mappings. *J. Funct. Anal.*, 261:1799–1834, 2011.
59. A. Kriegl, P. W. Michor, and A. Rainer. The convenient setting for Denjoy–Carleman differentiable mappings of Beurling and Roumieu type. *Revista Matematica Complutense*, 2015.
60. A. Kriegl, P. W. Michor, and A. Rainer. An exotic zoo of diffeomorphism groups on $\mathbb{R}^n$. *Ann. Glob. Anal. Geom.*, 47(2):179–222, 2015.
61. A. Kriegl, P. W. Michor, and A. Rainer. The exponential law for spaces of test functions and diffeomorphism groups. *Indagationes Mathematicae*, 2015.
62. A. Kriegl and L. D. Nel. A convenient setting for holomorphy. *Cahiers Top. Géo. Diff.*, 26:273–309, 1985.
63. J. Leslie. On a differential structure for the group of diffeomorphisms. *Topology*, 6:264–271, 1967.
64. T. Marquis and K.-H. Neeb. Half-Lie groups. *Transform. Groups*, 23(3):801–840, 2018. arXiv:1607.07728.
65. R. Meise and D. Vogt. *Introduction to functional analysis*, volume 2 of *Oxford Graduate Texts in Mathematics*. The Clarendon Press, Oxford University Press, New York, 1997. Translated from the German by M. S. Ramanujan and revised by the authors.
66. R. B. Melrose. *Differential Analysis on Manifolds with Corners*. 1996. http://www-math.mit.edu/~rbm/book.html.
67. M. Micheli, P. W. Michor, and D. Mumford. Sectional curvature in terms of the cometric, with applications to the Riemannian manifolds of landmarks. *SIAM J. Imaging Sci.*, 5(1):394–433, 2012.
68. M. Micheli, P. W. Michor, and D. Mumford. Sobolev metrics on diffeomorphism groups and the derived geometry of spaces of submanifolds. *Izvestiya: Mathematics*, 77(3):541–570, 2013.
69. P. W. Michor. *Manifolds of differentiable mappings*. Shiva Mathematics Series 3, Orpington, 1980.
70. P. W. Michor. Manifolds of smooth maps III: The principal bundle of embeddings of a non compact smooth manifold. *Cahiers Top. Geo. Diff.*, 21:325–337, 1980. MR 82b:58022c.
71. P. W. Michor. Manifolds of smooth mappings IV: Theorem of De Rham. *Cahiers Top. Geo. Diff.*, 24:57–86, 1983.

72. P. W. Michor. A convenient setting for differential geometry and global analysis I, II. *Cahiers Topol. Geo. Diff.*, 25:63–109, 113–178., 1984.
73. P. W. Michor. *Topics in differential geometry*, volume 93 of *Graduate Studies in Mathematics*. American Mathematical Society, Providence, RI, 2008.
74. P. W. Michor and D. Mumford. Vanishing geodesic distance on spaces of submanifolds and diffeomorphisms. *Doc. Math.*, 10:217–245, 2005.
75. P. W. Michor and D. Mumford. Riemannian geometries on spaces of plane curves. *J. Eur. Math. Soc.*, 8:1–48, 2006.
76. P. W. Michor and D. Mumford. An overview of the Riemannian metrics on spaces of curves using the Hamiltonian approach. *Appl. Comput. Harmon. Anal.*, 23(1):74–113, 2007.
77. P. W. Michor and D. Mumford. A zoo of diffeomorphism groups on $\mathbb{R}^n$. *Ann. Glob. Anal. Geom.*, 44:529–540, 2013.
78. J. Milnor. Remarks on infinite-dimensional Lie groups. In *Relativity, groups and topology, II (Les Houches, 1983)*, pages 1007–1057. North-Holland, Amsterdam, 1984.
79. B. S. Mitjagin. Approximate dimension and bases in nuclear spaces. *Uspehi Mat. Nauk*, 16(4 (100)):63–132, 1961.
80. D. Mumford and P. W. Michor. On Euler's equation and 'EPDiff'. *Journal of Geometric Mechanics*, 5:319–344, 2013.
81. E. Nelson. *Topics in dynamics. I: Flows*. Mathematical Notes. Princeton University Press, Princeton, N.J.; University of Tokyo Press, Tokyo, 1969.
82. D. Nenning. *ODE-closedness of function spaces and almost analytic extensions of ultradifferentiable functions*. PhD thesis, 2019. Universität Wien.
83. D. Nenning. On time-dependent Besov vector fields and the regularity of their flows, 2019. arXiv:1804.07595.
84. D. N. Nenning and A. Rainer. On groups of Hölder diffeomorphisms and their regularity. *Trans. Amer. Math. Soc.*, 370(8):5761–5794, 2018.
85. D. N. Nenning and A. Rainer. The Trouvé group for spaces of test functions. *Rev. R. Acad. Cienc. Exactas Fis. Nat. Ser. A Mat. RACSAM*, 113(3):1799–1822, 2019.
86. H. Omori, Y. Maeda, and A. Yoshioka. On regular Fréchet Lie groups I. Some differential geometric expressions of Fourier integral operators on a Riemannian manifold. *Tokyo J. Math.*, 3:353–390, 1980.
87. H. Omori, Y. Maeda, and A. Yoshioka. On regular Fréchet Lie groups II. Composition rules of Fourier Integral operators on a Riemannian manifold. *Tokyo J. Math.*, 4:221–253, 1981.
88. H. Omori, Y. Maeda, and A. Yoshioka. On regular Fréchet Lie groups III. *Tokyo J. Math.*, 4:255–277, 1981.
89. H. Omori, Y. Maeda, and A. Yoshioka. On regular Fréchet Lie groups IV. Definitions and fundamental theorems. *Tokyo J. Math.*, 5:365–398, 1982.
90. H. Omori, Y. Maeda, and A. Yoshioka. On regular Fréchet Lie groups V. Several basic properties. *Tokyo J. Math.*, 6:39–64, 1983.
91. H. Omori, Y. Maeda, A. Yoshioka, and O. Kobayashi. On regular Fréchet Lie groups VI. Infinite dimensional Lie groups which appear in general relativity. *Tokyo J. Math.*, 6:217–246, 1983.
92. R. S. Palais. *Foundations of global non-linear analysis*. W. A. Benjamin, Inc., New York-Amsterdam, 1968.
93. B. Riemann. Über die Hypothesen, welche der Geometrie zu Grunde liegen. *Abhandlungen der Königlichen Gesellschaft der Wissenschaften zu Göttingen*, XIII:1–20, 1868. Habilitationsvortrag 10. Juni 1854.
94. B. Riemann. On the hypotheses which lie at the bases of geometry. *Nature*, 8:14–17, 36–37, 1873. translated into English by William Kingdon Clifford.
95. D. M. Roberts and A. Schmeding. Extending Whitney's extension theorem: nonlinear function spaces, 2018. arXiv:1801.04126.
96. L. Schwartz. *Théorie des distributions*. Hermann, Paris, 1966. Nouvelle édition.
97. R. T. Seeley. Extension of C^∞-functions defined in a half space. *Proc. AMS*, 15:625–626, 1964.

98. N. E. Steenrod. A convenient category for topological spaces. *Michigan Math. J.*, 14:133–152, 1967.
99. E. M. Stein. *Singular integrals and differentiability properties of functions*. Princeton Mathematical Series, No. 30. Princeton University Press, Princeton, N.J., 1970.
100. S. T. Swift. Natural bundles. III. Resolving the singularities in the space of immersed submanifolds. *J. Math. Phys.*, 34(8):3841–3855, 1993.
101. M. Tidten. Fortsetzungen von C^∞-Funktionen, welche auf einer abgeschlossenen Menge in $\mathbb{R}^n$ definiert sind. *Manuscripta Math.*, 27(3):291–312, 1979.
102. M. Tidten. A geometric characterization for the property $(\underline{DN})$ of $\mathcal{E}(K)$ for arbitrary compact subsets K of $\mathbb{R}$. *Arch. Math. (Basel)*, 77(3):247–252, 2001.
103. A. Trouvé. An infinite dimensional group approach for physics based models in pattern recognition. pages 1–35, 1995. http://www.cis.jhu.edu/publications/papers_in_database/alain/trouve1995.pdf.
104. G. Valette. Stokes' formula for stratified forms. *Ann. Polon. Math.*, 114(3):197–206, 2015.
105. D. Vogt. Sequence space representations of spaces of test functions and distributions. In *Functional analysis, holomorphy, and approximation theory (Rio de Janeiro, 1979)*, volume 83 of *Lecture Notes in Pure and Appl. Math.*, pages 405–443. Dekker, New York, 1983.
106. J. Watts. *Diffeologies, Differential Spaces, and Symplectic Geometry*. ProQuest LLC, Ann Arbor, MI, 2012. Thesis (Ph.D.)–University of Toronto (Canada).
107. H. Whitney. Analytic extensions of differentiable functions defined in closed sets. *Trans. Amer. Math. Soc.*, 36(1):63–89, 1934.
108. L. Younes. *Shapes and diffeomorphisms*, volume 171 of *Applied Mathematical Sciences*. Springer-Verlag, Berlin, 2010.

Notes on Global Stress and Hyper-Stress Theories

Reuven Segev

Contents

Abstract The fundamental ideas and tools of the global geometric formulation of stress and hyper-stress theory of continuum mechanics are introduced. The proposed framework is the infinite dimensional counterpart of the statics of a system having a finite number of degrees of freedom, as viewed in the geometric approach to analytical mechanics. For continuum mechanics, the configuration space is the manifold of embeddings of a body manifold into the space manifold. Generalized velocity fields are viewed as elements of the tangent bundle of the configuration space and forces are continuous linear functionals defined on tangent vectors, elements of the cotangent bundle. It is shown, in particular, that a natural

---

R. Segev (✉)
Department of Mechanical Engineering, Ben-Gurion University of the Negev, Beer-Sheva, Israel
e-mail: rsegev@bgu.ac.il

R. Segev, M. Epstein (eds.), *Geometric Continuum Mechanics*, Advances in Mechanics and Mathematics 43, https://doi.org/10.1007/978-3-030-42683-5_2

choice of topology on the configuration space implies that force functionals may be represented by objects that generalize the stresses of traditional continuum mechanics.

1 Introduction

These notes provide an introduction to the fundamentals of global analytic continuum mechanics as developed in [6, 21, 32–34, 39]. The terminology "global analytic" is used to imply that the formulation is based on the notion of a configuration space of the mechanical system as in analytic classical mechanics. As such, this review is complementary to that of [38], which describes continuum mechanics on differentiable manifolds using a generalization of the Cauchy approach to flux and stress theory.

The general setting for the basics of kinematics and statics of a mechanical system is quite simple and provides an elegant geometric picture of mechanics. Consider the configuration space $\mathbb{Q}$ containing all admissible configuration of the system. Then, construct a differentiable manifold structure on the configuration space, define generalized (or virtual) velocities as tangent vectors, elements of $T\mathbb{Q}$, and define generalized forces as linear functions defined on the space of generalized velocities, elements of $T^*\mathbb{Q}$. The result of the action of a generalized force F on a generalized velocity w is interpreted as mechanical power. Thus, such a structure may be used to encompass both classical mechanics of mass particles and rigid bodies as well as continuum mechanics. The difference is that the configuration space for continuum mechanics and other field theories is infinite dimensional.

It is well known that the transition from the mechanics of mass particles and rigid bodies to continuum mechanics is not straightforward and requires the introduction of new notions and assumptions. The global analytic formulation explains this observation as follows. Linear functions, and forces in particular, are identically continuous when defined on a finite dimensional space. However, in the infinite dimensional situation, one has to specify exactly the topology on the infinite dimensional space of generalized velocities with respect to which forces should be continuous. Then, the properties of force functionals are deduced from the continuity requirement through a representation theorem. In other words, the properties of forces follow directly from the kinematics of the theory.

For continuum mechanics of a body $\mathcal{X}$ in space $\mathcal{S}$, the basic kinematic assumption is traditionally referred to as the axiom of material impenetrability. A configuration of the body in space is specified by a mapping $\kappa : \mathcal{X} \to \mathcal{S}$ which is assumed to be injective and of full rank at each point—an embedding. Hence, the configuration space for continuum mechanics should be the collection of embeddings of the body manifold into the space manifold. It turns out that the C^1-topology is the natural one to use in order to endow the collection of embeddings with a differentiable structure of a Banach manifold. The C^r-topologies for $r > 1$ are admissible also.

It follows that forces are continuous linear functionals on the space of vector fields over the body of class C^r, $r \geq 1$, equipped with the C^r-topology with a special role for the case $r = 1$. A standard procedure based on the Hahn–Banach theorem leads to a representation theorem for a force functional in terms of vector valued measures.

The measures representing a force generalize the stress and hyper-stress objects of continuum mechanics. The proof of the representation theorem implies that a stress measure is not determined uniquely by a force. This is in accordance with the inherent static indeterminacy of continuum mechanics. While the case $r = 1$ leads to continuum mechanics of order one, the cases $r > 1$ are extensions of higher-order continuum mechanics. Thus, an existence theorem for hyper-stresses follows naturally. The relation between a force and a representing stress object is a generalization of the principle of virtual work in continuum mechanics and so it is analogous to the equilibrium equations.

The representation of forces by stress measures is significant for two reasons. First, the existence of the stress object as well as the corresponding equilibrium condition are obtained for stress distributions that may be as singular as Radon measures. In addition, while force functionals cannot be restricted to subsets of a body, measures may be restricted to subsets. This reflects a fundamental feature of stress distributions—they induce force systems on bodies. It is emphasized that no further assumptions of mathematical or physical nature are made.

The framework described above applies to continuum mechanics on general differentiable manifolds without any additional structure such as a Riemannian metric or a connection. The body manifold is assumed here to be a compact manifold with corners. However, as described in [23], it is now possible to extend the applicability of this framework to a wider class of geometric object—Whitney manifold germs.

Starting with the introduction of notation used in the manuscript in Sect. 2, we continue with the construction of the manifold structure on the space of embeddings. Thus, Sect. 3 describes the Banachable vector spaces used to construct the infinite dimensional manifold structure on the configuration space and Sect. 4 is concerned with the Banach manifold structure on the set of C^r-sections of a fiber bundle $\xi : \mathcal{Y} \to \mathcal{X}$. This includes, as a special case, the space $C^r(\mathcal{X}, \mathcal{S})$ of C^r-mappings of the body into space and also provides a natural extension to continuum mechanics of generalized media. After describing the topology in $C^r(\mathcal{X}, \mathcal{S})$ in Sect. 5, we show in Sect. 6, that the set of embeddings is open in $C^r(\mathcal{X}, \mathcal{S})$, $r \geq 1$. As such, it is a Banach manifold also and the tangent bundle is inherited from that of $C^r(\mathcal{X}, \mathcal{S})$. In Sect. 7 we outline the framework for the suggested force and stress theory as described roughly above. Sections 8, 9 and 10 introduce relevant spaces of linear functionals on manifolds, and present some of their properties. These include some standard classes of functionals such as de Rham currents and Schwartz distributions on manifolds. The representation theorem of forces by stress measures in considered in Sect. 11. Section 12 discusses the natural situation of simple forces and stress, that is, the case $r = 1$.

Concluding remarks and references to further studies are made in Sect. 13.

2 Notation and Preliminaries

2.1 General Notation

A collection of indices $i_1 \cdots i_k$, $i_r = 1, \dots, n$ will be represented as a multi-index I and we will write $|I| = k$, the length of the multi-index. In general, multi-indices will be denoted by upper-case roman letters and the associated indices will be denoted by the corresponding lowercase letters. Thus, a generic element in a k-multilinear mapping $A \in \otimes^k \mathbb{R}^n$ is given in terms of the array (A_I), $|I| = k$. In what follows, we will use the summation convention for repeated indices as well as repeated multi-indices. Whenever the syntax is violated, e.g., when a multi-index appears more than twice in a term, it is understood that summation does not apply.

A multi-index I induces a sequence $(I_1, \dots, I_n)$ in which I_r is the number of times the index r appears in the sequence $i_1 \cdots i_k$. Thus, $|I| = \sum_r I_r$. Multi-indices may be concatenated naturally such that $|IJ| = |I| + |J|$.

In case an array A is symmetric, the independent components of the array may be listed as $A_{i_1 \cdots i_k}$ with $i_l \leq i_{l+1}$. A non-decreasing multi-index, that is, a multi-index that satisfies the condition $i_l \leq i_{l+1}$, will be denoted by boldface, upper-case roman characters so that a symmetric tensor A is represented by the components $(A_{\boldsymbol{I}})$, $|\boldsymbol{I}| = k$. In particular, for a function $u : \mathbb{R}^n \to \mathbb{R}$, a particular partial derivative of order k is written in the form

$$u_{,\boldsymbol{I}} := \partial_{\boldsymbol{I}} u := \frac{\partial^{|\boldsymbol{I}|} u}{(\partial X^1)^{I_1} \cdots (\partial X^n)^{I_n}}, \tag{1}$$

where $\boldsymbol{I}$ is a non-decreasing multi-index with $|\boldsymbol{I}| = k$.

The notation $\partial_i = \partial / \partial X^i$, will be used for both the partial derivatives in $\mathbb{R}^n$ and for the elements of a chart-induced basis of the tangent space $T_X \mathcal{X}$ of a manifold $\mathcal{X}$ at a point X. The corresponding dual basis for $T^*_X \mathcal{X}$ will be denoted by $\{\mathrm{d}X^i\}$.

Greek letters, λ, μ, ν, will be used for strictly increasing multi-indices used in the representation of alternating tensors and forms, e.g.,

$$\omega = \omega_\lambda \mathrm{d}X^\lambda := \omega_{\lambda_1 \cdots \lambda_{|\lambda|}} \mathrm{d}X^{\lambda_1} \wedge \cdots \wedge \mathrm{d}X^{\lambda_{|\lambda|}}. \tag{2}$$

Given a strictly increasing multi-index λ with $|\lambda| = p$, we will denote the strictly increasing $(n-p)$-multi-index that complements λ to $1, \dots, n$ by $\hat{\lambda}$. In this context, $\hat{\mu}$, $\hat{\nu}$, etc. will indicate generic increasing $(n-p)$ multi-indices. The Levi-Civita symbol will be denoted as ε_I or ε^I, $|I| = n$ so that, for example, $\mathrm{d}X^\lambda \wedge \mathrm{d}X^{\hat{\lambda}} = \varepsilon^{\lambda\hat{\lambda}} \mathrm{d}X$, where we also set

$$\partial_X := \partial_1 \wedge \cdots \wedge \partial_n, \qquad \mathrm{d}X := \mathrm{d}X^1 \wedge \cdots \wedge \mathrm{d}X^n. \tag{3}$$

Note that we view λ and $\hat{\lambda}$ as two distinct indices so summation is not implied in a term such as $\mathrm{d}X^{\lambda} \wedge \mathrm{d}X^{\hat{\lambda}}$. In particular, in such an expression, the indices λ and $\hat{\lambda}$ are not a superscript and a subscript.

The following identifications will be implied for tensor products of vector spaces and vector bundles:

$$V^* \otimes U \cong L(V, U), \qquad (\mathrm{V}\otimes U)^* \cong V^* \otimes U^*. \tag{4}$$

For vector bundles V and U over a manifold $\mathcal{X}$, let S be a section of $V^* \otimes U$ and χ a section of V. The notation $S \cdot \chi$ is used for the section of U given by

$$(S \cdot \chi)(X) = S(X)(\chi(X)). \tag{5}$$

For two manifolds $\mathcal{X}$ and $\mathcal{Y}$, $C^r(\mathcal{X}, \mathcal{Y})$ will denote the collection of C^r-mappings from $\mathcal{X}$ to $\mathcal{Y}$. If $\xi : \mathcal{Y} \to \mathcal{X}$ is a fiber bundle, $C^r(\xi) := C^r(\mathcal{X}, \mathcal{Y})$ is the space of C^r-sections $\mathcal{X} \to \mathcal{Y}$.

2.2 *Manifolds with Corners*

Our basic object will be a fiber bundle $\xi : \mathcal{Y} \longrightarrow \mathcal{X}$ where $\mathcal{X}$ is assumed to be an oriented manifold with corners. We recall (e.g., [3, 18, 20, 22–24]) that an n-dimensional manifold with corners is a manifold whose charts assume values in the n-quadrant of $\mathbb{R}^n$, that is, in

$$\overline{\mathbb{R}}^n_+ := \{X \in \mathbb{R}^n \mid X^i \geq 0,\ i = 1, \ldots, n\}. \tag{6}$$

In the construction of the manifold structure, it is understood that a function defined on a quadrant is said to be differentiable if it is the restriction to the quadrant of a differentiable function defined on $\mathbb{R}^n$. If $\mathcal{X}$ is an n-dimensional manifold with corners, a subset $\mathcal{Z} \subset \mathcal{X}$ is defined to be a k-dimensional, $k \leq n$, submanifold with corners of $\mathcal{X}$ if for any $Z \in \mathcal{Z}$ there is a chart (U, φ), $Z \in U$, such that $\varphi(\mathcal{Z} \cap U) \subset \{X \in \overline{\mathbb{R}}^n_+ \mid X^l = 0,\ k < l \leqslant n\}$.

With an appropriate natural definition of the integral of an $(n-1)$-form over the boundary of a manifold with corners, Stokes's theorem holds for manifolds with corners. (See [18, pp. 363–370] and [23, Section 3.5].)

Relevant to the subject at hand is the following result. (See [3, 20, 22] and [23, Section 3.2].) Every n-dimensional manifold with corners $\mathcal{X}$ is a submanifold with corners of a manifold $\tilde{\mathcal{X}}$ without boundary of the same dimension. In addition, if $\mathcal{X}$ is compact, it can be embedded as a submanifold with corners in a compact manifold without boundary $\tilde{\mathcal{X}}$ of the same dimension [20, pp. I.24–26]. Furthermore, C^k-forms defined on $\mathcal{X}$, may be extended continuously and linearly to forms defined on $\tilde{\mathcal{X}}$. Such a manifold $\tilde{\mathcal{X}}$ is referred to as an *extension* of $\mathcal{X}$. Each smooth vector bundle over $\mathcal{X}$ extends to a smooth vector bundle over $\tilde{\mathcal{X}}$. Each immersion (embedding) of $\mathcal{X}$ into a smooth manifold $\mathcal{Y}$ without boundary is the restriction of an immersion (embedding) of $\tilde{\mathcal{X}}$ into $\mathcal{Y}$.

It is emphasized that manifold with corners do not model some basic geometric shapes such as a pyramid with a rectangular base or a cone. However, much of material presented in this review is valid for a class of much more general objects, *Whitney manifold germs* as presented in [23].

2.3 Bundles, Jets, and Iterated Jets

We will consider a fiber bundle $\xi : \mathcal{Y} \to \mathcal{X}$, where $\mathcal{X}$ is n-dimensional and the typical fiber is m-dimensional. The projection ξ is represented locally by $(X^i, y^\alpha) \mapsto (X^i)$, $i = 1, \dots, n$, $\alpha = 1, \dots, m$. Let

$$T\xi : T\mathcal{Y} \longrightarrow T\mathcal{X} \tag{7}$$

be the tangent mapping represented locally by

$$(X^i, y^\alpha, \dot{X}^j, \dot{y}^\beta) \longmapsto (X^i, \dot{X}^j). \tag{8}$$

The *vertical sub-bundle* $V\mathcal{Y}$ of $T\mathcal{Y}$ is the kernel of $T\xi$. An element $v \in V\mathcal{Y}$ is represented locally as $(X^i, y^\alpha, 0, \dot{y}^\beta)$. With some abuse of notation, we will write both $\tau : T\mathcal{Y} \to \mathcal{Y}$ and $\tau : V\mathcal{Y} \to \mathcal{Y}$. For $v \in V\mathcal{Y}$ with $\tau(v) = y$ and $\xi(y) = X$, we may view v as an element of $T_y(\mathcal{Y}_X) = T_y(\xi^{-1}(X))$. In other words, elements of the vertical sub-bundle are tangent vectors to $\mathcal{Y}$ that are tangent to the fibers.

Let $\kappa : \mathcal{X} \to \mathcal{Y}$ be a section and let

$$\kappa^*\tau : \kappa^* V\mathcal{Y} \longrightarrow \mathcal{X} \tag{9}$$

be the pullback of the vertical sub-bundle. Then, we may identify $\kappa^* V\mathcal{Y}$ with the restriction of the vertical bundle to Image κ.

2.3.1 Jets

We will denote by $\xi^r : J^r(\mathcal{X}, \mathcal{Y}) \to \mathcal{X}$ the corresponding r-jet bundle of ξ. (See [17, 27] for general expositions of the theory of jets.) When no ambiguity may occur, we will often use the simpler notation $\xi^r : J^r\mathcal{Y} \to \mathcal{X}$ and refer to a section of ξ^r as a section of $J^r\mathcal{Y}$. One has the additional natural projections $\xi^r_l : J^r(\mathcal{X}, \mathcal{Y}) \to J^l(\mathcal{X}, \mathcal{Y})$, $l < r$, and in particular $\xi^r_0 : J^r(\mathcal{X}, \mathcal{Y}) \to \mathcal{Y} = J^0\mathcal{Y}$. The jet extension mapping associates with a C^r-section, κ, of ξ, a continuous section $j^r\kappa$ of the jet bundle ξ^r.

Let $\kappa : \mathcal{X} \to \mathcal{Y}$ be a section of ξ which is represented locally by

$$X \longmapsto (X, y = \kappa(X)), \qquad \text{or,} \qquad X^i \longmapsto (X^i, \kappa^\alpha(X^j)), \tag{10}$$

$i = 1, \dots, n$, $\alpha = 1, \dots, m$. Then, denoting the k-th derivative of $\boldsymbol{\kappa}$ by D^k, a local representative of $j^r\kappa$ is of the form

$$X \longmapsto (X, \kappa(X), \dots, D^r\kappa(X)), \quad \text{or,} \quad X^i \longmapsto (X^i, \kappa^\alpha_{,\boldsymbol{I}}(X^j)), \quad 0 \le |\boldsymbol{I}| \le r. \tag{11}$$

Accordingly, an element $A \in J^r(\mathcal{X}, \mathcal{Y})$ is represented locally by the coordinates

$$(X^i, A^\alpha_{\boldsymbol{I}}), \quad 0 \le |\boldsymbol{I}| \le r. \tag{12}$$

2.3.2 Iterated (Non-holonomic) Jets

Completely non-holonomic jets for the fiber bundle $\xi : \mathcal{Y} \to \mathcal{X}$ are defined inductively as follows. Firstly, one defines the fiber bundles

$$\hat{J}^0(\mathcal{X}, \mathcal{Y}) = \mathcal{Y}, \qquad \hat{J}^1(\mathcal{X}, \mathcal{Y}) := J^1(\mathcal{X}, \mathcal{Y}), \tag{13}$$

and projections

$$\hat{\xi}^1 = \xi^1 : \hat{J}(\mathcal{X}, \mathcal{Y}) \longrightarrow \mathcal{X}, \qquad \hat{\xi}^1_0 = \xi^1_0 : \hat{J}^1(\mathcal{X}, \mathcal{Y}) \longrightarrow \mathcal{Y}. \tag{14}$$

Then, we define the *iterated r-jet bundle* as

$$\hat{J}^r(\mathcal{X}, \mathcal{Y}) := J^1(\mathcal{X}, \hat{J}^{r-1}(\mathcal{X}, \mathcal{Y})), \tag{15}$$

with projection

$$\hat{\xi}^r = \hat{\xi}^{r-1} \circ \xi^{1,r}_{r-1} : \hat{J}^r(\mathcal{X}, \mathcal{Y}) \longrightarrow \mathcal{X}, \tag{16}$$

where

$$\xi^{1,r}_{r-1} : \hat{J}^r(\mathcal{X}, \mathcal{Y}) = J^1(\mathcal{X}, \hat{J}^{r-1}(\mathcal{X}, \mathcal{Y})) \longrightarrow \hat{J}^{r-1}(\mathcal{X}, \mathcal{Y}). \tag{17}$$

By induction, $\hat{\xi}^r : \hat{J}^r(\mathcal{X}, \mathcal{Y}) \to \mathcal{X}$ is a well-defined fiber bundle.

When the projections $\xi^{1,r}_{r-1}$ are used inductively l-times, we obtain a projection

$$\hat{\xi}^r_{r-l} : \hat{J}^r(\mathcal{X}, \mathcal{Y}) \longrightarrow \hat{J}^{r-l}(\mathcal{X}, \mathcal{Y}). \tag{18}$$

Let $\kappa : \mathcal{X} \to \mathcal{Y}$ be a C^r-section of ξ. The iterated jet extension mapping

$$\hat{j}^r : C^r(\xi) \longrightarrow C^0(\hat{\xi}^r) \tag{19}$$

is naturally defined by

$$\hat{j}^1 = j^1 : C^l(\xi) \longrightarrow C^{l-1}(\hat{\xi}^1), \qquad \text{and} \qquad \hat{j}^r = j^1 \circ \hat{j}^{r-1}. \tag{20}$$

Note that we use j^1 here as a generic jet extension mapping, omitting the indication of the domain.

There is a natural inclusion

$$\iota^r : J^r(\mathcal{X}, \mathcal{Y}) \longrightarrow \hat{J}^r(\mathcal{X}, \mathcal{Y}), \qquad \text{given by} \qquad j^r\kappa(X) \longmapsto \hat{j}^r\kappa(X). \tag{21}$$

Let $\pi : W \to \mathcal{X}$ be a vector bundle, then $\hat{\pi}^1 = \pi^1 : \hat{J}^1 W = J^1 W \to \mathcal{X}$ is a vector bundle. Continuing inductively,

$$\hat{\pi}^r : \hat{J}^r W \longrightarrow \mathcal{X} \tag{22}$$

is a vector bundle. In this case, the inclusion $\iota^r : J^r\pi \to \hat{J}^r\pi$ is linear. Naturally, elements in the image of ι^r are referred to as *holonomic*.

2.3.3 Local Representation of Iterated Jets[1]

The local representatives of iterated jets are also constructed inductively. Hence, at each step, G, to which we refer as *generation*, the number of arrays is multiplied. Hence, powers of two are naturally used below. We will use multi-indices of the form $I_\mathfrak{p}$, where $\mathfrak{p}$, $\mathfrak{q}$, etc. are binary numbers that enumerate the various arrays included in the representation. For example, a typical element of $\hat{J}^3(\mathcal{X}, \mathcal{Y})$, in the form

$$(X^j; y^\alpha; y_{i_1}^{1\beta_1}; y_{i_2}^{2\beta_2}, y_{i_3 i_4}^{3\beta_3}; y_{i_5}^{4\beta_4}, y_{i_6 i_7}^{5\beta_5}, y_{i_8 i_9}^{6\beta_6}, y_{i_{10} i_{11} i_{12}}^{7\beta_3}) \tag{23}$$

is written as

$$(X^j; y_0^{0\beta_0}; y_{I_1}^{1\beta_1}; y_{I_{10}}^{10\beta_{10}}, y_{I_{11}}^{11\beta_{11}}; y_{I_{100}}^{100\beta_{100}}, y_{I_{101}}^{101\beta_{101}}, y_{I_{110}}^{110\beta_{110}}, y_{I_{111}}^{111\beta_{111}}), \tag{24}$$

and for short

$$(X^j; y_{I_\mathfrak{p}}^{\mathfrak{p}\beta_\mathfrak{p}}), \qquad \text{for all } \mathfrak{p} \text{ with } 0 \le G_\mathfrak{p} \le 3. \tag{25}$$

Here, $G_\mathfrak{p}$ is the generation where the $\mathfrak{p}$-th array appears and it is given by

$$G_\mathfrak{p} = \lfloor \log_2 \mathfrak{p} \rfloor + 1, \tag{26}$$

[1]This section may be skipped without interrupting the reading of most of the following.

where $\lfloor \log_2 \mathfrak{p} \rfloor$ denotes the integer part of $\log_2 \mathfrak{p}$. In (23, 24) the generations are separated by semicolons. As indicated in the example above, with each $\mathfrak{p}$ we associate a multi-index $I_{\mathfrak{p}} = i_1 \cdots i_p$ as follows. For each binary digit 1 in $\mathfrak{p}$ there is an index i_l, $l = 1, \ldots, p$. Thus, the total number of digits 1 in $\mathfrak{p}$, which is denoted by $|\mathfrak{p}|$, is the total number of indices, p, in $I_{\mathfrak{p}}$. In other words, the length, $|I_{\mathfrak{p}}|$, of the induced multi-index $I_{\mathfrak{p}}$ satisfies

$$|I_{\mathfrak{p}}| = p = |\mathfrak{p}| \,. \tag{27}$$

Note also that the expression $\beta_{\mathfrak{p}}$, is not a multi-index since we use upper-case letters to denote multi-indices. Here, the subscript $\mathfrak{p}$ serves for the enumeration of the β indices. If no ambiguity may arise, we will often make the notation somewhat shorter and write $y_{I_{\mathfrak{p}}}^{\beta_{\mathfrak{p}}}$ for $y_{I_{\mathfrak{p}}}^{\mathfrak{p}\beta_{\mathfrak{p}}}$. Continuing by induction, let a section $\hat{A}$ of $\hat{J}^{r-1}(\mathcal{X}, \mathcal{Y})$ be represented locally by $(X^j;\, y_{I_{\mathfrak{p}}}^{\mathfrak{p}\alpha_{\mathfrak{p}}}(X^j))$, $G_{\mathfrak{p}} \leq r - 1$. Then, its 1-jet extension, a section of $\hat{J}^r(\mathcal{X}, \mathcal{Y})$, is of the form

$$(X^j;\, y_{I_{\mathfrak{p}}}^{\mathfrak{p}\alpha_{\mathfrak{p}}}(X^J);\, y_{I_{\mathfrak{p}},k_{\mathfrak{p}}}^{\mathfrak{p}\beta_{\mathfrak{p}}}(X^j)), \qquad G_{\mathfrak{p}} \leq r - 1, \tag{28}$$

or equivalently,

$$(X^j;\, y_{I_{\mathfrak{p}}}^{\mathfrak{p}\alpha_{\mathfrak{p}}}(X^j);\, y_{I_{\mathfrak{p}},k_{\mathfrak{p}}}^{1\mathfrak{p}\alpha_{1\mathfrak{p}}}(X^j)), \qquad G_{\mathfrak{p}} \leq r - 1, \tag{29}$$

where $1\mathfrak{p}$ is the binary representation of $2^r + p$. It is noted that the array $y^{1\mathfrak{p}}$ contains the derivatives of the array $y^{\mathfrak{p}}$, and that $G_{1\mathfrak{p}} = r$. Thus indeed, the number of digits 1 that appear in $\mathfrak{q}$, i.e., $|\mathfrak{q}|$, determine the length of the index $I_{\mathfrak{q}}$.

It follows that an element of $\hat{J}^r(\mathcal{X}, \mathcal{Y})$ may be represented in the form

$$(X^j;\, y_{I_{\mathfrak{p}}}^{\mathfrak{p}\alpha_{\mathfrak{p}}};\, y_{I_{\mathfrak{p}}k_{\mathfrak{p}}}^{\mathfrak{p}\alpha_{1\mathfrak{p}}}), \qquad G_{\mathfrak{p}} \leq r - 1 \tag{30}$$

or

$$(X^j;\, y_{I_{\mathfrak{q}}}^{\mathfrak{q}\alpha_{\mathfrak{q}}}), \quad \text{for all } \mathfrak{q} \text{ with } G_{\mathfrak{q}} \leq r. \tag{31}$$

That is, for each $\mathfrak{p}$ with $G_{\mathfrak{p}} \leq r - 1$, we have an index $\mathfrak{q} = \mathfrak{q}(\mathfrak{p})$ such that $\mathfrak{q} = \mathfrak{q}(\mathfrak{p}) = 1\mathfrak{p}$ if $G_{\mathfrak{q}} = r$ and $\mathfrak{q} = \mathfrak{q}(\mathfrak{p}) = \mathfrak{p}$ if $G_{\mathfrak{q}} < r$.

A similar line of reasoning leads to the expression for the local representatives of the iterated jet extension mapping. For a section $\kappa : \mathcal{X} \to \mathcal{Y}$, the iterated jet extension $B = \hat{j}^r\kappa$, a section of $\hat{\xi}^r$, the local representation $(X^j;\, B_{I_{\mathfrak{q}}}^{\mathfrak{q}\alpha_{\mathfrak{q}}})$, $G_{\mathfrak{q}} \leq r$, $|I_{\mathfrak{q}}| = |\mathfrak{q}|$, satisfies

$$B_{I_{\mathfrak{q}}}^{\mathfrak{q}\alpha_{\mathfrak{q}}} = \kappa_{,I_{\mathfrak{q}}}^{\alpha_{\mathfrak{q}}}, \quad |I_{\mathfrak{q}}| = |\mathfrak{q}| \,, \text{ independently of the particular value of } \mathfrak{q}. \tag{32}$$

Indeed, if $(X^j, B^{\mathfrak{q}\alpha_\mathfrak{q}}_{I_\mathfrak{q}})$, $G_\mathfrak{q} \le r-1$, with $B^{\mathfrak{q}\alpha_\mathfrak{q}}_{I_\mathfrak{q}} = \kappa^{\alpha_\mathfrak{q}}_{,I_\mathfrak{q}}$ represent $\hat{j}^{r-1}\kappa$, then, $\hat{j}^r\kappa$ is represented locally by

$$(X^i, B^{\mathfrak{q}\alpha_\mathfrak{q}}_{I_\mathfrak{q}}; B^{\mathfrak{q}\alpha_{1\mathfrak{q}}}_{I_\mathfrak{q},j_\mathfrak{q}}), \quad G_\mathfrak{q} \le r-1. \tag{33}$$

Thus, by induction, any $\mathfrak{p} = \mathfrak{p}(\mathfrak{q})$ with $G_\mathfrak{p} = r$ and $G_\mathfrak{q} < r$, may be written as $\mathfrak{p} = 2^{r-1} + \mathfrak{q}$, $I_\mathfrak{p} = I_\mathfrak{q} j_\mathfrak{q}$, so that $B^{\mathfrak{p}\alpha_\mathfrak{p}}_{I_\mathfrak{p}} = B^{\mathfrak{q}\alpha_{1\mathfrak{q}}}_{I_\mathfrak{q},j_\mathfrak{q}} = \kappa^{\alpha_\mathfrak{p}}_{,I_\mathfrak{p}}$.

Let an element $A \in \hat{J}^r(\mathcal{X}, \mathcal{Y})$ be represented by $(X^j; y^{\mathfrak{p}\alpha_\mathfrak{p}}_{I_\mathfrak{p}})$, $G_\mathfrak{p} \le r$, then $\hat{\xi}^r_l(A)$ is represented by $(X^i; y^{\mathfrak{p}\alpha_\mathfrak{p}}_{I_\mathfrak{p}})$, $G_\mathfrak{p} \le l$.

2.4 *Contraction*

The right and left contractions of a $(p + r)$-form and a p-vector are given respectively by

$$(\theta \llcorner \eta)(\eta') = \theta(\eta' \wedge \eta), \qquad (\eta \lrcorner \theta)(\eta') = \theta(\eta \wedge \eta'), \tag{34}$$

for every r-vector η'. We will use the notation

$$\mathsf{C}_\llcorner : \bigwedge^p T\mathcal{X} \otimes \bigwedge^{p+r} T^*\mathcal{X} \cong L(\bigwedge^p T^*\mathcal{X}, \bigwedge^{p+r} T^*\mathcal{X}) \longrightarrow \bigwedge^r T^*\mathcal{X}, \tag{35}$$

and

$$\mathsf{C}_\lrcorner : \bigwedge^p T\mathcal{X} \otimes \bigwedge^{p+r} T^*\mathcal{X} \cong L(\bigwedge^p T^*\mathcal{X}, \bigwedge^{p+r} T^*\mathcal{X}) \longrightarrow \bigwedge^r T^*\mathcal{X}, \tag{36}$$

for the mappings satisfying

$$\mathsf{C}_\llcorner(\xi \otimes \theta) = \theta \llcorner \xi, \qquad \text{and} \qquad \mathsf{C}_\lrcorner(\xi \otimes \theta) = \xi \lrcorner \theta, \tag{37}$$

respectively. The left and right contraction differ by a factor of $(-1)^{rp}$.

For the case $r + p = n$, $\dim(\bigwedge^p T\mathcal{X} \otimes \bigwedge^{p+r} T^*\mathcal{X}) = \dim \bigwedge^r T^*\mathcal{X}$; as the mappings $\mathsf{C}_\llcorner$ and $\mathsf{C}_\lrcorner$ are injective, they are invertible. Specifically, consider the mappings

$$e_\llcorner : \bigwedge^{n-p} T^*\mathcal{X} \longrightarrow \bigwedge^p T\mathcal{X} \otimes \bigwedge^n T^*\mathcal{X} \cong L(\bigwedge^p T^*\mathcal{X}, \bigwedge^n T^*\mathcal{X}), \tag{38}$$

and

$$e_\lrcorner : \bigwedge^{n-p} T^*\mathcal{X} \longrightarrow \bigwedge^p T\mathcal{X} \otimes \bigwedge^n T^*\mathcal{X} \cong L(\bigwedge^p T^*\mathcal{X}, \bigwedge^n T^*\mathcal{X}), \tag{39}$$

given by

$$e_\llcorner(\omega)(\psi) = \psi \wedge \omega, \qquad \text{and} \qquad e_\lrcorner(\omega)(\psi) = \omega \wedge \psi, \tag{40}$$

respectively. One can easily verify that these mappings are isomorphisms, and in fact, they are the inverses of the contraction mappings defined above.

For example,

$$\begin{aligned} e_{\lrcorner}(\mathsf{C}_{\llcorner}(\xi\otimes\theta))(\psi) &= \mathsf{C}_{\llcorner}(\xi\otimes\theta)\wedge\psi, \\ &= (\theta\llcorner\xi)\wedge\psi, \\ &= \psi(\xi)\theta, \\ &= (\xi\otimes\theta)(\psi), \end{aligned} \tag{41}$$

where we view ξ s an element of the double dual. Thus,

$$e_{\lrcorner} = \mathsf{C}_{\llcorner}^{-1}, \qquad \text{and} \qquad e_{\llcorner} = \mathsf{C}_{\lrcorner}^{-1}. \tag{42}$$

3 Banachable Spaces of Sections of Vector Bundles over Compact Manifolds

For a compact manifold $\mathcal{X}$, the infinite dimensional Banach manifold of mappings to a manifold $\mathcal{S}$ and the manifold of sections of the fiber bundle $\xi : \mathcal{Y} \to \mathcal{X}$, are modeled by Banachable spaces of sections of vector bundles over $\mathcal{X}$, as will be described in the next section. In this section we describe the Banachable structure of such a space of differentiable vector bundle sections and make some related observations. Thus, we consider in this section a vector bundle $\pi : W \to \mathcal{X}$, where $\mathcal{X}$ is a smooth compact n-dimensional manifold with corners and the typical fiber of W is an m-dimensional vector space. The space of C^r-sections $w : \mathcal{X} \to W$, $r \geq 0$, will be denoted by $C^r(\pi)$ or by $C^r(W)$ if no ambiguity may arise. A natural real vector space structure is induced on $C^r(\pi)$ by setting $(w_1 + w_2)(X) = w_1(X) + w_2(X)$ and $(cw)(X) = cw(X)$, $c \in \mathbb{R}$.

3.1 Precompact Atlases

Let K_a, $a = 1, \ldots, A$, be a finite collection of compact subsets whose interiors cover $\mathcal{X}$ such that for each a, K_a is a subset of a domain of a chart $\varphi_a : U_a \to \mathbb{R}^n$ on $\mathcal{X}$ and

$$(\varphi_a, \Phi_a) : \pi^{-1}(U_a) \longrightarrow \mathbb{R}^n \times \mathbb{R}^m, \quad v \longmapsto (X^i, v^\alpha) \tag{43}$$

is some given vector bundle chart on W. Such a covering may always be found by the compactness of $\mathcal{X}$ (using coordinate balls as, for example, in [18, p. 16] or [26, p. 10]). We will refer to such a structure as a *precompact atlas*. The same terminology will apply for the case of a fiber bundle.

3.2 The C^r-Topology on $C^r(\pi)$

For a section w of π and each $a = 1, \dots, A$, let

$$w_a : \varphi_a(K_a) \longrightarrow \mathbb{R}^m, \tag{44}$$

satisfying

$$w_a(\varphi_a(X)) = \Phi_a(w(X)), \quad \text{for all} \quad X \in K_a, \tag{45}$$

be a local representative of w.

Such a choice of a precompact vector bundle atlas and subsets K_a makes it possible to define, for a section w,

$$\|w\|^r = \sup_{a,\alpha,|\boldsymbol{I}|\leq r} \left\{ \sup_{X\in\varphi_a(K_a)} \left\{ \left|(w_a^\alpha)_{,\boldsymbol{I}}(X)\right| \right\} \right\}. \tag{46}$$

Palais [26, in particular, Chapter 4] shows that $\|\cdot\|^r$ is indeed a norm endowing $C^r(\pi)$ with a Banach space structure. The dependence of this norm on the particular choice of atlas and sets K_a, makes the resulting space Banachable, rather than a Banach space. Other choices will correspond to different norms. However, norms induced by different choices will induce equivalent topological vector space structures on $C^r(\pi)$ [23, Section 5].

3.3 The Jet Extension Mapping

Next, one observes that the foregoing may be applied, in particular, to the vector space $C^0(\pi^r) = C^0(J^r W)$ of continuous sections of the r-jet bundle $\pi^r : J^r W \to \mathcal{X}$ of π. As a continuous section B of π^r is locally of the form

$$(X^i) \longmapsto (X^i, B_{\boldsymbol{I}}^\alpha(X^i)), \qquad |\boldsymbol{I}| \leq r, \tag{47}$$

the analogous expression for the norm induced by a choice of a precompact vector bundle atlas is

$$\|B\|^0 = \sup_{a,\alpha,|\boldsymbol{I}|\leq r} \left\{ \sup_{X\in\varphi_a(K_a)} \left\{ \left|B_{a\boldsymbol{I}}^\alpha(X)\right| \right\} \right\}. \tag{48}$$

Once, the topologies of $C^r(\pi)$ and $C^0(\pi^r)$ have been defined, one may consider the jet extension mapping

$$j^r : C^r(\pi) \longrightarrow C^0(\pi^r). \tag{49}$$

For a section $w \in C^r(\pi)$, with local representatives w_a^α, $j^r w$ is represented by a section $B \in C^0(\pi^k)$, the local representatives of which satisfy,

$$B_{a\boldsymbol{I}}^{\alpha} = w_{a,\boldsymbol{I}}^{\alpha}. \tag{50}$$

Clearly, the mapping j^r is injective and linear. Furthermore, it follows that

$$\|j^r w\|^0 = \sup_{a,\alpha,|\boldsymbol{I}|\leq r} \left\{ \sup_{\boldsymbol{X}\in\varphi_a(K_a)} \left\{ \left| w_{a,\boldsymbol{I}}^{\alpha}(\boldsymbol{X}) \right| \right\} \right\}. \tag{51}$$

(Note that since we take the supremum of all partial derivatives, we could replace the non-decreasing multi-index $\boldsymbol{I}$ by a regular multi-index I.) Thus, in view of Eq. (46),

$$\|j^r w\|^0 = \|w\|^r \tag{52}$$

and we conclude that j^r is a linear embedding of $C^r(\pi)$ into $C^0(\pi^r)$. Evidently, j^r is not surjective as a section A of π^k need not be compatible, i.e., it need not satisfy (50), for some section w of π. As a result of the above observations, j^r has a continuous right inverse

$$(j^r)^{-1} : \text{Image}\, j^r \subset C^0(\pi^r) \longrightarrow C^r(\pi). \tag{53}$$

3.4 *The Iterated Jet Extension Mapping*

In analogy, we now consider the iterated jet extension mapping

$$\hat{j}^r : C^r(\pi) \longrightarrow C^0(\hat{\pi}^r). \tag{54}$$

Specializing Eq. (31) for the case of the non-holonomic r-jet bundle

$$\hat{\pi}^r : \hat{J}^r W \longrightarrow \mathcal{X}, \tag{55}$$

a section B of $\hat{\pi}^r$ is represented locally in the form

$$X^i \longmapsto (X^i, B_{aI_{\mathfrak{q}}}^{\mathfrak{q}\alpha_{\mathfrak{q}}}(X^i)), \quad \text{for all } \mathfrak{q} \text{ with } G_{\mathfrak{q}} \leq r. \tag{56}$$

Thus, the induced norm on $C^0(\hat{\pi}^r)$ is given by

$$\|B\|^0 = \sup \left\{ \left| B_{aI_{\mathfrak{q}}}^{\mathfrak{q}\alpha_{\mathfrak{q}}}(X^i) \right| \right\}, \tag{57}$$

where the supremum is taken over all $X \in \varphi_a(K_a)$, $a = 1, \ldots, A$, $\alpha_{\mathfrak{q}} = 1, \ldots, m$, $I_{\mathfrak{q}}$ with $|I| = |\mathfrak{q}|$, and $\mathfrak{q}$ with $G_{\mathfrak{q}} \leq r$.

Specializing (32) for the case of a vector bundle, it follows that if the section B of $\hat{\pi}^r$, satisfies $B = \hat{j}^r w$, its local representatives satisfy

$$B_{aI_{\mathfrak{q}}}^{\mathfrak{q}\alpha_{\mathfrak{q}}} = w_{a,I_{\mathfrak{q}}}^{\alpha_{\mathfrak{q}}} \quad |I_{\mathfrak{q}}| = |\mathfrak{q}|\,, \text{ independently of the particular value of } \mathfrak{q}. \tag{58}$$

It follows that in

$$\|\hat{j}^r w\|^0 = \sup\left\{\left|w_{a,I_{\mathfrak{q}}}^{\alpha_{\mathfrak{q}}}(X^i)\right|\right\}, \tag{59}$$

(where the supremum is taken over all $X^i \in \varphi_a(K_a)$, $a = 1, \ldots, A$, $\alpha = 1, \ldots, m$, $I_{\mathfrak{q}}$ with $|I| = |\mathfrak{q}|$, and $\mathfrak{q}$ with $G_{\mathfrak{q}} \leq r$), it is sufficient to take simply all derivatives $w_{a,I}^{\alpha}(X^i)$, for $|I| \leq r$. Hence,

$$\|\hat{j}^r w\|^0 = \sup\left\{\left|w_{a,I}^{\alpha_{\mathfrak{q}}}(X^i)\right|\right\}, \tag{60}$$

where the supremum is taken over all $X^i \in \varphi_a(K_a)$, $a = 1, \ldots, A$, $\alpha = 1, \ldots, m$, and I with $|I| \leq r$. It is therefore concluded that

$$\|\hat{j}^r w\|^0 = \|j^r w\|^0 = \|w\|^r. \tag{61}$$

In other words, one has a sequence of linear embeddings

$$C^r(\pi) \xrightarrow{j^r} C^0(\pi^r) \xrightarrow{C^0(\iota^r)} C^0(\hat{\pi}^r), \qquad \hat{j}^r = C^0(\iota^r)\circ j^r \tag{62}$$

where $\iota^r : J^r W \to \hat{J}^r W$ is the natural inclusion (21) and $C^0(\iota^r)$ defined as $C^0(\iota^r)(A) := \iota^r \circ A$, for every continuous section A of $J^r W$, is the inclusion of sections. These embeddings are not surjective. In particular, sections of $\hat{J}^r W$ need not have the symmetry properties that hold for sections of $J^r W$.

4 The Construction of Charts for the Manifold of Sections

In this section, we outline the construction of charts for the Banach manifold structure on the collection of sections $C^r(\xi)$ as in [26]. (See a detailed presentation of the subject in this volume [23, Section 5.9].)

Let κ be a C^r-section of ξ. Similarly to the construction of tubular neighborhoods, the basic idea is to identify points in a neighborhood of Image κ with vectors at Image κ which are tangent to the fibers. This is achieved by defining a second order differential equation, so that a neighboring point y in the same fiber as $\kappa(X)$

is represented through the solution $c(t)$ of the differential equations with the initial condition $v \in T_{\kappa(X)}(\mathcal{Y}_X)$ by $y = c(t = 1)$. In other words, y is the image of v under the exponential mapping.

To ensure that the image of the exponential mapping is located on the same fiber, $\mathcal{Y}_X$, the spray inducing the second order differential equation is a vector field

$$\omega : V\mathcal{Y} \longrightarrow T(V\mathcal{Y}) \tag{63}$$

which is again tangent to the fiber in the sense that for

$$T\tau_{\mathcal{Y}} : T(V\mathcal{Y}) \longrightarrow T\mathcal{Y}, \quad \text{one has,} \quad T\tau_{\mathcal{Y}} \circ \omega \in V\mathcal{Y}. \tag{64}$$

This condition, together with the analog of the standard condition for a second order differential equation, namely,

$$T\tau_{\mathcal{Y}} \circ \omega(v) = v, \quad \text{for all} \quad v \in V\mathcal{Y}, \tag{65}$$

imply that ω is represented locally in the form

$$(X^i, y^\alpha, 0, \dot{y}^\beta) \longmapsto (X^i, y^\alpha, 0, \dot{y}^\beta, 0, \dot{y}^\gamma, 0, \tilde{\omega}^\alpha(X^i, y^\alpha, \dot{y}^\beta)). \tag{66}$$

Finally, ω is a *bundle spray* so that

$$\tilde{\omega}^\alpha(X^i, y^\alpha, a_0\dot{y}^\beta) = a_0^2\tilde{\omega}^\alpha(X^i, y^\alpha, \dot{y}^\beta). \tag{67}$$

Bundle sprays can always be defined on compact manifolds using partitions of unity and the induced exponential mappings have the required properties.

The resulting structure makes it possible to identify an open neighborhood U—*a vector bundle neighborhood*—of Image κ in $\mathcal{Y}$ with

$$V\mathcal{Y}|_{\text{Image}\,\kappa} \cong \kappa^* V\mathcal{Y}. \tag{68}$$

(We note that a rescaling is needed if U is to be identified with the whole of $\kappa^* V\mathcal{Y}$. Otherwise, only an open neighborhood of the zero section of $\kappa^* V\mathcal{Y}$ will be used to parametrize U.)

Once the identification of U with $\kappa^* V\mathcal{Y}$ is available, the collection of sections $C^r(\mathcal{X}, U)$ may be identified with $C^r(\kappa^* V\mathcal{Y})$, $\kappa \in C^r(\xi)$. Thus, a chart into a Banachable space is constructed where κ is identified with the zero section (Fig. 1).

The construction of charts on the manifold of sections, implies that curves in $C^r(\xi)$ in a neighborhood of κ are represented locally by curves in the Banachable space $C^r(\kappa^* V\mathcal{Y})$. Thus, tangent vectors $w \in T_\kappa C^r(\xi)$ may be identified with elements of $C^r(\kappa^* V\mathcal{Y})$. We therefore make the identification

$$T_\kappa C^r(\xi) = C^r(\kappa^* V\mathcal{Y}). \tag{69}$$

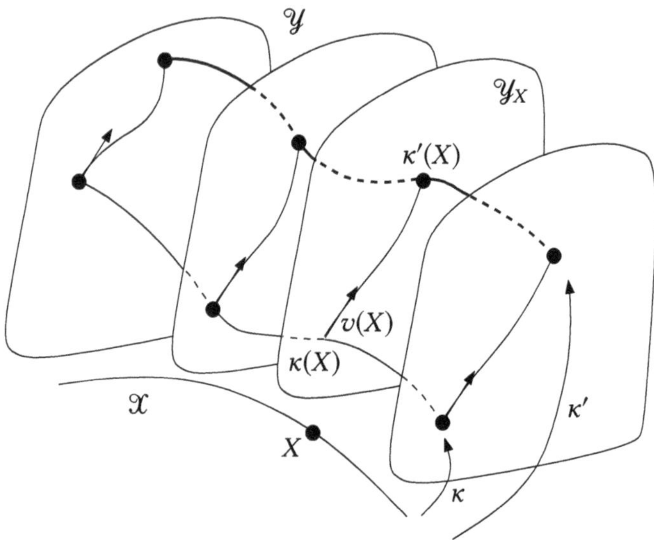

Fig. 1 Constructing the manifold of sections, a rough illustration

5 The C^r-Topology on the Space of Sections of a Fiber Bundle

The topology on the space of sections of fiber bundles is conveniently described in terms of filters of neighborhoods (e.g., [44]).

5.1 *Local Representatives of Sections*

We consider a fiber bundle $\xi : \mathcal{Y} \to \mathcal{X}$, where $\mathcal{X}$ is assumed to be a compact manifold with corners and the typical fiber is a manifold $\mathcal{S}$ without a boundary. Let $\{(U_a, \varphi_a, \Phi_a)\}$, $a = 1, \ldots, A$, and $K_a \subset U_a$, be a precompact (as in Sect. 3.1) fiber bundle trivialization on $\mathcal{Y}$. That is, the interiors of K_a cover $\mathcal{X}$, and $(\varphi_a, \Phi_a) : \xi^{-1}(U_a) \to \mathbb{R}^n \times \mathcal{S}$. Let $\{(V_b, \psi_b)\}$, $b = 1, \ldots, B$, be an atlas on $\mathcal{S}$ so that $\{V_b\}$ cover $\mathcal{S}$.

Consider a C^k-section $\kappa : \mathcal{X} \to \mathcal{Y}$. For any $a = 1, \ldots, A$, we can set

$$\tilde{\kappa}_a : U_a \longrightarrow \mathcal{S}, \quad \text{by } \tilde{\kappa}_a := \Phi_a \circ \kappa|_{U_a}. \tag{70}$$

Let

$$U_{ab} := U_a \cap \tilde{\kappa}_a^{-1}(V_b), \tag{71}$$

so that $\kappa(U_{ab}) \subset V_b$, and so, the local representatives of κ are

$$\kappa_{ab} := \psi_b \circ \tilde{\kappa}_a \circ \varphi_a^{-1}|_{\varphi_a(U_{ab})} : \varphi_a(U_{ab}) \longrightarrow \psi_b(V_b) \subset \mathbb{R}^m. \tag{72}$$

Thus, re-enumerating the subsets $\{U_{ab}\}$ and $\{V_b\}$ we may assume that we have a precompact trivialization $\{(U_a, \varphi_a, \Phi_a)\}$, $a = 1, \ldots, A$, $K_a \subset U_a$, on $\mathcal{Y}$, and an atlas $\{(V_a, \psi_a)\}$ on $\mathcal{S}$ such that $\tilde{\kappa}_a(U_a) \subset V_a$. The local representatives of κ relative to these atlases are

$$\kappa_a := \psi_a \circ \tilde{\kappa}_a \circ \varphi_a^{-1}|_{\varphi_a(U_a)} : \varphi_a(U_a) \longrightarrow \psi_a(V_a) \subset \mathbb{R}^m. \tag{73}$$

5.2 *Neighborhoods for $C^r(\xi)$ and the C^r-Topology*

Let $\kappa \in C^r(\xi)$ be given. Consider sets of sections of the form $U_{\kappa,\varepsilon}$ induced by the collection of local representatives as above and numbers $\varepsilon > 0$ in the form

$$U_{\kappa,\varepsilon} = \left\{\kappa' \in C^r(\xi) \mid \sup\left\{\left|((\kappa_a')^\alpha - \kappa_a^\alpha)_{,I}(\boldsymbol{X})\right|\right\} < \varepsilon\right\}, \tag{74}$$

where the supremum is taken over all

$$\boldsymbol{X} \in \varphi_a(K_a),\ \alpha = 1, \ldots, m,\ |I| \leq r,\ a = 1, \ldots, A.$$

The C^r-topology on $C^r(\xi)$ uses all such sets as a basis of neighborhoods. Using the transformation rules for the various variables, it may be shown that other choices of precompact trivialization and atlas will lead to equivalent topologies. It is noted that we use here the compactness of $\mathcal{X}$ which implies that the weak and strong C^r-topologies (see [13, p. 35]) become identical.

Remark 1 The collection of neighborhoods $\{U_{\kappa,\varepsilon}\}$ for the various values of ε generate a basis of neighborhoods for the topology of $C^r(\xi)$. If one keeps the value of $a = 1, \ldots, A$, fixed, then the collection of sections

$$U_{\kappa,\varepsilon_a,a} = \left\{\kappa' \in C^r(\xi) \mid \sup\left\{\left|((\kappa_a')^\alpha - \kappa_a^\alpha)_{,I}(\boldsymbol{X})\right|\right\} < \varepsilon_a\right\},$$

where the supremum is taken over all

$$\boldsymbol{X} \in \varphi_a(K_a),\ \alpha = 1, \ldots, m,\ |I| \leq r,$$

is a neighborhood as it contains the open neighborhood U_{κ,ε_a}. In fact, since $\mathcal{X}$ is assumed to be compact, the collection of sets of the form $\{U_{\kappa,\varepsilon_a,a}\}$ is a sub-basis of neighborhoods of κ for the topology on $C^r(\xi)$.

5.3 Open Neighborhoods for $C^r(\xi)$ Using Vector Bundle Neighborhoods

In order to specialize the preceding constructions for the case where a vector bundle neighborhood is used, we first consider local representations of sections.

Let $\kappa \in C^r(\xi)$ be a section and let

$$\kappa^*\tau_{\mathcal{Y}} : \kappa^* V\mathcal{Y} \longrightarrow \mathcal{X} \tag{75}$$

be the vector bundle identified with an open sub-bundle U of $\mathcal{Y}$. (We will use the two aspects of the vector bundle neighborhood, interchangeably.) Since the typical fiber of $\kappa^* V\mathcal{Y}$ is $\mathbb{R}^m$, one may choose a precompact vector bundle atlas $\{(U_a, \varphi_a, \Phi_a)\}$, $K_a \subset U_a$, $a = 1, \dots, A$, on $\kappa^* V\mathcal{Y}$, such that

$$(\varphi_a, \Phi_a) : (\kappa^*\tau_{\mathcal{Y}})^{-1}(U_a) \longrightarrow \mathbb{R}^n \times \mathbb{R}^m. \tag{76}$$

Thus, if we identify all open subsets V_a in Sect. 5.1 above with the typical fiber $\mathbb{R}^m$, the representatives of a section are of the form

$$\kappa_a := \Phi_a \circ \kappa|_{U_a} \circ \varphi_a^{-1}|_{\varphi_a(U_a)} : \varphi_a(U_a) \longrightarrow \mathbb{R}^m. \tag{77}$$

A basic neighborhood of κ is given by Eq. (74). However, from the point of view of a vector bundle neighborhood, κ is represented by the zero section and each κ' is viewed as a section of the vector bundle $\kappa^*\tau_{\mathcal{Y}}$, which we may denote by w'. Thus

$$U_{\kappa,\varepsilon} \cong \left\{ w' \in C^r(\kappa^*\tau_{\mathcal{Y}}) \mid \sup \left\{ \left| (w'_a)^\alpha_{,I}(X^i) \right| \right\} < \varepsilon \right\}. \tag{78}$$

In other words, using the structure of a vector bundle neighborhood we have

$$U_{\kappa,\varepsilon} \cong \left\{ w' \in C^r(\kappa^*\tau_{\mathcal{Y}}) \mid \|w'\|^r < \varepsilon \right\}, \tag{79}$$

so that $U_{\kappa,\varepsilon}$ is identified with a ball of radius ε in $C^r(\kappa^*\tau_{\mathcal{Y}})$ at the zero section.

It is concluded that the charts on $C^r(\xi)$ induced by the vector bundle neighborhood are compatible with the C^r-topology on $C^r(\xi)$.

6 The Space of Embeddings

The kinematic aspect of the Lagrangian formulation of continuum mechanics is founded on the notion of a configuration, an embedding of a body manifold $\mathcal{X}$ into the space manifold $\mathcal{S}$. The restriction of configurations to be embeddings, rather than generic C^r-mappings of the body into space follows from the traditional principle of material impenetrability which requires that configurations be injective and that infinitesimal volume elements are not mapped into elements of zero volume.

It is noted that any configuration $\kappa : \mathcal{X} \to \mathcal{S}$ may be viewed as a section of the trivial fiber bundle $\xi : \mathcal{X} \to \mathcal{Y} = \mathcal{X} \times \mathcal{S}$. Thus, the constructions described above apply immediately to configurations in continuum mechanics. In this particular case, we will write $C^r(\mathcal{X}, \mathcal{S})$ for the collection of all C^r-mappings. Our objective in this section is to describe how the set of embeddings $\mathrm{Emb}^r(\mathcal{X}, \mathcal{S})$ constitutes an open subset of $C^r(\mathcal{X}, \mathcal{S})$ for $r \geq 1$. In particular, it will follow that at each configuration κ, $T_\kappa \mathrm{Emb}^r(\mathcal{X}, \mathcal{S}) = T_\kappa C^r(\mathcal{X}, \mathcal{S})$. Since the C^r-topologies, for $r > 1$, are finer than the C^1-topology, it is sufficient to prove that $\mathrm{Emb}^1(\mathcal{X}, \mathcal{S})$ is open in $C^1(\mathcal{X}, \mathcal{S})$. This brings to light the special role that the case $r = 1$ plays in continuum mechanics.

6.1 The Case of a Trivial Fiber Bundle: Manifolds of Mappings

It is observed that the definitions of Sects. 5.1 and 5.2 hold with natural simplifications for the case of the trivial bundle. Thus, we use a precompact atlas $\{(U_a, \varphi_a)\}$, $a = 1, \dots, A$, and $K_a \subset U_a$, in $\mathcal{X}$ (the interiors, K_a°, cover $\mathcal{X}$). Given $\kappa \in C^1(\mathcal{X}, \mathcal{S})$, we can find an atlas $\{(V_a, \psi_a)\}$ on $\mathcal{S}$ such that $\kappa(U_a) \subset V_a$. The local representatives of κ are of the form

$$\kappa_a = \psi_a \circ \kappa|_{U_a} \circ \varphi_a^{-1} : \varphi_a(U_a) \longrightarrow \psi_a(V_a) \subset \mathbb{R}^m. \tag{80}$$

For the case $r = 1$, Eq. (74) reduces to

$$U_{\kappa,\varepsilon} = \left\{ \kappa' \in C^1(\mathcal{X}, \mathcal{S}) \;\middle|\; \sup \left\{ \left| (\kappa'_a)^\alpha(\boldsymbol{X}) - \kappa_a^\alpha(\boldsymbol{X}) \right|, \left| (\kappa'_a)^\alpha_{,j}(\boldsymbol{X}) - \kappa^\alpha_{a,j}(\boldsymbol{X}) \right| \right\} < \varepsilon \right\}, \tag{81}$$

where the supremum is taken over all

$$\boldsymbol{X} \in \varphi_a(K_a),\ \alpha = 1, \dots, m,\ a = 1, \dots, A.$$

Remark 2 It is noted that in analogy with Remark 1, for a fixed $a = 1, \dots, A$, a subset of mappings of the form

$$U_{\kappa,\varepsilon,a} = \left\{ \kappa' \in C^1(\mathcal{X}, \mathcal{S}) \;\middle|\; \sup \left\{ \left| (\kappa'_a)^\alpha(\boldsymbol{X}) - \kappa_a^\alpha(\boldsymbol{X}) \right|, \left| (\kappa'_a)^\alpha_{,j}(\boldsymbol{X}) - \kappa^\alpha_{a,j}(\boldsymbol{X}) \right| \right\} < \varepsilon \right\}, \tag{82}$$

where the supremum is taken over all

$$\boldsymbol{X} \in \varphi_a(K_a),\ j = 1, \dots, n,\ \alpha = 1, \dots, m,$$

is a neighborhood of κ as it contains a neighborhood as defined above. The collection of such sets for various values of a and ε form a sub-basis of neighborhoods for the topology on $C^1(\mathcal{X}, \mathcal{S})$.

6.2 The Space of Immersions

Let $\kappa \in C^1(\mathcal{X}, \mathcal{S})$ be an immersion, so that $T_X\kappa : T_X\mathcal{X} \to T_{\kappa(X)}\mathcal{S}$ is injective for every $X \in \mathcal{X}$. We show that there is a neighborhood $U_\kappa \subset C^1(\mathcal{X}, \mathcal{S})$ of κ such that all $\kappa' \in U_\kappa$ are immersions.

Note first that since the evaluation of determinants of $n \times n$ matrices is a continuous mapping, the collection of $m \times n$ matrices for which all $n \times n$ minors vanish is a closed set. Hence, the collection $L_{\text{In}}(\mathbb{R}^n, \mathbb{R}^m)$ of all injective $m \times n$ matrices is open in $L(\mathbb{R}^n, \mathbb{R}^m)$. Let κ be an immersion with representatives κ_a as above. For each a, the derivative

$$D\kappa_a : \varphi_a(U_a) \longrightarrow L(\mathbb{R}^n, \mathbb{R}^m), \quad \boldsymbol{X} \longmapsto D\kappa_a(\boldsymbol{X}) \tag{83}$$

is continuous, hence, $D\kappa_a(K_a)$ is a compact set of injective linear mappings. Choosing any norm in $L(\mathbb{R}^n, \mathbb{R}^m)$, one can cover $D\kappa_a(K_a)$ by a finite number of open balls all containing only injective mappings. In particular, setting

$$\|T\| = \max_{i,\alpha} \left\{ \left| T_i^\alpha \right| \right\}, \qquad T \in L(\mathbb{R}^n, \mathbb{R}^m), \tag{84}$$

let ε_a be the least radius of balls in this covering. Thus, we are guaranteed that any linear mapping T, such that $\|T - D\kappa_a(\boldsymbol{X})\| < \varepsilon_a$ for some $\boldsymbol{X} \in \varphi_a(K_a)$, is injective. Specifically, for any $\kappa' \in C^1(\mathcal{X}, \mathcal{S})$, if

$$\sup_{\boldsymbol{X} \in \varphi_a(K_a)} \left| (\kappa'_a)^\alpha_{,j}(\boldsymbol{X}) - \kappa^\alpha_{a,j}(\boldsymbol{X}) \right| \leqslant \varepsilon_a, \tag{85}$$

$D\kappa'_a$ is injective everywhere in $\varphi_a(K_a)$. Letting $\varepsilon = \min_a \varepsilon_a$, any configuration in $U_{\kappa,\varepsilon}$ as in (81) is an immersion.

6.3 Open Neighborhoods of Local Embeddings

Let $\kappa \in C^1(\mathcal{X}, \mathcal{S})$ and $X \in \mathcal{X}$. It is shown below that if $T_X\kappa$ is injective, then there is a neighborhood of mappings $U_{\kappa,X}$ of κ such that every $\kappa' \in U_{\kappa,X}$ is injective in some fixed neighborhood of X. Specifically, there is a neighborhood W_X of X, and a neighborhood $U_{\kappa,X}$ of κ such that for each $\kappa' \in U_{\kappa,X}$, $\kappa'|_{W_X}$ is injective.

Let (U, φ) and (V, ψ) be coordinate neighborhoods of X and $\kappa(X)$, respectively, such that $\kappa(U) \subset V$. Let $\boldsymbol{X}$ and $\boldsymbol{\kappa}$ be the local representative of X and κ relative to these charts. Thus, we are guaranteed that

$$M := \inf_{|\boldsymbol{v}|=1} |D\boldsymbol{\kappa}(\boldsymbol{X})(\boldsymbol{v})| > 0. \tag{86}$$

By a standard corollary of the inverse function theorem, due to the injectivity of $T_X\kappa$, we can choose U to be small enough so that the restrictions of κ and $\boldsymbol{\kappa}$ to U and its image under φ, respectively, are injective. Next, let W_X be a neighborhood of X such that $\varphi(W_X)$ is convex and its closure, $\overline{W}_X$, is a compact subset of U. Thus, define the neighborhood $U_{\kappa,X} \subset C^1(\mathcal{X}, \mathcal{S})$ whose elements, κ', satisfy the conditions

$$\kappa'(\overline{W}_X) \subset V, \quad \text{and} \quad \left|D\boldsymbol{\kappa}'(\boldsymbol{X}') - D\boldsymbol{\kappa}(\boldsymbol{X})\right| < \frac{M}{2}, \quad \text{for all} \quad \boldsymbol{X}' \in \varphi(\overline{W}_X). \tag{87}$$

By the definition of neighborhoods in $C^1(\mathcal{X}, \mathcal{S})$ in (81), $U_{\kappa,X}$ contains a neighborhood of κ, hence, it is also a neighborhood.

Next, it is shown that the fact that the values of the derivatives of elements of $U_{\kappa,X}$ are close to the injective $D\boldsymbol{\kappa}(\boldsymbol{X})$ everywhere in $\overline{W}_X$, implies that these mappings are close to the linear approximation using $D\boldsymbol{\kappa}(\boldsymbol{X})$, which in turn, implies injectivity in $\overline{W}_X$ of these elements. Specifically, for and $\boldsymbol{X}_1, \boldsymbol{X}_2 \in \overline{W}_X$, since

$$\boldsymbol{\kappa}'(\boldsymbol{X}_2) - \boldsymbol{\kappa}'(\boldsymbol{X}_1) = \boldsymbol{\kappa}'(\boldsymbol{X}_2) - \boldsymbol{\kappa}'(\boldsymbol{X}_1) - D\boldsymbol{\kappa}(\boldsymbol{X})(\boldsymbol{X}_2 - \boldsymbol{X}_1) + D\boldsymbol{\kappa}(\boldsymbol{X})(\boldsymbol{X}_2 - \boldsymbol{X}_1), \tag{88}$$

the triangle inequality implies that

$$\begin{aligned} \left|\boldsymbol{\kappa}'(\boldsymbol{X}_2) - \boldsymbol{\kappa}'(\boldsymbol{X}_1)\right| &\geq |D\boldsymbol{\kappa}(\boldsymbol{X})(\boldsymbol{X}_2 - \boldsymbol{X}_1)| - \left|\boldsymbol{\kappa}'(\boldsymbol{X}_2) - \boldsymbol{\kappa}'(\boldsymbol{X}_1) - D\boldsymbol{\kappa}(\boldsymbol{X})(\boldsymbol{X}_2 - \boldsymbol{X}_1)\right|, \\ &\geq M\,|\boldsymbol{X}_2 - \boldsymbol{X}_1| - \left|\boldsymbol{\kappa}'(\boldsymbol{X}_2) - \boldsymbol{\kappa}'(\boldsymbol{X}_1) - D\boldsymbol{\kappa}(\boldsymbol{X})(\boldsymbol{X}_2 - \boldsymbol{X}_1)\right|. \end{aligned} \tag{89}$$

Using the mean value theorem, there is a point $\boldsymbol{X}_0 \in \varphi(\overline{W}_X)$ such that

$$\boldsymbol{\kappa}'(\boldsymbol{X}_2) - \boldsymbol{\kappa}'(\boldsymbol{X}_1) = D\boldsymbol{\kappa}'(\boldsymbol{X}_0)(\boldsymbol{X}_2 - \boldsymbol{X}_1). \tag{90}$$

Hence,

$$\begin{aligned} \left|\boldsymbol{\kappa}'(\boldsymbol{X}_2) - \boldsymbol{\kappa}'(\boldsymbol{X}_1) - D\boldsymbol{\kappa}(\boldsymbol{X})(\boldsymbol{X}_2 - \boldsymbol{X}_1)\right| &= \left|(D\boldsymbol{\kappa}'(\boldsymbol{X}_0) - D\boldsymbol{\kappa}(\boldsymbol{X}))(\boldsymbol{X}_2 - \boldsymbol{X}_1)\right|, \\ &\leq \left|D\boldsymbol{\kappa}'(\boldsymbol{X}_0) - D\boldsymbol{\kappa}(\boldsymbol{X})\right| |\boldsymbol{X}_2 - \boldsymbol{X}_1|, \\ &< \frac{M}{2}\,|\boldsymbol{X}_2 - \boldsymbol{X}_1|. \end{aligned} \tag{91}$$

It follows that

$$\left|\boldsymbol{\kappa}'(\boldsymbol{X}_2) - \boldsymbol{\kappa}'(\boldsymbol{X}_1)\right| > \frac{M}{2}\,|\boldsymbol{X}_2 - \boldsymbol{X}_1|, \tag{92}$$

which proves the injectivity.

6.4 Open Neighborhoods of Embeddings

Finally, it is shown how every $\kappa \in \mathrm{Emb}^1(\mathcal{X}, \mathcal{S}) \subset C^1(\mathcal{X}, \mathcal{S})$ has a neighborhood consisting of embeddings only. It will follow that $\mathrm{Emb}^1(\mathcal{X}, \mathcal{S})$ is an open subset of $C^1(\mathcal{X}, \mathcal{S})$. This has far-reaching consequences in continuum mechanics and it explains the special role played by the C^1-topology in continuum mechanics.

Let κ be a given embedding. Using the foregoing result, for each $X \in \mathcal{X}$ there is an open neighborhood W_X of X and a neighborhood $U_{\kappa,X}$ of κ, such that for each $\kappa' \in U_{\kappa,X}$, $\kappa'|_{\overline{W}_X}$ is injective. The collection of neighborhoods $\{W_X\}$, $X \in \mathcal{X}$, is an open cover of $\mathcal{X}$ and by compactness, it has a finite sub-cover. Denote the finite number of open sets of the form W_x as above by W_a, $a = 1, \dots, A$, so that $K_a := \overline{W}_a$ is a compact subset of U_a, $\kappa(U_a) \subset V_a$. For each a, we have a neighborhood $U_{\kappa,a}$ of κ such that each $\kappa' \in U_{\kappa,a}$ satisfies the condition that $\kappa'|_{K_a}$ is injective. Let $\mathcal{N}_1 = \bigcap_{a=1}^{A} U_{\kappa,a}$ so that for each $\kappa' \in \mathcal{N}_1$, $\kappa'|_{K_a}$ is injective for all a. Let $\mathcal{N}_2$ be a neighborhood of κ which contains only immersions as in Sect. 6.2. Thus, $\mathcal{N}_0 = \mathcal{N}_1 \cap \mathcal{N}_2$ contains immersions which are locally injective.

Let κ be an embedding and $\mathcal{N}_0$ as above. If there is no neighborhood of κ that contains only injective mappings, then, for each $\nu = 1, 2, \dots$, there is a $\kappa_\nu \in U_{\kappa,\varepsilon_\nu}$, $\varepsilon_\nu = 1/\nu$, and points $X_\nu, X'_\nu \in \mathcal{X}$, $X_\nu \neq X'_\nu$, such that $\kappa_\nu(X_\nu) = \kappa_\nu(X'_\nu)$. As $\mathcal{N}_0$ is a neighborhood of κ, we may assume that $\kappa_\nu \in \mathcal{N}_0$ for all ν. By the compactness of $\mathcal{X}$ and $\mathcal{X} \times \mathcal{X}$, we can extract a converging subsequence from the sequence $((X_\nu, X'_\nu))$ in $\mathcal{X} \times \mathcal{X}$. We keep the same notation for the converging subsequences and let

$$(X_\nu, X'_\nu) \longrightarrow (X, X'), \qquad \text{as} \quad \nu \longrightarrow \infty. \tag{93}$$

We first exclude the possibility that $X = X'$. Assume $X = X' \in K_{a_0}$, for some $a_0 = 1, \dots, A$. Then, for any neighborhood $U_{\kappa,\varepsilon_\nu}$ of κ and any neighborhood of $X = X'$, there is a configuration κ_ν such that κ_ν is not injective. This contradicts the construction of local injectivity above.

Thus, one should consider the situation for which $X \neq X'$. Assume $X \in K_{a_0}$ and $X' \in K_{a_1}$ for $a_0, a_1 = 1, \dots, A$. By the definition of $U_{\kappa,\varepsilon_\nu}$, the local representatives of $\kappa_\nu|K_{a_0}$ and $\kappa_\nu|K_{a_1}$ converge uniformly to the local representatives of $\kappa|K_{a_0}$ and $\kappa|K_{a_1}$, respectively. This implies that

$$\kappa_\nu(X_\nu) \longrightarrow \kappa(X), \quad \kappa_\nu(X'_\nu) \longrightarrow \kappa(X'), \qquad \text{as} \quad \nu \longrightarrow \infty. \tag{94}$$

However, since for each ν, $\kappa_\nu(X_\nu) = \kappa_\nu(X'_\nu)$, it follows that $\kappa(X) = \kappa(X')$, which contradicts the assumption that κ is an embedding.

It is finally noted that the set of Lipschitz embeddings equipped with the Lipschitz topology may be shown to be open in the manifold of all Lipschitz mappings $\mathcal{X} \to \mathcal{S}$. See [8] and an application in continuum mechanics in [9].

7 The General Framework for Global Analytic Stress Theory

The preceding section implied that for the case where the kinematics of a material body $\mathcal{X}$ is described by its embeddings in a physical space $\mathcal{S}$, the collection of configurations—the *configuration space*

$$\mathbb{Q} := \mathrm{Emb}^r(\mathcal{X}, \mathcal{S}) \tag{95}$$

—is an open subset of the manifold of mappings $C^r(\mathcal{X}, \mathcal{S})$, for $r \geq 1$. As a result, the configuration space is a Banach manifold in its own right and

$$T_\kappa \mathbb{Q} = T_\kappa C^r(\mathcal{X}, \mathcal{S}) = T_\kappa C^r(\xi), \tag{96}$$

where $\xi : \mathcal{X} \times \mathcal{S} \to \mathcal{X}$ is the natural projection of the trivial fiber bundle.

In view (69), $T_\kappa \mathbb{Q} = C^r(\kappa^* V\mathcal{Y})$, where now

$$V\mathcal{Y} = \{v \in T\mathcal{Y} = T\mathcal{X} \times T\mathcal{S} \mid T\xi(v) = 0 \in T\mathcal{X}\}. \tag{97}$$

Hence, one may make the identifications

$$V\mathcal{Y} = \mathcal{X} \times T\mathcal{S} \tag{98}$$

and

$$(\kappa^* V\mathcal{Y})_X = (V\mathcal{Y})_{\kappa(X)} = T_{\kappa(X)}\mathcal{S}. \tag{99}$$

A section w of $\kappa^*\tau : \kappa^* V\mathcal{Y} \to \mathcal{X}$ is of the form

$$X \longmapsto w(X) \in T_{\kappa(X)}\mathcal{S} \tag{100}$$

and may be viewed as a *vector field along* κ, i.e., a mapping

$$w : \mathcal{X} \longrightarrow T\mathcal{S}, \qquad \text{such that}, \qquad \tau \circ w = \kappa. \tag{101}$$

Thus, a tangent vector to the configuration space at the configuration κ may be viewed as a C^r-vector field along κ. This is a straightforward generalization of the standard notion of a virtual velocity field and we summarize these observations by

$$T_\kappa \mathbb{Q} = C^r(\kappa^* V\mathcal{Y}) = \{w \in C^r(\mathcal{X}, T\mathcal{S}) \mid \tau \circ w = \kappa\}. \tag{102}$$

In the case of generalized continua, where $\xi : \mathcal{Y} \to \mathcal{X}$ need not be a trivial vector bundle, this simplification does not apply of course. However, the foregoing discussion motivates the definition of the *configuration space* for a general continuum mechanical system specified by the fiber bundle $\xi : \mathcal{Y} \to \mathcal{X}$ as

$$\mathbb{Q} = C^r(\xi), \qquad r \geq 1. \tag{103}$$

We note that the condition that configurations are embeddings is meaningless in the case of generalized continua.

The general framework for global analytic stress theory adopts the geometric structure for the statics of systems having a finite number of degrees of freedom. Once a configuration manifold $\mathbb{Q}$ is specified, *generalized* or *virtual velocities* are defined to be elements of the tangent bundle, $T\mathbb{Q}$, and *generalized forces* are defined to be elements of the cotangent bundle $T^*\mathbb{Q}$. The action of a force $F \in T^*_\kappa\mathbb{Q}$ on a virtual velocity $w \in T_\kappa\mathbb{Q}$ is interpreted as *virtual power* and as such, the notion of power has a fundamental role in this formulation.

The foregoing discussion, implies that a force at a configuration $\kappa \in C^r(\xi)$ is an element of $C^r(\kappa^* V\mathcal{Y})^*$—a continuous and linear functional on the Banachable space of C^r-section of the vector bundle. Thus, in the following sections we consider the properties of linear functionals on the space of C^r-sections of a vector bundle W. Of particular interest is the fact that our base manifold, or body manifold, is a manifold with corners rather than a manifold without boundary. The relation between such functionals, on the one hand, and Schwartz distribution and de Rham currents, on the other hand, is described. In Sect. 11 we show that the notions of stresses and hyper-stresses emerge from a representation theorem for such functionals and in Sect. 12 we study further the properties of stresses.

8 Duals to Spaces of Differentiable Sections of a Vector Bundle: Localization of Sections and Functionals

As follows from the foregoing discussion, generalized forces are modeled mathematically as elements of the dual space $C^r(\pi)^* = C^r(W)^*$ of the space of C^r-sections of a vector bundle $\pi : W \to \mathcal{X}$. This section reviews the basic notions corresponding to continuous linear functional in the dual space with particular attention to localization properties. While we assume that our base manifold $\mathcal{X}$ is compact with corners, we want to relate the nature of functionals defined on sections over $\mathcal{X}$ with analogous settings where $\mathcal{X}$ is a manifold without boundary. Thus, one can make a connection of the properties of generalized forces and objects like distributions, de Rham currents and generalized sections on manifolds. (See, in particular, Sect. 8.5.) As an additional motivation for considering sections over manifolds without boundaries, it is observed that in both the Eulerian formulation of continuum mechanics and in classical field theories, the base manifold, either space or space-time, is usually taken as manifold without boundary. We start with the case where $\mathcal{X}$ is a manifold without a boundary and continue with the case where bodies are modeled by compact manifolds with corners.

8.1 Spaces of Differentiable Sections over a Manifold Without Boundary and Linear Functionals

A comprehensive introduction to the subject considered here is available in the Ph.D thesis [43] and the corresponding [10, Chapter 3]. See also [12, Chapter VI] and [15].

Consider the space of C^r-sections of a vector bundle $\pi : W \to \mathcal{X}$, for $0 \leq r \leq \infty$. For manifolds without boundary that are not necessarily compact, the setting of Sect. 3.2 will not give a norm on the space of sections. Thus, one extends the settings used for Schwartz distributions and de Rham currents to sections of a vector bundle (see also [4, Chapter XVII]). Specifically, we turn our attention to $C^r_c(\pi)$, the space of *test sections*—C^r-sections of π having compact supports in $\mathcal{X}$.

Let $\{(U_a, \varphi_a, \Phi_a)\}_{a \in A}$, be a vector bundle atlas so that

$$(\varphi_a, \Phi_a) : \pi^{-1}(U_a) \longrightarrow \mathbb{R}^n \times \mathbb{R}^m, \qquad v \longmapsto (X^i, v^\alpha). \tag{104}$$

and let K be a compact subset of $\mathcal{X}$. Consider the vector subspace $C^r_{c,K}(\pi) \subset C^r_c(\pi)$ of sections, the supports of which are contained in K. Let $a_l \in A$, indicate a finite collection of charts such that $\{U_{a_l}\}$ cover K, and for each U_{a_l} let $K_{a_l,\mu}$, $\mu = 1, 2, \ldots$, be a fundamental sequence of compact sets, i.e., $K_{a_l,\mu} \subset K^{\mathrm{o}}_{a_l,\mu+1}$, covering $\varphi_{a_l}(U_{a_l}) \subset \mathbb{R}^n$. Then, for a section $w \in C^r_{c,K}(\pi)$, the collection of semi-norms

$$\|w\|^r_{K,\mu} = \sup_{a_l,\alpha,|\boldsymbol{I}|\leq r} \left\{ \sup_{X \in K_{a_l,\mu}} \left\{ \left|(w^\alpha_{a_l})_{,\boldsymbol{I}}(\boldsymbol{X})\right| \right\} \right\}, \tag{105}$$

induces a Fréchet space structure for $C^\infty_{c,K}(\pi)$. Since for each compact subset K, one has the inclusion mapping $\iota_K : C^r_{c,K}(\pi) \to C^r_c(\pi)$, one may define the topology on $C^r_c(\pi)$ as the inductive limit topology generated by these inclusions, i.e., the strongest topology on $C^r_c(\pi)$ for which all the inclusions are continuous. A sequence of sections in $C^r_c(\pi)$ converges to zero, if there is a compact subset $K \subset \mathcal{X}$ such that the supports of all sections in the sequence are contained in K and the r-jets of the sections converge uniformly to zero in K.

A linear functional $T \in C^r_c(\pi)^*$ is continuous when it satisfies the following condition. Let (χ_j) be a sequence of sections of π all of which are supported in a compact subset $K \subset U_a$ for some $a \in A$. In addition, assume that the local representatives of χ_j and their derivatives of all orders $k \leq r$ converge uniformly to zero in K. Then,

$$\lim_{j\to\infty} T(\chi_j) = 0. \tag{106}$$

Functionals in $C^r_c(\pi)^*$ for a finite value of r are referred to as functionals of order r.

For a linear functional T, the *support*, $\operatorname{supp} T$ is defined as follows. An open set $U \subset \mathcal{X}$ is termed a null set of T if $T(\chi) = 0$ for any section of π with $\operatorname{supp}\chi \subset U$. The union of all null sets, U_0 is an open set which is a null set also. Thus, one defines

$$\operatorname{supp} T := \mathcal{X} \setminus U_0. \tag{107}$$

8.2 Localization of Sections and Linear Functionals for Manifolds Without Boundaries

Let $\{(U_a, \varphi_a, \Phi_a)\}_{a\in A}$ be a locally finite vector bundle atlas on W and consider

$$E_{U_a} : C^r_c(\pi|_{U_a}) \longrightarrow C^r_c(\pi), \tag{108}$$

the natural zero extension of sections supported in compact subsets of U_a to the space of sections that are compactly supported in $\mathcal{X}$. This is evidently a linear and continuous injection of the subspace. On its image, the subspace of sections χ with $\operatorname{supp}\chi \subset U_a$, we have a left inverse, the natural restriction

$$\rho_{U_a} : \text{Image}\, E_{U_a} \longrightarrow C^r_c(\pi|_{U_a}), \tag{109}$$

a surjective mapping. However, it is well known (e.g., [28, 44, pp. 245–246]) that the inverse ρ_{U_a} is not continuous.

The dual,

$$E^*_{U_a} : C^r_c(\pi)^* \longrightarrow C^r_c(\pi|_{U_a})^*, \tag{110}$$

is the restriction of functionals on $\mathcal{X}$ to sections supported on U_a, and as ρ_{U_a} is not continuous, $E^*_{U_a}$ is not surjective (*loc. cit.*). We will write

$$T|_{U_a} := \tilde{T}_a := E^*_{U_a} T. \tag{111}$$

We also note that the restrictions $\{\tilde{T}_a\}$ satisfy the condition

$$\tilde{T}_a(\chi|_{U_a}) = \tilde{T}_b(\chi|_{U_b}) = T(\chi) \tag{112}$$

for any section χ supported in $U_a \cap U_b$.

Consider the mapping

$$s : \bigoplus_{a\in A} C^r_c(\pi|_{U_a}) \longrightarrow C^r_c(\pi) \tag{113}$$

given by

$$s(\chi_1, \ldots, \chi_a, \ldots) := \sum_{a\in A} E_{U_a}(\chi_a). \tag{114}$$

Due to the overlapping between domains of definition, the mapping s is not injective. However, s is surjective because using a partition of unity, $\{u_a\}$, which subordinate to this atlas, for each section, χ, $u_a\chi$ is a compactly supported in U_a and $\chi = \sum_a u_a\chi$. Hence, the dual mapping,

$$s^* : C^r_c(\pi)^* \longrightarrow \bigoplus_{a\in A} C^r_c(\pi|_{U_a})^*, \tag{115}$$

given by,

$$(s^*T)_a := E^*_{U_a}T = T|_{U_a}, \qquad s^*T(\chi_1, \ldots) = T\Big(\sum_{a\in A} E_{U_a}(\chi_a)\Big), \tag{116}$$

is injective. In other words, a functional is determined uniquely by the collection of its restrictions. Note that no compatibility condition is imposed above on the local sections $\{\chi_a\}$.

Since $\{\tilde{T}_a\} \in \operatorname{Image} s^*$ satisfy the compatibility condition (112), s^* is not surjective. However, it is easy to see that $\operatorname{Image} s^*$ is exactly the subspace of $\bigoplus_{a\in A} C^r_c(\pi|_{U_a})^*$ containing the compatible collections of local functionals. For let $\{\tilde{T}_a\}$ be local functionals that satisfy (112) and $\{u_a\}$ a partition of unity. Consider the functional $T \in C^r_c(\pi)^*$ given by

$$T(\chi) = \sum_{a\in A} \tilde{T}_a(u_a\chi). \tag{117}$$

If χ is supported in U_b for $b \in A$, then

$$\begin{aligned} T(\chi) &= \sum_{a\in A} \tilde{T}_a(u_a\chi), && U_a \cap U_b \neq \varnothing, \\ &= \sum_{a\in A} \tilde{T}_b(u_a\chi), && \text{by (112)}, \\ &= \tilde{T}_b\Big(\sum_{a\in A} u_a\chi\Big), \\ &= \tilde{T}_b(\chi). \end{aligned} \tag{118}$$

Thus, T is a well-defined functional on π and it is uniquely determined by the collection $\{\tilde{T}_a\}$—its restrictions, independently of the partition of unity chosen.

As mentioned above, a partition of unity induces an injective right inverse to s in the form

$$p : C^r_c(\pi) \longrightarrow \bigoplus_{a\in A} C^r_c(\pi|_{U_a}), \qquad p(\chi) = \{(u_a\chi)|_{U_a}\}, \tag{119}$$

that evidently satisfies $s \circ p = \iota$. It is noted that p is not a left inverse. In particular, for a section χ_a supported in U_a, with $\chi_b = 0$ for all $b \neq a$,

$$(p \circ s\{\chi_1, \dots\})_a = u_a \chi_a \tag{120}$$

which need not be equal to χ_a. Thus, p depends on the partition of unity.

For the surjective dual mapping

$$p^* : \bigoplus_{a \in A} C^r_c(\pi|_{U_a})^* \longrightarrow C^r_c(\pi)^*, \tag{121}$$

$$p^*(T_1, \dots) = \sum_{a \in A} u_a(\rho^*_{U_a} T_a), \qquad p^*(T_1, \dots)(\chi) = \sum_{a \in A} T_a((u_a \chi)|_{U_a}), \tag{122}$$

we note that $p^* \circ s^* = \mathrm{Id}$, while $s^* \circ p^* \neq \mathrm{Id}$, in general. The surjectivity of p^* implies that every functional T may be represented by a non-unique collection $\{T_a\}$ in the form

$$T(\chi) = \sum_{a \in A} T_a((u_a \chi)|_{U_a}), \qquad T = \sum_{a \in A} u_a T_a. \tag{123}$$

which depends on the partition of unity. Here, $u_a T$ denotes the functional defined by $u_a T(\chi) = T(u_a \chi)$.

Nevertheless, we may restrict p^* to the subspace of compatible local functionals, Image s^*, i.e., those satisfying (112). Thus, the restriction

$$p^*|_{\mathrm{Image}\, s^*} : \mathrm{Image}\, s^* \longrightarrow C^r_c(\pi)^*, \tag{124}$$

is an isomorphism (which depends on the partition of unity). It follows that

$$s^* \circ p^* = \mathrm{Id} : \mathrm{Image}\, s^* \longrightarrow \mathrm{Image}\, s^*. \tag{125}$$

(For additional details, see [4, pp. 244–245], which is restricted to the case of de Rham currents, and [10, pp. 234–235].)

8.3 *Localization of Sections and Linear Functionals for Manifolds with Corners*

In analogy with Sect. 8.2, we consider the various aspects of localization relevant to the case of compact manifolds with corners. Thus, the base manifold for the vector bundle $\pi : W \longrightarrow \mathcal{X}$ is assumed to be a manifold with corners and we are concerned with elements of $C^r(\pi)^*$ acting on sections that need not necessarily vanish together with their first r jets on the boundary of $\mathcal{X}$.

In [26, pp. 10–11], Palais proves what he refers to as the "Mayer–Vietoris Theorem." Adapting the notation and specializing the theorem to the C^r-topology, the theorem may be stated as follows.

Theorem 1 *Let $\mathcal{X}$ be a compact smooth manifold and let $K_1, \ldots, K_A$ be compact n-dimensional submanifolds of $\mathcal{X}$ whose interiors cover $\mathcal{X}$ (such as in a precompact atlas). Given the vector bundle π, set*

$$\tilde{C}^r(\pi) := \left\{ (\chi_1, \ldots, \chi_A) \in \bigoplus_{a=1}^{A} C^r(\pi|_{K_a}) \,\middle|\, \chi_a|_{K_b} = \chi_b|_{K_a} \right\}, \tag{126}$$

and define

$$\iota : C^r(\pi) \longrightarrow \tilde{C}^r(\pi), \qquad by \qquad \iota(\chi) = (\chi|_{K_1}, \ldots, \chi|_{K_A}). \tag{127}$$

Then, ι is an isomorphism of Banach spaces.

We will refer to the condition in (126) as the *compatibility condition* for local representatives of sections. The most significant part of the proof is the construction of ι^{-1}. Thus, one has to construct a field w when a collection $(w_1, \ldots, w_A)$, satisfying the compatibility condition, is given. This is done using a partition of unity which is subordinate to the interiors of $K_1, \ldots, K_A$.

It is noted that the situation may be viewed as "dual" to that described in Sect. 8.2. For functionals on spaces of sections with compact supports defined on a manifold without boundary, there is a natural restriction of functionals, $E^*_{U_a}$, and the images $\{\tilde{T}_a\}$ of a functional T under the restrictions satisfy the compatibility condition (112). The collection of restrictions determine T uniquely. Here, it follows from Theorem 1 that we have a natural restriction of sections, and the restricted sections satisfy the compatibility condition (126). The restrictions $\{\chi|_{K_a}\}$ also determine the global section χ, uniquely.

In Sect. 8.2, we observed that sections with compact supports on $\mathcal{X}$ cannot be "restricted" naturally to sections with compact supports on the various U_a. Such restrictions depend on the chosen partition of unity. The analogous situation for functionals on manifolds with corners is described below.

Corollary 1 *Let $T \in C^r(\pi)^*$, then T may be represented (non-uniquely) by $(T_1, \ldots, T_A)$, $T_a \in C^r(\pi|_{K_a})^*$, in the form*

$$T(w) = \sum_{a=1}^{A} T_a(w|_{K_a}). \tag{128}$$

Indeed, as ι in Theorem 1 is an embedding of $C^r(\pi)$ into a subspace of $\bigoplus_{a=1}^{A} C^r(\pi|_{K_a})$, one has a surjective

$$\iota^* : \bigoplus_{a=1}^{A} C^r(\pi|_{K_a})^* \longrightarrow C^r(\pi)^*, \tag{129}$$

given by

$$\iota^*(T_1, \ldots, T_A)(w) = \sum_{a=1}^{A} T_a(w|_{K_a}). \tag{130}$$

8.4 Supported Sections, Static Indeterminacy and Body Forces

The foregoing observations are indicative of the fundamental problem of continuum mechanics—that of static indeterminacy. Given a force F on a body as an element of $C^r(\pi)^*$ for some vector bundle $\pi : W \to \mathcal{X}$, and a sub-body $\mathcal{R} \subset \mathcal{X}$, there is no unique restriction of F to a force on $\mathcal{R}$ in $C^r(\pi|_{\mathcal{R}})^*$. This problem is evident for standard continuum mechanics in Euclidean spaces and continues all the way to continuum mechanics of higher order on differentiable manifolds.

Adopting the notation of [20], denote by $\dot{C}^r(\pi)$ the space of sections of π, the r-jet extensions of which vanish on all the components of the boundary $\partial\mathcal{X}$. Let $\tilde{\mathcal{X}}$ be a manifold without boundary extending $\mathcal{X}$ and let

$$\tilde{\pi} : \tilde{W} \longrightarrow \tilde{\mathcal{X}} \tag{131}$$

be an extension of π. Then, we may use zero extension to obtain an isomorphism

$$\dot{C}^r(\pi) \cong \{\chi \in C^r_c(\tilde{\pi}) \mid \operatorname{supp}\chi \subset \mathcal{X}). \tag{132}$$

If $\mathcal{R}$ is a sub-body of $\mathcal{X}$, then, one has the inclusion

$$\dot{C}^r(\pi|_{\mathcal{R}}) \hookrightarrow \dot{C}^r(\pi). \tag{133}$$

The dual $\dot{C}^r(\pi)^*$ to the space of sections supported in $\mathcal{X}$ is the space of *extendable functionals*. From [20, Proposition 3.3.1] it follows that the restriction

$$\rho : C^r(\pi)^* \longrightarrow \dot{C}^r(\pi)^*. \tag{134}$$

is surjective and its kernel is the space of functionals on $\tilde{\mathcal{X}}$ supported in $\partial\mathcal{X}$.

Thus, if we interpret $T \in C^r(\pi)^*$ as a force, $\rho(T) \in \dot{C}^r(\pi)^*$ is interpreted as the corresponding *body force*. For a sub-body $\mathcal{R}$, using the dual of (133), one has

$$\rho_{\mathcal{R}} : \dot{C}^r(\pi)^* \longrightarrow \dot{C}^r(\pi|_{\mathcal{R}})^*. \tag{135}$$

We conclude that even in this very general settings, body forces of any order may be restricted naturally to sub-bodies.

8.5 Supported Functionals

Distributions on closed subsets of $\mathbb{R}^n$ have been considered by Glaeser [11], Malgrange [19, Chapter 7] and Oksak [25]. The basic tool in the analysis of distributions on closed sets is Whitney's extension theorem [45] (see also [14, 31]) which guarantees that a differentiable function on a closed subset of $\mathbb{R}^n$ may be extended to a compactly supported smooth function on $\mathbb{R}^n$. For the case of manifolds with corners, the extension mapping between the corresponding function spaces is continuous. (See discussion and counterexamples in [23, Section 4.3].) The extension theorem implies that restriction of functions is surjective and so, the dual of the restriction mapping associates a unique distribution in an open subset of $\mathbb{R}^n$ with a linear functional defined on the given closed set. Distributions and functionals on manifold with corners have been considered by Melrose [20, Chapter 3], whom we follow below.

Thus, let $T \in C^r(\pi)^*$ and let $\tilde{\pi} : \tilde{W} \to \tilde{\mathcal{X}}$ be an extension of the vector bundle $\pi : W \to \mathcal{X}$, where $\tilde{\mathcal{X}}$ is a manifold without a boundary. The Whitney–Seeley extension

$$E : C^r(\pi) \longrightarrow C^r_c(\tilde{\pi}) \tag{136}$$

is a continuous injection. It follows that the natural restriction

$$\rho_{\mathcal{X}} : C^r_c(\tilde{\pi}) \longrightarrow C^r(\pi), \tag{137}$$

its left inverse satisfying $\rho_{\mathcal{X}} \circ E = \mathrm{Id}$, is surjective and the inclusion

$$\rho^*_{\mathcal{X}} : C^r(\pi)^* \longrightarrow C^r_c(\tilde{\pi})^* \tag{138}$$

is injective. In other words, each functional $T \in C^r(\pi)^*$, determines uniquely a functional $\tilde{T} = \rho^*_{\mathcal{X}} T$ satisfying

$$\tilde{T}(\tilde{\chi}) = T(\tilde{\chi}|_{\mathcal{X}}). \tag{139}$$

The last equation implies also that $\tilde{T}(\tilde{\chi}) = 0$ for any section $\tilde{\chi}$ supported in $\tilde{\mathcal{X}} \setminus \mathcal{X}$. Hence, $\tilde{T}$ is supported in $\mathcal{X}$.

Conversely, every $\tilde{T} \in C^r_c(\tilde{\pi})^*$, with $\operatorname{supp} \tilde{T} \subset \mathcal{X}$ represents a functional $T \in C^r(\pi)^*$, i.e., $\tilde{T} = \rho^*_{\mathcal{X}} T$. This may be deduced as follows. For any such $\tilde{T}$, consider $T = E^* \tilde{T}$. One needs to show that $\tilde{T} = \rho^*_{\mathcal{X}} \circ E^*(\tilde{T})$. Let $\tilde{\chi} \in C^r_c(\tilde{\pi})$, then,

$$\begin{aligned} (\tilde{T} - \rho^*_{\mathcal{X}} \circ E^*(\tilde{T}))(\tilde{\chi}) &= \tilde{T}(\tilde{\chi}) - \tilde{T}(E \circ \rho_{\mathcal{X}}(\tilde{\chi})), \\ &= \tilde{T}(\tilde{\chi} - E \circ \rho_{\mathcal{X}}(\tilde{\chi})). \end{aligned} \tag{140}$$

It is observed that $\tilde{\chi} - E \circ \rho_{\mathcal{X}}(\tilde{\chi})$ vanishes on $\mathcal{X}$ so that $\operatorname{supp}(\tilde{\chi} - E \circ \rho_{\mathcal{X}}(\tilde{\chi}) \subset \overline{\tilde{\mathcal{X}} \setminus \mathcal{X}}$. Since $\operatorname{supp} \tilde{T} \subset \mathcal{X}$, $\tilde{T}(\chi') = 0$ for any section χ' supported in $\tilde{\mathcal{X}} \setminus \mathcal{X}$. However, approximating the section $\tilde{\chi} - E \circ \rho_{\mathcal{X}}(\tilde{\chi})$, supported in the closure, $\overline{\tilde{\mathcal{X}} \setminus \mathcal{X}}$, by sections supported in $\tilde{\mathcal{X}} \setminus \mathcal{X}$, one concludes that $\tilde{T}(\tilde{\chi} - E \circ \rho_{\mathcal{X}}(\tilde{\chi})) = 0$ also.

Due to this construction, Melrose [20, Chapter 3] refers to such functionals (distributions) as *supported*. It is noted that such functionals of compact support are of a finite order r.

8.6 Density Dual and Smooth Functionals

A simple example for functionals on spaces of sections of a vector bundle $\pi : W \to \mathcal{X}$ is provided by *smooth functionals*. Consider, in analogy with the dual of a vector bundle, the vector bundle of linear mappings into another one-dimensional vector bundle, that of n-alternating tensors. Thus, for a given vector bundle, W, we use the notation (see Atiyah and Bott [1])

$$W' = L\big(W, \textstyle\bigwedge^n T^*\mathcal{X}\big) \cong W^* \otimes \textstyle\bigwedge^n T^*\mathcal{X}. \tag{141}$$

Let $A : W_1 \to W_2$ be a vector bundle morphism over $\mathcal{X}$. Then, in analogy with the dual mapping, one may consider

$$A' : W_2' \longrightarrow W_1', \qquad \text{given by} \qquad f \longmapsto f \circ A. \tag{142}$$

It is also noted that we have

$$\begin{aligned} (W')' &= (W^* \otimes \textstyle\bigwedge^n T^*\mathcal{X})' \\ &= (W^* \otimes \textstyle\bigwedge^n T^*\mathcal{X})^* \otimes \textstyle\bigwedge^n T^*\mathcal{X}, \\ &= W \otimes \textstyle\bigwedge^n T\mathcal{X} \otimes \textstyle\bigwedge^n T^*\mathcal{X}, \end{aligned} \tag{143}$$

and as $\bigwedge^n T\mathcal{X} \otimes \bigwedge^n T^*\mathcal{X}$ is isomorphic with $\mathbb{R}$, one has a natural isomorphism

$$(W')' \cong W. \tag{144}$$

For the vector bundles W, U,

$$(W \otimes U)' \cong W^* \otimes U^* \otimes \textstyle\bigwedge^n T^*\mathcal{X} \cong W^* \otimes U'. \tag{145}$$

We will refer to W' as the *density-dual bundle* and to A' as the *density-dual mapping*.

As an example, for the case $W = \bigwedge^p T^*\mathcal{X}$, we have an isomorphism (see Sect. 2.4),

$$e_\lrcorner : \textstyle\bigwedge^{n-p} T^*\mathcal{X} \longrightarrow (\textstyle\bigwedge^p T^*\mathcal{X})', \tag{146}$$

given by

$$e_{\lrcorner}(\omega)(\psi) = \omega \wedge \psi. \tag{147}$$

Smooth functionals may be induced by smooth sections of W'. For a section S of W', and a section χ of W, let $S \cdot \chi$ be the n-form

$$S \cdot \chi(X) = S(X)(\chi(X)). \tag{148}$$

The smooth functional T_S induced by S is defined by

$$T_S(\chi) := \int_{\mathcal{X}} S \cdot \chi. \tag{149}$$

8.7 Generalized Sections and Distributions

Let $\pi_0 : W_0 \to \mathcal{X}$ be a vector bundle and consider the case where the vector bundle π above is set to be

$$\pi : W_0' := L\big(W_0, \textstyle\bigwedge^n T^*\mathcal{X}\big) \cong W_0^* \otimes \textstyle\bigwedge^n T^*\mathcal{X} \longrightarrow \mathcal{X}. \tag{150}$$

Thus, the corresponding functionals on sections of π are elements of

$$C^r(\pi)^* = C^r\left(L\big(W_0, \textstyle\bigwedge^n T^*\mathcal{X}\big)\right)^* \cong C^r\left(W_0^* \otimes \textstyle\bigwedge^n T^*\mathcal{X}\right)^*. \tag{151}$$

In this case, smooth functionals are represented by smooth sections of

$$\begin{aligned} L\big(W, \textstyle\bigwedge^n T^*\mathcal{X}\big) &= L\big(W_0^* \otimes \textstyle\bigwedge^n T^*\mathcal{X}, \textstyle\bigwedge^n T^*\mathcal{X}\big), \\ &\cong \left(W_0^* \otimes \textstyle\bigwedge^n T^*\mathcal{X}\right)^* \otimes \textstyle\bigwedge^n T^*\mathcal{X}, \\ &\cong W_0 \otimes \textstyle\bigwedge^n T\mathcal{X} \otimes \textstyle\bigwedge^n T^*\mathcal{X}, \\ &\cong W_0. \end{aligned} \tag{152}$$

One concludes that smooth functionals in $C^r(W_0^* \otimes \bigwedge^n T^*\mathcal{X})^*$ are represented by sections of W_0. It is natural therefore to refer to elements of

$$C^{-r}(W_0) := C^r(W_0')^* \tag{153}$$

as *generalized sections* of W_0 (see [1, 10, 12], [15, p. 676]).

In the particular case where $W_0 = \mathcal{X} \times \mathbb{R}$ is the natural line bundle, smooth functionals are represented by real valued functions on $\mathcal{X}$. Consequently, elements of

$$C^r\left(\mathbb{R} \otimes \textstyle\bigwedge^n T^*\mathcal{X}\right)^* = C^r\left(\textstyle\bigwedge^n T^*\mathcal{X}\right)^* \tag{154}$$

are referred to as *generalized functions*.

The apparent complication in the definition of generalized sections using the density dual is justified in the sense that each element in $C^{-k}(W_0)$ may be approximated by a sequence of smooth functionals induced by sections of W_0'. (See [10, p. 241].)

In the literature, the term *section distributions* is used in different ways in this context. For example, in [10] and [15, p. 676], W_0-valued distributions are defined as elements of $C^r\left(W_0'\right)^*$, i.e., what are referred to here as generalized sections of W_0. (In [1, 2] they are referred to as *distributional sections*.) On the other hand, in [12], distributions are defined as generalized sections of $\bigwedge^n T^*\mathcal{X}$—elements of $C^r(\mathcal{X})^*$. See further comments on this issue and the corresponding terms *section distributional densities* and *generalized densities* in [10, 12, 15].

9 de Rham Currents

For a manifold without boundary $\mathcal{X}$, de Rham currents (see [5, 7, 29]) are functionals corresponding to the case of the vector bundle

$$\pi : \bigwedge^p T^*\mathcal{X} \longrightarrow \mathcal{X} \tag{155}$$

so that test sections are smooth p-forms having compact supports. Thus, a *p-current of order r* on $\mathcal{X}$ is a continuous linear functional on $C_c^r(\bigwedge^p T^*\mathcal{X})$.

A particular type of p-currents, smooth currents, are induced by differential $(n-p)$-forms. Such an $(n-p)$ form, ω, induces the currents ${}_{\omega}T$ and $T_\omega = (-1)^{p(n-p)}{}_{\omega}T$ by

$${}_{\omega}T(\psi) = \int_{\mathcal{X}} \omega \wedge \psi, \qquad T_\omega(\psi) = \int_{\mathcal{X}} \psi \wedge \omega. \tag{156}$$

Another simple p-current, $T_{\mathcal{Z}}$ is induced by an oriented p-dimensional submanifold $\mathcal{Z} \subset \mathcal{X}$. It is naturally defined by

$$T_{\mathcal{Z}}(\psi) = \int_{\mathcal{Z}} \psi. \tag{157}$$

These two examples illustrate the two points of views on currents. On the one hand, the example of the current ${}_{\omega}T$ suggests that a current in $C_c^r(\bigwedge^p T^*\mathcal{X})^*$ is viewed as a generalized $(n-p)$-form. With this point of view in mind, elements of $C_c^r(\bigwedge^p T^*\mathcal{X})$ are referred to as *currents of degree $n-p$*. Consequently, the space of p-currents on $\mathcal{X}$ is occasionally denoted by

$$C^{-r}(\bigwedge^{n-p} T^*\mathcal{X}) = C_c^r(\bigwedge^p T^*\mathcal{X})^*. \tag{158}$$

On the other hand, the example of the current $T_{\mathfrak{T}}$ induced by a p-dimensional manifolds $\mathfrak{T}$, suggests that currents be viewed as a geometric object of dimension p. Thus, an element of $C^r_c(\bigwedge^p T^*\mathfrak{X})^*$ is referred to as a *p-dimensional current*.

9.1 Basic Operations with Currents

The contraction operations of a $(p+q)$-current T and a q-form ω, yields the p-currents defined by

$$(T \llcorner \omega)(\psi) = T(\psi \wedge \omega) \qquad \text{and} \qquad (\omega \lrcorner T)(\psi) = T(\omega \wedge \psi), \tag{159}$$

so that

$$T \llcorner \omega = (-1)^{pq} \omega \lrcorner T. \tag{160}$$

Note that our notation is different from that of [5] and different in sign form that of [7]. In particular, given a p-current T, any p-form ψ induces naturally a zero-current

$$T \cdot \psi = T \llcorner \psi, \qquad \text{so that} \qquad (T \cdot \psi)(u) = T(u\psi). \tag{161}$$

The p-current ${}_\omega T$ defined above can be expressed using contraction in the form

$${}_\omega T = \omega \lrcorner T_{\mathfrak{X}}. \tag{162}$$

For a p-current T and a q-multi-vector field ξ, the $(p+q)$-currents $\xi \wedge T$ and $T \wedge \xi$ are defined by

$$(\xi \wedge T)(\psi) := T(\xi \lrcorner \psi), \qquad (T \wedge \xi)(\psi) := T(\psi \llcorner \xi), \tag{163}$$

for a $(p+q)$-form ψ. Using, $\xi \lrcorner \psi = (-1)^{qp} \psi \llcorner \xi$, one has

$$\xi \wedge T = (-1)^{pq} T \wedge \xi, \tag{164}$$

in analogy with the corresponding expression for multi-vectors. Thus, the wedge product of a p-current and an r-multi-vector is an $(r+p)$-current. Note that a real valued function u defined on $\mathfrak{X}$ may be viewed both as a zero-form and as a zero-multi-vector. Hence, we may write uT for any of the four operations defined above so that $(uT)(\psi) = T(u\psi)$.

The boundary operator

$$\partial : C^{-r}(\textstyle\bigwedge^{n-p} T^*\mathfrak{X}) \longrightarrow C^{-(r+1)}(\textstyle\bigwedge^{n-(p-1)} T^*\mathfrak{X}), \tag{165}$$

defined by

$$\partial T(\psi) = T(\mathrm{d}\psi), \tag{166}$$

is a linear and continuous operator. In other words, the boundary of a p-current is a $(p-1)$-current. In particular, for a smooth current, ${}_{\omega}T$ represented by the $(n-p)$-form ω, one has

$$\begin{aligned} \partial\, {}_{\omega}T(\psi) &= \int_{\mathfrak{X}} \omega \wedge \mathrm{d}\psi, \\ &= (-1)^{n-p} \int_{\mathfrak{X}} \mathrm{d}(\omega \wedge \psi) - (-1)^{n-p} \int_{\mathfrak{X}} \mathrm{d}\omega \wedge \psi, \\ &= (-1)^{n-p+1} T_{\mathrm{d}\omega}(\psi). \end{aligned} \tag{167}$$

Hence,

$$\partial\, {}_{\omega}T = (-1)^{n-p+1} T_{\mathrm{d}\omega}. \tag{168}$$

Similarly,

$$\partial T_{\omega} = (-1)^{p+1} T_{\mathrm{d}\omega}. \tag{169}$$

In order to strengthen further the point of view that a p-current is a generalized $(n-p)$-form, the *exterior derivative of a p-current* $\mathrm{d}T$ is defined by

$$\mathrm{d}T = (-1)^{n-p+1} \partial T. \tag{170}$$

Thus, in the smooth case,

$$\mathrm{d}\, {}_{\omega}T = T_{\mathrm{d}\omega}. \tag{171}$$

In addition, Stokes's theorem implies that for the p-current $T_{\mathfrak{L}}$ induced by the p-dimensional submanifold with boundary $\mathfrak{L}$, the boundary, a $(p-1)$-current, is given by

$$\partial T_{\mathfrak{L}} = T_{\partial \mathfrak{L}}. \tag{172}$$

It is quite evident, therefore, that the notion of a boundary generalizes and unites both the exterior derivative of forms and the boundaries of manifolds.

9.2 Local Representation of Currents

We consider next the local representation of de Rham currents in coordinate neighborhoods.

9.2.1 Representation by 0-Currents

Let $R = E^*_{U_a} T$ be the restriction of a p-current T to forms supported in a particular coordinate neighborhood—a local representative of T. Writing

$$\begin{aligned} R(\psi) &= R(\psi_\lambda \mathrm{d}X^\lambda), \qquad |\lambda| = p, \\ &= (\mathrm{d}X^\lambda \lrcorner R)(\psi_\lambda) \end{aligned} \tag{173}$$

(we could have used $R \llcorner \mathrm{d}X^\lambda$ just the same as the ψ_λ are real valued functions), one notes that locally

$$R(\psi) = R^\lambda(\psi_\lambda), \qquad \text{where} \qquad R^\lambda := \mathrm{d}X^\lambda \lrcorner R. \tag{174}$$

Using the exterior product of a multi-vector field ξ and a current in (163), we may write

$$R(\psi) = R^\lambda(\partial_\lambda \lrcorner \psi) = \partial_\lambda \wedge R^\lambda(\psi), \tag{175}$$

and so a current may be represented locally in the form

$$R = \partial_\lambda \wedge R^\lambda. \tag{176}$$

This representation suggests that T be interpreted as a generalized multi-vector field (cf. [46]).

In the sequel, when we refer to local representative of a current T, we will often keep the same notation, T, and it will be implied that we consider the restriction of T to forms (or sections, in general) supported in a generic coordinate neighborhood.

9.2.2 Representation by *n*-Currents

Alternatively (cf. [5, p. 36]), for a p-current R defined in a coordinate neighborhood and $\hat{\lambda}$ with $|\hat{\lambda}| = n - p$, consider the n-currents

$$R_{\hat{\lambda}} := \partial_{\hat{\lambda}} \wedge R, \qquad \text{so that} \qquad R_{\hat{\lambda}}(\theta) = R(\partial_{\hat{\lambda}} \lrcorner \theta). \tag{177}$$

Then, for every p-form ω,

$$\begin{aligned} (\mathrm{d}X^{\hat{\lambda}} \lrcorner R_{\hat{\lambda}})(\omega) &= R_{\hat{\lambda}}(\mathrm{d}X^{\hat{\lambda}} \wedge \omega), \\ &= R(\partial_{\hat{\lambda}} \lrcorner (\mathrm{d}X^{\hat{\lambda}} \wedge \omega)), \\ &= R_{\hat{\lambda}}(\varepsilon^{\hat{\lambda}\mu} \omega_\mu \mathrm{d}X), \end{aligned} \tag{178}$$

where we used

$$\mathrm{d}X^{\hat{\lambda}} \wedge \omega = \varepsilon^{\hat{\lambda}\mu}\omega_\mu \mathrm{d}X. \tag{179}$$

Also,

$$\begin{aligned}(\partial_{\hat{\lambda}} \lrcorner \mathrm{d}X)(\partial_\mu) &= \mathrm{d}X(\partial_{\hat{\lambda}} \wedge \partial_\mu),\\ &= \varepsilon_{\hat{\lambda}\mu},\\ &= \varepsilon_{\hat{\lambda}\nu}\mathrm{d}X^\nu(\partial_\mu),\end{aligned} \tag{180}$$

implies

$$\partial_{\hat{\lambda}} \lrcorner \mathrm{d}X = \varepsilon_{\hat{\lambda}\nu}\mathrm{d}X^\nu, \tag{181}$$

and so,

$$\partial_{\hat{\lambda}} \lrcorner (\mathrm{d}X^{\hat{\lambda}} \wedge \omega) = \omega_\lambda \mathrm{d}X^\lambda = \omega, \tag{182}$$

as expected. Hence,

$$(\mathrm{d}X^{\hat{\lambda}} \lrcorner R_{\hat{\lambda}})(\omega) = R(\omega), \tag{183}$$

and we conclude that R may be represented by the n-currents

$$R_{\hat{\lambda}} := \partial_{\hat{\lambda}} \wedge R, \qquad \text{in the form} \qquad R = \mathrm{d}X^{\hat{\lambda}} \lrcorner R_{\hat{\lambda}}, \tag{184}$$

with

$$R(\omega) = R_{\hat{\lambda}}(\mathrm{d}X^{\hat{\lambda}} \wedge \omega) = \varepsilon^{\hat{\lambda}\mu}R_{\hat{\lambda}}(\omega_\mu \mathrm{d}X) = \varepsilon^{\hat{\lambda}\lambda}R_{\hat{\lambda}}(\omega_\lambda \mathrm{d}X). \tag{185}$$

(It is recalled that in the last expression, summation is implied where λ and $\hat{\lambda}$ are considered as distinct indices.) This representation suggests again that a p-current T be interpreted as a generalized $(n-p)$-form. In particular, an n-current is a generalized function and is often referred to as a distribution on the manifold (e.g., [20, Chapter 3]).

Remark 3 It is noted that one may set

$$R'_{\hat{\lambda}} := R \wedge \partial_{\hat{\lambda}}. \tag{186}$$

Using (164) for the $(n-p)$-multi-vector $\partial_{\hat{\lambda}}$

$$R'_{\hat{\lambda}} = (-1)^{p(n-p)}R_{\hat{\lambda}}. \tag{187}$$

In addition, by (184) and (160), for the n-current $R_{\hat{\lambda}}$ and the $(n-p)$-form $\mathrm{d}X^{\hat{\lambda}}$,

$$R = R'_{\hat{\lambda}} \llcorner \mathrm{d}X^{\hat{\lambda}}, \tag{188}$$

and

$$R(\omega) = R'_{\hat{\lambda}}(\omega \wedge \mathrm{d}X^{\hat{\lambda}}) = \sum_{\lambda} \varepsilon^{\lambda\hat{\lambda}} R'_{\hat{\lambda}}(\omega_{\lambda} \mathrm{d}X). \tag{189}$$

10 Vector Valued Currents

A natural extension of the notions of generalized sections and de Rham currents yields vector valued currents that will be used to model stresses. Vector valued currents and their local representations will be considered in this section.

10.1 Vector Valued Forms

Let $\pi : W \to \mathcal{X}$ be a given vector bundle whose typical fiber is m-dimensional. We will refer to sections of

$$L\big(W, \bigwedge^p T^*\mathcal{X}\big) \cong W^* \otimes \bigwedge^p T^*\mathcal{X}. \tag{190}$$

as *vector valued p-forms*, which is short for the more appropriate *vector bundle valued p-form* (cf. [29, p. 340]). Thus in particular, sections of the density dual, $W' = W^* \otimes \bigwedge^n T^*\mathcal{X}$ are vector valued forms. In the mechanical context, we will also be concerned with co-vector valued forms, that is, sections of

$$L\big(W^*, \bigwedge^p T^*\mathcal{X}\big) \cong W \otimes \bigwedge^p T^*\mathcal{X}. \tag{191}$$

The terminology follows from the observation that using the isomorphism induced by transposition, i.e., $\bigwedge^p T\mathcal{X} \otimes W \cong W \otimes \bigwedge^p T\mathcal{X}$, a co-vector valued form may be viewed as a section of

$$\bigwedge^p T^*\mathcal{X} \otimes W \cong \bigwedge^p (T\mathcal{X}, W) \cong L(\bigwedge^p T\mathcal{X}, W). \tag{192}$$

Given a co-vector valued p-form, χ, and a vector valued $(n-p)$-form, f, one can define the bilinear action $f \dot{\wedge} \chi$ by setting

$$(g \otimes \omega) \dot{\wedge} (w \otimes \psi) := g(w)\omega \wedge \psi, \tag{193}$$

for sections g, w, ω, ψ of W^*, W, $\bigwedge^{n-p}T^*\mathcal{X}$, $\bigwedge^p T^*\mathcal{X}$, respectively. Thus, $\dot{\wedge}$ induces a bilinear mapping

$$\dot{\wedge}_{\lrcorner} : (W^* \otimes \bigwedge^{n-p}T^*\mathcal{X}) \times (W \otimes \bigwedge^p T^*\mathcal{X}) \longrightarrow \bigwedge^n T^*\mathcal{X}, \tag{194}$$

or a linear

$$\dot{\wedge}_{\lrcorner} : W^* \otimes \bigwedge^p T^*\mathcal{X} \otimes W \otimes \bigwedge^p T^*\mathcal{X} \longrightarrow \bigwedge^n T^*\mathcal{X}. \tag{195}$$

The mapping $\dot{\wedge}_{\lrcorner}$ gives rise to an extension of the isomorphism $e_{\lrcorner}$ considered above to an isomorphism (we keep the same notation)

$$e_{\lrcorner} : W^* \otimes \bigwedge^{n-p}T^*\mathcal{X} \longrightarrow \left(W \otimes \bigwedge^p T^*\mathcal{X}\right)', \qquad e_{\lrcorner}(f)(\chi) = f \dot{\wedge} \chi. \tag{196}$$

Let $\{(U_a, \varphi_a, \Phi_a)\}_{a \in A}$ be a vector bundle trivialization for the vector bundle $\pi : W \to \mathcal{X}$ so that

$$\Phi_a : \pi^{-1}(U_a) \longrightarrow U_a \times \boldsymbol{W}, \tag{197}$$

where $\boldsymbol{W}$ is the m-dimensional typical fiber. Given a basis in $\boldsymbol{W}$, let $\{\boldsymbol{e}_\alpha\}_{\alpha=1}^m$ and $\{\boldsymbol{e}^\alpha\}_{\alpha=1}^m$ be the local bases and dual bases induced by Φ_a^{-1} on $\pi^{-1}(U_a)$. Then, a co-vector valued form χ and a vector valued p-form f are represented locally in the forms

$$\chi_\lambda^\alpha \boldsymbol{e}_\alpha \otimes \mathrm{d}X^\lambda, \qquad \text{and} \qquad f_{\alpha\lambda}\boldsymbol{e}^\alpha \otimes \mathrm{d}X^\lambda, \qquad |\lambda| = p, \tag{198}$$

respectively.

10.2 Vector Valued Currents

We now substitute the vector bundle $W^* \otimes \bigwedge^p T^*\mathcal{X}$ for the vector bundle W_0 in definition (153) of generalized sections. Thus,

$$C^{-r}(W^* \otimes \bigwedge^p T^*\mathcal{X}) = C^r((W^* \otimes \bigwedge^p T^*\mathcal{X})')^*. \tag{199}$$

Using the isomorphism $e_{\lrcorner}$ as defined above, it is concluded that we may make the identifications

$$C^{-r}(W^* \otimes \bigwedge^p T^*\mathcal{X}) = C^r\left(W \otimes \bigwedge^{n-p}T^*\mathcal{X}\right)^* \tag{200}$$

(see [29, p. 340]). Comparing the last equation to (190) we may refer to elements of these spaces as *generalized vector valued p-forms* or as *vector valued $(n-p)$-currents*.

A smooth vector valued $(n-p)$-current may be represented by a $W^*\otimes\bigwedge^{n-p}T\mathcal{X}$ valued n-form—a smooth section S of $W^*\otimes\bigwedge^{n-p}T\mathcal{X}\otimes\bigwedge^{n}T^*\mathcal{X}$ by

$$\chi\longmapsto\int_{\mathcal{X}}S\cdot\chi,\tag{201}$$

where it is noted that $S\cdot\chi$ is an n-form. Locally, for $|\mu|=n-p$,

$$S=S^{\mu}_{\alpha}\boldsymbol{e}^{\alpha}\otimes\partial_{\mu}\otimes\mathrm{d}X,\qquad S\cdot\chi=S^{\mu}_{\alpha}\chi^{\alpha}_{\mu}\mathrm{d}X.\tag{202}$$

Alternatively, a smooth element of $C^{-r}\left(W^*,\bigwedge^{p}T^*\mathcal{X}\right)$ is induced by a section $\widehat{S}$ of $W\otimes\bigwedge^{p}T^*\mathcal{X}$ in the form

$$\chi\longmapsto\int_{\mathcal{X}}\widehat{S}\dot{\wedge}\chi,\tag{203}$$

for every C^r-section χ of $W\otimes\bigwedge^{n-p}T^*\mathcal{X}$. Locally, for $\left|\hat{\mu}\right|=p$, $|\lambda|=n-p$,

$$\widehat{S}=\widehat{S}_{\alpha\hat{\mu}}\boldsymbol{e}^{\alpha}\otimes\mathrm{d}X^{\hat{\mu}},\qquad\widehat{S}\dot{\wedge}\chi=\widehat{S}_{\alpha\hat{\mu}}\chi^{\alpha}_{\lambda}\mathrm{d}X^{\hat{\mu}}\wedge\mathrm{d}X^{\lambda}=\varepsilon^{\hat{\lambda}\lambda}\widehat{S}_{\alpha\hat{\lambda}}\chi^{\alpha}_{\lambda}\mathrm{d}X.\tag{204}$$

Comparing the last two expressions for the resulting densities, one concludes that

$$\widehat{S}_{\alpha\hat{\lambda}}=\varepsilon_{\hat{\lambda}\lambda}S^{\lambda}_{\alpha}.\tag{205}$$

Globally, it follows that

$$\widehat{S}=\mathsf{C}_{\llcorner}(S),\tag{206}$$

where

$$\mathsf{C}_{\llcorner}:W^*\otimes\bigwedge^{n-p}T\mathcal{X}\otimes\bigwedge^{n}T^*\mathcal{X}\longrightarrow W^*\otimes\bigwedge^{p}T^*\mathcal{X}\tag{207}$$

is induced by the right contraction $\theta\llcorner\eta_1(\eta_2)=\theta(\eta_2\wedge\eta_1)$. What determined the direction of the contraction was the choice of action of $\widehat{S}$ in (196) as in Remark 3.

10.3 Local Representation of Vector Valued Currents

We now consider the local representation of the restriction of a vector valued p-current to vector valued forms supported in some given vector bundle chart. We introduce first some basic operations.

10.3.1 The Inner Product of a Vector Valued Current and a Vector Field

Given a vector valued current T in $C^r(W \otimes \bigwedge^p T^*\mathcal{X})^*$ and a C^r-section w of W, we define the (scalar) p-current $T \cdot w$ by

$$T \cdot w(\omega) = T(w \otimes \omega). \tag{208}$$

For local representation, one may consider the p-currents

$$T_\alpha := T \cdot \boldsymbol{e}_\alpha. \tag{209}$$

Thus, in analogy with (173) and (174) we have

$$\begin{aligned} T(w \otimes \omega) &= T(w^\alpha \boldsymbol{e}_\alpha \otimes \omega), \\ &= (T \cdot \boldsymbol{e}_\alpha)(w^\alpha \omega), \\ &= T_\alpha(w^\alpha \omega). \end{aligned} \tag{210}$$

10.3.2 The Tensor Product of a Current and a Co-vector Field

A (scalar) p-current, T, and a C^r-section of W^*, g, induce a vector valued current $g \otimes T \in C^r\big(W \otimes \bigwedge^p T^*\mathcal{X}\big)^*$ by setting

$$(g \otimes T)(w \otimes \omega) := T((g \cdot w)\omega). \tag{211}$$

In particular, locally,

$$(\boldsymbol{e}^\alpha \otimes T)(w \otimes \omega) := T(w^\alpha \omega). \tag{212}$$

Utilizing this definition, one may write for local representatives

$$\begin{aligned} \boldsymbol{e}^\alpha \otimes T_\alpha(w \otimes \omega) &= T_\alpha(w^\alpha \omega), \\ &= T(w^\alpha \boldsymbol{e}_\alpha \otimes \omega), \end{aligned} \tag{213}$$

and so, complementing (210), one has

$$T = \boldsymbol{e}^\alpha \otimes T_\alpha. \tag{214}$$

10.3.3 Representation by 0-Currents

Proceeding as in Sect. 9.2, the p-current T may be represented by the 0-currents

$$T^\lambda_\alpha := \mathrm{d}X^\lambda \lrcorner T_\alpha = \mathrm{d}X^\lambda \lrcorner (T \cdot \boldsymbol{e}_\alpha), \qquad \text{in the form} \qquad T(\chi) = T^\lambda_\alpha(\chi^\alpha_\lambda). \tag{215}$$

Using (163) and (176), we finally have

$$T = \boldsymbol{e}^\alpha \otimes (\partial_\lambda \wedge T^\lambda_\alpha). \tag{216}$$

In the case where the 0-currents T^λ_α are represented locally by smooth n-forms $S^\lambda_\alpha \mathrm{d}X$, one has

$$T(\chi) = \int_U S^\lambda_\alpha \chi^\alpha_\lambda \mathrm{d}X \tag{217}$$

in accordance with (202).

10.3.4 The Exterior Product of a Vector Valued Current and a Multi-Vector Field

Next, in analogy with Sect. 9.2, for a vector valued p-current T and a q-multi-vector η, $q \leq n - p$, consider the vector valued $(p+q)$-current $\eta \wedge T$ defined by

$$(\eta \wedge T)(w \otimes \omega) := T(w \otimes (\eta \lrcorner \omega)). \tag{218}$$

In particular, for multi-indices $\hat\lambda$, $|\hat\lambda| = n - p$, we define locally the vector valued n-currents

$$T_{\hat\lambda} := \partial_{\hat\lambda} \wedge T, \qquad T_{\hat\lambda}(w \otimes \theta) = T(w \otimes (\partial_{\hat\lambda} \lrcorner \theta)), \tag{219}$$

so that

$$T_{\hat\lambda}(w \otimes \mathrm{d}X) = T(w \otimes (\partial_{\hat\lambda} \lrcorner \mathrm{d}X)). \tag{220}$$

10.3.5 The Contraction of a Vector Valued Current and a Form

Also, for a vector valued p-current T and a q-form ψ, $q \leq p$, define the vector valued $(p-q)$-currents $\psi \lrcorner T$ and $T \llcorner \psi$ as

$$(\psi \lrcorner T)(w \otimes \omega) := T(w \otimes (\psi \wedge \omega)) \tag{221}$$

and

$$(T \llcorner \psi)(w \otimes \omega) := T(w \otimes (\omega \wedge \psi)), \tag{222}$$

so that $T \llcorner \psi = (-1)^{pq} \psi \lrcorner T$. In the case where $q = p$, we obtain an element $\omega \lrcorner T \in C^r(W)^*$, a vector valued 0-current, satisfying

$$(\omega \lrcorner T)(w) = T(w \otimes \omega). \tag{223}$$

Locally, one may consider the functionals—vector valued 0-currents,

$$T^\lambda := \mathrm{d}X^\lambda \lrcorner T, \qquad T^\lambda(w) = T(w \otimes \mathrm{d}X^\lambda), \qquad |\lambda| = p. \tag{224}$$

Hence,

$$T(w \otimes \omega) = T^\lambda(\omega_\lambda w). \tag{225}$$

is a local representation of the action of T using vector valued 0-currents. It is implied by the identity $T^\lambda(\omega_\lambda w) = \partial_\lambda \wedge T^\lambda(w \otimes \omega)$, that

$$T = \partial_\lambda \wedge T^\lambda. \tag{226}$$

10.3.6 Representation by *n*-Currents

Next, for a local basis $\mathrm{d}X^{\hat{\lambda}}$ of $\bigwedge^{n-p} T^*\mathfrak{X}$, using (219) and (223),

$$\begin{aligned} (\mathrm{d}X^{\hat{\lambda}} \lrcorner T_{\hat{\lambda}})(w \otimes \omega) &= T_{\hat{\lambda}}(w \otimes (\mathrm{d}X^{\hat{\lambda}} \wedge \omega)), \\ &= T(w \otimes (\partial_{\hat{\lambda}} \lrcorner (\mathrm{d}X^{\hat{\lambda}} \wedge \omega))). \end{aligned} \tag{227}$$

Following the same procedure as that leading to (184) and (185), one concludes that the vector valued p-current T may be represented locally by the vector valued n-currents $T_{\hat{\lambda}}$ in the form

$$T = \mathrm{d}X^{\hat{\lambda}} \lrcorner T_{\hat{\lambda}}, \tag{228}$$

and

$$T(w \otimes \omega) = T_{\hat{\lambda}}(w \otimes (\mathrm{d}X^{\hat{\lambda}} \wedge \omega)) = \varepsilon^{\hat{\lambda}\lambda} T_{\hat{\lambda}}(\omega_\lambda w \otimes \mathrm{d}X). \tag{229}$$

Using (208) and (214), we may define the (scalar) n-currents

$$T_{\alpha\hat{\lambda}} := T_{\hat{\lambda}} \cdot \boldsymbol{e}_\alpha, \qquad \text{so that} \qquad T_{\hat{\lambda}} = \boldsymbol{e}^\alpha \otimes T_{\alpha\hat{\lambda}}, \tag{230}$$

and

$$T_{\alpha\hat{\lambda}}(\theta) = T_{\hat{\lambda}}(\boldsymbol{e}_\alpha \otimes \theta) = T(\boldsymbol{e}_\alpha \otimes (\partial_{\hat{\lambda}} \lrcorner \theta)). \tag{231}$$

Considering the p-currents T_α in (209), the local components $(T_\alpha)_{\hat{\lambda}}$ are defined by $(T_\alpha)_{\hat{\lambda}} = \partial_{\hat{\lambda}} \wedge (T \cdot \boldsymbol{e}_\alpha)$, hence,

$$
\begin{aligned}
(T_\alpha)_{\hat\lambda}(\theta) &= \partial_{\hat\lambda} \wedge (T \cdot \boldsymbol{e}_\alpha)(\theta), \\
&= (T \cdot \boldsymbol{e}_\alpha)(\partial_{\hat\lambda} \lrcorner\, \theta), \\
&= T(\boldsymbol{e}_\alpha \otimes (\partial_{\hat\lambda} \lrcorner\, \theta)),
\end{aligned} \tag{232}
$$

and we conclude that

$$
T_{\alpha\hat\lambda} = (T_\alpha)_{\hat\lambda}. \tag{233}
$$

Thus, Eqs. (228) and (229) may be rewritten as

$$
T = \mathrm{d}X^{\hat\lambda} \lrcorner\, (\boldsymbol{e}^\alpha \otimes T_{\alpha\hat\lambda}), \tag{234}
$$

and

$$
\begin{aligned}
T(w \otimes \omega) &= T_{\alpha\hat\lambda}(w^\alpha(\mathrm{d}X^{\hat\lambda} \wedge \omega)), \\
&= \varepsilon^{\hat\lambda\lambda} T_{\alpha\hat\lambda}(\omega_\lambda w^\alpha \mathrm{d}X), \\
&= \varepsilon^{\hat\lambda\lambda} \mathrm{d}X \lrcorner\, T_{\alpha\hat\lambda}(\omega_\lambda w^\alpha),
\end{aligned} \tag{235}
$$

Comparing the last equation with (215) we arrive at

$$
\mathrm{d}X \lrcorner\, T_{\alpha\hat\lambda} = \varepsilon_{\hat\lambda\lambda} T_\alpha^\lambda, \qquad T_{\alpha\hat\lambda} = \varepsilon_{\hat\lambda\lambda} \partial_X \wedge T_\alpha^\lambda. \tag{236}
$$

In the smooth case, the n-currents $T_{\alpha\hat\lambda}$ are represented by functions $\widehat{S}_{\alpha\hat\lambda}$ that make up the vector valued $(n-p)$-form

$$
\widehat{S} = \widehat{S}_{\alpha\hat\mu} \boldsymbol{e}^\alpha \otimes \mathrm{d}X^{\hat\mu} \tag{237}
$$

as in (203) and (204).

Remark 4 In summary, the representation by zero currents (e.g., (217)) corresponds to viewing the vector valued current as an element of $C^r(W \otimes \bigwedge^p T^*\mathcal{X})^* =: C^{-r}(W^* \otimes \bigwedge^p T\mathcal{X} \otimes \bigwedge^n T^*\mathcal{X})$. On the other hand, the representation as in (217, 237) corresponds to the point of view that by the isomorphism of $\bigwedge^p T\mathcal{X} \otimes \bigwedge^n T^*\mathcal{X}$ with $\bigwedge^{n-p} T^*\mathcal{X}$,

$$
C^r(W \otimes \textstyle\bigwedge^p T^*\mathcal{X})^* \cong C^{-r}(W^* \otimes \textstyle\bigwedge^{n-p} T^*\mathcal{X}) \tag{238}
$$

Remark 5 In the foregoing discussion we have made special choices and used, for example, the definitions, $T_{\hat\lambda} := \partial_{\hat\lambda} \wedge T$ and $T^\lambda := \mathrm{d}X^\lambda \lrcorner\, T$ rather than $T'_{\hat\lambda} := T \wedge \partial_{\hat\lambda}$ and $T'^\lambda := T \llcorner\, \mathrm{d}X^\lambda$, respectively. The correspondence between the two schemes is a natural extension of Remark 3. In particular, $\varepsilon_{\hat\lambda\lambda}$ will be replaced by $\varepsilon_{\lambda\hat\lambda}$.

11 The Representation of Forces by Hyper-Stresses and Non-holonomic Stresses

11.1 Stresses and Non-holonomic Stresses

We recall that the tangent space $T_\kappa C^r(\xi)$ to the Banach manifold of C^r-sections of the fiber bundle $\xi : \mathcal{Y} \to \mathcal{X}$ at the section $\kappa : \mathcal{X} \to \mathcal{Y}$ may be identified with the Banachable space $C^r(\kappa^* V\mathcal{Y})$ of sections of the pullback vector bundle $\kappa^* \tau_{\mathcal{Y}} : \kappa^* V\mathcal{Y} \longrightarrow \mathcal{X}$. Elements of the tangent space at κ to the configuration manifold represent generalized velocities of the continuous mechanical system (cf. [23, Section 5.8]). Consequently, a generalized force is modeled mathematically by and element $F \in C^r(\kappa^* V\mathcal{Y})^*$. The central message of this section is that although such functionals cannot be restricted naturally to sub-bodies of $\mathcal{X}$, as discussed in Sect. 8.3, forces may be represented, non-uniquely, by stress objects that enable restriction of forces to sub-bodies. In order to simplify the notation, we will consider a general vector bundle $\pi : W \to \mathcal{X}$, as in Sect. 3, and the notation introduced there will be used throughout. The construction is analogous to the representation theorem for distributions of finite order (e.g., [29, p. 91] or [44, p. 259]).

Consider the jet extension linear mapping

$$j^r : C^r(\pi) \longrightarrow C^0(\pi^r) \tag{239}$$

as in Sect. 3.3. As noted, j^r is an embedding and under the norm induced by an atlas, it is even isometric. Evidently, due to the compatibility constraint, Image j^r is a proper subset of $C^0(\pi^r)$ and its complement is open. Hence, the inverse

$$(j^r)^{-1} : \text{Image}\, j^r \longrightarrow C^r(\pi) \tag{240}$$

is a well-defined linear homeomorphism. Given a force $F \in C^r(\pi)^*$, the linear functional

$$F \circ (j^r)^{-1} : \text{Image}\, j^r \longrightarrow \mathbb{R} \tag{241}$$

is a continuous and linear functional on Image j^r. Hence, by the Hahn–Banach theorem, it may be extended to a linear functional $\varsigma \in C^0(\pi^r)^*$. In other words, the linear mapping

$$j^{r*} : C^0(\pi^r)^* \longrightarrow C^r(\pi)^* \tag{242}$$

is surjective.

By the definition of the dual mapping, ς represents a force F, i.e.,

$$j^{r*}\varsigma = F, \tag{243}$$

if and only if,

$$F(w) = \varsigma(j^r w) \tag{244}$$

for all C^r-virtual velocity fields w. The object $\varsigma \in C^0(\pi^r)^*$ is interpreted as a generalization of the notion of hyper-stress in higher-order continuum mechanics and will be so referred to. For $r = 1$, ς is a generalization of the standard stress tensor. The condition (244), resulting from the representation theorem, is a generalization of the principle of virtual work as it states that the power expended by the force F for a virtual velocity field w is equal to the power expended by the hyper-stress for $j^r w$—containing the first r derivatives of the velocity field. Accordingly, Eq. (243) is a generalization of the equilibrium equation of continuum mechanics.

It is noted that ς is not unique. The non-uniqueness originates from the fact that the image of the jet extension mapping, containing the compatible jet fields, is not a dense subset of $C^0(\pi^r)$. Thus, the static indeterminacy of continuum mechanics follows naturally from the representation theorem.

In view of (61), the same procedure applies if we use the non-holonomic jet extension $\hat{j} : C^r(\pi) \to C^0(\hat{\pi}^r)$. A force F may then be represented by a non-unique, non-holonomic stress $\hat{\varsigma} \in C^0(\hat{\pi})^*$ in the form

$$F = \hat{j}^{r*}\hat{\varsigma}. \tag{245}$$

The mapping $C^0(\iota^r)$ of Sect. 3.4 is an embedding. Hence, a hyper-stress ς may be represented by some non-unique, non-holonomic stress $\hat{\varsigma}$ in the form

$$\varsigma = C^0(\iota^r)^*(\hat{\varsigma}), \tag{246}$$

and in the following commutative diagram all mappings are surjective.

$$C^r(\pi)^* \xleftarrow{j^{r*}} C^0(\pi^r)^* \xleftarrow{C^0(\iota^r)^*} C^0(\hat{\pi}^r)^*, \qquad (\hat{j}^r)^* : C^0(\hat{\pi}^r)^* \to C^r(\pi)^* \tag{247}$$

11.2 Smooth Stresses

In view of the discussion in Sect. 8.7, hyper-stresses are elements of

$$C^0(\pi^r)^* = C^{-0}((\pi^r)') = C^{-0}(\pi^{r*} \otimes \textstyle\bigwedge^n T^*\mathcal{X}), \tag{248}$$

and so they may be approximated by smooth sections of $\pi^{r*} \otimes \bigwedge^n T^*\mathcal{X}$, i.e., n-forms valued in the dual of the r-jet bundle.

Similarly, non-holonomic stresses are elements of

$$C^0(\hat{\pi}^r)^* = C^{-0}(\hat{\pi}^{r*} \otimes \bigwedge^n T^*\mathcal{X}), \tag{249}$$

and smooth non-holonomic stresses are n-forms valued in the dual of the r-iterated jet bundle.

11.3 Stress Measures

Analytically, stresses are vector valued zero currents that are representable by integration. (See [7, Section 4.1], for the scalar case.)

As noted in Sect. 8.2, given a vector bundle atlas $\{(U_a, \varphi_a, \Phi_a)\}_{a\in A}$, a linear functional is uniquely determined by its restrictions to sections supported in the various domains $\varphi_\alpha(U_\alpha)$—its local representatives. In particular, for the case of an m-dimensional vector bundle $\pi : W \to \mathcal{X}$, and a functional $T \in C^0(\pi)^*$, a typical local representative is an element $T_a \in \left(C^0_c(\varphi_a(U_a))^*\right)^m$. Thus, each component $(T_a)_\alpha \in C^0_c(\varphi_a(U_a))^*$ is a Radon measure or a distribution representable by integration. We will use the same notation for the measure. Consequently, for a section w_a compactly supported in $\varphi_a(U_a)$, we may write

$$T_a(w_\alpha) = \int_{\varphi_a(U_a)} w_a \cdot \mathrm{d}T_a := \int_{\varphi_a(U_a)} w_a^\alpha \mathrm{d}T_{a\alpha}. \tag{250}$$

Given a partition of unity $\{u_a\}$ subordinate to the atlas, one has

$$T(w) = \int_{\mathcal{X}} w \cdot \mathrm{d}T := \sum_{a\in A} \int_{\varphi_a(U_a)} \Phi_a(u_a w) \cdot \mathrm{d}T_a = \sum_{a\in A} \int_{\varphi_a(U_a)} u_a w_a^\alpha \mathrm{d}T_{a\alpha}. \tag{251}$$

For the case of stresses, one has to replace W by $J^r W$, T by ς, and $T_{a\alpha}$ by $\varsigma^I_{a\alpha}$, $|I| \leq r$. In addition, as $\mathcal{X}$ is a manifold with corners, representing measures may be viewed as measures on the extension $\tilde{\mathcal{X}}$ which are supported in $\mathcal{X}$. Thus, for a section χ of $J^r W$, represented locally by χ^α_{aI},

$$\varsigma(\chi) = \int_{\mathcal{X}} \chi \cdot \mathrm{d}\varsigma := \sum_{a\in A} \int_{\varphi_a(U_a)} \Phi_a(u_a \chi) \cdot \mathrm{d}\varsigma_a = \sum_{a\in A} \int_{\varphi_a(U_a)} u_a \chi^\alpha_{aI} \mathrm{d}\varsigma^I_{a\alpha}. \tag{252}$$

We note that the components $\varsigma^I_{a\alpha}$ have the same symmetry under permutations of I as sections of the jet bundle. If w is a section of a vector bundle W_0, then,

$$\varsigma(j^r w) = \int_{\mathcal{X}} j^r w \cdot \mathrm{d}\varsigma = \sum_{a\in A} \int_{\varphi_a(U_a)} u_a w^\alpha_{a,I} \mathrm{d}\varsigma^I_{a\alpha}. \tag{253}$$

The same reasoning applies to the representation by non-holonomic stresses, only here we consider sections $\hat{\chi}$ of the iterated jet bundle represented locally by $\hat{\chi}_{aI_{\mathfrak{p}}}^{\mathfrak{p}\alpha_{\mathfrak{p}}}$, $G_{\mathfrak{p}} \leq r$. The local non-holonomic stress measures have components $\hat{\varsigma}_{a\alpha_{\mathfrak{p}}}^{\mathfrak{p}I_{\mathfrak{p}}}$ and

$$\hat{\varsigma}(\hat{\chi}) = \int_{\mathcal{X}} \hat{\chi} \cdot \mathrm{d}\hat{\varsigma} := \sum_{a\in A} \int_{\varphi_a(U_a)} \Phi_a(u_a\hat{\chi}) \cdot \mathrm{d}\hat{\varsigma}_a = \sum_{a\in A} \int_{\varphi_a(U_a)} u_a \hat{\chi}_{aI_{\mathfrak{p}}}^{\mathfrak{p}\alpha_{\mathfrak{p}}} \mathrm{d}\hat{\varsigma}_{a\alpha_{\mathfrak{p}}}^{\mathfrak{p}I_{\mathfrak{p}}}, \tag{254}$$

$$\hat{\varsigma}(\hat{j}^r w) = \int_{\mathcal{X}} \hat{j}^r w \cdot \mathrm{d}\hat{\varsigma} = \sum_{a\in A} \int_{\varphi_a(U_a)} u_a w_{a,I_{\mathfrak{p}}}^{\mathfrak{p}\alpha_{\mathfrak{p}}} \mathrm{d}\varsigma_{a\alpha_{\mathfrak{p}}}^{\mathfrak{p}I_{\mathfrak{p}}}, \tag{255}$$

where summation is implied on all values of $\alpha_{\mathfrak{p}}$, $I_{\mathfrak{p}}$, for all values of $\mathfrak{p}$ such that $G_{\mathfrak{p}} \leq r$.

It is concluded that for a given force F, there is some non-unique vector valued hyper-stress measure ς and a non-holonomic stress measure $\hat{\varsigma}$, such that

$$F(w) = \int_{\mathcal{X}} j^r w \cdot \mathrm{d}\varsigma = \int_{\mathcal{X}} \hat{j}^r w \cdot \mathrm{d}\hat{\varsigma}. \tag{256}$$

11.4 Force System Induced by Stresses

It was noted in Sect. 8.3 that given a force on a body $\mathcal{X}$, a manifold with corners, there is no unique way to restrict it to an n-dimensional submanifold with corners, a sub-body $\mathcal{R} \subset \mathcal{X}$. We view this as the fundamental problem of continuum mechanics—static indeterminacy.

Stress, though not determined uniquely by a force, provide means for inducing a *force system*, the assignment of a force $F_{\mathcal{R}}$ to each sub-body $\mathcal{R}$. Indeed, once a stress measure is given, be it a hyper-stress or a non-holonomic stress, integration theory makes it possible to consider the force system given by

$$F_{\mathcal{R}}(w) = \int_{\mathcal{R}} j^r w \cdot \mathrm{d}\varsigma = \int_{\mathcal{R}} \hat{j}^r w \cdot \mathrm{d}\hat{\varsigma} \tag{257}$$

for any section w of $\pi|_{\mathcal{R}}$.

Further details on the relation between hyper-stresses and force systems are available in [30]. It is our opinion that the foregoing line of reasoning captures the essence of stress theory in continuum mechanics accurately and elegantly.

12 Simple Forces and Stresses

We restrict ourselves now to the most natural setting for continuum mechanics, the case $r = 1$—the first value for which the set of C^r-embeddings is open in the manifold of mappings. (See [9] for consideration of configurations modeled as Lipschitz mappings.) Evidently, hyper-stresses and non-holonomic stresses become identical now, and therefore, it is natural in this case to use the terminology *simple forces* and *stresses*.

12.1 Simple Stresses

A simple stress ς on a body $\mathfrak{X}$ is an element of

$$C^0(J^1W)^* =: C^{-0}((J^1W)^* \otimes \textstyle\bigwedge^n T^*\mathfrak{X}), \tag{258}$$

which implies that smooth stress distributions are sections of

$$(J^1W)^* \otimes \textstyle\bigwedge^n T^*\mathfrak{X} = L\big(J^1W, \bigwedge^n T^*\mathfrak{X}\big). \tag{259}$$

Following the discussion in Sect. 8.5, ς may be viewed as a generalized section of $(J^1\tilde{W})^* \otimes \bigwedge^n T^*\tilde{\mathfrak{X}}$, which is supported in $\mathfrak{X}$, where we use the extension of the vector bundle to a vector bundle $\tilde{\pi} : \tilde{W} \to \tilde{\mathfrak{X}}$ over a compact manifold without boundary $\tilde{\mathfrak{X}}$.

A typical local representative of a section of the jet bundle is of the form

$$\chi = \chi^\alpha \boldsymbol{e}_\alpha + \chi_i^\alpha \mathrm{d}X^i \otimes \boldsymbol{e}_\alpha \tag{260}$$

so that locally,

$$\begin{aligned} \varsigma(\chi) &= \varsigma(\chi^\alpha \boldsymbol{e}_\alpha + \chi_i^\alpha \mathrm{d}X^i \otimes \boldsymbol{e}_\alpha), \\ &= \varsigma_\alpha(\chi^\alpha) + \varsigma_\alpha^i(\chi_i^\alpha). \end{aligned} \tag{261}$$

Here, ς_α and ς_α^i are 0-currents defined by

$$\varsigma_\alpha(u) := (\varsigma \cdot \boldsymbol{e}_\alpha)(u) = \varsigma(u\boldsymbol{e}_\alpha), \tag{262}$$

$$\varsigma_\alpha^i(u) := (\varsigma \cdot (\mathrm{d}X^i \otimes \boldsymbol{e}_\alpha))(u) = \varsigma(u\mathrm{d}X^i \otimes \boldsymbol{e}_\alpha). \tag{263}$$

In the smooth case, ς is represented by a section S of $(J^1W)^* \otimes \bigwedge^n T^*\mathfrak{X}$ in the form

$$\varsigma(\chi) = \int_{\mathfrak{X}} S \cdot \chi. \tag{264}$$

Locally, such a vector valued form is represented as

$$S = (S_\alpha \boldsymbol{e}^\alpha + S^i_\alpha \partial_i \otimes \boldsymbol{e}^\alpha) \otimes \mathrm{d}X \tag{265}$$

so that, for the domain of a chart, U, and a section χ with $\operatorname{supp}\chi \subset U$,

$$\varsigma(\chi) = \int_U (S_\alpha \chi^\alpha + S^i_\alpha \chi^\alpha_i)\mathrm{d}X. \tag{266}$$

12.2 The Vertical Projection

The vertical sub-bundle

$$V\pi^1 : VJ^1W \longrightarrow \mathcal{X} \tag{267}$$

is kernel of the natural projection

$$\pi^1_0 : J^1W \longrightarrow W. \tag{268}$$

In other words, elements of the vertical sub-bundle at a point $X \in \mathcal{X}$ are jets of sections that vanish at X. Thus, if a typical element of J^1W is represented locally in the form $(\chi^\alpha, \chi^\beta_i)$, an element of the vertical sub-bundle, $VJ^1W \subset J^1W$ has the form $(0, \chi^\beta_i)$ in any adapted coordinate system. The vertical sub-bundle may be identified with the vector bundle $T^*\mathcal{X} \otimes W$. Denoting the natural inclusion by

$$\iota_V : VJ^1W \longrightarrow J^1W, \tag{269}$$

one has the induced inclusion

$$C^0(\iota_V) : C^0(V\pi^1) \longrightarrow C^0(\pi^1), \qquad \chi \longmapsto \iota_V \circ \chi. \tag{270}$$

Clearly, $C^0(\iota_V)$ is injective and a homeomorphism onto its image. Hence, its dual

$$C^0(\iota_V)^* : C^0(\pi^1)^* \longrightarrow C^0(V\pi^1)^* \cong C^0(T^*\mathcal{X}\otimes W)^* = C^{-0}(T\mathcal{X}\otimes W^*\otimes \textstyle\bigwedge^n T^*\mathcal{X}), \tag{271}$$

is a well-defined surjection. Simply put, $C^0(\iota_V)^*(\varsigma)$ is the restriction of the stress ς to sections of the vertical sub-bundle. Accordingly, we will refer to an element of $C^0(T^*\mathcal{X} \otimes W)^*$ as a *vertical stress* and to $C^0(\iota_V)^*$ as the *vertical projection*.

In case the stress ς is represented locally by the 0-currents $(\varsigma_\alpha, \varsigma^i_\beta)$ as in (261), then $C^0(\iota_V)^*(\varsigma)$ is represented by (ς^i_β).

Let $\varsigma^+ \in C^0(T^*\mathcal{X} \otimes W)^*$ be a vertical stress and let $w \in C^0(\pi)$. Then, $\varsigma^+ \cdot w$ defined by

$$(\varsigma^+ \cdot w)(\varphi) := \varsigma^+(\varphi \otimes w), \qquad \varphi \in C^0(T^*\mathcal{X}) \tag{272}$$

is a 1-current. This is an indication of the fact that ς^+ may be viewed as a vector valued 1-current.

We may use the local representation of currents as in Sect. 10.3 to represent ς^+ by the scalar 1-currents $\varsigma_\alpha^+ = (\varsigma^+ \cdot \boldsymbol{e}_\alpha)$ given as

$$\varsigma_\alpha^+(\varphi) = (\varsigma^+ \cdot \boldsymbol{e}_\alpha)(\varphi) = \varsigma^+(\varphi \otimes \boldsymbol{e}_\alpha) \tag{273}$$

so that

$$\begin{aligned} \varsigma^+(\varphi \otimes w) &= \varsigma^+(\varphi^i w^\alpha \mathrm{d}X^i \otimes \boldsymbol{e}_\alpha), \\ &= \varsigma_\alpha^+(\varphi_i w^\alpha \mathrm{d}X^i), \\ &= (w^\alpha \lrcorner\, \varsigma_\alpha^+)(\varphi). \end{aligned} \tag{274}$$

Similarly, the 1-current $\varsigma^+ \cdot w$ may be represented locally as in Sect. 9.2 by 0-currents $(\varsigma^+ \cdot w)^i$ given as

$$(\varsigma^+ \cdot w)^i(u) = (\mathrm{d}X^i \lrcorner\, (\varsigma^+ \cdot w))(u) = (\varsigma^+ \cdot w)(u\mathrm{d}X^i) \tag{275}$$

in the form

$$(\varsigma^+ \cdot w)(\varphi) = (\varsigma^+ \cdot w)^i(\varphi_i). \tag{276}$$

Evidently,

$$\begin{aligned} (\varsigma^+ \cdot w)^i(u) &= \varsigma^+(u w^\alpha \mathrm{d}X^i \otimes \boldsymbol{e}_\alpha), \\ &= \varsigma_\alpha^{+i}(u w^\alpha), \\ &= (w^\alpha \lrcorner\, \varsigma_\alpha^{+i})(u), \end{aligned} \tag{277}$$

and so

$$(\varsigma^+ \cdot w)^i = w^\alpha \lrcorner\, \varsigma_\alpha^{+i}. \tag{278}$$

It is concluded that

$$\varsigma^+ = \boldsymbol{e}^\alpha \otimes (\partial_i \wedge \varsigma_\alpha^{+i}), \qquad \varsigma_\alpha^{+i} = \mathrm{d}X^i \lrcorner\, (\varsigma^+ \cdot \boldsymbol{e}_\alpha), \tag{279}$$

$$\varsigma^+(\varphi \otimes w) = \varsigma_\alpha^{+i}(\varphi_i w^\alpha), \tag{280}$$

where it is recalled that in case $\varsigma^+ = C^0(\iota_V)^*(\varsigma)$, then, $\varsigma_\alpha^{+i} = \varsigma_\alpha^i$.

In the smooth case, the vertical projection of the stress is represented by a section $S^+$of $T\mathcal{X} \otimes W^* \otimes \bigwedge^n T^*\mathcal{X}$ so that

$$C^0(\iota_V)^*(\varsigma)(\chi) = \int_{\mathcal{X}} S^+ \cdot \chi, \tag{281}$$

where

$$(S^+ \cdot \chi)(v_1, \dots, v_n)(X) = S^+(X)(\chi(X) \otimes (v_1(X) \wedge \cdots \wedge v_n(X))). \tag{282}$$

If ς is represented by a section S of $(J^1W)^* \otimes \bigwedge^n T^*\mathcal{X}$, then, $C^0(\iota_V)^*(\varsigma)$ is represented locally by

$$S^+ = S^i_\alpha \partial_i \otimes \boldsymbol{e}^\alpha \otimes \mathrm{d}X. \tag{283}$$

For a vertical stress ς^+ which is represented by S^i_α as above and a field w, the 1-current $\varsigma^+ \cdot w$ is given locally by

$$\varphi \longmapsto \int_U S^i_\alpha w^\alpha \varphi_i \mathrm{d}X, \tag{284}$$

for a 1-form φ supported in U. In other words, if the vertical stress ς^+ is represented by the section S^+ of $T\mathcal{X} \otimes W \otimes \bigwedge^n T^*\mathcal{X}$, the 1-current $\varsigma^+ \cdot w$ is represented by the density $S^+ \cdot w$, a section of $T\mathcal{X} \otimes \bigwedge^n T^*\mathcal{X}$ given by

$$(S^+ \cdot w)(\varphi) = S^+(\varphi \otimes w). \tag{285}$$

12.3 Traction Stresses

Using the transposition $\mathrm{tr} : W \otimes T^*\mathcal{X} \to T^*\mathcal{X} \otimes W$, one has a mapping on the space of vertical stresses

$$C^0(\mathrm{tr})^* : C^0(T^*\mathcal{X} \otimes W)^* \longrightarrow C^0(W \otimes T^*\mathcal{X})^*. \tag{286}$$

We define *traction stress distributions* to be elements of

$$C^0(W \otimes T^*\mathcal{X})^* = C^{-0}(W^* \otimes \bigwedge^{n-1} T^*\mathcal{X}). \tag{287}$$

Thus, it is noted that a traction stress is not much different than a vertical stress distribution but transposition enables its representation as a vector valued current. Using local representation in accordance with Sect. 10.3.6, a traction stress σ is represented locally by n-currents $\sigma_{\alpha\hat{\imath}}$, $|\hat{\imath}| = n - 1$, in the form

$$\sigma = \mathrm{d}X^{\hat{\imath}} \lrcorner \sigma_{\hat{\imath}} = \boldsymbol{e}^\alpha \otimes \sigma_\alpha = \mathrm{d}X^{\hat{\imath}} \lrcorner (\boldsymbol{e}^\alpha \otimes \sigma_{\alpha\hat{\imath}}), \tag{288}$$

where

$$\sigma_{\hat{i}} := \partial_{\hat{i}} \wedge \sigma, \qquad \sigma_\alpha := \sigma \cdot \boldsymbol{e}_\alpha, \qquad \sigma_{\alpha\hat{i}} := \sigma_{\hat{i}} \cdot \boldsymbol{e}_\alpha = (\sigma_\alpha)_{\hat{i}} = (\partial_{\hat{i}} \wedge \sigma) \cdot \boldsymbol{e}_\alpha. \tag{289}$$

Hence,

$$\begin{aligned} \sigma(w \otimes \varphi) &= \varepsilon^{\hat{i}i} \sigma_{\alpha\hat{i}}(\varphi_i w^\alpha \mathrm{d}X), \\ &= \sum_i (-1)^{n-i} \sigma_{\alpha\hat{i}}(\varphi_i w^\alpha \mathrm{d}X), \\ &= \sigma_{\alpha\hat{i}}(w^\alpha \varphi \wedge \mathrm{d}X^{\hat{i}}), \\ &= \sigma_{\alpha\hat{i}} \llcorner \mathrm{d}X^{\hat{i}}(w^\alpha \varphi). \end{aligned} \tag{290}$$

Introducing the notation

$$p_\sigma := C^0(\mathrm{tr})^* \circ C^0(\iota_V)^* : C^0(\pi^1)^* \longrightarrow C^0(W \otimes T^*\mathfrak{X})^*, \tag{291}$$

a simple stress distribution ς induces a traction stress distribution σ by

$$\sigma = p_\sigma(\varsigma). \tag{292}$$

Let $\sigma = p_\sigma(\varsigma)$, then, comparing the last equation with (280), it is concluded that, in accordance with (236), locally,

$$(-1)^{n-i} \sigma_{\alpha\hat{i}}(u \mathrm{d}X) = \varsigma_\alpha^i(u) \tag{293}$$

for any function u, and so

$$\mathrm{d}X \lrcorner \sigma_{\alpha\hat{i}} = (-1)^{n-i} \varsigma_\alpha^i, \qquad \sigma_{\alpha\hat{i}} = (-1)^{n-i} \partial_X \wedge \varsigma_\alpha^i. \tag{294}$$

Remark 6 Continuing Remark 5, it is observed that one may consider

$$\sigma'_{\hat{i}} := \sigma \wedge \partial_{\hat{i}}, \qquad \sigma'_{\alpha\hat{i}} := \sigma'_{\hat{i}} \cdot \boldsymbol{e}_\alpha = (\sigma_\alpha)'_{\hat{i}} = (\sigma \wedge \partial_{\hat{i}}) \cdot \boldsymbol{e}_\alpha, \tag{295}$$

so that

$$\sigma = \sigma'_{\hat{i}} \llcorner \mathrm{d}X^{\hat{i}} = (\boldsymbol{e}^\alpha \otimes \sigma'_{\alpha\hat{i}}) \llcorner \mathrm{d}X^{\hat{i}}. \tag{296}$$

Thus,

$$\begin{aligned} \sigma(w \otimes \varphi) &= (\boldsymbol{e}^\alpha \otimes \sigma'_{\alpha\hat{i}}) \llcorner \mathrm{d}X^{\hat{i}}(w \otimes \varphi), \\ &= \sigma'_{\alpha\hat{i}}(w^\alpha \varphi \wedge \mathrm{d}X^{\hat{i}}), \\ &= \sum_i (-1)^{i-1} \sigma'_{\alpha\hat{i}}(\varphi_i w^\alpha \mathrm{d}X), \end{aligned} \tag{297}$$

and comparison with (280) implies that

$$\sigma'_{\alpha\hat{i}} \llcorner \mathrm{d}X = (-1)^{i-1} \varsigma^i_\alpha, \qquad \sigma'_{\alpha\hat{i}} = (-1)^{i-1} \varsigma^i_\alpha \wedge \partial_X = (-1)^{n-1} \sigma_{\alpha\hat{i}}, \tag{298}$$

where it is observed that as ς^i_α are zero currents, the order of the contraction and wedge product in the last equation is immaterial.

12.4 Smooth Traction Stresses

In the smooth case, we adapt (204) to the current context. The traction stress σ is represented by a section s of $W^* \otimes \bigwedge^{n-1} T^*\mathfrak{X}$ so that

$$\sigma(\chi) = \int_{\mathfrak{X}} s \dot{\wedge} \chi. \tag{299}$$

Locally,

$$s = s_{\alpha\hat{i}} e^\alpha \otimes \mathrm{d}X^{\hat{i}} = \varepsilon^{\hat{i}i} s_{\alpha\hat{i}} e^\alpha \otimes (\mathrm{d}X \llcorner \partial_i), \qquad s \dot{\wedge} \chi = \varepsilon^{\hat{i}i} s_{\alpha\hat{i}} \chi^\alpha_i \mathrm{d}X \tag{300}$$

so that the local components of s are the n-currents—functions—that represent σ locally.

Let ς be a smooth stress represented by the vector valued form S, a section of $(J^1 W)^* \otimes \bigwedge^n T^*\mathfrak{X}$ as in (265) and (266) and let S^+ be its vertical component as in (283). In view of (206) and (205), $p_\sigma(S)$ is represented by the section s of $W^* \otimes \bigwedge^{n-1} T^*\mathfrak{X}$ with

$$s = \mathsf{C} \llcorner (S^+ \circ \mathrm{tr}), \qquad s_{\alpha\hat{i}} = \varepsilon_{\hat{i}i} S^i_\alpha. \tag{301}$$

Hence, using

$$\mathrm{d}X^{\hat{i}} = \varepsilon^{i\hat{i}} \partial_i \lrcorner \mathrm{d}X = \varepsilon^{\hat{i}i} \mathrm{d}X \llcorner \partial_i, \qquad \varepsilon^{i\hat{i}} = (-1)^{i-1}, \quad \varepsilon^{\hat{i}i} = (-1)^{n-i}, \tag{302}$$

we may write explicitly

$$s = s_{\alpha\hat{i}} e^\alpha \otimes \mathrm{d}X^{\hat{i}} = S^i_\alpha e^\alpha \otimes (\mathrm{d}X \llcorner \partial_i) = (-1)^{n-1} S^i_\alpha e^\alpha \otimes (\partial_i \lrcorner \mathrm{d}X). \tag{303}$$

The terminology, traction stress, originates from the fact that a traction stress σ represented by a smooth section s of $W^* \otimes \bigwedge^{n-1} T^*\mathfrak{X}$, induces the analog of a traction field on oriented hypersurfaces in the body as follows. (See [38] for further details.) We consider, for any given section s of $W^* \otimes \bigwedge^{n-1} T^*\mathfrak{X}$ and a field w, the $(n-1)$-form $\sigma \cdot w$ given by

$$(s \cdot w)(\eta) = s(w \otimes \eta) \tag{304}$$

for sections η of $\bigwedge^{n-1} T\mathcal{X}$. Consider an $(n-1)$-dimensional oriented smooth submanifold $\mathfrak{X} \subset \mathcal{X}$. Let

$$\iota_{\mathfrak{X}} : \mathfrak{X} \longrightarrow \mathcal{X} \tag{305}$$

be the natural inclusion and

$$\iota_{\mathfrak{X}}^* : C^\infty(\bigwedge^{n-1} T^*\mathcal{X}) \longrightarrow C^\infty(\bigwedge^{n-1} T^*\mathfrak{X}), \tag{306}$$

the corresponding restriction of $(n-1)$-forms. Combining the above, one may define a linear mapping

$$\rho_{\mathfrak{X}} : C^\infty(W^* \otimes \bigwedge^{n-1} T^*\mathcal{X}) \longrightarrow C^\infty(W^* \otimes \bigwedge^{n-1} T^*\mathfrak{X}) \tag{307}$$

whereby

$$\rho_{\mathfrak{X}}(s) \cdot w = \iota_{\mathfrak{X}}^*(s \cdot w) \in \bigwedge^{n-1} T^*\mathfrak{X}. \tag{308}$$

A section $\boldsymbol{t}$ of $W^* \otimes \bigwedge^{n-1} T^*\mathfrak{X}$ is interpreted as a *surface traction distribution* on the hypersurface $\mathfrak{X}$. Its action $\boldsymbol{t} \cdot w$ is interpreted as the power density of the corresponding surface force, and may be integrated over $\mathfrak{X}$. In particular, the condition

$$\boldsymbol{t} = (-1)^{n-1} \rho_{\mathfrak{X}}(s) \tag{309}$$

is a generalization of Cauchy's formula for the relation between traction fields and stresses.

Remark 7 The factor $(-1)^{n-1}$ that appears in (309) above, and is absent in [38], originates from our choice to use exterior multiplication on the left as in Remarks 3 and 5. Evidently, if we represented σ by the smooth vector valued form s' such that

$$\sigma(\chi) = \int_{\mathcal{X}} \chi \dot{\wedge} s' \tag{310}$$

instead of (299), the factor $(-1)^{n-1}$ would not appear in the analogous computation and

$$\boldsymbol{t} = \rho_{\mathfrak{X}}(s'). \tag{311}$$

In addition, the second of Eq. (301), may be rewritten as

$$s'_{\alpha \hat{i}} = (-1)^{n-1} S^i_\alpha. \tag{312}$$

12.5 The Generalized Divergence of the Stress

Let ς be a stress distribution, $\sigma = p_\sigma(\varsigma)$, and $w \in C^1(\pi)$. We compute, using (290), a local expression for the boundary of the 1-current $\sigma \cdot w$ as

$$\begin{aligned}
\partial(\sigma \cdot w)(u) &= (\sigma \cdot w)(\mathrm{d}u),\\
&= \sigma(w \otimes \mathrm{d}u),\\
&= \varsigma^i_\alpha(u_{,i} w^\alpha),\\
&= \varsigma^i_\alpha((uw^\alpha)_{,i}) - \varsigma^i_\alpha(uw^\alpha_{,i}),\\
&= \varsigma^i_\alpha(\partial_i \lrcorner \mathrm{d}(uw^\alpha)) - \varsigma^i_\alpha(uw^\alpha_{,i}) - \varsigma_\alpha(uw^\alpha) + \varsigma_\alpha(uw^\alpha),\\
&= \partial_i \wedge \varsigma^i_\alpha(\mathrm{d}(uw^\alpha)) - (\varsigma \cdot j^1 w)(u) + (w^\alpha \lrcorner \varsigma_\alpha)(u),\\
&= \partial(\partial_i \wedge \varsigma^i_\alpha)(uw^\alpha) - (\varsigma \cdot j^1 w)(u) + (w^\alpha \lrcorner \varsigma_\alpha)(u),\\
&= [(\partial_i \varsigma^i_\alpha \cdot w^\alpha) - (\varsigma \cdot j^1 w) + (w^\alpha \lrcorner \varsigma_\alpha)](u).
\end{aligned} \tag{313}$$

Here, for a 0-current $T = \varsigma^i_\alpha$, $\partial_i T$ is the "partial boundary" operator or the dual to the partial derivative, a 0-current defined by

$$\begin{aligned}
\partial_i T(u) &:= \partial(\partial_i \wedge T)(u),\\
&= (\partial_i \wedge T)(du),\\
&= T(u_{,i}).
\end{aligned} \tag{314}$$

Since $\sigma \cdot w$ and σ_α are 1-currents, definition (170) implies that the exterior derivatives satisfy

$$\mathrm{d}(\sigma \cdot w) = (-1)^n \partial(\sigma \cdot w), \qquad \mathrm{d}\sigma_\alpha = (-1)^n \partial\sigma_\alpha, \tag{315}$$

and

$$\mathrm{d}(\sigma \cdot w)(u) = (\mathrm{d}\sigma_\alpha \cdot w^\alpha)(u) + (-1)^{n-1}[\varsigma \cdot j^1 w - w^\alpha \lrcorner \varsigma_\alpha](u). \tag{316}$$

The preceding framework may be formalized as follows. Let

$$\tilde{p} : C^1(W) \times C^1(\mathfrak{X}) \longrightarrow C^1(W), \qquad \text{be defined by} \qquad \tilde{p}(w, u) = uw. \tag{317}$$

The mapping $\tilde{p}$ is clearly bilinear, and as such, it induces a linear mapping

$$p : C^1(W) \otimes C^1(\mathfrak{X}) \longrightarrow C^1(W), \qquad p(w \otimes u) = \tilde{p}(w, u) = uw. \tag{318}$$

The mapping p has a natural right inverse

$$p^{-1} : C^1(W) \longrightarrow C^1(W) \otimes C^1(\mathfrak{X}), \qquad w \longmapsto w \otimes \mathbf{1}, \tag{319}$$

where $\mathbf{1}(X) = 1$, for all $X \in \mathfrak{X}$. Evidently, p^{-1}, and the mappings derived from it, exist only because we do not limit ourselves to functions with compact support in the interior $\mathfrak{X}^{\mathrm{o}}$. It is further noted that the mapping p^{-1} is related to the integral of distributional n-forms defined and studied in [10, pp. 249–250]. The dual mappings satisfy

$$p^* : C^1(W)^* \longrightarrow (C^1(W) \otimes C^1(\mathfrak{X}))^*, \quad p^*(F)(w \otimes u) := F(p(w \otimes u)) = F(uw), \tag{320}$$

and

$$p^{-1*} : (C^1(W) \otimes C^1(\mathfrak{X}))^* \longrightarrow C^1(W)^*, \qquad p^{-1*}(G)(w) = G(w \otimes \mathbf{1}). \tag{321}$$

With the preceding observations, the computations above imply that there are "dual" linear differential operators

$$\tilde{\partial} : C^0(W \otimes T^*\mathfrak{X})^* \longrightarrow (C^1(W) \otimes C^1(\mathfrak{X}))^*, \tag{322}$$

and

$$\tilde{\mathrm{d}} : C^0(W \otimes T^*\mathfrak{X})^* \longrightarrow (C^1(W) \otimes C^1(\mathfrak{X}))^*, \tag{323}$$

such that

$$\tilde{\partial}\sigma(w \otimes u) := \sigma(w \otimes \mathrm{d}u) = \partial(\sigma \cdot w)(u) = (\tilde{\partial}\sigma \cdot w)(u), \qquad \tilde{\mathrm{d}}\sigma := (-1)^n \tilde{\partial}\sigma. \tag{324}$$

Consequently, we define the generalized divergence, a differential operator

$$\mathrm{div} : C^{-0}((J^1 W)^* \otimes \textstyle\bigwedge^n T^*\mathfrak{X}) \longrightarrow (C^1(W) \otimes C^1(\mathfrak{X}))^*, \tag{325}$$

by

$$\mathrm{div}\, \varsigma = -\tilde{\partial}(p_\sigma \varsigma) - j^{1*}\varsigma. \tag{326}$$

As mentioned above, we view the various terms as elements of $(C^1(W) \otimes C^1(\mathfrak{X}))^*$, so that each may be contracted with w to give an element of $C^{-1}(\bigwedge^n T^*\mathfrak{X})$. Thus,

$$\begin{aligned}(\mathrm{div}\, \varsigma \cdot w)(u) &= -(\tilde{\partial}(p_\sigma \varsigma) \cdot w)(u) - (\varsigma \cdot j^1 w)(u), \\ &= -p_\sigma \varsigma(w \otimes \mathrm{d}u) - (\varsigma \cdot j^1 w)(u).\end{aligned} \tag{327}$$

The local expression for the generalized divergence in a coordinate neighborhood U is obtained using (313). For a smooth function u having a compact support in U,

$$(\operatorname{div} \varsigma \cdot w)(u) = [-\partial_i \varsigma^i_\alpha \cdot w^\alpha - w^\alpha \lrcorner \varsigma_\alpha](u), \tag{328}$$

so that

$$\operatorname{div} \varsigma \cdot w = -\partial_i \varsigma^i_\alpha \cdot w^\alpha - w^\alpha \lrcorner \varsigma_\alpha. \tag{329}$$

12.6 *The Divergence for the Smooth Case*

In the smooth case

$$\begin{aligned}
\partial_i \varsigma^i_\alpha \cdot w^\alpha(u) &= \partial_i \varsigma^i_\alpha (w^\alpha u), \\
&= \varsigma^i_\alpha((w^\alpha u)_{,i}), \\
&= \int_U S^i_\alpha (w^\alpha u)_{,i} \mathrm{d}X, \\
&= \int_U (S^i_\alpha w^\alpha u)_{,i} \mathrm{d}X - \int_U S^i_{\alpha,i} w^\alpha u \mathrm{d}X, \\
&= (-1)^{n-1} \int_{\partial U} s_{\alpha \hat{i}} w^\alpha u \mathrm{d}X^{\hat{i}} - \int_U S^i_{\alpha,i} w^\alpha u \mathrm{d}X,
\end{aligned} \tag{330}$$

where we have used (303) to write

$$\begin{aligned}
(-1)^{n-1} \int_{\partial U} s_{\alpha \hat{i}} w^\alpha u \mathrm{d}X^{\hat{i}} &= \int_{\partial U} S^i_\alpha w^\alpha u \varepsilon_{i\hat{i}} \mathrm{d}X^{\hat{i}}, \\
&= \int_U \mathrm{d}(S^i_\alpha w^\alpha u \varepsilon_{i\hat{i}} \mathrm{d}X^{\hat{i}}), \\
&= \int_U (S^i_\alpha w^\alpha u)_{,j} \varepsilon_{i\hat{i}} \mathrm{d}X^j \wedge \mathrm{d}X^{\hat{i}}, \\
&= \int_U (S^i_\alpha w^\alpha u)_{,j} \varepsilon_{i\hat{i}} \varepsilon^{j\hat{i}} \mathrm{d}X, \\
&= \int_U (S^i_\alpha w^\alpha u)_{,j} \delta^j_i \mathrm{d}X, \\
&= \int_U (S^i_\alpha w^\alpha u)_{,i} \mathrm{d}X.
\end{aligned} \tag{331}$$

Thus, noting that in the smooth case $\varsigma_\alpha \cdot w^\alpha$ are represented by the n-forms $S_\alpha w^\alpha \mathrm{d}X$, we conclude that locally

$$
\begin{aligned}
(\operatorname{div}\varsigma\cdot w)(u) &= \int_U (S^i_{\alpha,i} - S_\alpha) w^\alpha u \mathrm{d}X - (-1)^{n-1}\int_{\partial U} s_{\alpha\hat{\imath}} w^\alpha u \mathrm{d}X^{\hat{\imath}},\\
&= \int_U (S^i_{\alpha,i} - S_\alpha) w^\alpha u \mathrm{d}X - \int_{\partial U} S^i_\alpha \varepsilon_{i\hat{\imath}} w^\alpha u \mathrm{d}X^{\hat{\imath}}.
\end{aligned} \tag{332}
$$

In previous work, e.g., [37, 38], the divergence operator was defined specifically for the smooth case where both the stress densities and their derivatives have well-defined pointwise values. Thus, it is possible to define the divergence of the stress, denoted now by $\tilde{\operatorname{div}}$ in order to exhibit the distinction from (326) and (327), as a section of $L(W, \bigwedge^n T^*\mathcal{X})$, given pointwise by the condition

$$
\tilde{\operatorname{div}} S\cdot w = (-1)^{n-1}\mathrm{d}(s\cdot w) - S\cdot j^1 w. \tag{333}
$$

Thus, locally

$$
\begin{aligned}
(\tilde{\operatorname{div}} S)_\alpha w^\alpha \mathrm{d}X &= (-1)^{n-1}\mathrm{d}(\varepsilon_{\hat{\imath}i} S^i_\alpha w^\alpha \mathrm{d}X^{\hat{\imath}}) - S_\alpha w^a \mathrm{d}X - S^i_\alpha w^\alpha_{,i}\mathrm{d}X,\\
&= (-1)^{n-1}(\varepsilon_{\hat{\imath}i} S^i_\alpha w^\alpha)_{,j}\mathrm{d}X^j \wedge \mathrm{d}X^{\hat{\imath}} - S_\alpha w^a \mathrm{d}X - S^i_\alpha w^\alpha_{,i}\mathrm{d}X,\\
&= (-1)^{n-1}(\varepsilon_{\hat{\imath}i} S^i_\alpha w^\alpha)_{,j}\varepsilon^{j\hat{\imath}}\mathrm{d}X - S_\alpha w^a \mathrm{d}X - S^i_\alpha w^\alpha_{,i}\mathrm{d}X,\\
&= (S^i_\alpha w^\alpha)_{,i}\mathrm{d}X - S_\alpha w^a \mathrm{d}X - S^i_\alpha w^\alpha_{,i}\mathrm{d}X,\\
&= (S^i_{\alpha,i} - S_\alpha) w^\alpha \mathrm{d}X.
\end{aligned} \tag{334}
$$

It is noted now that the weak form of this relation includes only the first integral of (332). Thus, the definition of the divergence in the general case does not generalize directly the definitions of $\tilde{\operatorname{div}}$ in ([37, 38]). However, by restricting (332) to functions u that vanish on $U \cap \partial\mathcal{X}$, $\tilde{\operatorname{div}}$ is determined uniquely. As a result, the second integral of (332) is also well determined by $\operatorname{div}\varsigma$.

We conclude that the divergence operator div includes a boundary term in addition to the bulk term $(S^i_{\alpha,i} - S_\alpha)\boldsymbol{e}^\alpha \otimes \mathrm{d}X$.

12.7 *The Invariance of the Divergence*

Returning to the non-smooth case, consider the definition of the divergence in Eq. (327). It is apparent that for a given force F, the term $(\varsigma\cdot j^1 w)(u) = \varsigma(uj^1 w)$ depends on the particular stress representing F because the jet field $uj^1 w$ is, in general, non-holonomic, i.e., there is no vector field w' such that $uj^1 w = j^1(w')$. In fact, if it were possible to compute this term independently of ς, we could restrict the force to sub-bodies by an appropriate sequence of functions u. The other term

$p_\sigma\varsigma(w\otimes \mathrm{d}u)$ also depends, apparently, on the stress representing F. Nevertheless, we whom below that $\operatorname{div}\varsigma$ is independent of the particular stress ς that represents F.

Consider first the jet $j^1(uw)$, for a differentiable function u and a differentiable vector field w represented locally by $w = w^\alpha \boldsymbol{e}_\alpha$. Then,

$$j^1 w = w^\alpha \boldsymbol{e}_\alpha + w^\alpha_{,i}\mathrm{d}X^i \otimes \boldsymbol{e}_\alpha, \tag{335}$$

and so

$$j^1(uw) = uw^\alpha \boldsymbol{e}_\alpha + (uw^\alpha)_{,i}\mathrm{d}X^i \otimes \boldsymbol{e}_\alpha. \tag{336}$$

Hence,

$$\begin{aligned} j^1(uw) &= uw^\alpha \boldsymbol{e}_\alpha + u_{,i}w^\alpha \mathrm{d}X^i \otimes \boldsymbol{e}_\alpha + uw^\alpha_{,i}\mathrm{d}X^i \otimes \boldsymbol{e}_\alpha,\\ &= u(w^\alpha \boldsymbol{e}_\alpha + w^\alpha_{,i}\mathrm{d}X^i \otimes \boldsymbol{e}_\alpha)\\ &\quad + (0\boldsymbol{e}_\alpha + (u_{,i}\mathrm{d}X^i)\otimes(w^\alpha \boldsymbol{e}_\alpha)), \end{aligned} \tag{337}$$

and we conclude that

$$j^1(uw) = uj^1w + \iota_V(\mathrm{d}u\otimes w). \tag{338}$$

Let ς be some stress representing a given force F. We have

$$\begin{aligned} F(uw) &= \varsigma(j^1(uw)),\\ &= \varsigma(uj^1w + C^0(\iota_V)(\mathrm{d}u\otimes w)),\\ &= \varsigma(uj^1w) + C^0(\iota_V)^*(\varsigma)(\mathrm{d}u\otimes w), \end{aligned} \tag{339}$$

and using (291),

$$F(uw) = \varsigma(uj^1w) + p_\sigma(\varsigma)(w\otimes \mathrm{d}u). \tag{340}$$

The last equation may serve as an additional motivation for the introduction of the traction stress $\sigma = p_\sigma(\varsigma)$. In addition, comparing it to the definition (327), it is immediately concluded that

$$(\operatorname{div}\varsigma\cdot w)(u) = -F(uw). \tag{341}$$

Hence, $\operatorname{div}\varsigma$ depends only on the force F and is independent of the stress ς representing it.

For a 0-current T, it is customary to use the notation

$$\int_{\mathfrak{X}} T := T(\mathbf{1}_{\mathfrak{X}}), \qquad \text{where } \mathbf{1}_{\mathfrak{X}}(X) = 1, \tag{342}$$

as in [10, p. 249]. Then, Eq. (341) implies that

$$F(w) = -\int_{\mathcal{X}} \operatorname{div} \varsigma \cdot w. \tag{343}$$

However, this notation may be confusing as we also use integration relative to the stress measures. Thus, we will denote below the integration mapping as

$$\operatorname{int} : (C^1(W) \otimes C^1(\mathcal{X}))^* \longrightarrow C^1(W)^* \tag{344}$$

whereby

$$\operatorname{int}(G)(w) = G(w \otimes \mathbf{1}). \tag{345}$$

Thus, Eq. (343) is rewritten as

$$F(w) = -\operatorname{int}(\operatorname{div} \varsigma)(w), \qquad \text{or,} \qquad F = -\operatorname{int} \circ \operatorname{div} \varsigma \tag{346}$$

12.8 The Balance Equation

We define now the body force current b and the boundary force current t corresponding to the stress ς, elements of $(C^1(W) \otimes C^1(\mathcal{X}))^*$, by

$$b := -\operatorname{div} \varsigma, \qquad t := -\tilde{\partial}(p_\sigma \varsigma) = (-1)^{n-1}\tilde{\mathrm{d}}(p_\sigma \varsigma). \tag{347}$$

From the definition of the divergence in (327) we deduce

$$\varsigma \cdot j^1 w = b \cdot w + t \cdot w, \tag{348}$$

The last equation is yet another generalization of the principle of virtual work in continuum mechanics.

For the smooth case, σ is represented by the smooth vector valued form s as in (299) and we can compute, for any differentiable function u defined on $\mathcal{X}$,

$$\begin{aligned}
\partial(\sigma \cdot w)(u) &= \int_{\mathcal{X}} s \dot{\wedge} (w \otimes \mathrm{d}u), \\
&= \int_{\mathcal{X}} (s \cdot w) \wedge \mathrm{d}u, \\
&= (-1)^{n-1} \int_{\mathcal{X}} \mathrm{d}((s \cdot w) \wedge u) - (-1)^{n-1} \int_{\mathcal{X}} \mathrm{d}(s \cdot w) \wedge u, \qquad (349)\\
&= (-1)^{n-1} \int_{\partial\mathcal{X}} (s \cdot w) u - (-1)^{n-1} \int_{\mathcal{X}} \mathrm{d}(s \cdot w) u, \\
&= \int_{\partial\mathcal{X}} (t \cdot w) u - (-1)^{n-1} \int_{\mathcal{X}} \mathrm{d}(s \cdot w) u,
\end{aligned}$$

where Stokes's theorem was utilized in the fourth line and (309) was used in the fifth line. Thus, in the smooth case, $\partial(\sigma \cdot w)$ contains, upon appropriate choices of u, information regarding the action of the surface force.

12.9 Application to Non-holonomic Stresses

In spite of numerous attempts (see [33, 40–42]), for the general geometry of manifolds, we were not able to extend the foregoing analysis to hyper-stresses, even for the case of stresses represented by smooth densities. Yet, the introduction of non-holonomic stresses makes it possible to carry out one step of the reduction.

Let W_0 be a vector bundle over $\mathcal{X}$ and consider forces in $C^r(W_0)^*$. Using the representation by non-holonomic stresses as in (245) in Sect. 11.1, let

$$W_{r-1} := \hat{J}^{r-1} W_0. \tag{350}$$

Then, a force $F \in C^r(W_0)^*$ is represented by an element

$$\hat{\varsigma} \in C^0(\hat{J}^r W_0)^* = C^0(J^1 W_{r-1})^* \tag{351}$$

in the form

$$\begin{aligned} F(w) &= \hat{j}^{r*}(\hat{\varsigma})(w), \\ &= \hat{\varsigma}(\hat{j}^r w), \\ &= \hat{\varsigma}(j^1(\hat{j}^{r-1} w)), \\ &= j^{1*}\hat{\varsigma}(\hat{j}^{r-1} w). \end{aligned} \tag{352}$$

Thus, one may apply the foregoing analysis of simple stresses to the study the action $\hat{\varsigma}(j^1\chi) = j^{1*}\hat{\varsigma}(\chi)$ for elements

$$\chi \in C^1(\hat{\pi}^{r-1})^* = C^1(W_{r-1})^*. \tag{353}$$

In other words, the analysis of simple stresses is used where we substitute W_{r-1} and χ for W and w above respectively. In particular, the balance equations for this reduction will yield

$$F(w) = \hat{\varsigma}(\hat{j}^r w) = b(\hat{j}^{r-1} w) + t(\hat{j}^{r-1} w), \tag{354}$$

where

$$t, b \in C^{-1}(W^*_{r-1} \otimes \textstyle\bigwedge^n T^*\mathcal{X}) \tag{355}$$

are interpreted as hyper surface traction and hyper body force distributions, respectively.

13 Concluding Remarks

The foregoing text is meant to serve as an introduction to global geometric stress and hyper-stress theory. We used a simple geometric model of a mechanical system in which forces are modeled as elements of the cotangent bundle of the configuration space and outlined the necessary steps needed in order to use it in the infinite dimensional case of continuum mechanics. The traditional choice of configurations as embeddings of a body in space, led us to the natural C^1-topology which determined the properties of forces as linear functionals. In particular, the stress object emerges from a representation theorem for force functionals.

The general stress object we obtain preserves the basic feature of the stress tensor—it induces a force system on the body and its sub-bodies as described in Sect. 11.4. Further details of the relation between hyper-stresses and force systems are presented in [30] for the general case where stresses are as irregular as measures.

Generalizing continuum mechanics to differentiable manifolds implies that derivatives can no longer be decomposed invariantly from the values of vector fields and jets, combining the values of the field and its derivatives, are used. As a result, simple stresses mix both components dual to the values of the velocity fields, ς_α, and components dual to the derivatives, ς_α^i. This distinction from the classical stress tensor may be treated if additional mathematical structure is introduced. It is noted that no conditions of equilibrium, which are equivalent to invariance of the virtual power under the action of the Euclidean group, were imposed. In the general case, one may assume the action of a Lie group on the space manifold and obtain corresponding balance laws (see [36]).

Another subject that has been omitted here is that of constitutive relations. Constitutive relations, in particular the notion of locality have been considered from the global point of view in [35]. Roughly speaking, it is shown in [35] that a local constitutive relation, viewed form the global point of view as a mapping that assigns a stress distribution to a configuration, which is continuous relative to the C^r-topology is a constitutive relation for a material of grade r. Thus, the notion of locality is tied in with that of continuity.

A framework for the dynamics of a continuous body, for the geometry of differentiable manifolds, was proposed in [16]. The dynamics of the system is specified using a Riemannian metric on the infinite dimensional configuration space.

Acknowledgements This work has been partially supported by H. Greenhill Chair for Theoretical and Applied Mechanics and the Pearlstone Center for Aeronautical Engineering Studies at Ben-Gurion University.

References

1. M. Atiyah and R. Bott. A Lefschetz fixed point formula for elliptic complexes: I. *Annals of Mathematics*, 86:374–407, 1967.
2. M. Atiyah and I. Singer. The index of elliptic operators: I. *Annals of Mathematics*, 87:484–530, 1968.
3. A. Douady and L. Hérault. Arrondissement des variétés à coins in Appendix of A. Borel J-P. Serre: Corners and arithmetic groups. *Commentarii Mathematici Helvetici*, 48:484–489, 1973.
4. J. Dieudonné. *Treatise on Analysis*, volume III. Academic Press, 1972.
5. G. de Rham. *Differentiable Manifolds*. Springer, 1984.
6. M. Epstein and R. Segev. Differentiable Manifolds and the Principle of Virtual Work in Continuum Mechanics. *Journal of Mathematical Physics*, 21:1243–1245, 1980.
7. H. Federer. *Geometric Measure Theory*. Springer, 1969.
8. K. Fukui and T. Nakamura. A topological property of Lipschitz mappings. *Topology and its Applications*, 148:143–152, 2005.
9. L. Falach and R. Segev. The configuration space and principle of virtual power for rough bodies. *Mathematics and Mechanics of Solids*, 20:1049–1072, 2015.
10. M. Grosser, M. Kunzinger, M. Oberguggenberger, and R. Steinbauer. *Geometric Theory of Generalized Functions with Applications to General Relativity*. Springer, 2001.
11. G. Glaeser. Etude de quelques algebres tayloriennes. *Journal d'Analyse Mathématique*, 6:1–124, 1958.
12. V. Guillemin and S. Sternberg. *Geometric Asymptotics*. American Mathematical Society, 1977.
13. M. Hirsch. *Differential Topology*. Springer, 1976.
14. L. Hörmander. *The analysis of linear partial differential operators. I. Distribution theory and Fourier analysis*. Springer, 1990.
15. J. Korbas. *Handbook of Global Analysis*, chapter Distributions, vector distributions, and immersions of manifolds in Euclidean spaces, pages 665–724. Elsevier, 2008.
16. R. Kupferman, E. Olami, and R. Segev. Continuum dynamics on manifolds: Applications to Elasticity of residually stresses bodies. *Journal of Elasticity*, 126:61–84, 2017.
17. I. Kolár, J. Slovák, and P.W. Michor. *Natural operations in differential geometry*. Springer, 1993.
18. J.M. Lee. *Introduction to Smooth Manifolds*. Springer, 2002.
19. B. Malgrange. *Ideals of Differentiable functions*, volume 3 of *Tata Institute of Fundamental Research*. Oxford University Press, 1966.
20. R.B. Melrose. *Differential Analysis on Manifolds with Corners*. www-math.mit.edu/~rbm/book.html, 1996.
21. J. E. Marsden and T. J. R. Hughes. *Mathematical Foundations of Elasticity*, chapter An invariant theory of stress and equilibrium by Segev and Epstein, pages 169–175. Dover, 1994.
22. P. W. Michor. *Manifolds of Differentiable Mappings*. Shiva, 1980.
23. P. Michor. Manifolds of mappings for continuum mechanics. In: R. Segev and M. Epstein (eds.), *Geometric Continuum Mechanics*. Advances in Mechanics and Mathematics. Springer, New York, 2020. https://doi.org/10.1007/978-3-030-42683-5_1.
24. J. Margalef-Roig and E Outerelo Domíngues. *Handbook of Global Analysis*, chapter Topology of manifolds with corners, pages 983–1033. Elsevier, 2008.
25. A.I. Oksak. On invariant and covariant Schwartz distributions in the case of a compact linear Groups. *Communications in Mathematical Physics*, 46:269–287, 1976.
26. R. S. Palais. *Foundations of Global Non-Linear Analysis*. Benjamin, 1968.
27. D. J. Saunders. *The Geometry of Jet Bundles*, volume 142 of *London Mathematical Society: Lecture Notes Series*. Cambridge University Press, 1989.
28. L. Schwartz. *Lectures on Modern Mathematics (Volume 1)*, chapter Some applications of the theory of distributions, pages 23–58. Wiley, 1963.
29. L. Schwartz. *Théorie des Distributions*. Hermann, 1973.

30. R. Segev and G. DeBotton. On the consistency conditions for force systems. *International Journal of Nonlinear Mechanics*, 26:47–59, 1991.
31. R.T. Seeley. Extension of C^∞ functions defined in a half space. *Proceeding of the American Mathematical Society*, 15:625–626, 1964.
32. R. Segev. *Differentiable manifolds and some basic notions of continuum mechanics*. PhD thesis, U. of Calgary, 1981.
33. R. Segev. Forces and the existence of stresses in invariant continuum mechanics. *Journal of Mathematical Physics*, 27:163–170, 1986.
34. R. Segev. On the Existence of Stresses in Continuum Mechanics. *Israel Journal of Technology*, 23:75–78, 1986.
35. R. Segev. Locality and Continuity in Constitutive Theory. *Archive for Rational Mechanics and Analysis*, 101:29–37, 1988.
36. R. Segev. A Geometrical Framework for the Statics of Materials with Microstructure. *Mathematical Models and Methods in Applied Sciences*, 6:871–897, 1994.
37. R. Segev. Metric-independent analysis of the stress-energy tensor. *Journal of Mathematical Physics*, 43:3220–3231, 2002.
38. R. Segev. Notes on metric independent analysis of classical fields. *Mathematical Methods in the Applied Sciences*, 36:497–566, 2013.
39. R. Segev. Continuum Mechanics, Stresses, Currents and Electrodynamics. *Philosophical Transactions of the Royal Society of London A*, 374:20150174, 2016.
40. R. Segev. Geometric analysis of hyper-stresses. *International Journal of Engineering Science*, 120:100–118, 2017.
41. R. Segev and J. Śniatycki. Hyper-stresses in k-jet field theories. *Journal of Elasticity*, 2018.
42. R. Segev and J. Śniatycki. On jets, almost symmetric tensors, and traction hyper-stresses. *Mathematics and Mechanics of Complex Systems*, 6:101–124, 2018.
43. R. Steinbauer. *Distributional Methods in General Relativity*. PhD thesis, Universität Wien, 2000.
44. F. Trèves. *Topological Vector Spaces, Distributions and Kernels*. Academic Press, 1967.
45. H. Whitney. Analytic extensions of functions defined in closed sets. *Transactions of the American Mathematical Society*, 36:63–89, 1934.
46. H. Whitney. *Geometric Integration Theory*. Princeton University Press, 1957.

Applications of Algebraic Topology in Elasticity

Arash Yavari

Contents

Abstract In this chapter we discuss some applications of algebraic topology in elasticity. This includes the necessary and sufficient compatibility equations of nonlinear elasticity for non-simply-connected bodies when the ambient space is Euclidean. Algebraic topology is the natural tool to understand the topological obstructions to compatibility for both the deformation gradient **F** and the right Cauchy–Green strain **C**. We investigate the relevance of homology, cohomology, and homotopy groups in elasticity. We also use the relative homology groups in order to derive the compatibility equations in the presence of boundary conditions. The differential complex of nonlinear elasticity written in terms of the deformation gradient and the first Piola–Kirchhoff stress is also discussed.

1 Introduction

Compatibility equations of elasticity are more than 150 years old and according to Love [31] were first studied by Saint Venant in 1864. In nonlinear elasticity a given distribution of strain on a body $\mathcal{B}$ may not correspond to a deformation

A. Yavari (✉)
School of Civil and Environmental Engineering & The George W. Woodruff School of Mechanical Engineering, Georgia Institute of Technology, Atlanta, GA, USA
e-mail: arash.yavari@ce.gatech.edu

R. Segev, M. Epstein (eds.), *Geometric Continuum Mechanics*, Advances in Mechanics and Mathematics 43, https://doi.org/10.1007/978-3-030-42683-5_3

mapping. Similarly, in linear elasticity a given distribution of linearized strains may not correspond to a well-defined displacement field. Strain has to satisfy a set of integrability equations in order to correspond to some deformation field. These integrability equations are called compatibility equations in continuum mechanics. We provided a detailed history of the compatibility equations in nonlinear and linear elasticity in [58] and will not repeat it here. Compatibility equations for simply connected bodies are well understood and are a set of PDEs that depend on the measure of strain. For non-simply-connected bodies these "bulk" compatibility equations are only necessary. In other words, when the bulk compatibility equations are satisfied in a non-simply-connected body the strain field may still be incompatible; there may be topological obstructions to compatibility. A classical example of incompatible strain fields that satisfy the bulk compatibility equations are Volterra's "distortions" (dislocations and disclinations) [51]. For a strain field on a non-simply-connected body to be compatible, in addition to the bulk compatibility equations, some extra compatibility equations that explicitly depend on the topology of the body are needed [13, 27, 36, 46, 51]. We call these extra compatibility equations the *complementary compatibility equations* [49] or the *auxiliary compatibility equations*.

The natural mathematical tool for understanding the topological obstruction to compatibility is algebraic topology. Topological methods, and particularly algebraic topology have been used in fluid mechanics [7], and electromagnetism [24] for quite sometime. In the case of electromagnetism this goes back to the work of Maxwell [35] before the formal developments of algebraic topology that started in the work of Poincaré [40]. Algebraic topology has not been used systematically in solid mechanics until recently [58]. To motivate the present study consider the following problem. Having a solid sphere (a ball) with the different types of holes shown in Fig. 1, what are the compatibility equations for **F** and **C**? The necessary compatibility equations ("bulk" compatibility equations) are well understood and our focus will be on the sufficient conditions. We will see that in case (a) of a spherical hole no extra compatibility equations are needed. For (b), (c), and (d) one needs to impose some extra constraints on the (red) loops (generators of the first homology group) to ensure compatibility.

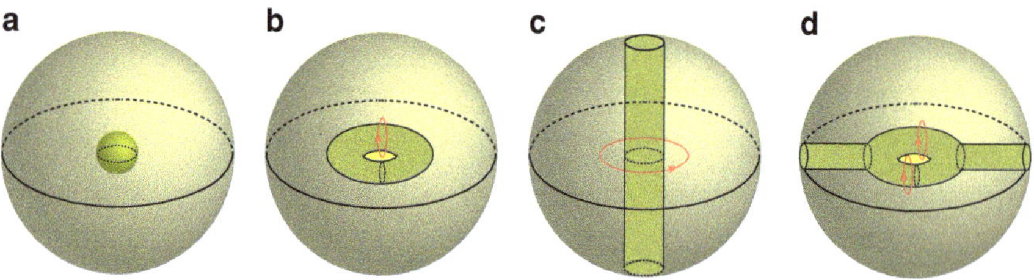

Fig. 1 Balls with (**a**) spherical, (**b**) toroidal, and (**c**) cylindrical holes. (**d**) A ball with a hole consisting of a solid torus attached to two solid cylinders. Betti numbers of these sets are zero, one, one, and two, respectively

This chapter is structured as follows. In Sect. 2 we tersely review differential geometry. This follows by short discussions of presentation of groups, homology and cohomology groups, relative homology groups, the idea of homotopy and the fundamental group, classification of 2-manifolds with boundary, knot theory and the fundamental group of their complements in $\mathbb{R}^3$, and the topology of 3-manifolds in Sect. 3. In Sect. 4 we discuss the kinematics of nonlinear elasticity. In Sect. 5, **F**-compatibility equations for non-simply-connected bodies are discussed. **F**-compatibility equations in the presence of essential (Dirichlet) boundary conditions are also derived. **C**-compatibility equations for non-simply-connected bodies are derived. Several examples are presented. Finally, the necessary and sufficient compatibility equations of linearized elasticity are derived. In Sect. 6, the differential complex of nonlinear elasticity written in terms of the deformation gradient and the first Piola–Kirchhoff stress is discussed. Some applications are also briefly mentioned.

2 Differential Geometry

In this section, we briefly review the differential geometry background needed in the kinematic description of nonlinear elasticity.

Consider a map $\pi : \mathcal{E} \to \mathcal{B}$, where $\mathcal{E}$ and $\mathcal{B}$ are sets. The fiber over $X \in \mathcal{B}$ is defined to be the set $\mathcal{E}_X := \pi^{-1}(X) \subset \mathcal{E}$. If the map π is onto, fibers are non-empty and $\mathcal{E} = \sqcup_{X \in \mathcal{B}} \mathcal{E}_X$, where $\sqcup$ denotes disjoint union of sets. Now assume that $\mathcal{E}$ and $\mathcal{B}$ are manifolds and for any $X \in \mathcal{B}$, there exists a neighborhood $\mathcal{U} \subset \mathcal{B}$ of X, a manifold $\mathcal{F}$, and a diffeomorphism $\psi : \pi^{-1}(\mathcal{U}) \to \mathcal{U} \times \mathcal{F}$ such that $\pi = \mathrm{pr}_1 \circ \psi$, where $\mathrm{pr}_1 : \mathcal{U} \times \mathcal{F} \to \mathcal{U}$ is projection onto the first factor. The triplet $(\mathcal{E}, \pi, \mathcal{B})$ is called a fiber bundle and $\mathcal{E}$, π, and $\mathcal{B}$ are called the total space, the projection, and the base space, respectively. If $\pi^{-1}(X)$ is a vector space, for any $X \in \mathcal{B}$, then $(\mathcal{E}, \pi, \mathcal{B})$ is called a vector bundle. The set of all smooth maps $\sigma : \mathcal{B} \to \mathcal{E}$ such that $\sigma(X) \in \mathcal{E}_X, \ \forall X \in \mathcal{B}$, is called the set of sections of this bundle, and is denoted by $\Gamma(\mathcal{E})$. The tangent bundle of a manifold is an example of a vector bundle for which $\mathcal{E} = T\mathcal{B}$.

A vector field on a manifold $\mathcal{B}$ is a section of the tangent bundle $T\mathcal{B}$ of $\mathcal{B}$. The set of all C^r vector fields on $\mathcal{B}$ is denoted by $\mathfrak{X}^r(\mathcal{B})$ and the set of all C^∞ vector fields by $\mathfrak{X}(\mathcal{B})$. A vector field on $\mathcal{B}$ is an assignment, to each $X \in \mathcal{B}$, of a tangent vector $\mathbf{W}_X \in T_X\mathcal{B}$. Note that for an N-dimensional manifold $\mathcal{B}$, $T_X\mathcal{B}$ is an N-dimensional vector space with a local basis $\left\{\frac{\partial}{\partial X^1}, \ldots, \frac{\partial}{\partial X^N}\right\}$ induced from a local chart $\{X^A\}$. Given a vector field $\mathbf{W}$, for each point $X \in \mathcal{B}$, $\mathbf{W}$ is locally described as

$$\mathbf{W}(X) = \sum_{A=1}^{N} W^A(X) \frac{\partial}{\partial X^A}, \tag{1}$$

where W^A are C^∞ maps. One important role of tangent vectors is the directional differentiation of functions. In other words, a vector field acts on functions by taking their directional derivative, i.e.,

$$\mathbf{W}[f] := \sum_{A=1}^{N} W^A(X) \frac{\partial f(X)}{\partial X^A}. \tag{2}$$

This is the directional or Lie derivative of f along $\mathbf{W}$ and is denoted by $\mathfrak{L}_{\mathbf{W}} f$. Thus, $\mathfrak{L}_{\mathbf{W}} f(X) := \mathbf{W}[f](X) = df(X) \cdot \mathbf{W}(X)$. This is the reason $\mathfrak{L} f = df$ belongs to the cotangent space of $\mathcal{B}$, where the cotangent space $T^*\mathcal{B}$ is defined as $T^*\mathcal{B} := \{\varphi : T\mathcal{B} \to \mathbb{R},\ \varphi \text{ is linear and bounded}\}$.

A linear (affine) connection on a manifold $\mathcal{B}$ is an operation $\nabla : \mathcal{X}(\mathcal{B}) \times \mathcal{X}(\mathcal{B}) \to \mathcal{X}(\mathcal{B})$, where $\mathcal{X}(\mathcal{B})$ is the set of vector fields on $\mathcal{B}$, such that $\forall\, \mathbf{X}, \mathbf{Y}, \mathbf{X}_1, \mathbf{X}_2, \mathbf{Y}_1, \mathbf{Y}_2 \in \mathcal{X}(\mathcal{B}), \forall\, f, f_1, f_2 \in C^\infty(\mathcal{B}), \forall\, a_1, a_2 \in \mathbb{R}$:

1. $\nabla_{f_1\mathbf{X}_1 + f_2\mathbf{X}_2}\mathbf{Y} = f_1 \nabla_{\mathbf{X}_1}\mathbf{Y} + f_2 \nabla_{\mathbf{X}_2}\mathbf{Y}$,
2. $\nabla_{\mathbf{X}}(a_1\mathbf{Y}_1 + a_2\mathbf{Y}_2) = a_1\nabla_{\mathbf{X}}\mathbf{Y}_1 + a_2\nabla_{\mathbf{X}}\mathbf{Y}_2$,
3. $\nabla_{\mathbf{X}}(f\mathbf{Y}) = f\nabla_{\mathbf{X}}\mathbf{Y} + (\mathbf{X}f)\mathbf{Y}$.

$\nabla_{\mathbf{X}}\mathbf{Y}$ is called the covariant derivative of $\mathbf{Y}$ along $\mathbf{X}$. In a local chart $\{X^A\}$, $\nabla_{\partial_A}\partial_B = \Gamma^C{}_{AB}\partial_C$, where $\Gamma^C{}_{AB}$ are the Christoffel symbols of the connection, and $\partial_A = \frac{\partial}{\partial x^A}$ are the natural bases for the tangent space corresponding to a coordinate chart $\{x^A\}$. A linear connection is said to be compatible with a metric $\mathbf{G}$ of the manifold if

$$\nabla_{\mathbf{X}}\langle\!\langle \mathbf{Y}, \mathbf{Z} \rangle\!\rangle_{\mathbf{G}} = \langle\!\langle \nabla_{\mathbf{X}}\mathbf{Y}, \mathbf{Z} \rangle\!\rangle_{\mathbf{G}} + \langle\!\langle \mathbf{Y}, \nabla_{\mathbf{X}}\mathbf{Z} \rangle\!\rangle_{\mathbf{G}}, \tag{3}$$

where $\langle\!\langle ., . \rangle\!\rangle_{\mathbf{G}}$ is the inner product induced by the metric $\mathbf{G}$. A connection ∇ is $\mathbf{G}$-compatible if and only if $\nabla\mathbf{G} = \mathbf{0}$, or in components, $G_{AB|C} = G_{AB,C} - \Gamma^D{}_{CA}G_{DB} - \Gamma^D{}_{CB}G_{AD} = 0$. We consider an N-dimensional manifold $\mathcal{B}$ with the metric $\mathbf{G}$ and a $\mathbf{G}$-compatible connection ∇. The torsion of a connection is a map $\boldsymbol{T} : \mathcal{X}(\mathcal{B}) \times \mathcal{X}(\mathcal{B}) \to \mathcal{X}(\mathcal{B})$ defined by

$$\boldsymbol{T}(\mathbf{X}, \mathbf{Y}) = \nabla_{\mathbf{X}}\mathbf{Y} - \nabla_{\mathbf{Y}}\mathbf{X} - [\mathbf{X}, \mathbf{Y}], \tag{4}$$

where $[\mathbf{X}, \mathbf{Y}] = \mathbf{XY} - \mathbf{YX}$ is the commutator of the vector fields $\mathbf{X}$ and $\mathbf{Y}$. For an arbitrary scalar field f, $[\mathbf{X}, \mathbf{Y}][f] = \mathbf{X}[f]\mathbf{Y} - \mathbf{Y}[f]\mathbf{X}$. In components in a local chart $\{X^A\}$, $T^A{}_{BC} = \Gamma^A{}_{BC} - \Gamma^A{}_{CB}$. The connection ∇ is symmetric if it is torsion-free, i.e., $\nabla_{\mathbf{X}}\mathbf{Y} - \nabla_{\mathbf{Y}}\mathbf{X} = [\mathbf{X}, \mathbf{Y}]$. It can be shown that on any Riemannian manifold $(\mathcal{B}, \mathbf{G})$ there is a unique linear connection (the Levi-Civita connection) ∇, which is compatible with $\mathbf{G}$ and is torsion-free with the Christoffel symbols $\Gamma^C{}_{AB} = \frac{1}{2}G^{CD}(G_{BD,A} + G_{AD,B} - G_{AB,D})$. In a manifold with a connection, the curvature is a map $\boldsymbol{\mathcal{R}} : \mathcal{X}(\mathcal{B}) \times \mathcal{X}(\mathcal{B}) \times \mathcal{X}(\mathcal{B}) \to \mathcal{X}(\mathcal{B})$ defined by

$$\boldsymbol{\mathcal{R}}(\mathbf{X}, \mathbf{Y})\mathbf{Z} = \nabla_{\mathbf{X}}\nabla_{\mathbf{Y}}\mathbf{Z} - \nabla_{\mathbf{Y}}\nabla_{\mathbf{X}}\mathbf{Z} - \nabla_{[\mathbf{X}, \mathbf{Y}]}\mathbf{Z}, \tag{5}$$

or in components, $\mathcal{R}^A{}_{BCD} = \Gamma^A{}_{CD,B} - \Gamma^A{}_{BD,C} + \Gamma^A{}_{BM}\Gamma^M{}_{CD} - \Gamma^A{}_{CM}\Gamma^M{}_{BD}$.

An N-dimensional Riemannian manifold is locally flat if it is isometric to Euclidean space. This is equivalent to vanishing of the curvature tensor [9, 28]. Ricci curvature is defined as $R_{AB} = \mathcal{R}^{C}{}_{ACB}$. The trace of Ricci curvature is called scalar curvature: $\mathsf{R} = R_{AB}G^{AB}$. In dimensions two and three Ricci curvature algebraically determines the entire curvature tensor. In dimension three [25]:

$$\mathcal{R}_{ABCD}=G_{AC}R_{BD}-G_{AD}R_{BC}-G_{BC}R_{AD}+G_{BD}R_{AC}-\frac{1}{2}\mathsf{R}\left(G_{AC}G_{BD}-G_{AD}G_{BC}\right). \tag{6}$$

In dimension two $R_{AB} = \mathsf{R}g_{AB}$, and hence, scalar curvature completely characterizes the curvature tensor and is twice the Gauss curvature.[1]

2.1 Exterior Calculus

We introduce differential forms on an arbitrary manifold $\mathcal{B}$ following [1]. The permutation group on N elements consists of all bijections $\tau : \{1,\ldots,N\} \to \{1,\ldots,N\}$ and is denoted by S_N. For Banach spaces $\mathbb{E}$ and $\mathbb{F}$, a k-multilinear mapping $t \in L^k(\mathbb{E};\mathbb{F})$, i.e., $t : \mathbb{E}\times\mathbb{E}\times\ldots\times\mathbb{E} \to \mathbb{F}$ is called skew-symmetric if

$$t(e_1,\ldots,e_k) = (\text{sign}\,\tau)t(e_{\tau(1)},\ldots,e_{\tau(k)}), \qquad \forall e_1,\ldots,e_k \in \mathbb{E},\ \tau \in S_k, \tag{7}$$

where sign τ is $+1$ (-1) if τ is an even (odd) permutation. The subspace of skew-symmetric elements of $L^k(\mathbb{E};\mathbb{F})$ is denoted by $\Lambda^k(\mathbb{E},\mathbb{F})$. Elements of $\Lambda^k(\mathbb{E},\mathbb{F})$ are called exterior k-forms. Wedge product of two exterior forms $\alpha \in \Lambda^k(\mathbb{E},\mathbb{F})$ and $\beta \in \Lambda^l(\mathbb{E},\mathbb{F})$ is a $(k+l)$-form $\alpha\wedge\beta \in \Lambda^{k+l}(\mathbb{E},\mathbb{F})$, which is defined in components as

$$(\alpha\wedge\beta)_{i_1\ldots i_{k+l}} = \sum_{(k,l)\in S_{K+l}} (\text{sign}\,\tau)\alpha_{\tau(i_1)\ldots\tau(i_k)}\beta_{\tau(i_{k+1})\ldots\tau(i_{k+l})}. \tag{8}$$

For a manifold $\mathcal{B}$, the vector bundle of exterior k-forms on $T\mathcal{B}$ is denoted by $\Lambda^k_{\mathcal{B}} : \Lambda^k(\mathcal{B}) \to \mathcal{B}$. In a local coordinate chart a differential k-form α has the following representation

$$\omega = \sum_{I_1<I_2<\ldots<I_k} \omega_{I_1 I_2\ldots I_k} dX^{I_1}\wedge\ldots\wedge dX^{I_k}, \qquad I_1, I_2\ldots, I_k \in \{1,2,\ldots,N\}, \tag{9}$$

where $\omega_{I_1 I_2\ldots I_k}$ are C^∞ maps. The space of k-forms on $\mathcal{B}$ is denoted $\Omega^k(\mathcal{B})$. Let

[1] It is known that the necessary compatibility equations for the right Cauchy–Green strain $\mathbf{C}^\flat$ in 2D and 3D are written as $\mathsf{R}(\mathbf{C}^\flat) = 0$ and $\boldsymbol{R}(\mathbf{C}^\flat) = \mathbf{0}$, respectively, i.e., in 2D there is only one compatibility equation while in 3D there are six. Note also that the Bianchi identities do not reduce the number of compatibility equations.

$$\Omega(\mathcal{B}) = \bigoplus_{k=0,1,\ldots} \Omega^k(\mathcal{B}), \tag{10}$$

with its structure as a real vector space and multiplication $\wedge$. $\Omega(\mathcal{B})$ is called the algebra of exterior differential forms on $\mathcal{B}$.

Let U be an open subset of an N-manifold $\mathcal{B}$. Consider the unique family of mappings $d^k(U) : \Omega^k(U) \to \Omega^{k+1}(U)$ $(k = 0, 1, \ldots, N)$ merely denoted d with the following properties:

1. $d(\alpha \wedge \beta) = d\alpha \wedge \beta + (-1)^k \alpha \wedge d\beta$, $\forall \alpha \in \Omega^k(U)$, $\beta \in \Omega^l(U)$,
2. If $f \in \Omega^0(U)$, df is the (usual) differential of f,
3. $d^2 = d \circ d = 0$ (i.e., $d^{k+1}(U) \circ d^k(U) = 0$),
4. d is a local operator (natural with respect to restrictions), i.e., if $U \subset V \subset \mathcal{B}$ are open and $\alpha \in \Omega^k(V)$, then $d(\alpha|_U) = (d\alpha)|_U$.

In component form, for the differential form in (9) one writes

$$d\omega = \frac{\partial\, \omega_{I_1 I_2 \ldots I_k}}{\partial X^J} dX^J \wedge dX^{I_1} \wedge \ldots \wedge dX^{I_k}, \tag{11}$$

where summation over repeated indices is implied.

For an N-manifold $\mathcal{B}$, $\dim[\Lambda^k(\mathcal{B})] = \binom{N}{k} = \binom{N}{N-k} = \dim[\Lambda^{N-k}(\mathcal{B})]$. This shows that $\Lambda^k(\mathcal{B})$ and $\Lambda^{N-k}(\mathcal{B})$ should be isomorphic to each other. The natural isomorphism is the Hodge star operator. Hodge star is the unique isomorphism $* : \Lambda^k(\mathcal{B}) \to \Lambda^{N-k}(\mathcal{B})$ satisfying

$$\alpha \wedge *\beta = \langle\!\langle \alpha, \beta \rangle\!\rangle_{\mathbf{G}}\, \mu, \quad \forall\, \alpha, \beta \in \Lambda^k(\mathcal{B}), \tag{12}$$

where $\langle\!\langle , \rangle\!\rangle_{\mathbf{G}}$ and μ are the standard Riemannian inner product and the standard volume element on $\mathcal{B}$, respectively. As an example, $\Lambda^1(\mathbb{R}^3)$ and $\Lambda^2(\mathbb{R}^3)$ are both three dimensional and $* : \Lambda^1(\mathbb{R}^3) \to \Lambda^2(\mathbb{R}^3)$ is defined by

$$e^1 \mapsto e^2 \wedge e^3, \;\; e^2 \mapsto e^3 \wedge e^1, \text{ and } e^3 \mapsto e^1 \wedge e^2. \tag{13}$$

The codifferential operator $\boldsymbol{\delta} : \Omega^{k+1}(\mathcal{B}) \to \Omega^k(\mathcal{B})$ is defined by

$$\begin{aligned} &\boldsymbol{\delta}(\Omega^0(\mathcal{B})) = 0, \\ &\boldsymbol{\delta}\alpha = (-1)^{Nk+1} * d * \alpha, \quad \forall\, \alpha \in \Omega^{k+1}(\mathcal{B}),\ k = 0, 1, \ldots, N-1. \end{aligned} \tag{14}$$

This is the adjoint of d with respect to $\langle\!\langle , \rangle\!\rangle_{\mathbf{G}}$. For an oriented smooth N-manifold $\mathcal{B}$ with boundary $\partial\mathcal{B}$ and $\alpha \in \Omega^{N-1}(\mathcal{B})$, Stokes' theorem states that

$$\int_{\partial\mathcal{B}} \alpha = \int_{\mathcal{B}} d\alpha, \tag{15}$$

assuming that both integrals exist.

3 Algebraic Topology

To make this chapter self-contained, we next tersely review some notation and facts from algebraic topology and also refer the reader to the relevant literature for more details.

3.1 *Homology and Cohomology Groups*

An r-form ω is closed if $d\omega = 0$ and it is exact if there exists an $(r-1)$-form α such that $\omega = d\alpha$. An exact differential form is closed, and from Poincaré's lemma a closed form is locally exact. However, globally a closed differential form may not be exact. Cohomology aims in finding the topological obstructions to exactness. This turns out to be directly related to the compatibility equations of elasticity. In the following we mainly follow [18, 22, 24, 37, 50].

3.1.1 Group Theory

For two Abelian groups $(G_1, .)$ and $(G_2, .)$, a map $f : G_1 \to G_2$ is a homomorphism if

$$f(x.y) = f(x).f(y), \quad \forall x, y \in G_1. \tag{16}$$

Our notation is flexible here; we use $x.y$ and xy interchangeably. If in addition f is a bijection, it is an isomorphism, G_1 and G_2 are said to be isomorphic, and this is denoted by $G_1 \cong G_2$. Let $H \subset G$ be a subgroup. If $xy^{-1} \in H$, then $x, y \in G$ are called equivalent and we write $x \sim y$. The equivalence class of x is denoted by $[x]$. G/H is the quotient space—the set of equivalence classes—and $[x].[y] = [xy]$. If $ghg^{-1} \in H, \forall g \in G, h \in H$, H is called a normal subgroup. For a normal subgroup H, G/H is always a subgroup called the quotient group. For a homomorphism $f : G_1 \to G_2$, $\operatorname{Ker} f$ and $\operatorname{Im} f$ are subgroups of G_1 and G_2, respectively, where

$$\operatorname{Ker} f = \{x \in G_1 | f(x) = 1\}, \quad \operatorname{Im} f = \{x \in G_2 | x \in f(G_1) \subset G_2\}, \tag{17}$$

and 1 is the identity element of G_2. The isomorphism theorem of group theory tells us that $G_1/\operatorname{Ker} f \cong \operatorname{Im} f$.

Let $(G, .)$ be an Abelian group, i.e., $x.y = y.x, \ \forall\, x, y \in G$. If there exist $g_1, \ldots, g_n \in G$ such that

$$g = g_1^{\lambda_1} \ldots g_n^{\lambda_n}, \quad \forall\, g \in G, \lambda_i \in \mathbb{Z}, \tag{18}$$

then G is called a finitely generated Abelian group with generators $g_1, \ldots, g_n$. If in addition

$$g = g_1^{\lambda_1} \ldots g_n^{\lambda_n} = 1 \;\Rightarrow\; \lambda_1 = \ldots = \lambda_n = 0, \tag{19}$$

G is called a free finitely generated Abelian group, and $g_1, \ldots, g_n$ are called free generators or a basis. It can be shown that $(G, .)$ is a free finitely generated Abelian group if and only if every g has a unique representation with respect to the basis $\{g_1, \ldots, g_n\}$.

Suppose $S = \{s_1, \ldots, s_k\}$ is a set of distinct elements. Let $\tilde{S}$ be the set of expressions of the form $\tilde{s} = \prod_{i=1}^k s_i^{\lambda_i}$, where $\lambda_i \in \mathbb{Z}$. Then $\prod_{i=1}^k s_i^{\lambda_i} = \prod_{i=1}^k s_i^{\mu_i}$ if and only if $\lambda_i = \mu_i, i = 1, \ldots, k$. Multiplication is defined as

$$\prod_i s_i^{\lambda_i} \prod_i s_i^{\mu_i} = \prod_i s_i^{\lambda_i + \mu_i}. \tag{20}$$

$\tilde{S}$ is a free finitely generated Abelian group with basis$\{s_1^1 s_2^0 \ldots s_k^0, \ldots, s_1^0 \ldots s_{k-1}^0 s_k^1\}$. $\tilde{S}$ is called the free finitely generated Abelian group on S. If G is an Abelian group, $g \in G$ has finite order if $g^n = 1$ for some $n \in \mathbb{N}$. The set of all elements of finite order in G is a subgroup called the torsion subgroup T of G. If T is trivial, i.e., $T = \{1\}$, G is called torsion-free. Any free Abelian group is torsion-free. For $x, y \in G$, and G a group, $[x, y] = xyx^{-1}y^{-1} \in G$ is called the commutator of x and y. $[G, G]$ is a normal subgroup of G generated by all commutators. Note that $G/[G, G]$ is an Abelian group.

The direct sum of two groups A and B is the set of pairs $(a, b), a \in A, b \in B$ and is denoted by $A \oplus B$. Group multiplication in $A \oplus B$ is defined as

$$(a_1, b_1).(a_2, b_2) = (a_1 a_2, b_1 b_2), \quad \forall a_1, a_2 \in A, \; \forall b_1, b_2 \in B. \tag{21}$$

Generalization of this to any finite number of groups is straightforward.

3.1.2 Combinatorial Group Theory

In combinatorial group theory one studies groups that are described by generators and some defining relations. Here we mainly follow [8] and [50]. If $X \subset G$, the smallest subgroup of G containing X is denoted by $\langle X \rangle$ and is characterized as

$$\langle X \rangle = \{g \in G |\; g = x_1^{\epsilon_1} x_2^{\epsilon_2} \ldots x_k^{\epsilon_k},\; x_i \in X, \epsilon_i = \pm 1\}. \tag{22}$$

$x_1^{\epsilon_1} x_2^{\epsilon_2} \ldots x_k^{\epsilon_k}$ is called an X-word or simply a word. A word is reduced if $x_i = x_{i+1}$ implies that $\epsilon_i + \epsilon_{i+1} \neq 0,\; i = 1, \ldots, k-1$. For example, the word $x_1^{-1} x_1^{-1} x_2 x_2^{-1} x_1 x_1 x_1 x_2$ is not reduced while $x_1 x_2$ is reduced. If $G = \langle X \rangle$ and every non-empty reduced X-word $w \neq_G 1$, X is called a free group. In this case, two reduced X-words have equal values in G if and only if they are identical. A group is finitely generated if it can be generated by a finite set. If G is a freely generated group by X, then for any group H and map $\psi : X \to H$, there is a unique homomorphism $\varphi : G \to H$ such that $\varphi|_X = \psi$. For a group G, and $X \subset G$, the normal closure of X in G (the smallest normal subgroup of G containing X) is defined as

$$\mathsf{gp}_G(X) = \Big\langle \{g^{-1}xg \mid g \in G, x \in X\} \Big\rangle. \tag{23}$$

If F is a free group on $X \subset G$ and $\psi : X \to G$, a map such that $G = \langle \psi(X) \rangle$, then the extension of this map $\varphi : F \to G$ has kernel $K = \mathsf{gp}_F(R)$, where $R \subset F$. Then one writes $G = \langle X; R \rangle$ and this is called a presentation for G, which comes with an implicit map $\psi : X \to G$, the presentation map. Elements of R are called defining relators. A group is finitely presented if it has a finite presentation, i.e., if both X and R are finite.

Any normal subgroup of a group G consists of elements expressed by words of the following form

$$\prod_{i=1}^{n} g_i x_{j_i}^{\epsilon_i} g_i^{-1}, \quad g_i, x_{j_i} \in G, \ \epsilon_i = \pm 1. \tag{24}$$

This normal subgroup is said to be generated by $x_1, x_2, \ldots \in G$ and is denoted by $\mathsf{gp}_G(\{x_1, x_2, \ldots\})$ as in (23). Dyck's theorem says that the group $\langle X, R \rangle$ is the quotient of $F = \langle X \rangle$ by its normal subgroup $\mathsf{gp}_G(R)$.

3.1.3 Chain Complexes and Homology Groups

Let $\{v_0, \ldots, v_k\}$ be a geometrically independent set in $\mathbb{R}^N$, i.e., $\{v_1 - v_0, \ldots, v_k - v_0\}$ is a set of linearly independent vectors in $\mathbb{R}^N$. A k-simplex σ^k is defined as

$$\sigma^k = \left\{ x \in \mathbb{R}^N \,\middle|\, x = \sum_{i=0}^{k} t_i v_i, \text{ where } 0 \leq t_i \leq 1, \sum_{i=0}^{k} t_i = 1 \right\}. \tag{25}$$

The numbers t_i are uniquely determined by x and are called barycentric coordinates of the point x of σ with respect to vertices $v_0, \ldots, v_k$. The number k is the dimension of σ^k. A simplicial complex K in $\mathbb{R}^N$ is a collection of simplices in $\mathbb{R}^N$ such that (1) every face of a simplex of K is in K, and (2) the intersection of any two simplices is either empty or a face of each of them. The largest dimension of the simplices of K is called the dimension of K. A subcomplex of K is a subcollection of K that contains all faces of its elements.

Suppose K is an oriented simplicial complex of dimension n. Let α_p be the number of p-simplices of K, $0 \leq p \leq n$. Let $\{\sigma_p^1, \ldots, \sigma_p^{\alpha_p}\}$ be the set of p-

simplices of K. The pth chain group of K with integer coefficients is denoted by $C_p(K)$ and is a free Abelian group on the set $\{\sigma_p^1, \ldots, \sigma_p^{\alpha_p}\}$, i.e.,[2]

$$\sigma \in C_p(K), \quad \sigma = \sum_{i=1}^{\alpha_p} \lambda_i \sigma_p^i, \quad \lambda_i \in \mathbb{Z}. \tag{26}$$

For $p > n$ or $p < 0$, $C_p(K) = 0$. Let $\sigma = (v^0, \ldots, v^p)$ be an oriented p-simplex of K. Then, the boundary of σ is defined as

$$\partial\sigma = \partial_p\sigma = \sum_{i=0}^{p} (-1)^i (v^0, \ldots, \hat{v^i}, \ldots, v^p), \tag{27}$$

where hat over v^i indicates omission of v^i. The boundary homomorphism $\partial_p : C_p(K) \to C_{p-1}(K)$ is defined as

$$\partial_p \left(\sum \lambda_i \sigma_p^i\right) = \sum_i \lambda_i \partial_p(\sigma_p^i). \tag{28}$$

Note that for any p, $\partial \circ \partial = \partial_{p-1} \circ \partial_p = 0$. Note also that $\operatorname{Im} \partial_{p+1} \subset \operatorname{Ker} \partial_p$. $Z_p = \operatorname{Ker} \partial_p$ is the set of p-cycles and $B_p = \operatorname{Im} \partial_{p+1}$ is the set of p-boundaries. $H_p(K) = Z_p(K)/B_p(K)$ is a finitely generated Abelian group and quantifies the non-bounding p-cycles of K. This is called the pth homology group of K (with integer coefficients). Note that $H_n(K) = Z_n(K)$ is free Abelian. Two p-cycles z and $z' \in Z_p(K)$ are homologous ($z \sim z'$) if $z - z' \in B_p(K)$. It is a fact that homology groups are topological invariants, i.e., two homeomorphic topological spaces have isomorphic homology groups. For a simplicial complex, the set of simplices as subsets of $\mathbb{R}^m$ ($m \leq n$) is called the polyhedron $|K|$ of K. For a topological space X, if there exists a simplicial complex K and a homeomorphism $f : |K| \to X$, X is said to be triangulable and (K, f) is called a triangulation of X. For a triangulable topological space X, given an arbitrary triangulation (K, f), $H_r(X) := H_r(K), r = 0, 1, \ldots$.[3]

Example 3.1 Circle S^1 is not the boundary of any 2-chain, and hence, $H_1(S^1)$ is generated by the circle itself (only one generator), i.e., $H_1(S^1) = \mathbb{Z}$. S^1 is connected, and hence, $H_0(S^1) = \mathbb{Z}$. A similar example is the punctured plane $\mathbb{R}^2\backslash(0, 0)$, which is connected and its first homology group is generated by any simple closed curve circling the origin once.

---

[2]Here, we find it more convenient to use an additive notation. Also, to be more specific we should denote the group by $C_p(K; \mathbb{Z})$ to emphasis that it has integer coefficients.

[3]Note that the homology groups are independent of triangulations. Note also that not every space can be triangulated. For such spaces one can still define homology, e.g., singular and Čech homologies.

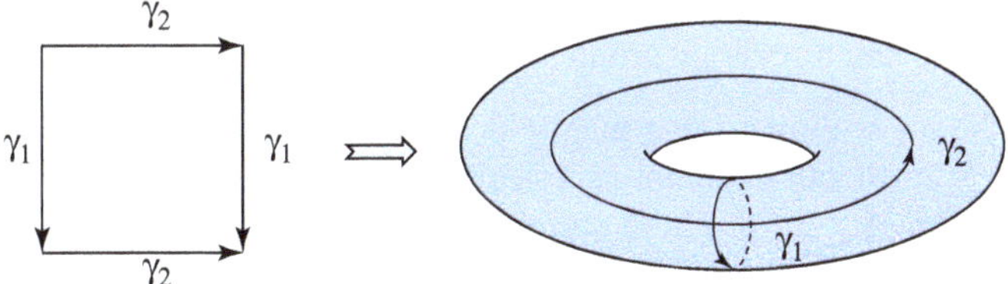

Fig. 2 A torus can be constructed from a square by the identifications shown above. γ_1 and γ_2 are generators of the first homology and first homotopy groups

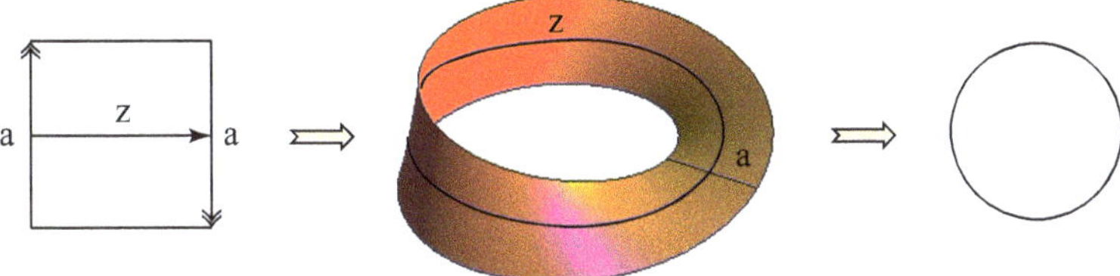

Fig. 3 Möbius band and its deformation retract to a circle

Example 3.2 Torus T^2 is not a boundary of any 3-chain. Thus, $H_2(T^2)$ is freely generated by one generator, the surface itself, i.e., $H_2(T^2) \cong \mathbb{Z}$. T^2 is connected, and hence, $H_0(T^2) \cong \mathbb{Z}$. $H_1(T^2)$ is freely generated by the loops γ_1 and γ_2 (see Fig. 2), and hence, $H_1(T^2) \cong \mathbb{Z} \oplus \mathbb{Z}$. The group presentation can be written as $\pi_1(T^2) = \langle \gamma_1, \gamma_2 \rangle$. For a torus of genus g (the number of closed cuts that leave the torus path-connected)

$$H_1(\Sigma_g) \cong \underbrace{\mathbb{Z} \oplus \mathbb{Z} \oplus \ldots \oplus \mathbb{Z}}_{2g}. \tag{29}$$

Example 3.3 Möbius band is constructed from a square by the identification shown in Fig. 3. z is a generator of the first homology group $H_1(\mathcal{M}, \mathbb{Z})$, i.e., $H_1(\mathcal{M}, \mathbb{Z}) = \mathbb{Z}$.

Remark 3.4 Note that $Z_r(K)$ and $B_r(K)$ are both free Abelian groups as they are both subgroups of a free Abelian group $C_r(K)$. However, this does not imply that $H_r(K)$ is also free Abelian. From the fundamental theorem of finitely generated Abelian groups one has

$$H_1(K; \mathbb{Z}) \cong \underbrace{\mathbb{Z} \oplus \mathbb{Z} \oplus \ldots \oplus \mathbb{Z}}_{f} \oplus \underbrace{\mathbb{Z}_{k_1} \oplus \ldots \oplus \mathbb{Z}_{k_p}}_{\text{torsion subgroup}}, \tag{30}$$

where $k_1, \ldots, k_p$ are integers, k_{i+1} divides k_i ($i = 1, \ldots, p-1$), and $\mathbb{Z}_{k_i} = \mathbb{Z}/k_i\mathbb{Z}$ is the set of integers modulo k_i. f is called the rank of $H_1(K; \mathbb{Z})$ or the first Betti number and p is called the torsion number. The torsion subgroup contains all the elements of the first homology group that have finite order.

Let M be an m-dimensional manifold and let σ_r be an r-simplex in $\mathbb{R}^m$, and $f : \sigma_r \to M$ a smooth map, not necessarily invertible. $s_r = f(\sigma_r) \subset M$ is called a singular r-simplex in M (these simplices do not provide a triangulation of M). Given the set of r-simplices $\{s_i^r\}$ in M, an r-chain in M is defined as

$$c = \sum_i a_i s_i^r, \quad a_i \in \mathbb{R}. \tag{31}$$

The r-chains in M form the chain group $C_r(M)$ with real coefficients. The boundary of a singular r-simplex s_r is defined as $\partial s_r := f(\partial \sigma_r)$. The boundary and cycle groups $B_r(M)$ and $Z_r(M)$ are defined similar to those of simplicial complexes. The singular homology group is defined as $H_r(M) := Z_r(M)/B_r(M)$. The singular homology group is isomorphic to the corresponding simplicial homology group with $\mathbb{R}$-coefficients.

3.1.4 Cohomology Groups

Integration of an r-form ω over an r-chain in M is defined as

$$\int_{s_r} \omega = \int_{\sigma_r} f^* \omega, \tag{32}$$

where $f^*\omega$ is the pull-back of ω under f. For $c = \sum_i a_i s_i^r \in C_r(M)$:

$$\int_c \omega = \sum_i a_i \int_{s_i^r} \omega. \tag{33}$$

The set of closed r-forms (rth cocycle group) is denoted by $Z^r(M)$. The set of exact r-forms (the rth coboundary group with real coefficients) is denoted by $B^r(M)$. The rth de Rham cohomology group of M is defined as

$$H^r(M; \mathbb{R}) := Z^r(M)/B^r(M). \tag{34}$$

For $\omega \in Z^r(M)$, $[w] \in H^r(M)$ (the equivalence class of ω) is defined as

$$[\omega] = \{\omega' \in Z^r(M) | \omega' = \omega + d\psi,\ \psi \in \Omega^{r-1}(M)\}, \tag{35}$$

where $\Omega^{r-1}(M)$ is the set of $(r-1)$-forms on M.

Example 3.5 The first cohomology group of the unit circle $S^1 = \{e^{i\theta} | 0 \leq \theta < 2\pi\}$ is calculated as follows. Let ω and ω' be closed forms ($d\omega = d\omega' = 0$) that are not exact. Note that $\omega' - a\omega$ is exact when $a = \int_0^{2\pi} \omega' / \int_0^{2\pi} \omega$. Thus, given ω such that $d\omega = 0$, any closed 1-form ω' is cohomologous to $a\omega$ for some $a \in \mathbb{R}$. Hence, each cohomology class is given by a real number a. Therefore, $H^1(S^1) = \mathbb{R}$.

The period of a closed r-form ω over a cycle c is defined as $(c, \omega) = \int_c \omega$. For $[c] \in H_r(M)$, $[\omega] \in H^r(M)$ define

$$\Lambda([c], [\omega]) := (c, \omega) = \int_c \omega. \tag{36}$$

We note that both $\Lambda(., [\omega]) : H_r(M) \to \mathbb{R}$, and $\Lambda([c], .) : H^r(M) \to \mathbb{R}$ are linear maps. De Rham's theorem [16, 23] says that if M is a compact manifold, $H_r(M)$ and $H^r(M)$ are finite-dimensional and the map $\Lambda : H_r(M) \times H^r(M) \to \mathbb{R}$ is bilinear and non-degenerate. Hence, $H^r(M)$ is the dual vector space of $H_r(M)$. As a corollary of de Rham's theorem, for a compact manifold M, let $b_r = \dim H_r(M; \mathbb{R})$ be its rth Betti number. Let $c_1, \ldots, c_{b_r}$ be generators of $Z_r(M)$. Then, a closed r-form ψ is exact if and only if[4]

$$\int_{c_i} \psi = 0, \quad i = 1, \ldots, b_r. \tag{37}$$

Note that $\Lambda([c_i], .) : H^r(M) \to \mathbb{R}$ is non-degenerate, and hence, $\Lambda([c_i], [\omega]) = 0$ implies $[\omega] = 0$, i.e., the cohomology class of exact forms. Duff [20] generalized this theorem to manifolds with boundary.[5]

3.1.5 Relative Homology Groups

The relative homology groups were introduced by S. Lefschetz [29]. These are important in problems with boundary conditions and also appear in duality theorems. Let K be an oriented simplicial complex of dimension n and $L \subset K$. The pth chain group of K modulo L (the pth relative chain group) is the subgroup of $C_p(K)$ in which the coefficient of every simplex of L is zero. This is denoted by $C_p(K, L) \subset C_p(K)$. Let us define a homomorphism $j = j_q : C_q(K) \to C_q(K, L)$, which changes to zero the coefficient of every simplex in L. The relative boundary homomorphism $\tilde{\partial} = \tilde{\partial}_p : C_p(K, L) \to C_{p-1}(K, L)$ is defined as

$$\tilde{\partial} c = j_{p-1}(\partial_p c), \qquad \forall c \in C_p(K, L). \tag{38}$$

[4] This was conjectured by Cartan in 1928 and was proved later on by de Rham [18]. This theorem can be summarized as follows. If for a closed form ω, $(c, \omega) = 0$ for all p-cycles, then ω is exact. If for a p-cycle c, $(c, \omega) = 0$ for all closed p-forms, then c is a boundary.

[5] Duff [20] showed that a closed form with zero relative periods in $H_1(M, \partial M)$ is a closed relative form, i.e., a closed form with compact support in M.

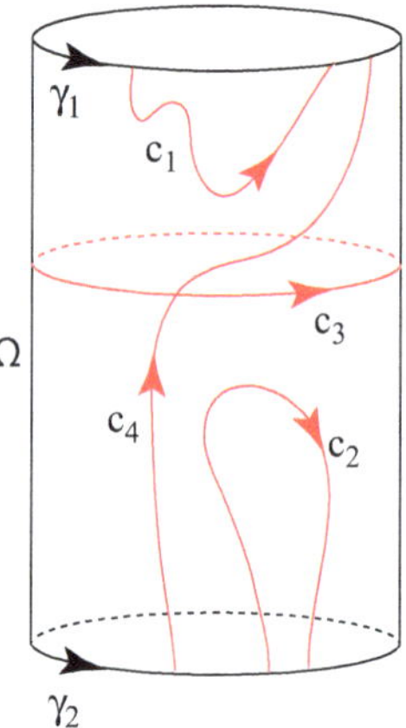

Fig. 4 A cylinder $\Omega = S^1 \times [0, 1]$. $S = \partial\Omega = \gamma_1 \cup \gamma_2$ has two components. c_1 and c_2 are relative boundaries, c_3 generates $H_1(\Omega)$, and c_4 is a relative cycle but not a relative boundary; it generates $H_1(\Omega, \partial\Omega)$.

Note that $\tilde{\partial}_p = j_{p-1} \circ \partial_p \circ i_p$, where $i_p : C_p \to C_p(K)$ is the inclusion map. Note also that for any p, $\tilde{\partial} \circ \tilde{\partial} = \tilde{\partial}_{p-1} \circ \tilde{\partial}_p = 0$.

Let Ω be a compact manifold and $S \subset \Omega$ a compact subset. $C_*(\Omega) = \{C_p(\Omega), \partial_p\}$ is the chain complex corresponding to Ω and for $S \subset \Omega, C_*(S) = \{C_p(S), \partial'_p\}$, where $C_p(S) \subset C_p(\Omega)$, $\forall p$, is the chain complex associated with S. The relative chain group is defined as

$$C_p(\Omega, S) := C_p(\Omega)/C_p(S) = \{c + C_p(S)\},\ c \in C_p(\Omega). \tag{39}$$

The induced boundary operator $\partial''_p : C_p(\Omega)/C_p(S) \to C_{p-1}(\Omega)/C_{p-1}(S)$ is defined the obvious way. $Z_p(\Omega, S) = \operatorname{Ker} \partial''_p$ is the group of relative p-cycles modulo S and $B_p(\Omega, S) = \operatorname{Im} \partial''_{p+1}$ is the group of relative p-boundaries of Ω modulo S. Note that z is a relative p-cycle if its boundary lies in S and b is a relative p-boundary if it is homologous to some p-chain in S. In Fig. 4, four paths on a cylinder are shown. c_1 and c_2 are relative boundaries, i.e., are elements of $B_1(\Omega, \partial\Omega)$, $c_3 \in H_1(\Omega)$, and $c_4 \in H_1(\Omega, \partial\Omega)$.

$C^p(\Omega, S)$ is defined to be the set of linear combinations of p-forms whose support lies in $\Omega\backslash S$. For $z \in Z^p_c(\Omega\backslash S)$, $\int_z \omega$ is the relative period of ω on z, where $Z^p_c(\Omega\backslash S)$ is the set of closed p-forms with compact support in $\Omega\backslash S$. Suppose M is a manifold with boundary ∂M. If a closed p-form has zero relative periods in M, then α is an exact relative p-form [20].

3.1.6 Duality Theorems in Algebraic Topology

The following duality theorems are useful in nonlinear elasticity applications.

- Poincaré duality: For an orientable n-manifold M without boundary, $H^p_c(M) \cong H_{n-p}(M)$, where $H^p_c(M) := Z^p_c(M)/B^p_c(M)$, and $Z^p_c(M)$ and $B^p_c(M)$ are the closed and exact p-forms with compact supports in M, respectively. For compact manifolds from de Rham's theorem $H_p(M) \cong H_{n-p}(M)$.

- Lefschetz duality: For a compact n-manifold M, $H_c^{n-p}(M) \cong H_p(M, \partial M)$. From de Rham's theorem, $H_{n-p}(M) \cong H_c^p(M \backslash \partial M)$. Therefore, $H_{n-p}(M) \cong H_p(M, \partial M)$.[6] Thus, $b_{n-p}(M) = b_p(M, \partial M)$.
- Poincaré–Lefschetz duality: For a compact, orientable n-manifold M with boundary (for $0 \leq k \leq n$), $H^k(M; \mathbb{Z}) \cong H_{n-k}(M, \partial M; \mathbb{Z})$. This holds for any Abelian coefficient group as well.
- Alexander duality: For a closed subset M of an n-manifold Q, $H^p(M) \cong H_{n-p}(Q, Q \backslash M)$. In elasticity applications, $Q = \mathbb{R}^3$. It can be shown that for $p \neq 2$, $H^p(M) \cong H_{2-p}(\mathbb{R}^3 \backslash M)$, and $\mathbb{R} \otimes H^2(M) \cong H_0(\mathbb{R}^3 \backslash M)$ [24]. Thus, for $p \neq 2$, $b_p(M) = b_{2-p}(\mathbb{R}^3 \backslash M)$, and $1 + b_2(M) = b_0(\mathbb{R}^3 \backslash M)$.

Let us now restrict ourselves to embedded 3-submanifolds of $\mathbb{R}^3$,[7] which model our three-dimensional deformable bodies in elasticity. $H_0(M)$ is generated by equivalence classes of points in M; two points are in the same equivalence class if they can be connected to each other by a continuous path in M. $H_1(M)$ is generated by equivalent classes of oriented loops; two loops are in the same equivalence class if their "difference" is the boundary of an oriented surface in M. $H_1(M, \partial M)$ is generated by the equivalence class of oriented paths with end points on ∂M; two paths are equivalent if their "difference" (augmented by paths on ∂M if necessary) is the boundary of an oriented surface in M. From Poincaré duality we know that

$$H_0(M) \cong H_3(M, \partial M), \tag{40}$$

$$H_1(M) \cong H_2(M, \partial M), \tag{41}$$

$$H_2(M) \cong H_1(M, \partial M), \tag{42}$$

$$H_3(M) \cong H_0(M, \partial M). \tag{43}$$

Define $M^c = \mathbb{R}^3 \backslash M$. From Alexander duality one has

$$H_0(M) \cong H_2(M^c), \quad H_1(M) \cong H_1(M^c), \quad H_0(M^c) \cong \mathbb{R} \otimes H_2(M). \tag{44}$$

Let $\Sigma_1, \ldots, \Sigma_k$ be a family of surfaces in M with boundaries on ∂M such that they generate $H_2(M, \partial M)$. As an example, consider the solid torus with two holes shown in Fig. 5 for which $k = 2$. Let $\gamma_1, \ldots, \gamma_k$ be loops in the interior of M that generate

[6] Love [31] in Article 156 writes: "Now suppose the multiply-connected region to be reduced to a simply-connected one by means of a system of barriers." A "barrier" Ω in a three-dimensional body $\mathcal{B}$ is a generator of $H_2(\mathcal{B}, \partial \mathcal{B}) \cong H_1(\mathcal{B})$, and in a two-dimensional body it is a generator of $H_2(\mathcal{B}, \partial \mathcal{B}) \cong H_1(\mathcal{B})$.

[7] Cantarella et al. [11] present an elementary exposition of homology theory with applications to vector calculus. The reader may find their exposition useful.

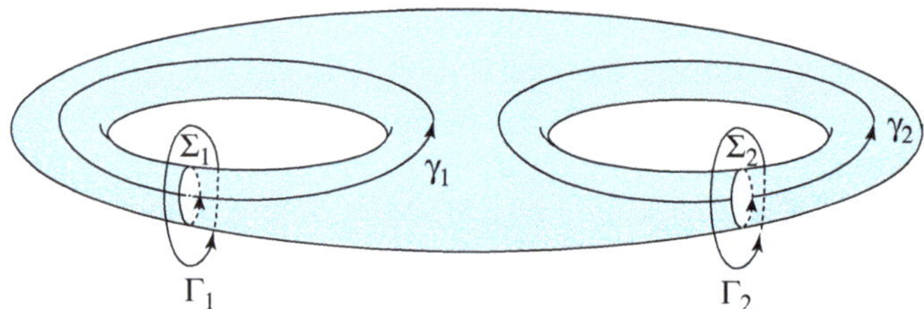

Fig. 5 A two-hole solid torus M. The closed curves γ_1 and γ_2 are generators of $H_1(M)$. Γ_1 and Γ_2 are generators of $H_1(\mathbb{R}^3\backslash M)$

$H_1(M)$ chosen such that intersection number of c_i with Σ_j is δ_{ij}.[8] These loops can be chosen to be disjoint. If one pushes the boundaries of $\Sigma_1, \ldots, \Sigma_k$ slightly into M^c, one obtains the loops $\Gamma_1, \ldots, \Gamma_k$ that generate $H_1(M^c)$.

3.2 *Homotopy and the Fundamental Group*

Fundamental group was introduced by Poincaré in 1895 and plays an important role in understanding compatibility equations. It is much easier to define compared to homology groups but it is much harder to calculate, in general. A path in a topological space X is a map $c : [0, 1] \to X$. It is simple if it is one-to-one. A closed path (loop) has the same end points, i.e., $c(0) = c(1)$, which is called the base point of the loop. A cycle is a continuous map $\gamma : S^1 \to X$. It is different from a loop because in a cycle no end points are distinguished. Two paths c_1 and c_2 having the same end points are homotopic if there is a continuous family of paths whose end points are the same as those of c_1 and c_2. Roughly speaking, the set of equivalent paths based at x_0 constitute the fundamental group $\pi_1(X, x_0)$. An isotopy between c_1 and c_2 is a homotopy for which the curves remain simple during the whole deformation process from c_1 to c_2. Note that two homotopic simple paths are not necessarily isotopic. We make these notions more precise in the following.

Consider a topological space X and a base point $x_0 \in X$. Two loops based at x_0 are equivalent if one loop can be continuously deformed to the other loop. A loop based at x_0 is a continuous map $f : I = [0, 1] \to X$ such that $f(0) = f(1) = x_0$. Two loops f, g are called homotopic if there is a continuous function $F : I \times I \to X$ such that $F(s, 0) = f(s)$, $F(s, 1) = g(s)$, $F(0, t) = F(1, t) = x_0$. F is a homotopy between f and g and this is denoted by $f \sim_F g$. It can be shown that homotopy gives an equivalence relation on loops based at x_0. The equivalence class of f is denoted by $[f]$ and the equivalence classes are elements of the fundamental group $\pi_1(X, x_0)$. Group multiplication is defined as $[f][g] = [fg]$, where fg is defined

[8]This is possible as a consequence of Poincaré duality.

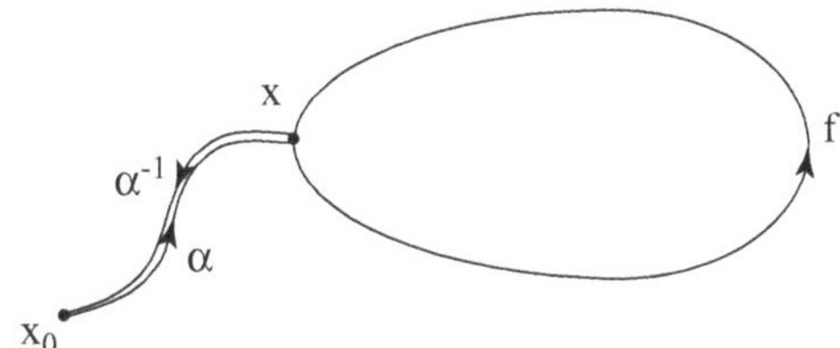

Fig. 6 Having a loop f based at x a loop $\alpha\gamma\alpha^{-1}$ based at x_0 is constructed

by first going along the loop f and then along the loop g. Inverse of a loop f, f^{-1} is the same loop with the opposite orientation and $[f]^{-1} = [f^{-1}]$. Identity loop at x_0 is a loop $f : [0, 1] \to X$ such that $f(s) = x_0, \forall\, s \in [0, 1]$. For a path-connected topological space X fundamental groups at two distinct points x_0 and x are isomorphic. A path α connecting x_0 to x $(\alpha(0) = x_0, \alpha(1) = x)$, induces an isomorphism $\alpha_* : \pi_1(X, x) \to \pi_1(X, x_0)$ defined as $\alpha_*([f]) = [\alpha f \alpha^{-1}]$ (see Fig. 6).

A path-connected space X is simply connected if $\pi_1(X, x_0) \cong \{1\}$. $\mathbb{R}^n$ is an example of a simply connected space. Another example is the 2-sphere S^2. In a simply connected and path-connected space any closed path can be continuously shrunk to any point in the space.

Example 3.6 The fundamental group of the unit circle S^1 is $\pi_1(S^1) = \mathbb{Z}$. Homotopy class of a loop is determined by the number of times it winds around. In other words, any closed path in the circle can be tightened through homotopy into the product of n standard circular paths. Torus $T^2 = S^1 \times S^1$ has the fundamental group $\pi_1(T^2) \cong \mathbb{Z} \oplus \mathbb{Z}$ and is Abelian.

Consider two paths $f, g : I \to X$, $f(0) = a_0$, $f(1) = a_1$, and $g(0) = b_0$, $g(1) = b_1$. f and g are said to be freely homotopic if there exists a continuous map $F : I \times I \to X$ such that $F(s, 0) = f(s)$, $F(s, 1) = g(s)$. In addition to this if $F(0, t) = a_0$, $F(1, t) = b_0$, f and g are called homotopic. Two loops $f, g : I \to X$ are freely homotopic if there is a continuous map $F : I \times I \to X$ such that $F(s, 0) = f(s)$, $F(s, 1) = g(s)$ and $F(0, t) = F(1, t) = \alpha(t)$ is a path between $f(0) = f(1) = a$ and $g(0) = g(1) = b$.

Let Y be a topological space. $X \subset Y$ is a retract of Y if there exists a continuous map $r : Y \to X$ such that $r(x) = x$ for all $x \in X$. X is a deformation retract of Y if it is a retract of Y and there is a continuous map $h : [0, 1] \times Y \to Y$ such that: (1) $h(0, y) = y$, $h(1, y) = r(y)$, $\forall y \in Y$, and (2) $h(t, x) = x$, $\forall x \in X$, $\forall t \in [0, 1]$. A deformation retract $r : Y \to X$ induces an isomorphism $r_* : \pi_1(Y) \to \pi_1(X)$. One can visualize deformation retraction as a continuous collapse of Y onto X in such a way that each point of X remains fixed during the deformation process.

Example 3.7 The Möbius band $\mathcal{M}$ is constructed from a square by the identification shown in Fig. 3. This is an example of a non-orientable surface. The circle S^1 is a deformation retract of the Möbius band, and hence, $\pi_1(\mathcal{M}) = \mathbb{Z}$.

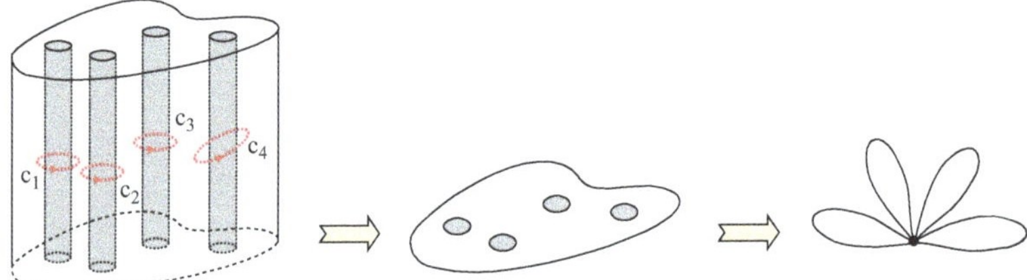

Fig. 7 A solid cylinder with four tubular holes and its deformation retract to a bouquet of four circles. c_1, c_2, c_3, c_4 are the generators of the fundamental group

Example 3.8 Consider the solid cylinder Ω with four tubular holes shown in Fig. 7. As is shown schematically Ω has a deformation retract to a bouquet of four circles, and hence, $\pi_1(\Omega) = \mathbb{Z}\otimes\mathbb{Z}\otimes\mathbb{Z}\otimes\mathbb{Z}$, i.e., the free group with four generators. If this is a solid body, e.g., a hollow bar under torsion and bending, we will see in the next section that because c_i's are free generators of the fundamental group, each would require an additional (vectorial) compatibility equation.

H. F. F. Tietze (1908) showed that the fundamental group of any compact, finite-dimensional, path-connected manifold is finitely presented. One forms the Abelization of a group by taking the quotient over the subgroup generated by all commutators $g^{-1}h^{-1}gh$. The Poincaré isomorphism theorem tells us that (Poincaré, 1895)[9]

$$\pi_1(M)/[\pi_1(M), \pi_1(M)] \cong H_1(M, \mathbb{Z}). \tag{45}$$

Given a group G with the presentation

$$G = \langle a_1, \ldots, a_m;\ r_1, \ldots, r_n \rangle, \tag{46}$$

its Abelianization is obtained by adding the relations $a_i a_j = a_j a_i$ and it is independent of the presentation of G.

3.3 Classification of Compact 2-Manifolds with Boundary

Let M_1 and M_2 be compact manifolds with boundary. Assume that their boundaries have the same number of components. M_1 and M_2 are homeomorphic if and only if the manifolds M_1^* and M_2^* obtained by gluing a disk to each boundary component are homeomorphic. Any compact surface is either homeomorphic to a sphere, a connected sum of tori, or a connected sum of projective planes. Any compact orientable 2-manifold with boundary is homeomorphic to a sphere with n handles and k holes, see Fig. 8.

[9] If $\gamma_1^{n_1}\gamma_2^{n_2}\ldots\gamma_k^{n_k} = 1$, Poincaré observed that $n_1\gamma_1 + n_2\gamma_2 + \ldots + n_k\gamma_k$ is null-homologous [18].

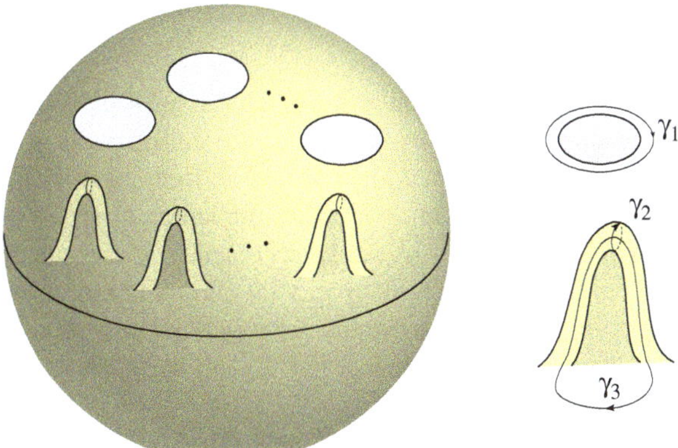

Fig. 8 A sphere with k holes and n handles. γ_1, γ_2, and γ_3 are typical generators for the first homology group. Note that a sphere with a single hole is simply connected, i.e., there are $k - 1$ generators corresponding to the k holes

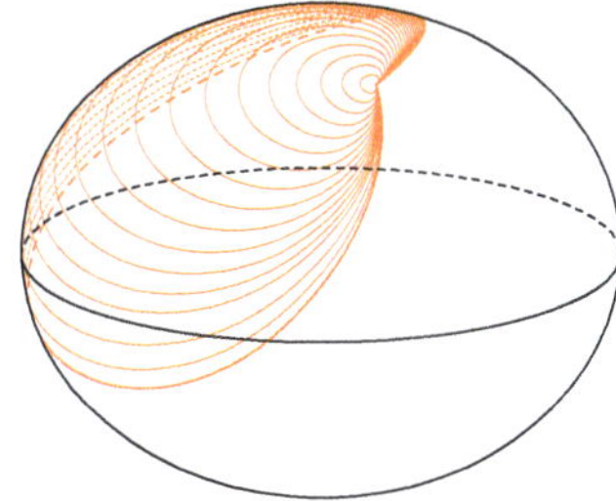

Fig. 9 A null-homotopic curve on an orientable surface bounds a region

3.4 Curves on Oriented Surfaces

R. Baer in 1928 showed that simple closed curves on a 2-manifold are isotopic if and only if they are homotopic [50]. Epstein [21] showed that any two simple, homotopic, non-contractible loops on an orientable surface are isotopic. If c is a simple, null-homotopic (contractible) loop on a surface, then it is the boundary of a topological disk (a genus zero surface with one boundary curve) [30, 32], see Fig. 9. A zero-genus surface with two boundary curves is called a cylinder. Any two non-contractible, non-intersecting, and freely homotopic curves on a closed surface bound a cylinder [30]. We will use these facts to derive the "bulk" compatibility equations.[10]

[10]These topological results are implicitly assumed in the literature of compatibility equations.

3.5 Theory of Knots

Topology of subsets of $\mathbb{R}^3$ with tubular holes can, at least partially, be understood using the complementary spaces of knots. For the background in knot theory we mainly follow [3, 15, 50]. A knot $\mathcal{K}$ is a simple closed curve in $\mathbb{R}^3$. A knot $\mathcal{K}$ is trivial if it is isotopic to the circle in $\mathbb{R}^3$. The fundamental group of a trivial knot $\mathbb{R}^3\backslash\mathcal{K}$ is infinite cyclic. Any knot $\mathcal{K}$ can be represented by a projection on a plane with no multiple points higher than double, with an indication of the upper branch of each crossing point (each of the double points). A projection of the trefoil knot (the simplest non-trivial knot) is shown in Fig. 10. A link is a set of knots tangled up together.

If the lower branch (under crossing) of each crossing is broken, one obtains a finite number of arcs α_i. It turns out that $\pi_1(\mathbb{R}^3\backslash\mathcal{K})$ is generated by loops c_i that pass around these arcs (this is rigorously proved using the Seifert–van Kampen theorem). This means that the number of generators of $\pi_1(\mathbb{R}^3\backslash\mathcal{K})$ is equal to the number of crossing points. Given the crossing point shown in Fig. 11a, the three generators of the fundamental group corresponding to the arcs α_i, α_{i+1}, and α_j are c_i, c_{i+1}, and c_j, respectively, and are oriented using the right-hand rule. It can be shown that $c_i c_j^{-1} c_{i+1}^{-1} c_j$ is null-homotopic, or equivalently, at this crossing point we have the relation $c_{i+1}c_j = c_j c_i$ [50]. All the four possibilities and their corresponding group relations are shown in Fig. 11b.

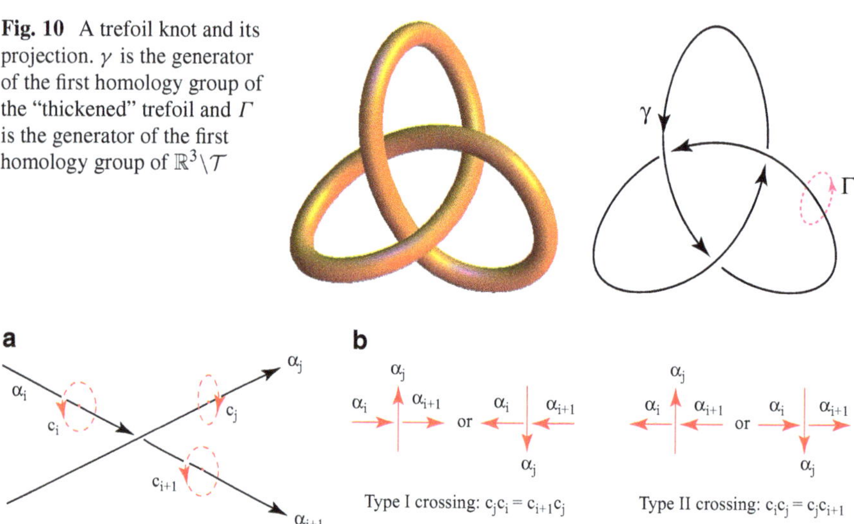

Fig. 10 A trefoil knot and its projection. γ is the generator of the first homology group of the "thickened" trefoil and Γ is the generator of the first homology group of $\mathbb{R}^3\backslash\mathcal{T}$

Fig. 11 (**a**) A crossing point. α_j corresponds to the over crossing and α_i and α_{i+1} correspond to the under crossing. Their corresponding loops c_j, c_i, and c_{i+1} are oriented using the right-hand rule. (**b**) Two types of crossing points and their corresponding group relations. Note that c_i is the fundamental group generator corresponding to the arc α_i, etc.

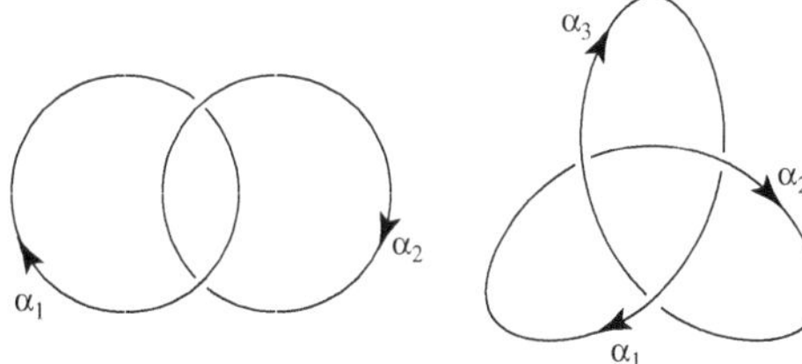

Fig. 12 The double link and the trefoil knot and their corresponding arcs α_i

Next, as examples, we find the fundamental groups of the complements of the two-crossing link, and the trefoil knot (see Fig. 12). Using the diagrams of Fig. 11b, it is straightforward to see that the fundamental groups of the complements of the two-crossing link $\mathcal{T}_1$ and the trefoil knot $\mathcal{T}_2$ are, respectively:

$$\begin{aligned} \pi_1\left(\mathbb{R}^3\backslash\mathcal{T}_1\right) &= \langle c_1, c_2;\ c_1c_2 = c_2c_1\rangle, \\ \pi_1\left(\mathbb{R}^3\backslash\mathcal{T}_2\right) &= \langle c_1, c_2, c_3;\ c_3c_1 = c_2c_3, c_2c_3 = c_1c_2, c_1c_3 = c_2c_1\rangle. \end{aligned} \tag{47}$$

Note that if the two circles are unlinked then $\pi_1\left(\mathbb{R}^3\backslash\mathcal{T}_1\right) = \langle c_1, c_2\rangle$, i.e., a free group with c_1 and c_2 as generators.

Remark 3.9 In the case of knots, Abelianization always gives an infinite cyclic group [50]. A handlebody $\mathcal{H}_n$ is a solid body bounded by an orientable surface of genus n embedded in $\mathbb{R}^3$. $\pi_1\left(\mathbb{R}^3\backslash\mathcal{H}_n\right)$ is the free group of rank n.

3.6 Topology of 3-Manifolds

Material manifold—the natural configuration of a body—may be non-Euclidean in many applications [38, 48, 54–57]. However, for most applications the ambient space is the Euclidean 3-space. We consider a body that has a non-trivial topology, i.e., it has "holes." We assume that the body is elastic and the material manifold is an embedded 3-submanifold of $\mathbb{R}^3$. There is a complete classification of 3-manifolds [26, 34], but it is not known what 3-manifolds can be embedded in $\mathbb{R}^3$. However, a large class of embedded 3-submanifolds can be constructed by thickened knots and their complements in $\mathbb{R}^3$. The important thing to note is the complexity of embedded 3-manifolds and the importance of algebraic topology in deriving their necessary and sufficient compatibility equations for non-simply-connected bodies.

For an embedded 3-manifold with boundary in $\mathbb{R}^3$, its boundary is an embedded closed (orientable) 2-manifold, which has a complete classification. If the boundary of the 3-manifold is the two-sphere, then its topology is uniquely determined by the genus of the boundary, i.e., the manifold is simply the compact region bounded by the boundary in $\mathbb{R}^3$ (by the generalized Jordan–Brouwer separation theorem, any closed embedded 2-manifold in $\mathbb{R}^3$ divides $\mathbb{R}^3$ into a pair of regions, and precisely one of these regions has compact closure). If the boundary is not connected, then things are more complicated. For instance even when the boundary consists of a single torus, the compact region that it bounds in $\mathbb{R}^3$ is not uniquely determined, but it is known that it must be either a solid torus or a knot complement. Things get more complicated when the boundary has genus larger than one. The only simple case is when the boundary is a sphere, in which case the manifold must necessarily be a ball by Jordan's theorem. To summarize, while 3-manifolds with boundary have been completely classified, it is not known which ones can be embedded in $\mathbb{R}^3$. The answer certainly depends on both the topology of the boundary and its isotopy (or knotting) in $\mathbb{R}^3$. As to what types of "holes" can occur in a 3-dimensional solid, consider the following example: Put a knot in the solid body, then "thicken" it to obtain a (knotted) solid torus, and then remove the interior of that torus. This way one can construct as many different types of holes (or topological types for the solid) as there are knots. Now consider doing the same construction with multiple tori or higher genus surfaces, which may be linked with each other.

4 Kinematics of Nonlinear Elasticity

In this section we review the kinematics of nonlinear elasticity. A body $\mathcal{B}$ is identified with a Riemannian manifold $(\mathcal{B}, \mathbf{G})$[11] and a configuration of $\mathcal{B}$ is a mapping $\varphi : \mathcal{B} \to \mathcal{S}$, where $(\mathcal{S}, \mathbf{g})$ is another Riemannian manifold. The set of all configurations of $\mathcal{B}$ is denoted by $\mathcal{C}$. A motion is a curve $c : \mathbb{R} \to \mathcal{C}; t \mapsto \varphi_t$ in $\mathcal{C}$. The material manifold is, by construction, the natural configuration of the body. For a fixed t, $\varphi_t(X) = \varphi(X, t)$ and for a fixed X, $\varphi_X(t) = \varphi(X, t)$, where X is the position of a material point in the undeformed configuration $\mathcal{B}$. The material velocity is given by $\mathbf{V}_t(X) = \mathbf{V}(X, t) = \frac{\partial \varphi(X,t)}{\partial t}$. Similarly, the material acceleration is defined by $\mathbf{A}_t(X) = \mathbf{A}(X, t) = \frac{\partial \mathbf{V}(X,t)}{\partial t}$. In components, $A^a = \frac{\partial V^a}{\partial t} + \gamma^a{}_{bc} V^b V^c$, where $\gamma^a{}_{bc}$ is the Christoffel symbol of the local coordinate chart $\{x^a\}$. The spatial velocity of a regular motion φ_t is defined as $\mathbf{v}_t : \varphi_t(\mathcal{B}) \to T_{\varphi_t(X)}\mathcal{S}$, $\mathbf{v}_t = \mathbf{V}_t \circ \varphi_t^{-1}$, and the spatial acceleration $\mathbf{a}_t$ is defined as $\mathbf{a} = \dot{\mathbf{v}} = \frac{\partial \mathbf{v}}{\partial t} + \nabla_{\mathbf{v}} \mathbf{v}$. In components, $a^a = \frac{\partial v^a}{\partial t} + \frac{\partial v^a}{\partial x^b} v^b + \gamma^a{}_{bc} v^b v^c$.

Let $\varphi : \mathcal{B} \to \mathcal{S}$ be a C^1 configuration of $\mathcal{B}$ in $\mathcal{S}$, where $\mathcal{B}$ and $\mathcal{S}$ are manifolds. The deformation gradient is the tangent map of φ and is denoted by $\mathbf{F} = T\varphi$. Thus, at each point $X \in \mathcal{B}$, it is a linear map $\mathbf{F}(X) : T_X\mathcal{B} \to T_{\varphi(X)}\mathcal{S}$. If $\{x^a\}$ and $\{X^A\}$ are local coordinate charts on $\mathcal{S}$ and $\mathcal{B}$, respectively, the components of $\mathbf{F}$ are $F^a{}_A(X) = \frac{\partial \varphi^a}{\partial X^A}(X)$. $\mathbf{F}$ has the local representation $\mathbf{F} = F^a{}_A \frac{\partial}{\partial x^a} \otimes dX^A$. $\mathbf{F}$ can be thought of a vector-valued 1-form with the representation $\mathbf{F} = \frac{\partial}{\partial x^a} \otimes \vartheta^a$, with the coframes $\vartheta^a = F^a{}_A dX^A$. The adjoint of $\mathbf{F}$ is defined by

[11] In general, $(\mathcal{B}, \mathbf{G})$ is the underlying Riemannian manifold of the material manifold, i.e., its natural state. See [54–57] for more details.

$$\mathbf{F}^{\mathsf{T}} : T_x\mathcal{S} \to T_X\mathcal{B}, \quad \langle\!\langle \mathbf{FW}, \mathbf{w} \rangle\!\rangle_{\mathbf{g}} = \langle\!\langle \mathbf{W}, \mathbf{F}^{\mathsf{T}}\mathbf{w} \rangle\!\rangle_{\mathbf{G}}, \quad \forall \mathbf{W} \in T_X\mathcal{B}, \ \mathbf{w} \in T_X\mathcal{S}. \tag{48}$$

In components, $(F^{\mathsf{T}}(X))^A{}_a = g_{ab}(x)F^b{}_B(X)G^{AB}(X)$, where $\mathbf{g}$ and $\mathbf{G}$ are metric tensors on $\mathcal{S}$ and $\mathcal{B}$, respectively. The right Cauchy–Green deformation tensor is defined as

$$\mathbf{C}(X) : T_X\mathcal{B} \to T_X\mathcal{B}, \qquad \mathbf{C}(X) = \mathbf{F}^{\mathsf{T}}(X)\mathbf{F}(X). \tag{49}$$

In components, $C^A{}_B = (F^{\mathsf{T}})^A{}_a F^a{}_B$. It is straightforward to show that, $\mathbf{C}^\flat = \varphi^*(\mathbf{g}) = \mathbf{F}^\star\mathbf{g}\mathbf{F}$, i.e., $C_{AB} = F^a{}_A(g_{ab} \circ \varphi)F^b{}_B$, where the dual of the deformation gradient is defined as $\mathbf{F}^\star = F^a{}_A dX^A \otimes \frac{\partial}{\partial x^a}$. The Finger tensor is defined as $\mathbf{b} = \mathbf{c}^{-1}$, where $\mathbf{c} = \varphi_{t*}\mathbf{G}$. In components, $b^a{}_b = F^a{}_A g_{bc} F^c{}_B G^{AB} = F^a{}_A \left(F^{\mathsf{T}}\right)^A{}_b$. Thus

$$\mathbf{b}(x) : T_x\mathcal{S} \to T_x\mathcal{S}, \qquad \mathbf{b}(x) = \mathbf{F}(X)\mathbf{F}^{\mathsf{T}}(X). \tag{50}$$

Polar decomposition theorem states that $\mathbf{F} = \mathbf{RU}$ [45]. In components it reads $F^a{}_A = R^a{}_B U^B{}_A$, where $\mathbf{R}(X) : T_X\mathcal{B} \to T_{\varphi_t(X)}\mathcal{S}$ is a $(\mathbf{G}, \mathbf{g})$-orthogonal transformation, i.e., $G_{AB} = R^a{}_A R^b{}_B g_{ab}$, and $\mathbf{U}(X) : T_X\mathcal{B} \to T_X\mathcal{B}$ is the material stretch tensor. Note that $\mathbf{G} = \mathbf{R}^*\mathbf{g}$ and $\mathbf{C} = \mathbf{U}^*\mathbf{G}$.

5 Compatibility Equations in Nonlinear Elasticity

In this section we summarize the results of [4, 58], and [5]. We assume a finite body, and hence, the material manifold $(\mathcal{B}, \mathbf{G})$ is a compact Riemannian manifold. We also assume that the first homology and homotopy groups $H_1(\mathcal{B})$ and $\pi_1(\mathcal{B})$ are given. In the presence of boundary conditions we will use the relative homology groups, which are also assumed to be given. In [58] we derived the compatibility equations for the deformation gradient $\mathbf{F}$ using a generalization of de Rham's theorem. The $\mathbf{F}$-compatibility equations can be derived using the fundamental group as well. It turns out that understanding the role of homotopy in compatibility equations is crucial in formulating the $\mathbf{C}$-compatibility equations [58].

5.1 Compatibility Equations for the Deformation Gradient F

The following old questions in vector calculus are relevant to the compatibility equations: (1) Given a vector field defined on some bounded domain in the Euclidean 3-space, is it the gradient of some function defined on the same domain? (2) Is it the curl of another vector field? It turns out that the topology of the domain of definition of the vector field plays a crucial role. The $\mathbf{F}$-compatibility problem is stated as: Given a body $\mathcal{B} \subset \mathbb{R}^3$, find the condition(s) that guarantee existence of a map $\varphi : \mathcal{B} \to \mathbb{R}^3$ such that $\mathbf{F} = T\varphi$. Question (1) is related to compatibility equations while question (2) is related to the existence of stress functions in elasticity. The following proposition summarizes the **F**-compatibility equations, which is a simple extension of de Rham's theorem to $\mathbb{R}^3$-valued forms.

Proposition 5.1 (Yavari [58]) *The necessary and sufficient* **F***-compatibility equations are*[12]

$$d\mathbf{F} = \mathbf{0}, \quad \text{and} \quad \int_{c_i} \mathbf{F}d\mathbf{X} = \mathbf{0}, \quad i = 1, \ldots, b_1(\mathcal{B}), \tag{51}$$

where c_i, $i = 1, \ldots, b_1(\mathcal{B})$ *are the generators of* $H_1(\mathcal{B}; \mathbb{R})$.

Instead of using de Rham's theorem, one may follow a different path using the fundamental group. Let us assume that the position of a material point $X_0 \in \mathcal{B}$ in the deformed configuration $x_0 \in \mathcal{S}$ is given. The position of an arbitrary material point $X \in \mathcal{B}$ in the deformed configuration is given as

$$\mathbf{x} = \mathbf{x}_0 + \int_\gamma \mathbf{F}d\mathbf{X}. \tag{52}$$

Note that the ambient space is Euclidean, and hence, integrating vector fields makes sense. **F** is compatible if and only if the above integral is path-independent, which is equivalent to

$$\int_\gamma \mathbf{F}d\mathbf{X} = \mathbf{0}, \tag{53}$$

for any closed path γ based at X_0.

Suppose $\pi_1(\mathcal{B})$ has the generators $\{\gamma_i\}_{i=1,\ldots,m}$. For a compact material manifold $\mathcal{B}$, i.e., a finite body, the fundamental group has a finite presentation [50]

$$\pi_1(\mathcal{B}) = \langle \gamma_1, \ldots, \gamma_m; r_1, \ldots, r_n \rangle, \tag{54}$$

where

$$r_i = \gamma_{i_1}^{\epsilon_{i_1}} \cdots \gamma_{j_i}^{\epsilon_{j_i}} = 1, \quad i = 1, \ldots, n, \ \epsilon_k = \pm 1, \tag{55}$$

are the relators of the fundamental group. If γ is a contractible (null-homotopic) curve that lies on a 2-submanifold $\mathcal{P} \subset \mathcal{B}$, then

$$\int_\gamma \mathbf{F}d\mathbf{X} = \int_{\partial\mathcal{U}} \mathbf{F}d\mathbf{X} = \int_{\mathcal{U}} d(\mathbf{F}d\mathbf{X}) = \mathbf{0}, \tag{56}$$

[12] The exterior derivative of the deformation gradient $d\mathbf{F}$ can be identified with $\mathsf{Curl}\,\mathbf{F}$. Note that $d\mathbf{F} = \mathbf{0}$ is equivalent to $\nabla^{\mathbf{G}}\mathbf{F} = \mathbf{0}$, where $\nabla^{\mathbf{G}}$ is the Levi-Civita connection corresponding to the material metric **G** [59]. In components, $F^a{}_{A,B} = F^a{}_{B,A}$, or equivalently, $F^a{}_{A|B} = F^a{}_{B|A}$.

where $\gamma = \partial\mathcal{U} \subset \mathcal{P}$ [21, 30]. Because $\mathcal{P}$ is arbitrary one concludes that $d\mathbf{F} = \mathbf{0}$ in $\mathcal{B}$, which is a necessary compatibility condition. Note that from (53)

$$\int_{\gamma_i} \mathbf{F}d\mathbf{X} = \mathbf{0}, \quad i = 1, \ldots, m. \tag{57}$$

Therefore, $d\mathbf{F} = \mathbf{0}$, and $\int_{\gamma_i} \mathbf{F}d\mathbf{X} = \mathbf{0},\ i = 1, \ldots, m$ subjected to $\int_{r_i} \mathbf{F}d\mathbf{X} = \mathbf{0},\ i = 1, \ldots, n$ are necessary for compatibility of $\mathbf{F}$. It turns out that they are sufficient as well. Given a null-homotopic curve γ, $\gamma = \partial\Omega$, and hence, from $d\mathbf{F} = \mathbf{0}$, one can write

$$\int_{\gamma} \mathbf{F}d\mathbf{X} = \int_{\partial\mathcal{U}} \mathbf{F}d\mathbf{X} = \int_{\mathcal{U}} d(\mathbf{F}d\mathbf{X}) = \int_{\mathcal{U}} d\mathbf{F} \wedge d\mathbf{X} = \mathbf{0}. \tag{58}$$

If γ is non-contractible, in terms of the group generators it has the representation $\gamma = \left(w_1\gamma_1 w_1^{-1}\right)^{\epsilon_1} \ldots \left(w_p\gamma_p w_p^{-1}\right)^{\epsilon_p}$, where w_i is a curve joining X_0 to a point on γ_i and $\{\gamma_1, \ldots, \gamma_p\}$ is a subset of the group generators with possible relabelings. Using the relations $\int_{w_i\gamma_i w_i^{-1}} \mathbf{F}d\mathbf{X} = \int_{\gamma_i} \mathbf{F}d\mathbf{X}$, one has

$$\int_{\gamma} \mathbf{F}d\mathbf{X} = \epsilon_1 \int_{\gamma_1} \mathbf{F}d\mathbf{X} + \ldots + \epsilon_p \int_{\gamma_p} \mathbf{F}d\mathbf{X} = \mathbf{0}. \tag{59}$$

The relators of the group representation impose the following constraints

$$\int_{r_i} \mathbf{F}d\mathbf{X} = \mathbf{0}, \quad i = 1, \ldots, n. \tag{60}$$

This implies that the conditions $\int_{\gamma_i} \mathbf{F}d\mathbf{X} = \mathbf{0},\ i = 1, \ldots, m$, may not all be independent.

Proposition 5.2 (Yavari [58]) *The necessary and sufficient* **F***-compatibility conditions are:*

(i) $d\mathbf{F} = \mathbf{0}$ *in* $\mathcal{B}$,
(ii) If $\pi_1(\mathcal{B}) = \langle \gamma_1, \ldots, \gamma_m; r_1, \ldots, r_n \rangle$, *then*

$$\int_{\gamma_i} \mathbf{F}d\mathbf{X} = \mathbf{0}, \quad i = 1, \ldots, m, \tag{61}$$

$$\text{subjected to} \int_{r_i} \mathbf{F}d\mathbf{X} = \mathbf{0}, \quad i = 1, \ldots, n. \tag{62}$$

For a path-connected set $\mathcal{B}$ the first homology group is the Abelianization of the fundamental group [10]. One Abelianizes $\pi_1(\mathcal{B})$ by adding the relations $\gamma_i\gamma_j = \gamma_j\gamma_i$, which do not lead to any new compatibility equations.

One should note that the generators of the torsion subgroup do not contribute to the **F**-compatibility equations because for γ an element of the torsion subgroup $\gamma^n = 1$ for some $n \in \mathbb{N}$, and thus, $\int_\gamma \mathbf{F}d\mathbf{X} = \mathbf{0}$ is trivially satisfied. This means that it is sufficient to have $\int_\gamma \mathbf{F}dX = \mathbf{0}$ only on each generator of the first homology group with real coefficients. Therefore, the number of the complementary compatibility equations is $Nb_1(\mathcal{B})$, where $N = \dim \mathcal{S}$.

Example 5.3 Let us assume that $\dim \mathcal{B} = 1$ and $\mathcal{S} = \mathbb{R}^2$. The bulk compatibility equations are trivially satisfied. It is known that when $\mathcal{B}$ is a graph its fundamental group is freely generated. Assuming that $\gamma_1, \ldots, \gamma_k$ are the free generators of $\pi_1(\mathcal{B})$, there are $2k$ compatibility equations. As an example, let us assume that $\mathcal{B} = S^1(R)$, i.e., the circle with radius R and let $X = \Theta$ be the standard parametrization of S^1. Compatibility equations read

$$\int_0^{2\pi} \mathbf{F}d\Theta = \mathbf{0}. \tag{63}$$

As examples, $\mathbf{F} = (\kappa_1\Theta, \kappa_2)^\mathsf{T}$, where κ_1 and κ_2 are arbitrary constants, is not compatible, while $\mathbf{F} = (\kappa_1 \sin\Theta, \kappa_2 \cos\Theta)^\mathsf{T}$ is compatible.

Remark 5.4 One should note that the **F**-compatibility equations derived here are valid for anelastic bodies as well. In other words, we have not assumed a flat material manifold $(\mathcal{B}, \mathbf{G})$; the compatibility equations have the same form even in problems for which the material manifold is non-flat. As an example, see the discussion of universal deformations and eigenstrains in compressible solids in [59].

5.2 *Examples of Non-simply-connected Bodies and Their F-Compatibility Equations*

We next look at a few examples of 2D and 3D non-simply-connected bodies and derive their compatibility equations.

5.2.1 2D Elasticity on a Torus and a Punctured Torus

The first homology groups of both torus and punctured torus (handle) are generated by the loops γ_1 and γ_2 in Figs. 2 and 13. Hence, the **F**-compatibility equations read

$$d\mathbf{F} = \mathbf{0}, \quad \int_{\gamma_1} \mathbf{F}d\mathbf{X} = \int_{\gamma_2} \mathbf{F}d\mathbf{X} = \mathbf{0}. \tag{64}$$

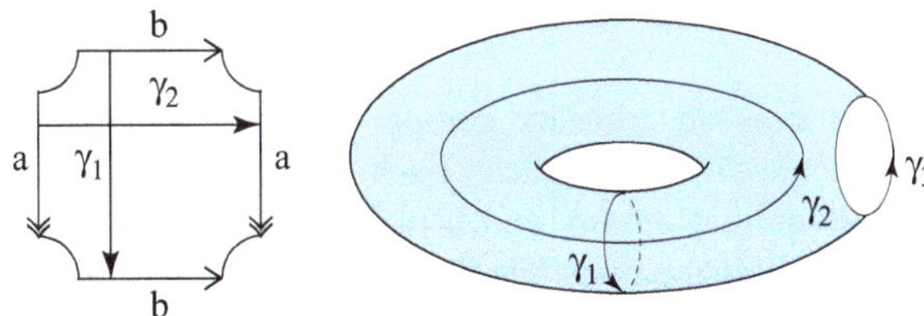

Fig. 13 A punctured torus. γ_1, γ_2, and γ_3 are generators of the fundamental group

The fundamental group of torus (see Fig. 2) has the presentation

$$\pi_1(T^2) = \langle \gamma_1, \gamma_2; \gamma_1\gamma_2 = \gamma_2\gamma_1 \rangle. \tag{65}$$

Therefore, the group relator is written as $r_1 = \gamma_1\gamma_2\gamma_1^{-1}\gamma_2^{-1} = 1$. Note that

$$\int_{r_1} \mathbf{F}d\mathbf{X} = \int_{\gamma_1\gamma_2\gamma_1^{-1}\gamma_2^{-1}} \mathbf{F}d\mathbf{X} = \int_{\gamma_1} \mathbf{F}d\mathbf{X} + \int_{\gamma_2} \mathbf{F}d\mathbf{X} - \int_{\gamma_1} \mathbf{F}d\mathbf{X} - \int_{\gamma_2} \mathbf{F}d\mathbf{X} = \mathbf{0}. \tag{66}$$

This means that (64) are the necessary and sufficient **F**-compatibility equations.

For a punctured torus (see Fig. 13) the fundamental group has three generators and the following presentation [50]

$$\pi_1(\mathcal{H}) = \langle \gamma_1, \gamma_2, \gamma_3; \gamma_3 = \gamma_1\gamma_2\gamma_1^{-1}\gamma_2^{-1} \rangle. \tag{67}$$

Therefore, the group relator is written as $r_1 = \gamma_3\gamma_2\gamma_1\gamma_2^{-1}\gamma_1^{-1} = 1$. Note that

$$\begin{aligned}
\mathbf{0} &= \int_{r_1} \mathbf{F}d\mathbf{X} \\
&= \int_{\gamma_3\gamma_2\gamma_1\gamma_2^{-1}\gamma_1^{-1}} \mathbf{F}d\mathbf{X} \\
&= \int_{\gamma_3} \mathbf{F}d\mathbf{X} + \int_{\gamma_2} \mathbf{F}d\mathbf{X} + \int_{\gamma_1} \mathbf{F}d\mathbf{X} - \int_{\gamma_2} \mathbf{F}d\mathbf{X} - \int_{\gamma_1} \mathbf{F}d\mathbf{X} \\
&= \int_{\gamma_3} \mathbf{F}d\mathbf{X}.
\end{aligned} \tag{68}$$

Therefore, the necessary and sufficient **F**-compatibility equations read

$$d\mathbf{F} = \mathbf{0}, \quad \int_{\gamma_1} \mathbf{F}d\mathbf{X} = \int_{\gamma_2} \mathbf{F}d\mathbf{X} = \mathbf{0}. \tag{69}$$

One observes that γ_3 is a generator of the fundamental group but does not have a corresponding complementary compatibility equation. The boundary of the hole in a punctured torus is null-homologous path but not null-homotopic.

5.2.2 2D Elasticity on Arbitrary Compact Orientable 2-Manifolds

When $\mathcal{B}$ is an arbitrary compact orientable 2-manifold, it is homeomorphic to a sphere with n handles. Each handle corresponds to two generators of the first homology group, and hence, there are $3 \times 2n$ complementary compatibility equations. If the body has boundaries they correspond to k holes, which introduce another $k-1$ generators of the first homology group (see Fig. 8). The total number of complementary compatibility equations are $3(2n+k-1)$.

5.2.3 3D Elastic Bodies with Holes

A 3D solid with internal cavities has a trivial $H_1(\mathcal{B})$. As an example, a solid with a spherical hole (see Fig. 1a) has a trivial first homology group, and hence, $d\mathbf{F} = \mathbf{0}$ is both necessary and sufficient for compatibility of $\mathbf{F}$. The first homology group of a solid torus has only one generator. The body shown in Fig. 1c is homeomorphic to a solid torus and the (red) closed curve generates its first homology group. The body shown in Fig. 1d has Betti number two and the two (red) loops generate its first homology group. The complement of a solid torus has Betti number one (see Fig. 1b). The first homology group of a 2-holed solid torus has the two generators γ_1 and γ_2 shown in Fig. 5. Its complement has Betti number two and the generators Γ_1 and Γ_2 are shown in Fig. 5. A thick torus with two tubular holes has Betti number three. The generators of the first homology group are shown in Fig. 14.

A solid trefoil knot has Betti number one and a generator of its first homology group is γ shown in Fig. 10. Its complement in $\mathbb{R}^3$ has Betti number one as well and Γ in Fig. 10 is its generator. A cylinder and an annulus are homeomorphic. The first homology group is generated by the loop c_3 in Fig. 4. A solid with tubular holes shown in Fig. 7 has a fundamental group freely generated by the four loops $c_i, i = 1, 2, 3, 4$. Each c_i corresponds to three extra compatibility equations for $\mathbf{F}$ (six extra compatibility equations for $\mathbf{C}^\flat$). A thick hollow cylinder is the special case of this example when there is only one hole. The Betti number of both the Möbius band $\mathcal{M}$ and the thick Möbius band $\mathcal{M} \times [0, 1]$ is one. The knotted ball shown in Fig. 15 (left) has Betti number one. Note that its fundamental group has

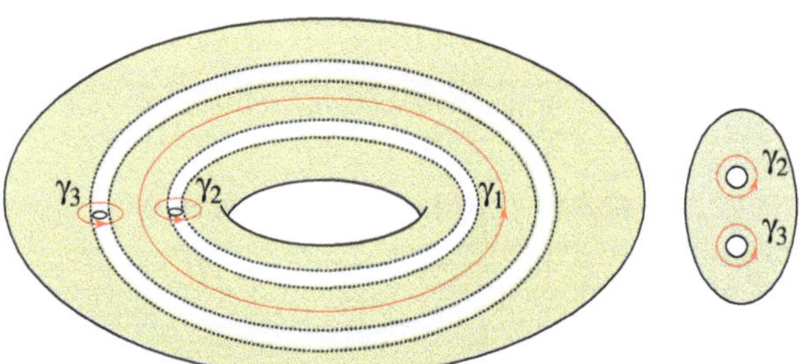

Fig. 14 A solid torus with two tubular holes

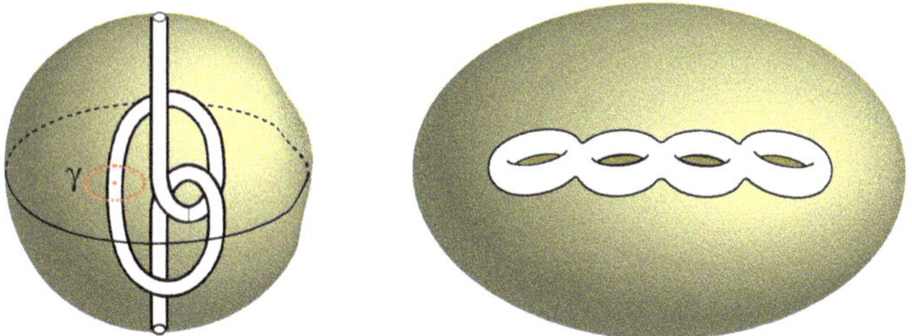

Fig. 15 Left: A knotted ball. γ is a generator of the first homology group. Right: A ball with a toroidal hole of genus four. This body has Betti number four

four generators but only one requires complementary compatibility equations. The ball shown in Fig. 15 (right) has a hole, which is a genus four handlebody. Its Betti number is four.

5.3 F-Compatibility Equations in the Presence of Dirichlet Boundary Conditions

In [5], the compatibility equations in the presence of Dirichlet boundary conditions were derived using some Hodge-type orthogonal decompositions. Here, we follow [20] and find the **F**-compatibility equations when deformation mapping (or displacement field) is prescribed on part of the boundary $\partial_D \mathcal{B} \subset \partial \mathcal{B}$.

Consider a k-form ω ($k \geq 1$) on $\mathcal{B}$. Using the inclusion map $\iota : \partial \mathcal{B} \hookrightarrow \mathcal{B}$, the tangential component of ω is defined as $\mathsf{t}\omega = \iota^* \omega$ [41]. This can equivalently be defined using the decomposition of vector fields on $\partial \mathcal{B}$ into tangential and normal parts. Given $\mathbf{X} \in \Gamma(T\mathcal{B}|_{\partial \mathcal{B}})$, $\mathbf{X} = \mathbf{X}^{\parallel} + \mathbf{X}^{\perp}$, and the tangential part of the k-form ω is defined as

$$\mathsf{t}\omega(\mathbf{X}_1, \ldots, \mathbf{X}_k) = \omega(\mathbf{X}_1^{\parallel}, \ldots, \mathbf{X}_k^{\parallel}), \quad \forall \mathbf{X}_1, \ldots, \mathbf{X}_k \in \Gamma(T\mathcal{B}|_{\partial \mathcal{B}}). \tag{70}$$

The normal part is defined as $\mathsf{n}\omega = \omega - \mathsf{t}\omega$. For $k = 0$, $\mathsf{t}\omega = \omega$. The deformation mapping can be thought of as an $\mathbb{R}^3$-valued 0-form. The Dirichlet boundary conditions are given as $\varphi^a|_{\partial_D \mathcal{B}} = \hat{\varphi}^a$, where φ^a, $a = 1, 2, 3$, are 0-forms defined on $\mathcal{B}$, and $\hat{\varphi}^a$, $a = 1, 2, 3$, are 0-forms defined on $\partial_D \mathcal{B}$.

The following result is a simple corollary of [20, Theorem 6].

Proposition 5.5 *Suppose* $\mathbf{F}$ *is an* $\mathbb{R}^3$*-valued* 1*-form in* $\mathcal{B}$*. Also assume that* $\mathsf{t}\,\mathbf{F} = d\hat{\varphi}$ *on* $\partial_D \mathcal{B}$.[13] *The necessary and sufficient conditions for compatibility of* $\mathbf{F}$*, i.e., the existence of an* $\mathbb{R}^3$*-valued* 0*-form* φ *such that* $\mathbf{F} = d\varphi$*, and* $\varphi|_{\partial_D \mathcal{B}} = \hat{\varphi}$ *are:*

[13] $\mathsf{t}\mathbf{F} = d\hat{\varphi}$ means that for any vector $\mathbf{W} \in T\mathcal{B}|_{\partial_D \mathcal{B}}$, $\mathbf{F}\mathbf{W}^{\parallel} = \langle d\hat{\varphi}, \mathbf{W}^{\parallel} \rangle$.

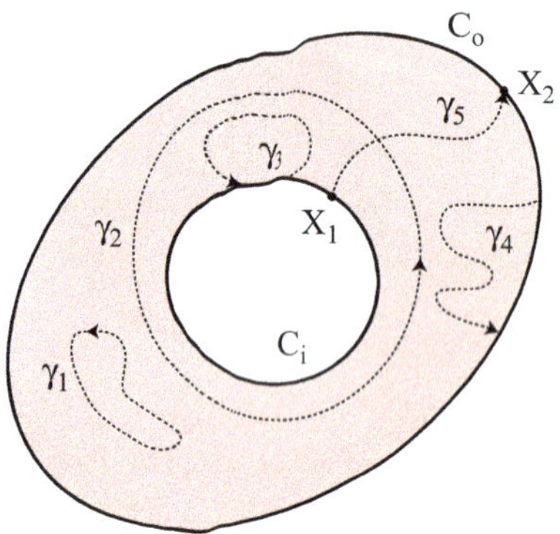

Fig. 16 The boundary of $\mathcal{B}$ is the union of the inner circle C_i and the outer ellipse C_o

$$d\mathbf{F} = \mathbf{0}, \quad \text{and} \int_{c_i} \mathbf{F}d\mathbf{X} = \int_{\partial c_i} \hat{\varphi} = \hat{\varphi}(\boldsymbol{X}_2^i) - \hat{\varphi}(\boldsymbol{X}_1^i), \quad i = 1, \ldots, b_1(\mathcal{B}, \partial_D\mathcal{B}), \tag{71}$$

where c_i's are the generators of the first relative singular homology group $H_1(\mathcal{B}, \partial_D\mathcal{B}; \mathbb{R})$. Note that each $\partial c_i = [\boldsymbol{X}_1^i, \boldsymbol{X}_2^i]$ is an oriented pair of points $(\boldsymbol{X}_1^i, \boldsymbol{X}_2^i)$ such that $\boldsymbol{X}_1^i$ and $\boldsymbol{X}_2^i$ lie on $\partial_D\mathcal{B}$.

Example 5.6 Let us consider the body shown in Fig. 16. We consider the following four cases of boundary conditions.

- $\partial_D\mathcal{B} = \varnothing$: In this case the auxiliary compatibility equation reads

$$\int_{\gamma_2} \mathbf{F}d\mathbf{X} = \mathbf{0}, \tag{72}$$

  where γ_2 is the generator of the first de Rham cohomology group (see Fig. 16).
- $\partial_D\mathcal{B} = C_i$: In this case there are no auxiliary compatibility equations. Note that γ_2 and γ_3 are relative boundaries.
- $\partial_D\mathcal{B} = C_o$: In this case there are no auxiliary compatibility equations. Note that γ_2 and γ_4 are relative boundaries.
- $\partial_D\mathcal{B} = \partial\mathcal{B}$: In this case a generator of $H_1(\mathcal{B}, \partial_D\mathcal{B}; \mathbb{R})$ is γ_5, and the auxiliary compatibility equation reads

$$\int_{\gamma_5} \mathbf{F}d\mathbf{X} = \hat{\varphi}(X_2) - \hat{\varphi}(X_1). \tag{73}$$

  Note that γ_2, γ_3, and γ_4 are relative boundaries.

Our calculations in this example are consistent with [5, Example 10] in which the Dirichlet boundary was assumed fixed.

5.4 Compatibility Equations for the Right Cauchy–Green Strain $\mathbf{C}^\flat$

Consider a motion of a body $\varphi_t : \mathcal{B} \to \mathcal{S}$ and assume that $\dim \mathcal{B} = \dim \mathcal{S}$. The right Cauchy–Green deformation tensor is defined as $\mathbf{C}^\flat = \varphi_t^* \mathbf{g}$. For a Euclidean ambient space $\mathcal{R}(\mathbf{g}) = \mathbf{0}$. Thus

$$\mathbf{0} = \varphi_t^* \mathcal{R}(\mathbf{g}) = \mathcal{R}(\varphi_t^* \mathbf{g}) = \mathcal{R}(\mathbf{C}^\flat), \tag{74}$$

i.e., a necessary condition for $\mathbf{C}^\flat$ to be compatible is vanishing of its Riemann curvature, or equivalently local flatness of the Riemannian manifold $(\mathcal{B}, \mathbf{C}^\flat)$. Note that this is independent of the geometry of $(\mathcal{B}, \mathbf{G})$. In other words, even for a non-flat material manifold $\mathcal{R}(\mathbf{C}^\flat) = \mathbf{0}$ is a necessary compatibility equation for $\mathbf{C}^\flat$. Marsden and Hughes [33] showed that this condition is locally sufficient as well. In the case of simply connected elastic bodies this condition guarantees the existence of a global deformation mapping [14].

Suppose $\{X^A\}$, $\{x^a\}$ are coordinate charts for $\mathcal{B}$, and $\mathcal{S}$, respectively. The Levi-Civita connection coefficients of $\mathbf{g}$ and $\mathbf{C}^\flat = \varphi^* \mathbf{g}$ are denoted by $\gamma^a{}_{bc}$ and $\Gamma^A{}_{BC}$, respectively. They are related as

$$\Gamma^A{}_{BC} = \frac{\partial X^A}{\partial x^a} \frac{\partial x^b}{\partial X^B} \frac{\partial x^c}{\partial X^C} \gamma^a{}_{bc} + \frac{\partial^2 x^m}{\partial X^B \partial X^C} \frac{\partial X^A}{\partial x^m}. \tag{75}$$

Assuming that $\{x^a\}$ is a Cartesian coordinate chart for the Euclidean ambient space, $\gamma^a{}_{bc} = 0$, and hence

$$\Gamma^A{}_{BC} = \frac{\partial^2 x^m}{\partial X^B \partial X^C} \frac{\partial X^A}{\partial x^m}. \tag{76}$$

Therefore

$$\frac{\partial^2 x^a}{\partial X^B \partial X^C} = \frac{\partial}{\partial X^C} F^a{}_B = F^a{}_A \Gamma^A{}_{BC}. \tag{77}$$

Using the polar decomposition in (77) one obtains[14]

$$R^a{}_{A,B} = R^a{}_C \Omega^C{}_{AB}, \tag{78}$$

where

$$\Omega^C{}_{AB} = \left(\Gamma^M{}_{BN} U^C{}_M - U^C{}_{N,B} \right) U_A{}^N, \; \Gamma^C{}_{AB} = \frac{1}{2} C^{CD} (C_{BD,A} + C_{AD,B} - C_{AB,D}), \tag{79}$$

[14]Note that Eq. (78) is identical to Shield [42]'s Eq. (8).

and $U_A{}^N$ are components of $\mathbf{U}^{-1}$. Note that the material manifold is assumed to be embedded in the Euclidean ambient space. Choosing Cartesian coordinates for $\mathcal{B}$, $G_{AB} = \delta_{AB}$. For a path γ that connects $X_0,\ X \in \mathcal{B}$ and is parametrized by $s \in I$, one obtains the following system of linear ODEs governing the rotation tensor

$$\frac{d}{ds}\mathbf{R} = \mathbf{R}\mathbf{K}, \tag{80}$$

where

$$K^C{}_A(s) = \Omega^C{}_{AB}(s)\dot{X}^B(s). \tag{81}$$

Note that $K_{BA} = -K_{AB}$. Therefore, (80) is a linear ODE for $\mathbf{R} \in SO(3)$, and $\mathbf{K} \in \mathfrak{so}(3)$, the Lie algebra of the Lie group $SO(3)$. For each a

$$\frac{dR^a{}_A}{ds} - \Omega^C{}_{AB}R^a{}_C\dot{X}^B(s) = 0. \tag{82}$$

This is the equation of parallel transport of $R^a{}_A$ along the curve γ when $\mathcal{B}$ is equipped with the connection $\boldsymbol{\Omega}$. Let us assume that $\mathbf{R}(0) = \mathbf{R}_0$. We see that rotation tensor at s is the parallel transport of $\mathbf{R}_0$. It can be shown that in a simply connected body the integrality conditions of (82) are equivalent to vanishing of curvature tensor of $\mathbf{C}^\flat$ [39]. For solving (80) in [58] we used product integration and wrote the solution as

$$\mathbf{R}(s) = \mathbf{R}_0 \prod_0^s(\gamma)\, e^{\mathbf{K}(\xi)d\xi}, \tag{83}$$

where $\mathbf{R}_0 = \mathbf{R}(s)$ is assumed to be given and $\prod_0^s(\gamma)\, e^{\mathbf{K}(\xi)d\xi}$ is the product integral of $\mathbf{K}$ along the path γ from 0 to s. For more details on product integration see [58], and [19, 47].

For a compatible $\mathbf{C}^\flat$, the rotation tensor calculated from (83) must be independent of the path γ. Therefore, for any closed path γ in $\mathcal{B}$

$$\prod_\gamma e^{\mathbf{K}(s)ds} = \mathbf{I}. \tag{84}$$

It was shown in [58] that a necessary and sufficient condition is

$$\int_0^1 \mathbf{K}(s)ds = \mathbf{0}, \tag{85}$$

where $\gamma : [0, 1] \to \mathcal{B}$ is any closed path.

$\mathbf{C}^\flat$-compatibility is formulated as follows. Given $\mathbf{C}$, $\mathbf{U} = \sqrt{\mathbf{C}}$ is determined uniquely. The system of ODEs (78) govern the rotation $\mathbf{R}$. The calculated rotation is path-independent if and only if the curvature tensor of $\mathbf{C}^\flat$ vanishes, and (85) are satisfied over each generator of the first homology group.

Proposition 5.7 (Yavari [58]) *The necessary and sufficient* $\mathbf{C}^\flat$*-compatibility conditions in a non-simply-connected body* $\mathcal{B}$ *are:*

(i) $\mathcal{R}(\mathbf{C}^\flat) = \mathbf{0}$ *in* $\mathcal{B}$,
(ii) $\int_{c_i} \mathbf{K}(s)ds = \mathbf{0},\ i = 1, \ldots, b_1(\mathcal{B})$, *where* c_i*'s are generators of* $H_1(\mathcal{B}; \mathbb{R})$,
(iii) The above two conditions guarantee that deformation gradient $\mathbf{F} = \mathbf{R}\sqrt{\mathbf{C}}$ *is uniquely determined. For the deformation gradient to be compatible, one must have,* $\int_{\gamma_i} \mathbf{F}d\mathbf{X} = \mathbf{0},\ i = 1, \ldots, b_1(\mathcal{B})$.

5.5 Compatibility Equations in Linearized Elasticity

Suppose φ_ϵ is a 1-parameter family of deformations around a reference motion $\mathring{\varphi}$, and let $\epsilon = 0$ correspond to the reference motion. The displacement field is defined as [53, 60]:

$$\mathbf{U}(X) = \delta\varphi(X) = \left.\frac{d\varphi_\epsilon(X)}{d\epsilon}\right|_{\epsilon=0}. \tag{86}$$

The linearization of the deformation gradient is written as [33, 60]: $\mathscr{L}(\mathbf{F}) = \nabla\mathbf{U}$. In components, $\mathscr{L}(\mathbf{F})^a{}_A = U^a{}_{|A} = \frac{\partial U^a}{\partial X^A} + \gamma^a{}_{bc} F^b{}_A U^c$, where $\gamma^a{}_{bc}$ are the connection coefficients of the Riemannian manifold $(\mathcal{S}, \mathbf{g})$. The spatial and material strain tensors are defined, respectively, as $\mathbf{e} = \frac{1}{2}(\mathbf{g} - \varphi_{t*}\mathbf{G})$, and $\mathbf{E} = \frac{1}{2}(\varphi_t^*\mathbf{g} - \mathbf{G})$ [33]. In components, $e_{ab} = \frac{1}{2}\left(g_{ab} - G_{AB}F_a{}^A F_b{}^B\right)$, and $E_{AB} = \frac{1}{2}(C_{AB} - G_{AB})$. It can be shown that $\mathscr{L}(\mathbf{C})_{AB} = 2F^a{}_A F^b{}_B\, \epsilon_{ab}$, where $\epsilon_{ab} = \frac{1}{2}(u_{a|b} + u_{b|a})$ is the linearized strain, and $\mathbf{u} = \mathbf{U} \circ \varphi^{-1}$. Thus, $\mathscr{L}(\mathbf{C}) = 2\varphi_t^*\boldsymbol{\epsilon}$, and hence, $\boldsymbol{\epsilon} = \varphi_{t*}\mathscr{L}(\mathbf{E})$. When the ambient space is Euclidean and one uses Cartesian coordinates the covariant derivatives reduce to partial derivatives and the classical definition of linear strain in terms of partial derivatives is recovered, i.e.,

$$\epsilon_{ab} = \frac{1}{2}\left(\frac{\partial u_a}{\partial x^b} + \frac{\partial u_b}{\partial x^a}\right). \tag{87}$$

The necessary and sufficient conditions for compatibility in terms of $\mathbf{F}$ are $\int_\gamma \mathbf{F}d\mathbf{X} = \mathbf{0}$, for every loop γ in $\mathcal{B}$. The linearization of this condition reads $\int_\gamma \nabla\mathbf{U}d\mathbf{X} = \mathbf{0}$. In components, $\int_\gamma u^a{}_{,B}dX^B = 0$, where $\{X^A\}$ and $\{x^a\}$ are coordinate charts for $\mathcal{B}$ and $\mathcal{S}$, respectively. Linearization is assumed about the standard embedding of $\mathcal{B}$ in $\mathbb{R}^N$, i.e., $F^a{}_A = \delta^a_A$. This implies that $dX^B = \frac{\partial X^B}{\partial x^b}dx^b = \delta^B_b dx^b$, and thus

$$\int_\gamma u_{a,B} dx^B = \int_\gamma u_{a,b} dx^b = \int_\gamma (e_{ab} + \omega_{ab})\, dx^b = 0, \tag{88}$$

where $e_{ab} = u_{(a,b)} = \frac{1}{2}(u_{a,b} + u_{b,a})$, and $\omega_{ab} = u_{[a,b]} = \frac{1}{2}(u_{a,b} - u_{b,a})$, are the linearized strain and rotation tensors, respectively. Note that

$$\int_\gamma \omega_{ab} dx^b = \int_\gamma \left[(x^c \omega_{ac})_{,b} - x^c \omega_{ac,b}\right] dx^b = -\int_\gamma x^c \omega_{ac,b} dx^b. \tag{89}$$

The gradient of the rotation tensor is rewritten as

$$\begin{aligned} \omega_{ac,b} &= \frac{1}{2}\left(u_{a,cb} - u_{c,ab}\right) + \frac{1}{2}\left(u_{b,ac} - u_{b,ac}\right) \\ &= \frac{1}{2}\left(u_{a,bc} + u_{b,ac}\right) - \frac{1}{2}\left(u_{c,abc} + u_{b,ac}\right) \\ &= e_{ab,c} - e_{bc,a}. \end{aligned} \tag{90}$$

For a given e_{ab}, ω_{ab} is calculated by integrating $\omega_{ab,c} = e_{ac,b} - e_{cb,a}$ along an arbitrary curve. To ensure that the rotation field is well-defined one must have $\int_\gamma \left(e_{ac,b} - e_{cb,a}\right) dx^c = 0$, for any closed path $\gamma \in \mathcal{B}$. When γ is null-homotopic, $\gamma = \partial\Omega$, on a 2-submanifold of $\mathcal{B}$, and hence

$$\begin{aligned} \int_\gamma \left(e_{ac,b} - e_{cb,a}\right) dx^c &= \int_\Omega d\left(e_{ac,b} - e_{cb,a}\right) \wedge dx^c \\ = \int_\Omega &\left(e_{ad,bc} + e_{bc,ad} - e_{ac,bd} - e_{bd,ac}\right)(dx^c \wedge dx^d) = 0, \end{aligned} \tag{91}$$

where $\{(dx^c \wedge dx^d)\} = \{dx^c \wedge dx^d\}_{c<d}$ is a basis of 2-forms. Note that (91) is equivalent to $\mathsf{curl} \circ \mathsf{curl}\, \mathbf{e} = \mathbf{0}$, which is the classical bulk compatibility equation of linear elasticity [31]. From (90), one writes

$$\int_\gamma u_{a,b} dx^b = \int_\gamma \mathsf{C}_{ab} dx^b = 0, \tag{92}$$

where $\mathsf{C}_{ab} = e_{ab} - x^c(e_{ab,c} - e_{bc,a})$ is called the Cesàro tensor. The above representation is called the Cesàro integral [13]. For a null-homotopic path γ that lies on a surface $\mathcal{P} \subset \mathcal{B}$, $\gamma = \partial\Omega$ for some $\Omega \subset \mathcal{P}$. Hence, using Stokes' theorem one writes

$$\begin{aligned} \int_\gamma \mathsf{C}_{ab} dx^b &= \int_\Omega d\mathsf{C}_{ab} \wedge dx^b = \int_\Omega \mathsf{C}_{ab,c} dx^c \wedge dx^b \\ &= \int_\Omega \left[e_{bc,a} - x^d\left(e_{ab,cd} - e_{bd,ac}\right)\right] dx^c \wedge dx^b. \end{aligned} \tag{93}$$

Due to symmetry of strain $e_{bc,a}dx^c \wedge dx^b = 0$, and hence

$$\int_\gamma u_{a,b}dx^b = \int_\Omega x^d \left(e_{bd,ac} - e_{ab,cd}\right) dx^c \wedge dx^b$$
$$= \int_\Omega x^d \left(e_{ab,cd} + e_{cd,ab} - e_{ac,bd} - e_{bd,ac}\right) dx^b \wedge dx^c = 0. \tag{94}$$

One should note that (94) are equivalent to $\mathsf{curl} \circ \mathsf{curl}\, \mathbf{e} = \mathbf{0}$, i.e., the classical bulk compatibility equations [31].

Proposition 5.8 (Yavari [58]) *The necessary and sufficient conditions for compatibility conditions for the linearized strain* $\mathbf{e} = \mathfrak{L}_\mathbf{u}\mathbf{g}$ *in* $\mathcal{B}$ *are:*

(i) $\mathsf{curl} \circ \mathsf{curl}\, \mathbf{e} = \mathbf{0}$ *in* $\mathcal{B}$*,*
(ii) For each generator of $H_1(\mathcal{B}; \mathbb{R})$

$$\int_{c_i} \mathcal{C}d\mathbf{X} = \mathbf{0}, \quad \& \quad \int_{c_i} \left(e_{ac,b} - e_{cb,a}\right) dx^c = 0, \qquad i = 1, \ldots, b_1(\mathcal{B}). \tag{95}$$

We call $(95)_1$ and $(95)_2$ the Cesàro and the *rotation compatibility* equations, respectively. Note that in dimension n ($n = 2$ or 3) for each c_i, there are n Cesàro and $n(n-1)/2$ rotation compatibility equations. Hence, each c_i has $n(n+1)$ complementary compatibility equations. In dimension three, there are six bulk compatibility equations, and six auxiliary compatibility equations for each generator of the first homology group. We should mention that this is consistent with Weingarten's theorem [52] that says that if a body is cut along a surface the jump in the displacement field is a rigid-body motion, see Love [31] for a detailed discussion (he calls homotopic paths, "reconcilable circuits" and a null-homotopic path, a "evanescible circuit"). Zubov [61] and Casey [12] demonstrated the validity of Weingarten's theorem for finite strains (see also Acharya [2]). In [58] it was pointed out that the discussion in [46] regarding sufficient compatibility equations in linear elasticity is flawed as Skalak et al. missed the rotation compatibility conditions $(95)_2$. In [58, Example 29] the rotation compatibility conditions were trivially satisfied. Next, we provide an example of an incompatible strain field for which the Cesàro compatibility conditions are satisfied while the rotation compatibility conditions are not satisfied.

Example 5.9 Consider a single wedge disclination [17, 56] along the z-axis in an infinite linear elastic body. The linearized strain field of the disclination in the Cartesian coordinates (x, y, z) reads [17]

$$
\begin{aligned}
e_{11} &= \frac{\Omega}{4\pi(1-\nu)}\left[(1-2\nu)\ln\sqrt{x^2+y^2}+\frac{y^2}{x^2+y^2}\right],\\
e_{22} &= \frac{\Omega}{4\pi(1-\nu)}\left[(1-2\nu)\ln\sqrt{x^2+y^2}+\frac{x^2}{x^2+y^2}\right],\\
e_{12} &= -\frac{\Omega}{4\pi(1-\nu)}\frac{xy}{x^2+y^2},\qquad e_{33}=e_{13}=e_{23}=0,
\end{aligned}
\tag{96}
$$

where Ω is the Frank vector of the disclination. For this strain field the bulk compatibility equation $e_{11,yy}+e_{22,xx}-2e_{12,xy}=0$ is satisfied in $\mathbb{R}^3\setminus$ z-axis. Eq. $(95)_1$ gives the following two Cesàro compatibility equations

$$
\begin{aligned}
&\int_\gamma \left[e_{11}-y(e_{11,y}-e_{12,x})\right]dx+\left[e_{12}-y(e_{12,y}-e_{22,x})\right]dy=0,\\
&\int_\gamma \left[e_{12}-x(e_{12,x}-e_{11,y})\right]dx+\left[e_{22}-x(e_{22,x}-e_{12,y})\right]dy=0,
\end{aligned}
\tag{97}
$$

where γ is any closed curve lying in a plane normal to the z-axis and enclosing the origin. Using a square path with corners $(a,-a,0)$, $(a,a,0)$, $(-a,a,0)$, and $(-a,-a,0)$, where $a>0$, it is straightforward to show that the two Cesàro compatibility equations are trivially satisfied. For this strain field there is only one rotation compatibly equation, which is not satisfied, namely

$$
\int_\gamma (e_{11,y}-e_{12,x})dx+(e_{12,y}-e_{22,x})dy=-\Omega\neq 0.
\tag{98}
$$

6 Differential Complexes in Nonlinear Elasticity

For a flat 3-manifold $\mathcal{B}$, let $\Omega^k(\mathcal{B})$ be the space of smooth k-forms on $\mathcal{B}$, i.e., $\boldsymbol{\alpha}\in\Omega^k(\mathcal{B})$ is an anti-symmetric $\binom{0}{k}$-tensor with smooth components $\alpha_{i_1\cdots i_k}$. The exterior derivatives $d_k:\Omega^k(\mathcal{B})\to\Omega^{k+1}(\mathcal{B})$ are linear differential operators that satisfy $d_{k+1}\circ d_k=0$. In order to simplify the notation we drop the subscript k in d_k. The following sequence of spaces and operators

$$
0\longrightarrow\Omega^0(\mathcal{B})\xrightarrow{d}\Omega^1(\mathcal{B})\xrightarrow{d}\Omega^2(\mathcal{B})\xrightarrow{d}\Omega^3(\mathcal{B})\longrightarrow 0
\tag{99}
$$

is denoted by $(\Omega(\mathcal{B}),d)$ and is called the de Rham complex. Note that each operator is linear and the composition of any two successive operators vanishes. Also the first operator on the left sends 0 to the zero function and the last operator on the right sends $\Omega^3(\mathcal{B})$ to zero. The property $d\circ d=0$ implies that $\operatorname{im}d_k\subset\ker d_{k+1}$, where $\operatorname{im}d_k$ is the image of d_k and $\ker d_{k+1}$ is the kernel of d_{k+1}. If $\operatorname{im}d_k=\ker d_{k+1}$, the complex is exact. The de Rham cohomology groups are defined as $\mathrm{H}^k_{dR}(\mathcal{B}) = \ker d_k / \mathrm{im}\, d_{k-1}$. A complex is exact if and only if all $\mathrm{H}^k_{dR}(\mathcal{B})$ are the trivial group $\{0\}$. Cohomology groups quantify non-exactness of a complex.

For $\boldsymbol{\beta} \in \Omega^k(\mathcal{B})$, the necessary and sufficient condition for the existence of a solution for the PDE $d\boldsymbol{\alpha} = \boldsymbol{\beta}$ is $\boldsymbol{\beta} \in \mathrm{im}\, d$. If $(\Omega(\mathcal{B}), d)$ is exact, $\boldsymbol{\beta} \in \mathrm{im}\, d$ if and only if $d\boldsymbol{\beta} = 0$. Assuming that $\mathrm{H}^k_{dR}(\mathcal{B})$ is finite-dimensional, de Rham's theorem states that $\dim \mathrm{H}^k_{dR}(\mathcal{B}) = b_k(\mathcal{B})$, where $b_k(\mathcal{B})$ is the k-th Betti number—a purely topological property of $\mathcal{B}$. Thus, $\boldsymbol{\beta} \in \mathrm{im}\, d$ if and only if

$$d\boldsymbol{\beta} = 0, \text{ and } \int_{c_k} \boldsymbol{\beta} = 0, \quad k = 1, \ldots, b_k(\mathcal{B}), \tag{100}$$

where c_k are the generators of the kth homology group of $\mathcal{B}$.

Sometimes one may be able to establish a connection between a given complex and the de Rham complex. A complex closely related to the de Rham complex is the grad-curl-div complex of vector analysis. Let $C^\infty(\mathcal{B})$ and $\mathfrak{X}(\mathcal{B})$ be the spaces of smooth real-valued functions and smooth vector fields on $\mathcal{B}$, an open subset of $\mathbb{R}^3$. Consider the three operators grad : $C^\infty(\mathcal{B}) \to \mathfrak{X}(\mathcal{B})$, curl : $\mathfrak{X}(\mathcal{B}) \to \mathfrak{X}(\mathcal{B})$, and div : $\mathfrak{X}(\mathcal{B}) \to C^\infty(\mathcal{B})$. The classical identities curl $\circ$ grad $= 0$, and div $\circ$ curl $= 0$, allow one to write the following complex

$$0 \longrightarrow C^\infty(\mathcal{B}) \xrightarrow{\text{grad}} \mathfrak{X}(\mathcal{B}) \xrightarrow{\text{curl}} \mathfrak{X}(\mathcal{B}) \xrightarrow{\text{div}} C^\infty(\mathcal{B}) \longrightarrow 0 \tag{101}$$

which is called the grad-curl-div complex or simply the gcd complex. One can show that the gcd complex is equivalent to the de Rham complex, or more precisely is isomorphic to the de Rham complex [4]. As an example of the application of this isomorphism, one can show that a vector field $\mathbf{w}$ is the gradient of a function if and only if

$$\operatorname{curl} \mathbf{w} = \mathbf{0}, \text{ and } \int_\gamma \mathbf{w} \cdot \mathbf{t}_\gamma \, ds = 0, \quad \forall \gamma \subset \mathcal{B}, \tag{102}$$

where γ is an arbitrary closed curve in $\mathcal{B}$, $\mathbf{t}_\gamma$ is the unit tangent vector field along γ, and $\mathbf{w} \cdot \mathbf{t}_\gamma$ is the standard inner product of $\mathbf{w}$ and $\mathbf{t}_\gamma$ in $\mathbb{R}^3$.

It turns out that when using deformation gradient $\mathbf{F}$ and its corresponding stress, i.e., the first Piola–Kirchhoff stress $\mathbf{P}$, the differential complex of nonlinear elasticity is isomorphic to the $\mathbb{R}^3$-valued de Rham complex [4]. Let us assume that the ambient space is Euclidean, i.e., $\mathcal{S} = \mathbb{R}^3$ with Cartesian coordinates $\{x^i\}$. Suppose $\varphi : \mathcal{B} \to \mathcal{S}$ is a smooth map and define the following operators for two-point tensors in $\Gamma(T\varphi(\mathcal{B}))$ and $\Gamma(T\varphi(\mathcal{B}) \otimes T\mathcal{B})$

$$\begin{aligned}
\mathbf{Grad} &: \Gamma(T\varphi(\mathcal{B})) \to \Gamma(T\varphi(\mathcal{B}) \otimes T\mathcal{B}), && (\mathbf{Grad\,U})^{iI} = U^i{}_{,I}, \\
\mathbf{Curl} &: \Gamma(T\varphi(\mathcal{B}) \otimes T\mathcal{B}) \to \Gamma(T\varphi(\mathcal{B}) \otimes T\mathcal{B}), && (\mathbf{CurlF})^{iI} = \varepsilon_{IKL} F^{iL}{}_{,K}, \\
\mathbf{Div} &: \Gamma(T\varphi(\mathcal{B}) \otimes T\mathcal{B}) \to \Gamma(T\varphi(\mathcal{B})), && (\mathbf{Div\,F})^i = F^{iI}{}_{,I}.
\end{aligned}$$

Note that **Curl** $\circ$ **Grad** $= \mathbf{0}$, and **Div** $\circ$ **Curl** $= \mathbf{0}$. Therefore, the GCD complex or the *nonlinear elasticity complex* is written as:

$$\mathbf{0} \longrightarrow \Gamma(T\varphi(\mathcal{B})) \xrightarrow{\mathbf{Grad}} \Gamma(T\varphi(\mathcal{B}) \otimes T\mathcal{B}) \xrightarrow{\mathbf{Curl}} \Gamma(T\varphi(\mathcal{B}) \otimes T\mathcal{B}) \xrightarrow{\mathbf{Div}} \Gamma(T\varphi(\mathcal{B})) \longrightarrow \mathbf{0}$$

Any $\mathbb{R}^3$-valued k-form $\boldsymbol{\alpha} \in \Omega^k(\mathcal{B};\mathbb{R}^3)$ can be written as $\boldsymbol{\alpha} = (\alpha^1, \alpha^2, \alpha^3)$, where $\alpha^i \in \Omega^k(\mathcal{B})$, $i = 1, 2, 3$. The exterior derivative $\mathbf{d} : \Omega^k(\mathcal{B};\mathbb{R}^3) \to \Omega^{k+1}(\mathcal{B};\mathbb{R}^3)$ is defined as $\mathbf{d}\boldsymbol{\alpha} = (d\alpha^1, d\alpha^2, d\alpha^3)$. From $d \circ d = 0$, one concludes that $\mathbf{d} \circ \mathbf{d} = \mathbf{0}$, which gives the $\mathbb{R}^3$-valued de Rham complex $\big(\Omega(\mathcal{B};\mathbb{R}^3), \mathbf{d}\big)$.

Let us define the following isomorphisms

$$\begin{aligned}
&\boldsymbol{I}_0 : \Gamma(T\varphi(\mathcal{B})) \to \Omega^0(\mathcal{B};\mathbb{R}^3), && [\boldsymbol{I}_0(\mathbf{U})]^i = U^i,\\
&\boldsymbol{I}_1 : \Gamma(T\varphi(\mathcal{B}) \otimes T\mathcal{B}) \to \Omega^1(\mathcal{B};\mathbb{R}^3), && [\boldsymbol{I}_1(\mathbf{F})]^i{}_J = F^{iJ},\\
&\boldsymbol{I}_2 : \Gamma(T\varphi(\mathcal{B}) \otimes T\mathcal{B}) \to \Omega^2(\mathcal{B};\mathbb{R}^3), && [\boldsymbol{I}_2(\mathbf{F})]^i{}_{JK} = \varepsilon_{JKL}F^{iL},\\
&\boldsymbol{I}_3 : \Gamma(T\varphi(\mathcal{B})) \to \Omega^3(\mathcal{B};\mathbb{R}^3), && [\boldsymbol{I}_3(\mathbf{U})]^i{}_{123} = U^i,
\end{aligned}$$

where ε_{JKL} is the permutation symbol. The following diagram commutes [4]

$$\begin{array}{ccccccccc}
0 \longrightarrow & \Gamma(T\varphi(\mathcal{B})) & \xrightarrow{\text{Grad}} & \Gamma(T\varphi(\mathcal{B})\otimes T\mathcal{B}) & \xrightarrow{\text{Curl}} & \Gamma(T\varphi(\mathcal{B})\otimes T\mathcal{B}) & \xrightarrow{\text{Div}} & \Gamma(T\varphi(\mathcal{B})) & \longrightarrow 0\\
& \downarrow I_0 & & \downarrow I_1 & & \downarrow I_2 & & \downarrow I_3 & \\
0 \longrightarrow & \Omega^0(\mathrm{B};\mathbb{R}^3) & \xrightarrow{\mathrm{d}} & \Omega^1(\mathrm{B};\mathbb{R}^3) & \xrightarrow{\mathrm{d}} & \Omega^2(\mathrm{B};\mathbb{R}^3) & \xrightarrow{\mathrm{d}} & \Omega^3(\mathrm{B};\mathbb{R}^3) & \longrightarrow 0
\end{array}$$

The above isomorphisms induce a cohomology isomorphism $\mathrm{H}^k_{\mathsf{GCD}}(\mathcal{B}) \approx \oplus_{i=1}^3 \mathrm{H}^k_{dR}(\mathcal{B})$, where $\mathrm{H}^k_{\mathsf{GCD}}(\mathcal{B})$ is the k-th cohomology group of the GCD complex. Let $\langle \mathbf{F}, \mathbf{W} \rangle := \sum_{i,I=1}^3 F^{iI} W^I \mathbf{e}_i$, where $\{\mathbf{e}_i\}$ is the standard basis of $\mathbb{R}^3$. The following result can be proved using the fact that the nonlinear elasticity and the $\mathbb{R}^3$-valued de Rham complexes are isomorphic.

Theorem 6.1 (Angoshtari and Yavari [4]) *Given* $\mathbf{F} \in \Gamma(T\varphi(\mathcal{B}) \otimes T\mathcal{B})$, *there exists* $\mathbf{U} \in \Gamma(T\varphi(\mathcal{B}))$ *such that* $\mathbf{F} = \mathbf{Grad}\,\mathbf{U}$, *if and only if*

$$\mathbf{Curl}\,\mathbf{F} = \mathbf{0}, \quad \text{and} \quad \int_\gamma \langle \mathbf{F}, \mathbf{t}_\gamma \rangle dS = \mathbf{0}, \quad \forall \gamma \subset \mathcal{B}, \tag{103}$$

where γ *is any closed curve in* $\mathcal{B}$, *and* $\mathbf{t}_\gamma$ *is the unit tangent vector field along* γ. *Moreover, there exists* $\boldsymbol{\Psi} \in \Gamma(T\varphi(\mathcal{B}) \otimes T\mathcal{B})$ *such that* $\mathbf{F} = \mathbf{Curl}\,\boldsymbol{\Psi}$, *if and only if*

$$\mathbf{Div}\,\mathbf{F} = \mathbf{0}, \quad \text{and} \quad \int_{\mathcal{C}} \langle \mathbf{F}, \mathbf{N}_{\mathcal{C}} \rangle dA = \mathbf{0}, \quad \forall \mathcal{C} \subset \mathcal{B}, \tag{104}$$

where $\mathcal{C}$ *is any closed surface in* $\mathcal{B}$ *and* $\mathbf{N}_{\mathcal{C}}$ *is the unit outward normal vector field of* $\mathcal{C}$.

Using the first Piola–Kirchhoff stress **P**, one defines a complex that describes both the kinematics and the kinetics of motion. The displacement field $\mathbf{U} \in \Gamma(T\varphi(\mathcal{B}))$ is defined as $\mathbf{U}(X) = \varphi(X) - X \in T_{\varphi(X)}\mathcal{S}$, $\forall X \in \mathcal{B}$. Then, $\mathbf{Grad\,U}$ is the displacement gradient, and $\mathbf{Curl} \circ \mathbf{Grad\,U} = \mathbf{0}$ expresses the compatibility of the displacement gradient. On the other hand, $\mathbf{P} = \mathbf{Curl}\,\boldsymbol{\Psi}$, where $\boldsymbol{\Psi}$ is a stress function. $\mathbf{DivP} = \mathbf{0}$ are the equilibrium equations. Therefore, the GCD complex or the nonlinear elasticity complex contains both the kinematics and the kinetics of motion as schematically shown below.

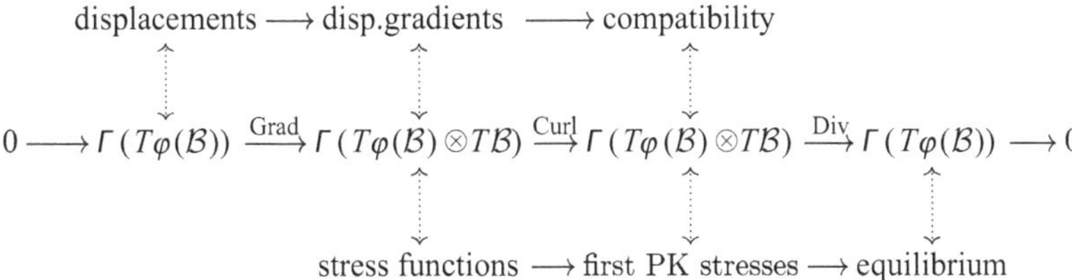

Using the differential complex of nonlinear elasticity, a new family of mixed finite elements—*compatible-strain mixed finite element methods* (CSFEMs)—has been introduced for both compressible and incompressible nonlinear elasticity [6, 43, 44]. CSFEMs are capable of capturing very large strains and accurately approximating stresses. CSFEMs, by construction, satisfy both the Hadamard jump conditions, and the continuity of traction at the discrete level. This makes them quite efficient for modeling heterogeneous solids. Moreover, CSFEMs seem to be free from numerical artifacts such as checkerboarding of pressure, hourglass instability, and locking.

References

1. Abraham, R., J.E. Marsden and T. Ratiu [1988], *Manifolds, Tensor Analysis, and Applications*, Springer-Verlag, New York.
2. Acharya, A. [2019], On Weingarten-Volterra defects. *Journal of Elasticity* **134**(1):79–101.
3. Adams, C.C. [2004], *The Knot Book*. American Mathematical Society, Providence.
4. Angoshtari, A. and A. Yavari, Differential complexes in continuum mechanics, *Archive for Rational Mechanics and Analysis* **216**, 2015, pp. 193–220.
5. Angoshtari, A. and A. Yavari, On the compatibility equations of nonlinear and linear elasticity in the presence of boundary conditions, *Zeitschrift für Angewandte Mathematik und Physik (ZAMP)* **66**(6), 2015, pp. 3627–3644.
6. Angoshtari, A., M.F. Shojaei, and A. Yavari, Compatible-strain mixed finite element methods for 2D nonlinear elasticity, *Computer Methods in Applied Mechanics and Engineering* **313**, 2017, pp. 596–631.
7. Arnold, V. I. and Khesin, B. A. [1999], *Topological Methods in Hydrodynamics*. Springer Science, New York.
8. Baumslag, G. [1993], *Topics in Combinatorial Group Theory*, Birkhäuser, Basel.
9. Berger, M. [2003], *A Panoramic View of Riemannian Geometry*, Springer-Verlag, New York.
10. Bredon, G. E. [1993], *Topology and Geometry*, Springer, New York.

11. Cantarella, J., DeTurck, D., and Gluck, H. [2002], Vector calculus and the topology of domains in 3-space. *American Mathematical Monthly* **109**(5):409–442.
12. Casey, J. [1995], On Volterra dislocations of finitely deforming continua. *Mathematics and mechanics of solids* **9**:473–492.
13. Cesàro, E. [1906], Sulle formole del Volterra, fondamentali nella teoria delle distorsioni elastiche. *Rendiconti d. Accademia R. di Napoli* **12**:311–321.
14. Ciarlet, P. G. and Laurent, F. [2002], On the recovery of a surface with prescribed first and second fundamental forms. *Journal de Mathématiques Pures et Appliquées* **81**(2):167–185.
15. Crowell, R.H. and Fox, R.H. [2008], *Introduction to Knot Theory*. Dover Publications, Mineola.
16. De Rham, G. [1931], Sur l'analysis situs des variétés à n dimensions. *Journal de Mathématiques Pures et Appliquées Sér 9* **10**:115–200.
17. de Wit, R. [1972], Theory of disclinations: IV. Straight disclinations, *Journal of Research of the National Bureau of Standards Section a-Physics and Chemistry A* **77**(5): 607–658.
18. Dieudonné, J. [1989], *A History of Algebraic and Differential Topology 1900–1960*. Birkhäuser, Boston.
19. Dollard, J.D. and Friedman, C.N. [1979], *Product Integration with Applications to Differential Equations*, Addison-Wesley, London.
20. Duff, G.F.D. [1966], Differential forms in manifolds with boundary. *Annals of Mathematics* **56**(1):115–127.
21. Epstein, D.B.A. [1966], Curves on 2-manifolds and isotopies. *Acta Mathematica* **115**(1):83–107.
22. Gibin, P. [2010], *Graphs, Surfaces and Homology*, Cambridge University Press, Cambridge.
23. Goldberg, S.I. [1998], *Curvature and Homology*, Dover Publications, Mineola.
24. Gross, P. and Kotiuga, P. R. [2004], *Electromagnetic Theory and Computation: A Topological Approach*, Cambridge University Press, Cambridge.
25. Hamilton, R. S. [1982], 3-manifolds with positive Ricci curvature. *Journal of Differential Geometry* **17**:255–306.
26. Hempel, J. [1976], *3-Manifolds*, AMS Chelsea Publishing.
27. Hetenyi, M. [1950], *Saint-Venant Theory of Torsion and Flexure* in Handbook of Experimental Stress Analysis, Editor M. Hetenyi, John Wiley & Sons, New York.
28. Lee, J.M. [1997], *Riemannian Manifold An Introduction to Curvature*, Springer-Verlag, New York.
29. Lefschetz, S. [1930], *Topology*, American Mathematical Society, New York.
30. Levine, H. I. [1963], Homotopic curves on surfaces. *Proceedings of the American Mathematical Society* **14**:986–990.
31. Love, A.E.H. [1927], *A Treatise on the Mathematical Theory of Elasticity*, Dover, New York.
32. Marden, A., Richards, I., and Rodin, B. [1966], On the regions bounded by homotopic curves. *Pacific Journal of Mathematics* **16**(2):337–339.
33. Marsden, J.E. and T.J.R. Hughes [1983], *Mathematical Foundations of Elasticity*, Dover, New York.
34. Matveev, S. [2003], *Algorithmic Topology and Classification of 3-Manifolds*, Springer, Berlin.
35. Maxwell, J.C. [1891], *A Treatise on Electricity and Magnetism*, Oxford University Press, Clarendon, England.
36. Michell, J. H. [1899], On the direct determination of stress in an elastic solid, with application to the theory of plates. *Proceedings of the London Mathematical Society* **s1-31**:100–124.
37. Munkres, J. (1984). *Elements of Algebraic Topology*, Springer, New York/Addison-Wesley, Menlo Park, CA.
38. Ozakin, A. and A. Yavari [2010], A geometric theory of thermal stresses, *Journal of Mathematical Physics* **51**, 032902.
39. Pietraszkiewicz, W. and Badur, J. [1983], Finite rotations in the description of continuum deformation. *International Journal of Engineering Science* **21**(9):1097–1115.
40. Poincaré, H. [1895], Analysis Situs. *Journal de l'École Polytechnique ser 2* **1**:1–123.

41. Schwarz, G. [2006], *Hodge Decomposition-A Method for Solving Boundary Value Problems*, Springer, Berlin.
42. Shield, R. T. [1973], Rotation associated with large strains. *SIAM Journal on Applied Mathematics* **25**(3):483–491.
43. Shojaei, M.F. and Yavari, A. [2018], Compatible-strain mixed finite element methods for incompressible nonlinear elasticity. *Journal of Computational Physics* **361**:247–279.
44. Shojaei, M.F. and Yavari, A. [2019], Compatible-strain mixed finite element methods for 3D compressible and incompressible nonlinear elasticity. *Computer Methods in Applied Mechanics and Engineering* **357**, 2019, 112610.
45. Simo, J. C. and J. E. Marsden [1984], On the rotated stress tensor and the material version of the Doyle-Ericksen formula. *Archive for Rational Mechanics and Analysis* **86**: 213–231.
46. Skalak, R. and Zargaryan, S. and Jain, R. K. and Netti, P. A. and Hoger, A. [1996], Compatibility and the genesis of residual stress by volumetric growth. *Journal of Mathematical Biology* **34**:889–914.
47. Slavík, A. [2007], *Product Integration, Its History and Applications*, Matfyzpress, Prague.
48. Sozio, F. and A. Yavari [2019], Nonlinear mechanics of accretion, *Journal of Nonlinear Science*, **29**(4):1813–1863
49. Sternberg, E. [1984], On Saint-Venant torsion and the plane problem of elastostatics for multiply connected domains. *Archive for Rational Mechanics and Analysis* **85**(4):295–310.
50. Stillwell, J. [1993], *Classical Topology and Combinatorial Group Theory*, Springer-Verlag, New York.
51. Volterra, V. [1907], Sur l'équilibre des corps élastiques multiplement connexes. *Annales Scientifiques de l'Ecole Normale Supérieure, Paris* **24**(3):401–518.
52. Weingarten, G. [1901], Sulle superficie di discontinuità nella teoria della elasticità dei corpi solidi. *Atti della Reale Accademia dei Lincei, Rendiconti, Series 5* **10**:57–60.
53. Yavari, A. and Ozakin, A. [2008]. Covariance in linearized elasticity. *ZAMP* **59**, 1081–1110.
54. Yavari, A. [2010], A geometric theory of growth mechanics. *Journal of Nonlinear Science* **20** (6), 2010, pp. 781–830.
55. Yavari, A. and A. Goriely [2012] Riemann-Cartan geometry of nonlinear dislocation mechanics. *Archive for Rational Mechanics and Analysis*, **205**(1), 2012, pp. 59–118.
56. Yavari, A. and A. Goriely [2012] Riemann-Cartan geometry of nonlinear disclination mechanics. *Mathematics and Mechanics of Solids*, **18**(1), 2013, pp. 91–102.
57. Yavari, A. and A. Goriely [2012] Weyl geometry and the nonlinear mechanics of distributed point defects, *Proceedings of the Royal Society A* **468**, 2012, pp. 3902–3922.
58. Yavari, A., Compatibility equations of nonlinear elasticity for non-simply-connected bodies, *Archive for Rational Mechanics and Analysis* **209**(1), 2013, pp. 237–253.
59. Yavari, A. and A. Goriely [2016] The anelastic Ericksen's problem: Universal eigenstrains and deformations in compressible isotropic elastic solids, *Proceedings of the Royal Society A* **472**, 2016, 20160690.
60. Yavari, A. and A. Golgoon, Nonlinear and linear elastodynamic transformation cloaking, *Archive for Rational Mechanics and Analysis* **234**(1), 2019, 211–316.
61. Zubov, L.M. [1997], *Nonlinear Theory of Dislocations and Disclinations in Elastic Bodies*. Springer, Berlin.

De Donder Construction for Higher Jets

Jędrzej Śniatycki and Reuven Segev

Contents

Abstract We construct generalizations of boundary forms of Kupferman, Olami and Segev and of the De Donder form to higher order Lagrangians. In principle, the resulting forms may depend on the coordinate system used. Nevertheless, it leads to many invariant conclusions. In some cases, like second order gravity, the construction leads to a unique De Donder form on the jet bundle that is independent of the choice of coordinates.

1 Introduction

In 1929, De Donder [5, 6], formulated an approach to study first order variational problems for several independent variables in terms of a differential form obtained from the Lagrangian by the Legendre transformation in all independent variables. His construction was generalized by Lepage [12] yielding a family of forms, each of which could be used in the same way as the De Donder form to reduce the original variational problem to a system of equations in exterior differential forms.

J. Śniatycki
Department of Mathematics and Statistics, University of Calgary, Calgary, AB, Canada

R. Segev (✉)
Department of Mechanical Engineering, Ben Gurion-University of the Negev, Beer-Sheva, Israel
e-mail: rsegev@bgu.ac.il

R. Segev, M. Epstein (eds.), *Geometric Continuum Mechanics*, Advances in Mechanics and Mathematics 43, https://doi.org/10.1007/978-3-030-42683-5_4

A geometric formulation of the De Donder construction in terms of jets was given by Śniatycki in 1970 [16]. It showed that the De Donder form depended only on the original Lagrangian and the canonical structure of the appropriate jet bundle. De Donder form, also called Poincaré-Cartan form, facilitated an invariant multisymplectic formulation of field theories [2, 7, 9, 10]. The De Donder construction was generalized in 2017 by Kupferman et al. [11] in the context of continuum mechanics of first order materials, to forms on the first jet bundle that need not be exact.

In 1977 Aldaya and Azcárraga [1] investigated generalization of Lepagean forms to higher order variational problems. For higher order Lagrangians, the natural generalization of the De Donder construction in terms of Ostrogradski's[1] Legendre transformation [14] in all independent variables leads to a form that depends on the adapted coordinate system used for its construction. This has led to search for additional geometric structures, which would ensure global existence of Poincaré-Cartan forms, see [3] and references quoted there. In the context of continuum mechanics, the analysis for higher order jets has been replaced by an analogous, yet further underdetermined, analysis on iterated jet bundles [15].

In this paper, we generalize De Donder approach to construct boundary forms that depend on the adapted coordinate system used in the construction. In continuum mechanics, use of boundary forms leads to splitting of the total force acting on the body into body force and surface traction. Moreover, this splitting is independent of the choice of the boundary form used. In calculus of variations, use of boundary forms leads to equations in exterior differential forms that are equivalent to the Euler–Lagrange equations. Infinitesimal symmetries of the theory lead to conservation laws valid for any choice of the boundary form. In an example, we show that the boundary conditions lead to independence of constants of motion of the choice of the boundary form.

2 Spaces of Smooth Sections

We are interested in geometric structure of calculus of variations with $m > 1$ independent variables and n dependent variables and its relation to continuum mechanics. Both theories deal with differentiation of functions on spaces of maps. There are several approaches to manifold structure of a space of maps. Here, we use the traditional approach of the calculus of variations flavoured by the insight from theory of differential spaces [17].

Consider a locally trivial fibration $\pi : N \to M$ with $\dim M = m$ and $\dim N = m + n$. Let K be an open relatively compact submanifold of M with smooth boundary ∂K. The closure $\bar{K} = K \cup \partial K$ is a manifold with boundary. A map $\sigma : \bar{K} \to N$ is a section of π if $\pi \circ \sigma$ is the identity on $\bar{K}$. In the spirit

[1]English transcription of the original Russian name is Ostrogradsky. However, Ostrogradski wrote in French and used the French transcription of his name.

of theory of differential spaces, we say that a section $\sigma : \bar{K} \to N$ is smooth if it extends to a smooth section of π defined on an open subset of M that contains $\bar{K}$. We denote by $S^{\infty}(\bar{K}, N)$ the space of smooth sections $\sigma : \bar{K} \to N$ of π. In the following, we assume that $\bar{K}$ is contained in the domain of a chart on M.

The next stage is to identify smooth functions on $S^{\infty}(\overline{K}, N)$. We use terminology of jet bundles reviewed in the Appendix. Let Λ be an m-form on the space $J^{\mathrm{k}}(M, N)$ of k-jets of sections of π. We say that Λ is semi-basic with respect to the source map $\pi^{\mathrm{k}} : J^{\mathrm{k}}(M, N) \to M$ if $X \lrcorner \Lambda = 0$ for every vector field X tangent to fibres of $\pi^{\mathrm{k}} : J^{\mathrm{k}}(M, N) \to M$, where $\lrcorner$ denotes the left interior product (contraction) of vectors and forms. The form Λ gives rise to the corresponding action functional

$$A : S^{\infty}(\bar{K}, N) \mapsto \mathbb{R} : \sigma \mapsto A(\sigma) = \int_K j^{\mathrm{k}}\sigma^{*}\Lambda. \tag{1}$$

Calculus of variations is concerned with study of critical points of action functionals. Let $\mathcal{A}$ denote the space of all action functionals on $S^{\infty}(\bar{K}, N)$. In other words, a function $F : S^{\infty}(\bar{K}, N) \to \mathbb{R}$ is in $\mathcal{A}$ if there exists an integer $\mathrm{k} \geq 0$, and an m-form Λ on $J^{\mathrm{k}}(M, N)$, semi-basic with respect to the source map $\pi^{\mathrm{k}} : J^{\mathrm{k}}(M, N) \to M$, such that

$$F(\sigma) = \int_K j^{\mathrm{k}}\sigma^{*}\Lambda, \ \ \forall \ \sigma \in S^{\infty}(\bar{K}, N). \tag{2}$$

Here, for $\mathrm{k} = 0$, we use identifications $J^{0}(M, N) = N$ and $\pi^{0} = \pi$.

The tangent space $T_{\sigma}S^{\infty}(\bar{K}, N)$ is the space of smooth maps $Y_{\sigma} : \bar{K} \to TN$ such that, for each $x \in \bar{K}$, $Y_{\sigma}(x) \in T_{\sigma(x)}N$ is tangent to the fibre $\pi^{-1}(x)$. It should be noted that every $Y_{\sigma} \in T_{\sigma}S^{\infty}(\bar{K}, N)$ can be extended to a vector field Y on N tangent to fibres of π and such that $Y_{\sigma}(x) = Y(\sigma(x))$ for every $x \in \bar{K}$. For $Y_{\sigma} \in T_{\sigma}S^{\infty}(\bar{K}, N)$ and $F(\sigma)$ given by Eq. (2), the derivative of F in direction Y_{σ} is

$$D_{Y_{\sigma}}F = \int_K j^{\mathrm{k}}\sigma^{*}(\mathcal{L}_{Y^{\mathrm{k}}}\Lambda), \tag{3}$$

where Y^{k} is the prolongation of an extension of Y_{σ} to a vertical vector field Y on N. It should be noted that the integral in (3) does not depend on the choice of the extension Y of Y_{σ}.

The next step is to identify vector fields on $S^{\infty}(\bar{K}, N)$. On manifolds, vector fields play two roles: they are global derivations of the differential structure, and they generate local one-parameter groups of diffeomorphisms. On manifolds with singularities, e.g. stratified spaces, global derivations need not generate local diffeomorphisms [17]. In this paper, we consider only vector fields on $S^{\infty}(\bar{K}, N)$ that are generated by global vertical vector fields Y on N as follows. A vertical vector field Y on N gives rise to a section $\boldsymbol{Y} : \bar{K} \to TS^{\infty}(\bar{K}, N)$ such that, $\boldsymbol{Y}(\sigma) = Y_{\sigma}$ for every $\sigma \in S^{\infty}(\bar{K}, N)$. In other words, for every F given by Eq. (2)

$$(YF)(\sigma) = D_{Y_\sigma} F = \int_K j^{\mathrm{k}}\sigma^*(\mathcal{L}_{Y^{\mathrm{k}}}\Lambda) \tag{4}$$

for every $\sigma \in S^\infty(\bar{K}, N)$. We denote by $\mathfrak{Y}(S^\infty(\bar{K}, N))$ the space of vector fields on $S^\infty(\bar{K}, N)$ defined above.

Now that we have vectors tangent to $S^\infty(\bar{K}, N)$, we can consider forms on $S^\infty(\bar{K}, N)$. Suppose that Φ is an $(m+1)$-form on $J^{\mathrm{k}}(M, N)$ such that $X \lrcorner \Phi$ is semi-basic with respect to the source map $\pi^{\mathrm{k}} : J^k(M, N) \to M$ for every vector field X on $J^{\mathrm{k}}(M, N)$ tangent to fibres of $\pi^{\mathrm{k}} : J^{\mathrm{k}}(M, N) \to M$. It gives rise to a 1-form $\mathbf{\Phi}$ on $S^\infty(\bar{K}, N)$ defined as follows. For every $\mathbf{Y} \in \mathfrak{Y}(S^\infty(\bar{K}, N))$ and $\sigma \in S^\infty(\bar{K}, N)$,

$$\langle \mathbf{\Phi} \mid \mathbf{Y} \rangle (\sigma) = \int_K j^{\mathrm{k}}\sigma^* \left(Y^{\mathrm{k}} \lrcorner \Phi \right). \tag{5}$$

If $Y_\sigma \in T_\sigma S^\infty(\bar{K}, N)$ is the restriction of Y to σ, then the restriction to $j^{\mathrm{k}}\sigma(K)$ of the prolongation Y^{k} of Y depends only on Y_σ and not on its extension off $\sigma(\bar{K})$. This shows that the 1-form $\mathbf{\Phi}$ restricts to a linear map $\mathbf{\Phi}_\sigma : T_\sigma S^\infty(\bar{K}, N) \to \mathbb{R}$ such that

$$\langle \mathbf{\Phi}_\sigma \mid Y_\sigma \rangle = \langle \mathbf{\Phi} \mid \mathbf{Y} \rangle (\sigma). \tag{6}$$

In applications to continuum mechanics, $\bar{K}$ represents the body manifold, sections $\sigma \in S^\infty(\bar{K}, N)$ are configurations of the body, and vectors $Y_\sigma \in T_\sigma S^\infty(\bar{K}, N)$ are virtual displacement fields. The form $\mathbf{\Phi}$ may be referred to as a force functional.

3 Boundary Forms

Let Φ be an $(m+1)$-form on $J^{\mathrm{k}}(M, N)$ such that $X \lrcorner \Phi$ is semi-basic with respect to the source map $\pi^{\mathrm{k}} : J^{\mathrm{k}}(M, N) \to M$ for every vector field X on $J^{\mathrm{k}}(M, N)$ tangent to fibres of $\pi^{\mathrm{k}} : J^{\mathrm{k}}(M, N) \to M$. Let $(x^i, y^a, z_i^a, \ldots, z^a_{i_1 \ldots i_k})$ be local coordinates on $J^{\mathrm{k}}(M, N)$. The corresponding local representation of Φ is

$$\Phi = \sum_{a=1}^{n} \left(\Phi_a \mathrm{d}y^a + \sum_{i=1}^{m} \Phi_a^i \mathrm{d}z_i^a + \ldots \sum_{i_1 \leq \ldots \leq i_l} \Phi_a^{i_1 \ldots i_l} \mathrm{d}z^a_{i_1 \ldots i_l} \right) \wedge \mathrm{d}_m x \tag{7}$$
$$+ \ldots + \sum_{a=1}^{n} \sum_{i_1 \leq \ldots \leq i_k} \Phi_a^{i_1 \ldots i_k} \mathrm{d}z^a_{i_1 \ldots i_k} \wedge \mathrm{d}_m x.$$

Note that, for every $l = 2, \ldots, \mathrm{k}$, coordinates $z^a_{i_1 \ldots i_l}$ are symmetric in indices $i_1, \ldots, i_l$ and so are $\Phi_a^{i_1 \ldots i_l}$. Therefore, the sum in Eq. (7) is taken only over the independent components. In the following, we modify the summation convention by the requirement that the indices $i_1, \ldots, i_l$ occurring in $z^a_{i_1,\ldots,i_l}$ are taken in a non-decreasing order. This allows us to rewrite Eq. (7) as follows,

$$\Phi = \left(\Phi_a \mathrm{d}y^a + \Phi^i_a \mathrm{d}z^a_i + \ldots \Phi^{i_1\ldots i_l}_a \mathrm{d}z^a_{i_1\ldots i_l} + \ldots + \Phi^{i_1\ldots i_k}_a \mathrm{d}z^a_{i_1\ldots i_k}\right) \wedge \mathrm{d}_m x. \tag{8}$$

An alternative approach would be use of multi-indices.

Theorem 1 *There exists locally a smooth m-form Ξ on $J^{2k-1}(M,N)$, which satisfies the following conditions.*

1. a. *Ξ is semi-basic with respect to the forgetful map $\pi^{2k-1}_{k-1} : J^{2k-1}(M,N) \to J^{k-1}(M,N)$. In other words, for any vector field X tangent to fibres of $\pi^{2k-1}_{k-1} : J^{2k-1}(M,N) \to J^{k-1}(M,N)$,*

$$X \lrcorner \Xi = 0.$$

   b. *For every vector field X on $J^{2k-1}(M,N)$ tangent to fibres of the source map $\pi^{2k-1} : J^{2k-1}(M,N) \to M$ the left interior product $X \lrcorner \Xi$ is semi-basic with respect to the source map. Thus, for every pair X_1, X_2 of vector fields on $J^{2k-1}(M,N)$ tangent to fibres of the source map $\pi^{2k-1} : J^{2k-1}(M,N) \to M$,*

$$X_2 \lrcorner\, (X_1 \lrcorner \Xi) = 0.$$

2. *For every section σ of $\pi : N \to M$,*

$$j^{2k-1}\sigma^* \Xi = 0,$$

where $j^{2k-1}\sigma^*\Xi = \Xi \circ \wedge^n T(j^{2k-1}\sigma)$ is the pull-back of Ξ by $j^{2k-1}\sigma : M \to J^{2k-1}(M,N)$.

3. *For every vector field X on $J^{2k-1}(M,N)$ tangent to fibres of the target map $\pi^{2k-1}_0 : J^{2k-1}(M,N) \to N$, and every section σ of $\pi : N \to M$,*

$$j^{2k-1}\sigma^* \left(X \lrcorner \left(\pi^{2k-1*}_{k} \Phi + \mathrm{d}\Xi\right)\right) = 0.$$

Proof The first and the second condition imply that Ξ is a linear combination of contact forms up to order k with coefficients given by forms that are semi-basic with respect to the source map. In local coordinates,

$$\begin{aligned}\Xi = & \, p^i_a (\mathrm{d}y^a - z^a_j \mathrm{d}x^j) \wedge \left(\frac{\partial}{\partial x^i} \lrcorner \mathrm{d}_m x\right) \\ & + p^{i_1 i_2}_a (\mathrm{d}z^a_{i_2} - z^a_{i_2 j} \mathrm{d}x^j) \wedge \left(\frac{\partial}{\partial x^{i_1}} \lrcorner \mathrm{d}_m x\right) \\ & + \ldots + p^{i_1 i_2 \ldots i_k}_a (\mathrm{d}z^a_{i_2\ldots i_k} - z^a_{i_2 i_2 \ldots i_k j} \mathrm{d}x^j) \wedge \left(\frac{\partial}{\partial x^{i_1}} \lrcorner \mathrm{d}_m x\right),\end{aligned} \tag{9}$$

where the coefficients $p_a^{i_1 i_2 \dots i_l}$ are symmetric in indices $i_2, \dots, i_l$, for $l = 3, \dots, k$ and the summation is taken over indices in non-decreasing order. Hence,

$$\Xi = \left(p_a^{i_1} \mathrm{d}y^a + p_a^{i_1 i_2} \mathrm{d}z_{i_2}^a + \dots + p_a^{i_1 i_2 \dots i_k} \mathrm{d}z_{i_2 \dots i_k}^a\right) \wedge \left(\frac{\partial}{\partial x^{i_1}} \lrcorner \mathrm{d}_m x\right) \tag{10}$$
$$- \left(p_a^{i_1} z_{i_1}^a + p_a^{i_1 i_2} z_{i_1 i_2}^a + \dots + p_a^{i_1 i_2 \dots i_k} z_{i_1 i_2 \dots i_k}^a\right) \mathrm{d}_m x.$$

Note that the symmetry of $z_{i_1 i_2 \dots i_l}^a$ in $i_1, \dots, i_l$ implies in Eq. (10) that only the fully symmetric parts of $p_a^{i_1 i_2 \dots i_l}$ contribute to the sum in the second term.

In order to use the third condition, we need the exterior differential of Ξ. Equation (9) yields

$$\mathrm{d}\Xi = \mathrm{d}p_a^i \wedge (\mathrm{d}y^a - z_j^a \mathrm{d}x^j) \wedge \left(\frac{\partial}{\partial x^i} \lrcorner \mathrm{d}_m x\right) + \tag{11}$$
$$\mathrm{d}p_a^{i_1 i_2} \wedge (\mathrm{d}z_{i_2}^a - z_{i_2 j}^a \mathrm{d}x^j) \wedge \left(\frac{\partial}{\partial x^{i_1}} \lrcorner \mathrm{d}_m x\right) + \dots +$$
$$+ \mathrm{d}p_a^{i_1 i_2 \dots i_k} \wedge (\mathrm{d}z_{i_2 \dots i_k}^a - z_{i_2 \dots i_k j}^a \mathrm{d}x^j) \wedge \left(\frac{\partial}{\partial x^{i_1}} \lrcorner \mathrm{d}_m x\right) +$$
$$- p_a^i \mathrm{d}z_j^a \wedge \mathrm{d}x^j \wedge \left(\frac{\partial}{\partial x^i} \lrcorner \mathrm{d}_m x\right) -$$
$$- p_a^{i_1 i_2} \mathrm{d}z_{i_2 j}^a \wedge \mathrm{d}x^j \wedge \left(\frac{\partial}{\partial x^{i_1}} \lrcorner \mathrm{d}_m x\right) + \dots +$$
$$- p_a^{i_1 i_2 \dots i_k} \mathrm{d}z_{i_2 \dots i_k j}^a \wedge \mathrm{d}x^j \wedge \left(\frac{\partial}{\partial x^{i_1}} \lrcorner \mathrm{d}_m x\right),$$

which can be simplified to

$$\mathrm{d}\Xi = \mathrm{d}p_a^i \wedge (\mathrm{d}y^a - z_j^a \mathrm{d}x^j) \wedge \left(\frac{\partial}{\partial x^i} \lrcorner \mathrm{d}_m x\right) + \tag{12}$$
$$\mathrm{d}p_a^{i_1 i_2} \wedge (\mathrm{d}z_{i_2}^a - z_{i_2 j}^a \mathrm{d}x^j) \wedge \left(\frac{\partial}{\partial x^i} \lrcorner \mathrm{d}_m x\right) + \dots +$$
$$+ \mathrm{d}p_a^{i_1 i_2 \dots i_k} \wedge (\mathrm{d}z_{i_2 \dots i_k}^a - z_{i_2 \dots i_k j}^a \mathrm{d}x^j) \wedge \left(\frac{\partial}{\partial x^i} \lrcorner \mathrm{d}_m x\right) +$$
$$- \left(p_a^i \mathrm{d}z_i^a + p_a^{i_1 i_2} \mathrm{d}z_{i_1 i_2}^a + \dots + p_a^{i_1 i_2 \dots i_k} \mathrm{d}z_{i_1 i_2 \dots i_k}^a\right) \wedge \mathrm{d}_m x.$$

Let X be a vector field on $J^{2k-1}(M, N)$ tangent to fibres of the target map $\pi_0^{2k-1} : J^{2k-1}(M, N) \to N$. In local coordinates,

$$X = X_i^a \frac{\partial}{\partial z_i^a} + \ldots + X_{i_1 \ldots i_{2k-1}}^a \frac{\partial}{\partial z_{i_1 \ldots i_{2k-1}}^a}. \tag{13}$$

Note that, for $l = 2, \ldots, 2k-1$, $X_{i_1 \ldots i_l}^a$ is symmetric in the indices $i_1, \ldots, i_l$. Then,

$$\begin{aligned} X \lrcorner \mathrm{d}\Xi = {} & (X p_a^i)(\mathrm{d}y^a - z_j^a \mathrm{d}x^j) \wedge \left(\frac{\partial}{\partial x^i} \lrcorner \mathrm{d}_m x \right) + \ldots + \\ & + (X p_a^{i_1 i_2 \ldots i_k})(\mathrm{d}z_{i_2 \ldots i_k}^a - z_{i_2 \ldots i_k j}^a \mathrm{d}x^j) \wedge \left(\frac{\partial}{\partial x^i} \lrcorner \mathrm{d}_m x \right) + \\ & - X_{i_2}^a \mathrm{d}p_a^{i_1 i_2} \wedge \left(\frac{\partial}{\partial x^i} \lrcorner \mathrm{d}_m x \right) - \ldots - \\ & - X_{i_2 \ldots i_k}^a \mathrm{d}p_a^{i_1 i_2 \ldots i_k} \wedge \left(\frac{\partial}{\partial x^i} \lrcorner \mathrm{d}_m x \right) + \\ & - \left(p_a^i X_i^a + p_a^{i_1 i_2} X_{i_1 i_2}^a + \ldots + p_a^{i_1 i_2 \ldots i_k} X_{i_1 i_2 \ldots i_k}^a \right) \mathrm{d}_m x, \end{aligned} \tag{14}$$

where $(X p_a^i)$ is the derivation of p_a^i in direction X, etc.

The first two lines of Eq. (14) do not contribute to $j^{2k-1}\sigma^*(X \lrcorner \mathrm{d}\Xi)$, because they are linear combinations of contact forms. Hence, Eq. (14) implies that

$$\begin{aligned} j^{2k-1}\sigma^* (X \lrcorner \mathrm{d}\Xi) = {} & -X_{i_2}^a P_{a,i_1}^{i_1 i_2} \mathrm{d}_m x - \ldots - X_{i_2 \ldots i_k}^a P_{a,i_1}^{i_1 i_2 \ldots i_k} \mathrm{d}_m x \\ & - \left(P_a^i X_i^a + P_a^{i_1 i_2} X_{i_1 i_2}^a + \ldots + P_a^{i_1 i_2 \ldots i_k} X_{i_1 i_2 \ldots i_k}^a \right) \mathrm{d}_m x, \end{aligned} \tag{15}$$

where

$$P_a^{i_1} = j^{2k-1}\sigma^* p_a^{i_1}, \;\; \ldots, \;\; P_a^{i_1 i_2 \ldots i_k} = j^{2k-1}\sigma^* p_a^{i_1 i_2 \ldots i_k}, \tag{16}$$

and the components of X are evaluated on the range of $j^{2k-1}\sigma$. Following the symmetry argument leading to Eq. (10), we can rewrite Eq. (15) in the form

$$\begin{aligned} j^{2k-1}\sigma^* (X \lrcorner \mathrm{d}\Xi) = {} & -X_{i_2}^a P_{a,i_1}^{i_1 i_2} d_n x - \ldots - X_{i_2 \ldots i_k}^a P_{a,i_1}^{i_1 i_2 \ldots i_k} \mathrm{d}_m x \\ & - \left(P_a^i X_i^a + P_a^{(i_1 i_2)} X_{i_1 i_2}^a + \ldots + P_a^{(i_1 i_2 \ldots i_k)} X_{i_1 i_2 \ldots i_k}^a \right) \mathrm{d}_m x, \end{aligned} \tag{17}$$

where $(i_1, \ldots, i_l)$ denotes symmetrization in indices $i_1, \ldots, i_l$. Therefore, the third condition of the theorem yields

$$\begin{aligned} 0 = {} & \Phi_a^i X_i^a + \ldots + \Phi_a^{i_1 \ldots i_k} X_{i_1 \ldots i_k}^a - X_{i_2}^a P_{a,i_1}^{i_1 i_2} - \\ & - \ldots - X_{i_2 \ldots i_k}^a P_{a,i_1}^{i_1 i_2 \ldots i_k} - \left(P_a^i X_i^a + P_a^{(i_1 i_2)} X_{i_1 i_2}^a + \ldots + P_a^{(i_1 i_2 \ldots i_k)} X_{i_1 i_2 \ldots i_k}^a \right), \end{aligned} \tag{18}$$

where $\Phi_a^i, \ldots, \Phi_a^{i_1 \ldots i_k}$ are evaluated on the range $j^k\sigma$. Equation (18) is equivalent to

$$\begin{aligned} 0 = {} & (\Phi_a^{i_1 \ldots i_k} - P_a^{(i_1 i_2 \ldots i_k)}) X^a_{i_1 i_2 \ldots i_k} + \\ & (\Phi_a^{i_2 \ldots i_k} - P_a^{(i_2 \ldots i_k)} - P_{a,i_1}^{i_1 i_2 \ldots i_k}) X^a_{i_2 \ldots i_k} + \ldots + \\ & + (\Phi_a^i - P_a^i - P_{a,i_1}^{i_1 i}) X^a_i . \end{aligned} \tag{19}$$

Since the components $X^a_{i_1 i_2 .. i_l}$ of X are arbitrary functions symmetric in the indices $i_1, \ldots, i_k$, it follows that

$$\begin{aligned} \Phi_a^{i_1 \ldots i_k} - P_a^{(i_1 i_2 \ldots i_k)} &= 0, \\ \Phi_a^{i_2 \ldots i_k} - P_a^{(i_2 \ldots i_k)} - P_{a,i_1}^{i_1 i_2 \ldots i_k} &= 0, \\ \Phi_a^{i_l \ldots i_k} - P_a^{(i_l \ldots i_k)} - P_{a,i_{l-1}}^{i_{l-1} i_l \ldots i_k} &= 0, \quad \text{for } l = 3, \ldots, k-2, \\ \Phi_a^i - P_a^i - P_{a,i_1}^{i_1 i} &= 0. \end{aligned} \tag{20}$$

This shows that there is no unique local form Ξ satisfying the conditions of our theorem. In order to prove existence, we are free to impose an additional condition on the coefficients $p_a^{i_1 i_2 \ldots i_l}$.

An obvious generalization of the De Donder construction corresponds to an additional condition that all $p_a^{i_1 i_2 \ldots i_l}$ are fully symmetric in all indices $i_1, \ldots, i_l$. With this additional assumption, Eq. (20) yields

$$\begin{aligned} \Phi_a^{i_1 \ldots i_k} - P_a^{i_1 i_2 \ldots i_k} &= 0, \\ \Phi_a^{i_2 \ldots i_k} - P_a^{i_2 \ldots i_k} - P_{a,i_1}^{i_1 i_2 \ldots i_k} &= 0, \\ \Phi_a^{i_l \ldots i_k} - P_a^{i_l \ldots i_k} - P_{a,i_{l-1}}^{i_{l-1} i_l \ldots i_k} &= 0, \quad \text{for } l = 3, \ldots, k-2, \\ \Phi_a^i - P_a^i - P_{a,i_1}^{i_1 i} &= 0. \end{aligned} \tag{21}$$

In Eq. (21) Φ depends on $j^k\sigma(x)$. In local coordinates, the section σ is given by $y^b = \sigma^b(x) = \sigma^b(x^1, \ldots, x^m)$, for $b = 1, ..n$, and $j^k\sigma(x)$ has coordinates

$$(x^i, y^b, z^b_{j_1}, \ldots, z^b_{j_1 \ldots j_k}) = (x^i, \sigma^b(x), \sigma^b_{,j_1}(x), \ldots, \sigma^b_{,j_1 \ldots j_k}(x)).$$

Hence,

$$P_a^{i_1 i_2 \ldots i_k}(x) = \Phi_a^{i_1 i_2 \ldots i_k}(x, \sigma^b(x), \sigma^b_{,j_1}(x), \ldots, \sigma^b_{,j_1 \ldots j_k}(x)). \tag{22}$$

Next,

$$
\begin{aligned}
P_a^{i_2\ldots i_k}(x) &= \Phi_a^{i_2\ldots i_k}(x,\sigma^b(x),\sigma^b_{,j_1}(x),\ldots,\sigma^b_{,j_1\ldots j_k}(x)) - P_{a,i_1}^{i_1 i_2\ldots i_k}(x) \\
&= \Phi_a^{i_2\ldots i_k}(x,\sigma^b(x),\sigma^b_{,j_1}(x),\ldots,\sigma^b_{,j_1\ldots j_k}(x)) + \\
&\quad -\frac{\partial}{\partial x^{i_1}}\left[\Phi_a^{i_1 i_2\ldots i_k}(x,\sigma^b(x),\sigma^b_{,j_1}(x),\ldots,\sigma^b_{,j_1\ldots j_k}(x))\right] \\
&= \Phi_a^{i_2\ldots i_k}(x,\sigma^b(x),\sigma^b_{,j_1}(x),\ldots,\sigma^b_{,j_1\ldots j_k}(x)) + \\
&\quad -\frac{\partial\Phi_a^{i_1 i_2\ldots i_k}}{\partial x^{i_1}}(j^k\sigma(x)) - \frac{\partial\Phi_a^{i_1 i_2\ldots i_k}}{\partial y^b}(j^k\sigma(x))\frac{\partial\sigma^b}{\partial x^{i_1}}(x) + \\
&\quad -\frac{\partial\Phi_a^{i_1 i_2\ldots i_k}}{\partial z^b_{j_1}}(j^k\sigma(x))\frac{\partial\sigma^b_{,j_1}}{\partial x^{i_1}}(x) - \ldots - \frac{\partial\Phi_a^{i_1 i_2\ldots i_k}}{\partial z^b_{j_1\ldots j_k}}(j^k\sigma(x))\frac{\partial\sigma^b_{,j_1\ldots j_k}}{\partial x^{i_1}}(x).
\end{aligned}
\tag{23}
$$

Since

$$
\frac{\partial\sigma^b_{,j_1\ldots j_k}}{\partial x^{i_1}}(x) = \sigma^b_{,j_1\ldots j_k i_1}(x),
$$

Eq. (23) implies that $P_a^{i_2\ldots i_k}(x)$ depends on $j^{k+1}\sigma(x)$.

Continuing, we get a complete solution of Eq. (21) in the form

$$
\begin{aligned}
&P_a^{i_l\ldots i_k}(x) = \Phi_a^{i_l\ldots i_k}(x,\sigma^b(x),\sigma^b_{,j_1}(x),\ldots,\sigma^b_{,j_1\ldots j_k}(x)) + \\
&+\sum_{j=1}^{l-1}(-1)^j\frac{\partial^j}{\partial x^{i_{l-j}}\ldots\partial x^{i_{l-1}}}\left[\Phi_a^{i_{l-j}i_{l-j+1}\ldots i_{l-1}i_l\ldots i_k}(x,\sigma^b(x),\sigma^b_{,j_1}(x),\ldots,\sigma^b_{,j_1\ldots j_k}(x))\right]
\end{aligned}
\tag{24}
$$

for $l = 1,\ldots,k$. It shows that $P_a^{i_l\ldots i_k}(x)$ depends on $j^{k+l-1}\sigma(x)$. In particular, $P_a^{i_k}(x)$ depends on $j^{2k-1}\sigma(x)$.

Recall that $P_a^{i_l\ldots i_k} = j^{2k-1}\sigma^* p_a^{i_l\ldots i_k}$, for $l = 1,\ldots,k-1$, where $p_a^{i_l\ldots i_k}$ is a function on $J^{2k-1}(M,N)$, see Eq. (16). Equation (24) gives $P_a^{i_l\ldots i_k}$ for all sections σ of π. Hence, we can use it to get an explicit expression for $p_a^{i_l\ldots i_k}$ as a function of coordinates $(x^i, y^a, z^a_{i_1}, \ldots z^a_{i_1\ldots i_{2k-1}})$. To this end we define a differential operator

$$
D_i = \frac{\partial}{\partial x^i} + z^a_i\frac{\partial}{\partial y^a} + z^a_{ij_1}\frac{\partial}{\partial z^a_{j_1}} + \ldots + z^a_{ij_1\ldots j_{2k-2}}\frac{\partial}{\partial z^a_{j_1\ldots j_{2k-2}}} + z^a_{ij_1\ldots j_{2k-1}}\frac{\partial}{\partial z^a_{j_1\ldots j_{2k-1}}}
\tag{25}
$$

acting on $C^\infty(J^{2\mathrm{k}-1}(M,N))$.

It enables us to write

$$
\begin{aligned}
&p_a^{i_l\ldots i_k}(x^i,y^b,z^b_{j_1},\ldots,z^b_{j_1\ldots j_{2k-1}}) = \Phi_a^{i_l\ldots i_k}(x^i,y^b,z^b_{j_1},\ldots,z^b_{j_1\ldots j_k}) + \\
&+\sum_{j=1}^{l-1}(-1)^j D_{i_{l-j}}\ldots D_{i_{l-1}}\left[\Phi_a^{i_{l-j}i_{l-j+1}\ldots i_{l-1}i_l\ldots i_k}(x^i,y^b,z^b_{j_1},\ldots,z^b_{j_1\ldots j_k})\right].
\end{aligned}
\tag{26}
$$

Hence, in terms of local coordinates $(x^i, y^b, z^b_{j_1}, \ldots, z^b_{j_1\ldots j_{2k-1}})$ on $J^{2k-1}(M, N)$, an obvious generalization of the De Donder construction yields form Ξ given by Eq. (9), where the coefficients $p_a^{i_1}, \ldots, p_a^{i_1 i_2 \ldots i_k}$ are given by Eq. (26).

Definition 1 Local forms

$$\Xi = p_a^i(\mathrm{d}y^a - z^a_j \mathrm{d}x^j) \wedge \left(\frac{\partial}{\partial x^i} \lrcorner \mathrm{d}_m x\right) + p_a^{i_1 i_2}(\mathrm{d}z^a_{i_2} - z^a_{i_2 j}\mathrm{d}x^j) \wedge \left(\frac{\partial}{\partial x^{i_1}} \lrcorner \mathrm{d}_m x\right) +$$
$$+ \ldots + p_a^{i_1 i_2 \ldots i_k}(\mathrm{d}z^a_{i_2 \ldots i_k} - z^a_{i_2 i_3 \ldots i_k j}\mathrm{d}x^j) \wedge \left(\frac{\partial}{\partial x^{i_1}} \lrcorner \mathrm{d}_m x\right),$$

are called boundary forms. If Ξ satisfies Condition 3 of Theorem 1, we say that Ξ is a *boundary form* of Φ.

In the following we discuss some properties of boundary forms. This means, we do not make additional assumptions on the symmetry properties of coefficients $p_a^{i_1 \ldots i_l}$, and do not specify the form Φ explicitly.

Lemma 1 *For each vector field Y on N, which projects to a vector field on M, and every section σ of $\pi : M \to N$,*

$$j^{2k-1}\sigma^* \left(\mathcal{L}_{Y^{2k-1}} \Xi\right) = 0, \tag{27}$$

where Y^{2k-1} is the prolongation of Y to $J^{2k-1}(M, N)$.

Proof By definition,

$$\mathcal{L}_{Y^{2k-1}} \Xi = \frac{d}{dt}(\mathrm{e}^{tY^{2k-1}} {}^* \Xi)_{|t=0} = \lim_{t \to 0}\left[\frac{1}{t}\left(\mathrm{e}^{tY^{2k-1}} {}^* \Xi - \Xi\right)\right].$$

Hence,

$$\begin{aligned} j^{2k-1}\sigma^* \left(\mathcal{L}_{Y^{2k-1}} \Xi\right) &= j^{2k-1}\sigma^* \lim_{t \to 0}\left[\frac{1}{t}\left(\mathrm{e}^{tY^{2k-1}} {}^* \Xi - \Xi\right)\right] \\ &= \lim_{t \to 0}\left[\frac{1}{t}\left(j^{2k-1}\sigma^* \mathrm{e}^{tY^{2k-1}} {}^* \Xi - j^{2k-1}\sigma^* \Xi\right)\right] \\ &= \lim_{t \to 0}\left\{\frac{1}{t}\left[(\mathrm{e}^{tY^{2k-1}} \circ j^{2k-1}\sigma)^* \Xi - j^{2k-1}\sigma^* \Xi\right]\right\} \\ &= \lim_{t \to 0}\left\{\frac{1}{t}\left[(j^{2k-1}(\mathrm{e}^{tY} {}_* \sigma))^* \Xi - j^{2k-1}\sigma^* \Xi\right]\right\}. \\ &= \lim_{t \to 0}\left[\frac{1}{t}\left(j^{2k-1}(\mathrm{e}^{tY} {}_* \sigma)^* \Xi - j^{2k-1}\sigma^* \Xi\right)\right] = 0. \end{aligned}$$

Condition 2 of Theorem 1 ensures that $j^{2k-1}\sigma^*\Xi$ and $(j^{2k-1}(\mathrm{e}^{tY}{}_*\sigma))^*\Xi = 0$ for every t in a neighbourhood of 0. Therefore, $j^{2k-1}\sigma^* \left(\mathcal{L}_{Y^{2k-1}} \Xi\right) = 0$, which completes the proof.

Proposition 1 *For every* $Y_\sigma \in T_\sigma S^\infty(\bar{K}, N)$, *a boundary form* Ξ *leads to a decomposition*

$$\int_K j^{\mathrm{k}}\sigma^*\left(Y^{\mathrm{k}} \lrcorner \Phi\right) = \int_K \left[j^{\mathrm{k}}\sigma^*(\Phi_a) - P^i_{a,i}\right] Y^a_\sigma \mathrm{d}_m x + \int_{j^{2\mathrm{k}-1}\sigma(\partial K)} (Y^{2\mathrm{k}-1}_\sigma \lrcorner \Xi), \tag{28}$$

where $P^i_a = j^{2k-1}\sigma^* p^i_a$, *as in Eq. (16).*

Proof Let Y be an extension of Y_σ to a vertical vector field on N. Clearly,

$$\begin{aligned}\int_K j^{\mathrm{k}}\sigma^*\left(Y^k \lrcorner \Phi\right) &= \int_K j^{2\mathrm{k}-1}\sigma^*\left(\pi^{2\mathrm{k}-1*}_{\mathrm{k}}\left(Y^{\mathrm{k}} \lrcorner \Phi\right)\right) \\ &= \int_K j^{2\mathrm{k}-1}\sigma^*\left(Y^{2\mathrm{k}-1} \lrcorner \pi^{2\mathrm{k}-1*}_{\mathrm{k}} \Phi\right) \\ &= \int_K j^{2\mathrm{k}-1}\sigma^*\left(Y^{2\mathrm{k}-1} \lrcorner \left(\pi^{2\mathrm{k}-1*}_{\mathrm{k}} \Phi + \mathrm{d}\Xi - d\Xi\right)\right).\end{aligned}$$

Thus,

$$\begin{aligned}\int_K j^{\mathrm{k}}\sigma^*\left(Y^{\mathrm{k}} \lrcorner \Phi\right) = \int_K j^{2\mathrm{k}-1}\sigma^*\left(Y^{2\mathrm{k}-1} \lrcorner \left(\pi^{2\mathrm{k}-1*}_{\mathrm{k}} \Phi + \mathrm{d}\Xi\right)\right) + \\ - \int_K j^{2\mathrm{k}-1}\sigma^*\left(Y^{2\mathrm{k}-1} \lrcorner d\Xi\right).\end{aligned} \tag{29}$$

By Condition 3 in Theorem 1,

$$j^{2\mathrm{k}-1}\sigma^*\left(X \lrcorner \left(\pi^{2\mathrm{k}-1*}_{\mathrm{k}} \Phi + \mathrm{d}\Xi\right)\right) = 0,$$

for every vector field X tangent to fibres of the target map $\pi^{2\mathrm{k}-1}_0 : J^{2\mathrm{k}-1}(M, N) \to N$. On the other hand, the prolongation $Y^{2\mathrm{k}-1}$ of a vertical vector field $Y = Y^a \frac{\partial}{\partial y^\alpha}$ on N is $\pi^{2\mathrm{k}-1}_0$-related to Y. Therefore, in local coordinates, treating $Y^a \frac{\partial}{\partial y^\alpha}$ as a vector field on $J^{2\mathrm{k}-1}(M, N)$, the difference $Y^{2\mathrm{k}-1} - Y^a \frac{\partial}{\partial y^\alpha}$ is tangent to fibres of the target map $\pi^{2\mathrm{k}-1}_0$, so that

$$\begin{aligned}& j^{2\mathrm{k}-1}\sigma^*\left(Y^{2\mathrm{k}-1} \lrcorner \mathrm{d}\left(\pi^{2\mathrm{k}-1*}_{\mathrm{k}} \Phi + d\Xi\right)\right) \\ &= j^{2\mathrm{k}-1}\sigma^*\left(\left[Y^{2\mathrm{k}-1} - Y^a \frac{\partial}{\partial y^\alpha} + Y^a \frac{\partial}{\partial y^\alpha}\right] \lrcorner \mathrm{d}\left(\pi^{2\mathrm{k}-1*}_{\mathrm{k}} \Phi + d\Xi\right)\right) \\ &= j^{2\mathrm{k}-1}\sigma^*\left(Y^a \frac{\partial}{\partial y^\alpha} \lrcorner \mathrm{d}\left(\pi^{2\mathrm{k}-1*}_{\mathrm{k}} \Phi + \mathrm{d}\Xi\right)\right).\end{aligned}$$

Taking into account equations (12) and (16), we get

$$j^{2k-1}\sigma^{*}\left(Y^{2k-1}\lrcorner\left(\pi_{k}^{2k-1*}\Phi+\mathrm{d}\Xi\right)\right)=(j^{k}\sigma^{*}\Phi_{a}-P_{a,i}^{i})Y_{\sigma}^{a}\mathrm{d}_{m}x. \tag{30}$$

Lemma 1 ensures that $\mathcal{L}_{Y^{2k-1}}\Xi=0$. Hence,

$$Y^{2k-1}\lrcorner d\,\Xi=-\mathrm{d}(Y^{2k-1}\lrcorner\,\Xi).$$

Therefore, the second line in Eq. (29) can be rewritten in the form

$$\begin{aligned}
-\int_{K}j^{2k-1}\sigma^{*}\left(Y^{2k-1}\lrcorner\mathrm{d}\Xi\right) &= -\int_{K}j^{2k-1}\sigma^{*}\left(\mathcal{L}_{Y^{2k-1}}\Xi-\mathrm{d}(Y^{2k-1}\lrcorner\,\Xi)\right)\\
&= -\int_{K}j^{2k-1}\sigma^{*}\mathcal{L}_{Y^{2k-1}}\Xi+\int_{K}j^{2k-1}\sigma^{*}\mathrm{d}(Y^{2k-1}\lrcorner\,\Xi)\\
&= \int_{K}\mathrm{d}\left(j^{2k-1}\sigma^{*}(Y^{2k-1}\lrcorner\,\Xi)\right)\\
&= \int_{\partial K}j^{2k-1}\sigma^{*}(Y^{2k-1}\lrcorner\,\Xi)\\
&= \int_{j^{2k-1}\sigma(\partial K)}Y_{\sigma}^{2k-1}\lrcorner\,\Xi.
\end{aligned}$$

This completes the proof.

Next, we want to show that the decomposition (28) is independent of the choice of boundary form Ξ. Let X be a vector field on $J^{2k-1}(M,N)$ tangent to fibres of the source map $\pi^{2k-1}:J^{2k-1}(M,N)\to M$. For any boundary form Ξ of Φ, Eqs. (12) and (16) yield

$$\begin{aligned}
&j^{2k-1}\sigma^{*}\left(X\lrcorner d\,\Xi\right)\\
&=j^{2k-1}\sigma^{*}\left[\left(-X^{a}\mathrm{d}p_{a}^{i_1}-X_{i_2}^{a}\mathrm{d}p_{a}^{i_1i_2}-\ldots-X_{i_2\ldots i_k}^{a}\mathrm{d}p_{a}^{i_1i_2\ldots i_k}\right)\wedge\left(\frac{\partial}{\partial x^{i_1}}\lrcorner\mathrm{d}_{m}x\right)\right]\\
&\quad-j^{2k-1}\sigma^{*}\left(p_{a}^{i_1}X_{i_1}^{a}+p_{a}^{i_1i_2}X_{i_1i_2}^{a}+p_{a}^{i_1i_2\ldots i_k}X_{i_1i_2\ldots i_k}^{a}\right)\mathrm{d}_{m}x\\
&=-\left(X^{a}P_{a,i_1}^{i_1}+X_{i_2}^{a}P_{a,i_1}^{i_1i_2}+\ldots+X_{i_2\ldots i_k}^{a}P_{a,i_1}^{i_1i_2\ldots i_k}\right)\mathrm{d}_{m}x\\
&\quad-\left(P_{a}^{i_1}X_{i_1}^{a}+P_{a}^{i_1i_2}X_{i_1i_2}^{a}+\ldots+P_{a}^{i_1i_2\ldots i_k}X_{i_1i_2\ldots i_k}^{a}\right)\mathrm{d}_{m}x.
\end{aligned}$$

Hence

$$\begin{aligned}
j^{2k-1}\sigma^{*}\left(X\lrcorner d\,\Xi\right)=&-\left[X^{a}P_{a,i_1}^{i_1}+X_{i_2}^{a}\left(P_{a,i_1}^{i_1i_2}+P_{a}^{i_2}\right)+\ldots\right]\mathrm{d}_{m}x\\
&-\left[X_{i_2\ldots i_k}^{a}\left(P_{a,i_1}^{i_1i_2\ldots i_k}+P_{a}^{i_2\ldots i_k}\right)+X_{i_1i_2\ldots i_k}^{a}P_{a}^{i_1i_2\ldots i_k}\right]\mathrm{d}_{m}x.
\end{aligned} \tag{31}$$

Let Ξ' be another boundary form of Φ such that

$$j^{2k-1}\sigma^*\left(X \lrcorner d\Xi'\right) = -\left[X^a P'^{i_1}_{a,i_1} + X^a_{i_2}\left(P'^{i_1 i_2}_{a,i_1} + P'^{i_2}_a\right) + \ldots\right] d_m x \qquad (32)$$
$$-\left[X^a_{i_2 \ldots i_k}\left(P'^{i_1 i_2 \ldots i_k}_{a,i_1} + P'^{i_2 \ldots i_k}_a\right) + X^a_{i_1 i_2 \ldots i_k} P'^{i_1 i_2 \ldots i_k}_a\right] d_m x,$$

where the coefficients $P'^{i_1 i_2}_a, \ldots, P'^{i_1 i_2 \ldots i_k}_a$ are symmetric in the upper indices. For the sake of simplicity, we introduce the notation

$$Q^{i_1 i_2 \ldots i_l}_a = P^{i_1 i_2 \ldots i_l}_a - P'^{i_1 i_2 \ldots i_l}_a \qquad (33)$$

for $l = 1, \ldots, k$. Then

$$j^{2k-1}\sigma^*\left(X \lrcorner d(\Xi - \Xi')\right) = -\left[X^a Q^{i_1}_{a,i_1} + X^a_{i_2}\left(Q^{i_1 i_2}_{a,i_1} + Q^{i_2}_a\right) + \ldots\right] d_m x \qquad (34)$$
$$-\left[X^a_{i_2 \ldots i_k}\left(Q^{i_1 i_2 \ldots i_k}_{a,i_1} + Q^{i_2 \ldots i_k}_a\right) + X^a_{i_1 i_2 \ldots i_k} Q^{i_1 i_2 \ldots i_k}_a\right] d_m x.$$

Since Ξ and Ξ' are boundary forms of the same form, and X is an arbitrary vector field tangent to fibres of the source map, Condition 3 of Theorem 1 yields

$$\begin{aligned} Q^{(i_1 i_2 \ldots i_k)}_a &= 0, \\ Q^{(i_2 \ldots i_k)}_a + Q^{i_1 i_2 \ldots i_k}_{a,i_1} &= 0, \\ Q^{(i_1 \ldots i_l)}_a + Q^{i_1 i_2 \ldots i_l}_{a,i_1} &= 0, \quad \text{for } l = 2, \ldots, k, \\ Q^i_a + Q^{i_1 i_2}_{a,i_1} &= 0. \end{aligned} \qquad (35)$$

Note that, by construction, $Q^{i_1 i_2 \ldots i_l}_a$ is symmetric in the indices $i_2, \ldots, i_l$. Hence,

$$j^{2k-1}\sigma^*\left(X \lrcorner d(\Xi - \Xi')\right) = -X^a Q^i_{a,i} d_m x. \qquad (36)$$

Lemma 2 *For boundary forms Ξ and Ξ', given by Eqs. (31) and (32), respectively,*

$$Q^i_{a,i} = (P^i_a - P'^i_a)_{,i} = 0. \qquad (37)$$

Proof We begin with the case when the difference $Q^i_a = (P^i_a - P'^i_a)$ is generated at the highest differential level. In other words, we consider ${}_0Q^{i_1 i_2 \ldots i_k}_a = P^{i_1 i_2 \ldots i_l}_a - P'^{i_1 i_2 \ldots i_l}_a \neq 0$ such that

$${}_0Q^{(i_1 i_2 \ldots i_k)}_a = 0, \qquad (38)$$

and, the remaining differences are symmetric and satisfy the equations

$$
\begin{aligned}
{}_0Q_a^{i_2\ldots i_l} + {}_0Q_{a,i_1}^{i_1 i_2\ldots i_l} &= 0, \quad \text{for } l = 3, \ldots, k-1, \\
{}_0Q_a^{i_2} + {}_0Q_{a,i_1}^{i_1 i_2} &= 0.
\end{aligned}
\tag{39}
$$

Therefore,

$$
{}_0Q_a^{i_k} = (-1)^{k-1}\, {}_0Q_{a,i_1 i_2\ldots i_{k-1}}^{i_1 i_2\ldots i_{k-1} i_k},
\tag{40}
$$

and

$$
\begin{aligned}
{}_0Q_{a,i_k}^{i_k} &= (-1)^{k-1}\, {}_0Q_{a,i_1 i_2\ldots i_{k-1} i_k}^{i_1 i_2\ldots i_{k-1} i_k} = (-1)^{k-1}\, {}_0Q_{a,(i_1 i_2\ldots i_{k-1} i_k)}^{i_1 i_2\ldots i_{k-1} i_k} \\
&= (-1)^{k-1}\, {}_0Q_{a,(i_1 i_2\ldots i_{k-1} i_k)}^{(i_1 i_2\ldots i_{k-1} i_k)} = (-1)^{k-1}\, {}_0Q_{a,i_1 i_2\ldots i_{k-1} i_k}^{(i_1 i_2\ldots i_{k-1} i_k)} = 0
\end{aligned}
\tag{41}
$$

because partial derivatives commute.

In the next step, we consider the situation when Ξ and Ξ' agree on the highest differential, that is we assume that ${}_1Q_a^{i_1\ldots i_k} = 0$. Moreover, we assume that

$$
\begin{aligned}
{}_1Q_a^{i_1\ldots i_{k-1}} &\neq 0, \\
{}_1Q_a^{(i_1\ldots i_{k-1})} &= 0, \\
{}_1Q_a^{i_2\ldots i_l} + {}_1Q_{a,i_1}^{i_1 i_2\ldots i_l} &= 0, \quad \text{for } l = 3, \ldots, k-1, \\
{}_1Q_a^{i_2} + {}_1Q_{a,i_1}^{i_1 i_2} &= 0.
\end{aligned}
\tag{42}
$$

The same arguments as above lead to

$$
{}_1Q_a^{i_k} = (-1)^{k-2}\, {}_1Q_{a,i_1\ldots i_{k-1}}^{i_1\ldots i_{k-1} i_k}
$$

so that

$$
{}_1Q_{a,i_k}^{i_k} = (-1)^{k-2}\, {}_1Q_{a,i_2\ldots i_{k-1} i_k}^{i_2\ldots i_{k-1} i_k} = 0
\tag{43}
$$

because ${}_1Q_a^{(i_2\ldots i_k)} = 0$. Continuing this procedure, for every $r = 2, , \ldots, k-1$, we consider ${}_rQ_a^{i_l\ldots i_k}$ such that,

$$
\begin{aligned}
{}_rQ_a^{i_1\ldots i_{k-l}} &= 0, \quad \text{for } l < r, \\
{}_rQ_a^{(i_1\ldots i_{k-r})} &= 0, \\
{}_rQ_a^{i_2\ldots i_l} + {}_rQ_{a,i_{l-1}}^{i_1 i_2\ldots i_l} &= 0, \quad \text{for } l = 3, \ldots, k-r, \\
{}_rQ_a^{i_k} + {}_rQ_{a,i_{k-1}}^{i_{k-1} i_k} &= 0.
\end{aligned}
\tag{44}
$$

As before, for this choice of ${}_rQ_a^{i_1\ldots i_l}$, we have

$$ {}_rQ^{i_k}_{a,i_k} = (-1)^{k-r}\ {}_rQ^{i_2\ldots i_{k-1}i_k}_{a,i_r\ldots i_{k-1}i_k} = 0. \tag{45}$$

The general $Q_a^{i_1\ldots i_k}$ can be expressed as the sum of terms ${}_rQ_a^{i_l\ldots i_k}$, for $r = 0,\ldots,k-1$. That is,

$$ Q_a^{i_l\ldots i_k} = {}_0Q_a^{i_l\ldots i_k} + {}_1Q_a^{i_l\ldots i_k} + \ldots + {}_{k-1}Q_a^{i_l\ldots i_k}. \tag{46}$$

The defining equations (44) for the terms ${}_rQ_a^{i_l\ldots i_k}$ ensure that the decomposition (46) satisfies Eq. (37). Taking into account equations (41), (43) and (45) we get

$$ Q^i_{a,i} = ({}_0Q^i_a + {}_1Q^i_a + \ldots + {}_{k-1}Q^i_a)_{,i} = {}_0Q^i_{a,i} + {}_1Q^i_{a,i} + \ldots + {}_{k-1}Q^i_{a,i} = 0. \tag{47}$$

We have shown that $Q^i_{a,i} = P^i_{a,i} - P'^i_{a,i} = 0$, under the assumption that Ξ' is the obvious choice of boundary form with fully symmetric coefficients and no additional assumptions on Ξ. Hence, Eq. (37) holds for any pair of boundary forms of the same form Φ. We have shown that $P'^i_{a,i} = P^i_{a,i}$ for any other boundary form Ξ'. If Ξ'' is still another boundary form of Φ, then $P''^i_{a,i} = P^i_{a,i}$.

This implies the following result.

Corollary 1

1. *If Ξ and Ξ' are boundary forms of the same form Φ, then*

$$ j^{2\mathrm{k}-1}\sigma^*\left(X \lrcorner \mathrm{d}\left(\Xi - \Xi'\right)\right) = 0 \tag{48}$$

   for every vector field X tangent to fibres of the source map $\pi^{2\mathrm{k}-1} : J^{2\mathrm{k}-1}(M,N) \to M$.
2. *The decomposition (28) is independent of the choice of boundary form Ξ for Φ such that $j^{2\mathrm{k}-1}\sigma(\bar{K})$ is in the domain of definition of Ξ.*

Proof Equation (48) is the consequence of Eqs. (36) and (37).
Equations (28), (36) and (37) yield

$$\begin{aligned}\int_K \left[j^{\mathrm{k}}\sigma^*(\Phi_a) - P^i_{a,i}\right] Y^a_\sigma \mathrm{d}_m x &= \int_K \left[j^{\mathrm{k}}\sigma^*(\Phi_a) - P'^i_{a,i} - (P^i_{a,i} - P'^i_{a,i})\right] Y^a_\sigma \mathrm{d}_m x \\ &= \int_K \left[j^{\mathrm{k}}\sigma^*(\Phi_a) - P'^i_{a,i}\right] Y^a_\sigma \mathrm{d}_m x.\end{aligned}$$

because $P^i_{a,i} - P'^i_{a,i} = 0$. Therefore, decompositions (28) for the boundary forms Ξ and Ξ' yield

$$\begin{aligned}\int_{j^{2\mathrm{k}-1}\sigma(\partial K)}(Y_\sigma^{2\mathrm{k}-1}\lrcorner\,\Xi) &= \int_K j^{\mathrm{k}}\sigma^*\left(Y^{\mathrm{k}}\lrcorner\,\Phi\right) - \int_K \left[j^{\mathrm{k}}\sigma^*(\Phi_a) - P^i_{a,i}\right] Y^a_\sigma \mathrm{d}_m x \\ &= \int_K j^{\mathrm{k}}\sigma^*\left(Y^{\mathrm{k}}\lrcorner\,\Phi\right) - \int_K \left[j^{\mathrm{k}}\sigma^*(\Phi_a) - P'^i_{a,i}\right] Y^a_\sigma \mathrm{d}_m x \\ &= \int_{j^{2\mathrm{k}-1}\sigma(\partial K)}(Y_\sigma^{2\mathrm{k}-1}\lrcorner\,\Xi').\end{aligned}$$

This shows that the decomposition (28) is independent of the choice of Ξ.

Since boundary forms are constructed in terms of adapted coordinate systems, non-uniqueness of the De Donder construction implies only local existence of the result. We see in Corollary 1 that decomposition (28) does not depend on the choice of boundary form with the same domain of definition. If boundary forms are globally defined, then decomposition (28) is unique and it holds for every section of π and each relatively compact open submanifold K of M with piece-wise smooth boundary ∂K. The existence of global boundary forms is a topological condition on the fibration $\pi : N \to M$. It is satisfied if the fibration is trivial and M and the typical fibre of π are diffeomorphic to open subsets of $\mathbb{R}^m$ and $\mathbb{R}^n$, respectively. In particular, it is satisfied in many problems in continuum mechanics. In a recent paper [18], the authors proved that the De Donder form is globally defined for every diffeomorphism invariant second order Lagrangian form in general relativity.

4 Application to Variational Problems

4.1 *Critical Points of Action Functionals*

In this section, we consider the case when $\Phi = \mathrm{d}\Lambda$, where Λ is a semi-basic m-form on $J^{\mathrm{k}}(M, N)$. Let $K \subseteq M$ be an open relatively compact submanifold of M with piece-wise smooth boundary ∂K. As in Sect. 2, we consider the space $S^\infty(\bar{K}, N)$ of smooth section $\sigma : \bar{K} \to N$. The form Λ defines an action functional A on $S^\infty(\bar{K}, N)$, given by

$$A(\sigma) = \int_K j^{\mathrm{k}}\sigma^*\Lambda = \int_{j^{\mathrm{k}}\sigma(K)} \Lambda. \tag{49}$$

Definition 2 A section $\sigma \in S^\infty(\bar{K}, N)$ is a *critical point* of A if $D_{Y_\sigma}A = 0$ for every $Y_\sigma \in T_\sigma S^\infty(\overline{K}, N)$, which vanishes on ∂K together with its partial derivatives up to order $\mathrm{k} - 1$.

Taking into account equation (3), we see that $\sigma \in S^\infty(\bar{K}, N)$ is a *critical point* of A if

$$\int_{j^{\mathrm{k}}\sigma(K)} \mathcal{L}_{Y^{\mathrm{k}}}\Lambda = 0 \tag{50}$$

for every $Y_\sigma \in \mathcal{T}_\sigma S^\infty(\bar{K}, N)$, which vanishes on ∂K together with its partial derivatives up to order k − 1. Here, Y^{k} is the prolongation to $J^{\mathrm{k}}(M, N)$ of an extension of Y_σ to a vertical vector field Y on N.

For every vector field Y on N,

$$\mathcal{L}_{Y^{\mathrm{k}}} \Lambda = Y^{\mathrm{k}} \lrcorner \mathrm{d}\Lambda + \mathrm{d}\left(Y^{\mathrm{k}} \lrcorner \Lambda\right). \tag{51}$$

The identity (51) and Stokes' theorem yield

$$\begin{aligned} \int_{j^{\mathrm{k}}\sigma(K)} \mathcal{L}_{Y^{\mathrm{k}}} \Lambda &= \int_{j^{\mathrm{k}}\sigma(K)} \left[Y^{\mathrm{k}} \lrcorner \mathrm{d}\Lambda + \mathrm{d}\left(Y^{\mathrm{k}} \lrcorner \Lambda\right)\right] \\ &= \int_{j^{\mathrm{k}}\sigma(K)} Y^{\mathrm{k}} \lrcorner \mathrm{d}\Lambda + \int_{\partial j^{\mathrm{k}}\sigma(K)} \left(Y^{\mathrm{k}} \lrcorner \Lambda\right) \\ &= \int_{j^{\mathrm{k}}\sigma(K)} Y^{\mathrm{k}} \lrcorner \mathrm{d}\Lambda \end{aligned}$$

because Λ is semi-basic with respect to the source map $\pi^{\mathrm{k}} : J^{\mathrm{k}}(M, N) \to M$. Hence Eq. (50) is equivalent to

$$\int_K j^{\mathrm{k}}\sigma^* \left(Y^{\mathrm{k}} \lrcorner \mathrm{d}\Lambda\right) = 0 \tag{52}$$

for every vertical vector field Y on N. Therefore, σ is a critical section of A if Eq. (52) holds for every vertical vector field Y on N such that $Y^{\mathrm{k}-1}$ vanishes on $\partial j^{\mathrm{k}-1}\sigma(K) = j^{\mathrm{k}-1}\sigma(\partial K)$.

4.2 Euler–Lagrange Equations

The Euler–Lagrange equations are obtained by using the coordinate description of Λ,

$$\Lambda = L(x^i, y^a, z^a_{i_1}, \ldots, z^a_{i_1,\ldots,i_k})\mathrm{d}_m x. \tag{53}$$

The usual rule that "variation of the derivative is the derivative of the variation" corresponds to the choice of extension of Y_σ to a vertical vector field $Y = Y^a(x^i)\frac{\partial}{\partial y^a}$ with components independent of y^a. Its prolongation to $J^{\mathrm{k}}(M, N)$ is

$$Y^{\mathrm{k}}(x^i, z^b_{j_1}, \ldots, z^b_{j_1\ldots j_k}) = Y^a(x^i)\frac{\partial}{\partial y^a} + Y^a_{,i}(x^i)\frac{\partial}{\partial z^a_i} + \ldots + Y^a_{,i_1\ldots i_k}(x^i)\frac{\partial}{\partial z^a_{i_1\ldots i_k}}. \tag{54}$$

Finally, the coordinate description of $j^{\mathrm{k}}\sigma$ is

$$j^{\mathrm{k}}\sigma : M \to J^{\mathrm{k}}(M, N) : (x^i) \mapsto (x^i, y^a(x), z^a_{i_1}(x), \ldots, z^a_{i_1,\ldots,i_k}(x)), \tag{55}$$

where

$$z^a_{i_1\ldots i_l}(x) = y^a(x)_{,i_1,\ldots,i_l} \tag{56}$$

for every positive integer l. With this notation,

$$j^{\mathrm{k}}\sigma^* \left(Y^{\mathrm{k}} \lrcorner \mathrm{d}(L\mathrm{d}_m x)\right) = \left(\frac{\partial L}{\partial y^a} Y^a + \frac{\partial L}{\partial z^a_i} Y^a_{,i} + \ldots + \frac{\partial L}{\partial z^a_{i_1\ldots i_k}} Y^a_{,i_1\ldots i_k}\right) \mathrm{d}_m x, \tag{57}$$

where all quantities on the right-hand side are expressed as functions of $(x^1, \ldots, x^n)$. Integrating this result over K and using Stokes' theorem yields

$$\begin{aligned}
\int_K j^{\mathrm{k}}\sigma^* \left(Y^{\mathrm{k}} \lrcorner \mathrm{d}\Lambda\right) &= \int_K \left(\frac{\partial L}{\partial y^a} Y^a + \frac{\partial L}{\partial z^a_{i_1}} Y^a_{,i_1} + \ldots + \frac{\partial L}{\partial z^a_{i_1\ldots i_k}} Y^a_{,i_1\ldots i_k}\right) \mathrm{d}_m x \\
&= \int_K \left[\frac{\partial L}{\partial y^a} Y^a + \frac{\partial}{\partial x^{i_1}} \left(\frac{\partial L}{\partial z^a_{i_1}} Y^a\right) - \left(\frac{\partial}{\partial x^{i_1}} \frac{\partial L}{\partial z^a_{i_1}}\right) Y^a + \ldots\right] \mathrm{d}_m x + \\
&\quad + \int_K \left[\frac{\partial}{\partial x^{i_1}} \left(\frac{\partial L}{\partial z^a_{i_1\ldots i_k}} Y^a_{,i_2\ldots i_k}\right) - \left(\frac{\partial}{\partial x^{i_1}} \frac{\partial L}{\partial z^a_{i_1\ldots i_k}}\right) Y^a_{,i_2\ldots i_k}\right] \mathrm{d}_m x, \\
&= \int_K \left[\frac{\partial L}{\partial y^a} Y^a - \left(\frac{\partial}{\partial x^i} \frac{\partial L}{\partial z^a_i}\right) Y^a - \ldots - \left(\frac{\partial}{\partial x^{i_k}} \frac{\partial L}{\partial z^a_{i_1\ldots i_k}}\right) Y^a_{,i_1\ldots i_{k-1}}\right] \mathrm{d}_m x \\
&\quad + \int_{\partial K} \left[\frac{\partial L}{\partial z^a_{i_1}} Y^a + \ldots + \frac{\partial L}{\partial z^a_{i_1\ldots i_k}} Y^a_{,i_2\ldots i_k}\right] \left(\frac{\partial}{\partial x^{i_1}} \lrcorner \mathrm{d}_m x\right) \\
&= \int_K \left[\frac{\partial L}{\partial y^a} Y^a - \left(\frac{\partial}{\partial x^i} \frac{\partial L}{\partial z^a_i}\right) Y^a - \ldots - \left(\frac{\partial}{\partial x^{i_k}} \frac{\partial L}{\partial z^a_{i_1\ldots i_k}}\right) Y^a_{,i_1\ldots i_{k-1}}\right] \mathrm{d}_m x
\end{aligned}$$

because $Y^{\mathrm{k}-1}$ vanishes on $j^{\mathrm{k}-1}\sigma(\partial K)$. Continuing integration by parts, we get

$$\begin{aligned}
\int_K j^{\mathrm{k}}\sigma^* \left(Y^{\mathrm{k}} \lrcorner d\Lambda\right) &= \int_K \left(\frac{\partial L}{\partial y^a} - \frac{\partial}{\partial x^{i_1}} \frac{\partial L}{\partial z^a_{i_1}} + \ldots + (-1)^k \frac{\partial^k}{\partial x^{i_1} \ldots \partial x^{i_k}} \frac{\partial L}{\partial z^a_{i_1\ldots i_k}}\right) Y^a \mathrm{d}_m x \\
&= \int_K \frac{\delta L}{\delta y^a} Y^a \mathrm{d}_m x,
\end{aligned} \tag{58}$$

where

$$\frac{\delta L}{\delta y^a} = \frac{\partial L}{\partial y^a} - \frac{\partial}{\partial x^{i_1}} \frac{\partial L}{\partial z^a_{i_1}} + \ldots + (-1)^k \frac{\partial^k}{\partial x^{i_1} \ldots \partial x^{i_k}} \frac{\partial L}{\partial z^a_{i_1 \ldots i_k}} \tag{59}$$

is called the *Lagrange derivative* of L. Comparing Eq. (58) with Eq. (28) observe that, if $\Phi = d(L d_m x)$, then $j^{\mathrm{k}}\sigma^*(\Phi_a) - P^i_{a,i} = \frac{\delta L}{\delta y^a}$.

Taking into account the Fundamental Theorem in the Calculus of Variations, we conclude that σ is a critical section of A_K if and only if, for every $a = 1, \ldots, n$,

$$\left(\frac{\partial L}{\partial y^a} - \frac{\partial}{\partial x^{i_1}} \frac{\partial L}{\partial z^a_{i_1}} + \ldots + (-1)^k \frac{\partial^k}{\partial x^{i_1} \ldots \partial x^{i_k}} \frac{\partial L}{\partial z^a_{i_1 \ldots i_k}} \right)_{|K} = 0. \tag{60}$$

Equation (60) is the *Euler–Lagrange equations* for critical points of the action functional corresponding to the Lagrangian L.

4.3 De Donder Equations

Let Ξ be the boundary form of $d\Lambda$, and let

$$\Theta = \pi_{\mathrm{k}}^{2\mathrm{k}-1*} \Lambda + \Xi. \tag{61}$$

Equation (61) generalizes the construction of De Donder [5] to $\mathrm{k} > 1$. We refer to Θ as a De Donder form of Λ. It follows from Theorem 1 that Θ satisfies the following conditions.

Corollary 2

1. *Θ is semi-basic with respect basic to the forgetful map $\pi_{\mathrm{k}-1}^{2\mathrm{k}-1} : J^{2\mathrm{k}-1}(M, N) \to J^{\mathrm{k}}(M, N)$. In other words, for any vector field X tangent to fibres of $\pi_{\mathrm{k}-1}^{2\mathrm{k}-1} : J^{2\mathrm{k}-1}(M, N) \to J^{\mathrm{k}}(M, N)$,*

$$X \lrcorner \Theta = 0. \tag{62}$$

2. *For every vector field X on $J^{2\mathrm{k}-1}(M, N)$ tangent to fibres of the source map $\pi^{2\mathrm{k}-1} : J^{2\mathrm{k}-1}(M, N) \to M$, the left interior product $X \lrcorner \Theta$ is semi-basic with respect to the source map. In other words, for every pair X_1, X_2 of vector fields on $J^{2\mathrm{k}-1}(M, N)$ tangent to fibres of the source map $\pi^{2\mathrm{k}-1} : J^{2\mathrm{k}-1}(M, N) \to M$,*

$$X_2 \lrcorner (X_1 \lrcorner \Theta) = 0. \tag{63}$$

3. *For every section σ of $\pi : N \to M$,*

$$j^{2\mathrm{k}-1}\sigma^* \Theta = j^{\mathrm{k}}\sigma^* \Lambda. \tag{64}$$

4. *For every vector field* X *on* $J^{2k-1}(M, N)$ *tangent to fibres of the target map* $\pi_0^{2k-1} : J^{2k-1}(M, N) \to N$, *and every section* σ *of* $\pi : N \to M$,

$$j^{2k-1}\sigma^* (X \lrcorner \mathrm{d}\Theta) = 0. \tag{65}$$

Theorem 2 *For* $\sigma \in S^\infty(\bar{K}, N)$, *suppose that* $j^{2k-1}\sigma(\bar{K})$ *is in the domain of a De Donder form* Θ. *Then* σ *is a critical section of the functional A, given by Eq. (49), if and only if*

$$j^{2k-1}\sigma^* (X \lrcorner \mathrm{d}\Theta) = 0 \tag{66}$$

for every vector field X *on* $J^{2k-1}(M, N)$ *that is tangent to fibres of the source map* $\pi^{2k-1} : J^{2k-1}(M, N) \to M$.

Proof Equation (64) implies that replacing $j^k\sigma^*\Lambda$ by $j^{2k-1}\sigma^*\Theta$ in Eq. (49) does not change the action functional,

$$A(\sigma) = \int_K j^k\sigma^*\Lambda = \int_{j^{2k-1}\sigma(K)} \Theta. \tag{67}$$

Moreover, if Y is a vertical vector field on N, then

$$j^{2k-1}\sigma^*\mathcal{L}_{Y^{2k-1}}\Theta = j^k\sigma^*\mathcal{L}_{Y^k}\Lambda, \tag{68}$$

where Y^{2k-1} is the prolongation of Y to $J^{2k-1}(M, N)$. Hence, σ is a critical section of the functional A if

$$\int_K j^{2k-1}\sigma^*\mathcal{L}_{Y^{2k-1}}\Theta = 0 \tag{69}$$

for every vertical vector field Y on N such that Y^{k-1} vanishes on $j^{k-1}\sigma(\partial K)$. The argument leading from Eq. (50) to Eq. (52) ensures that σ is a critical section of A if and only if

$$\int_K j^{2k-1}\sigma^* \left(Y^{2k-1} \lrcorner \mathrm{d}\Theta\right) = 0 \tag{70}$$

for all vertical vector fields Y on N, such that Y^{k-1} vanishes on $j^{k-1}\sigma^*(\partial K)$.

Equation (65) ensures that in Eq. (70), we can replace Y^{2k-1} by an arbitrary vector field X on $J^{2k-1}(M, N)$ that is tangent to fibres of the source map π^{2k-1} : $J^{2k-1}(M, N) \to M$ and satisfies the condition $T\pi_{k-1}^{2k-1} \circ X \circ j^{k-1}\sigma(\partial K) = 0$. In other words, we may omit the requirement that Y^{2k-1} is the prolongation of a vertical vector field Y on N. This proves that σ is a critical section of A_K if and only if

$$\int_K j^{2k-1}\sigma^* (X \lrcorner d\Theta) = 0 \tag{71}$$

for every vector field X on $J^{2k-1}(M, N)$ that is tangent to fibres of the source map $\pi^{2k-1} : J^{2k-1}(M, N) \to M$ and satisfies the condition $T\pi^{2k-1}_{k-1} \circ X \circ j^{k=1}\sigma(\partial K) = 0$.

Suppose that σ is a critical section of A. Equation (71) and the Fundamental Theorem in the Calculus of Variations ensure that

$$j^{2k-1}\sigma^* (X \lrcorner d\Theta) = 0 \tag{72}$$

for every vector field X on $J^{2k-1}(M, N)$ that is tangent to fibres of the source map $\pi^{2k-1} : J^{2k-1}(M, N) \to M$.

Conversely, assume that Eq. (72) is satisfied for all vector fields on $J^{2k-1}(M, N)$ that are tangent to fibres of the source map $\pi^{2k-1} : J^{2k-1}(M, N) \to M$. Then, Eq. (71) is satisfied for every vector field X on $J^{2k-1}(M, N)$ because the integrand is identically zero. In particular, Eq. (70) is satisfied for prolongations Y^{2k-1} of vertical vector fields Y on N that vanish on ∂K together with all derivatives up to order k. This ensures that σ is a critical point of A.

We refer to (66) and (72) as De Donder equations. They are a system of equations in differential forms that is equivalent to Euler–Lagrange equations.

Note that Condition 4 in Corollary 2 on De Donder form Θ, see Eq. (65), differs from Eq. (72) only by restriction on the range of the vector field X. We can combine these two conditions in the corollary below.

Corollary 3 *A section $\sigma \in S^\infty(\bar{K}, N)$ is a critical section of the functional A, given by Eq. (49), if there exists a boundary form Ξ such that*

$$j^{2k-1}\sigma^* \left(X \lrcorner d \left(\pi^{2k-1*}_{k} \Lambda + \Xi \right) \right) = 0 \tag{73}$$

for every vector field X on $J^{2k-1}(M, N)$ that is tangent to fibres of the source map $\pi^{2k-1} : J^{2k-1}(M, N) \to M$.

Equation (73) is a relation in the space of pairs (Ξ, σ). However, it is not in the form of the symplectic relation occurring in Tulczyjew triples. For a discussion of Tulczyjew triples in higher derivative field theory see reference [8].

Since boundary forms are defined only locally, the assumption in Theorem 2 appears to be quite restrictive. We show that this is not the case.

Proposition 2 *A section $\sigma \in S^\infty(\bar{K}, N)$ is a critical section of the functional A if there exists an open cover $\{U_\alpha\}$ of $\bar{K} \subset M$ such that $j^{2k-1}\sigma(U_\alpha)$ is in the domain of a De Donder form Θ_α, and*

$$j^{2k-1}\sigma^* (X \lrcorner d\Theta_\alpha)_{|K\cap U_\alpha} = 0 \tag{74}$$

for each α, and every vector field X on $J^{2k-1}(M, N)$.

Proof Corollary 1 ensures that, if Ξ and Ξ' are boundary forms of $\mathrm{d}\Lambda$ with the same domain and Θ and Θ' are the De Donder forms corresponding to Ξ and Ξ', respectively, then

$$j^{2k-1}\sigma^{*}\left(X \lrcorner \mathrm{d}\Theta'\right) = j^{2k-1}\sigma^{*}\left(X \lrcorner \mathrm{d}\Theta\right)$$

for each section σ of π and every vector field X on $J^{2k-1}(M, N)$. Hence, the choice of a De Donder form does not matter.

For each α, Eq. (74) is equivalent to Euler–Lagrange equations for σ in $K \cap U_{\alpha}$. Since Euler–Lagrange equations are local, the conditions of Proposition 2 imply that σ satisfies Euler–Lagrange equations in K.

4.4 Symmetries and Conservation Laws

Definition 3 A vector field Y on N is an infinitesimal symmetry of the Lagrangian system with Lagrangian $\Lambda = L\mathrm{d}_{m}x$ of differential order k if it projects to a vector field on M and $\mathcal{L}_{Y^{k}}\mathrm{d}\Lambda = 0$.

Let Y be an infinitesimal symmetry of Λ. For every boundary form Ξ of $\mathrm{d}\Lambda$, Lemma 1 ensures that $j^{2k-1}\sigma^{*}(\mathcal{L}_{Y^{2k-1}}\Xi) = 0$ for all sections σ of $\pi : M \to N$. Since $\Theta = \pi_{k}^{2k-1*}\Lambda + \Xi$ is the local De Donder form corresponding to Ξ, it follows that

$$j^{2k-1}\sigma^{*}(\mathcal{L}_{Y^{2k-1}}\Theta) = 0. \tag{75}$$

Hence,

$$j^{2k-1}\sigma^{*}(Y^{2k-1} \lrcorner \mathrm{d}\Theta) + j^{2k-1}\sigma^{*}[\mathrm{d}(Y^{2k-1} \lrcorner \Theta)] = 0. \tag{76}$$

If σ satisfies De Donder equations, we get the conservation law

$$\mathrm{d}\left[j^{2k-1}\sigma^{*}\left(Y^{2k-1} \lrcorner \Theta\right)\right] = 0. \tag{77}$$

If K is an open, relatively compact submanifold of M with boundary ∂K, which is contained in domain σ, then

$$\int_{\partial K} j^{2k-1}\sigma^{*}\left(Y^{2k-1} \lrcorner \Theta\right) = 0. \tag{78}$$

In other words, if $\partial K = \Sigma_{1} \cup \Sigma_{2}$, where Σ_{1} and Σ_{2} inherit outer orientation from ∂K, and $\Sigma_{1} \cap \Sigma_{2}$ is smooth of dimension $n-2$, then

$$\int_{\Sigma_{1}} j^{2k-1}\sigma^{*}\left(Y^{2k-1} \lrcorner \Theta\right) = \int_{\Sigma_{2}} j^{2k-1}\sigma^{*}\left(Y^{2k-1} \lrcorner \Theta\right). \tag{79}$$

If the De Donder equations are hyperbolic, and Σ_1 and Σ_2 are Cauchy surfaces, then the integrals in Eq. (79) are conserved quantities corresponding to the infinitesimal symmetry Y.

A priori, the integrals on each side of Eq. (79) depend on the choice of the boundary form Ξ. However, the difference between the left and the right-hand sides of Eq. (79) vanishes for every Ξ. In an example below, we show how boundary conditions lead to unique expressions for constants of motion.

5 Example

5.1 Cauchy Problem

Consider $M = \mathbb{R}^2$ with coordinates $\boldsymbol{x} = (x^1, x^2)$ and $N = T\mathbb{R}^2$ with coordinates $(\boldsymbol{x}, \boldsymbol{y}) = (x^1, x^2, y^1, y^2)$.

$$L = g_{ab} g^{ij} g^{kl} z^a_{ij} z^b_{kl}, \tag{80}$$

where g_{ab} is the Minkowski metric.

$$\frac{\partial L}{\partial z^a_{ij}} = 2 g_{ab} g^{ij} g^{kl} z^b_{kl}, \tag{81}$$

$$\frac{\partial L}{\partial z^a_i} = 0 \text{ and } \frac{\partial L}{\partial y^a} = 0.$$

Euler–Lagrange equations

$$\frac{\partial^2}{\partial x^i \partial x^j} \frac{\partial L}{\partial y^a_{,ij}} = 0$$

read

$$2 g_{ab} g^{ij} g^{kl} y^a_{,klij} = 0.$$

Writing $x^1 = t$, $x^2 = x$, we get

$$\left(\frac{\partial^2}{\partial t^2} - \frac{\partial^2}{\partial x^2}\right)\left(\frac{\partial^2}{\partial t^2} - \frac{\partial^2}{\partial x^2}\right) y^a(t, x) = 0.$$

$$\left(\frac{\partial^4}{\partial t^4} - 2\frac{\partial^2}{\partial t^2}\frac{\partial^2}{\partial x^2} + \frac{\partial^4}{\partial x^2}\right) y^a(t, x) = 0.$$

Set $y(t,x)$, $\dot{y}(t,x)$, $\ddot{y}(t,x)$, and $\dddot{y}(t,x)$ as the Cauchy data at t. Then

$$\begin{aligned}
\frac{\partial}{\partial t} y^a(t,x) &= \dot{y}^a(t,x),\\
\frac{\partial}{\partial t} \dot{y}^a(t,x) &= \ddot{y}^a(t,x),\\
\frac{\partial}{\partial t} \ddot{y}^a(t,x) &= \dddot{y}^a(t,x),\\
\frac{\partial}{\partial t} \dddot{y}^a(t,x) &= \frac{\partial^4}{\partial t^4} y^a(t,x) = 2\frac{\partial^2}{\partial x^2}\ddot{y}^a(t,x) - \frac{\partial^4}{\partial x^4} y^a(t,x).
\end{aligned}$$

Therefore,

$$\frac{\partial}{\partial t}\begin{pmatrix} y^a \\ \dot{y}^a \\ \ddot{y}^a \\ \dddot{y}^a \end{pmatrix} = \begin{pmatrix} \dot{y}^a \\ \ddot{y}^a \\ \dddot{y}^a \\ 2\frac{\partial^2}{\partial x^2}\ddot{y}^a - \frac{\partial^4}{\partial x^4} y^a \end{pmatrix} = A \begin{pmatrix} y^a \\ \dot{y}^a \\ \ddot{y}^a \\ \dddot{y}^a \end{pmatrix},$$

where

$$A = \begin{pmatrix} 0 & 1 & 0 & 0 \\ 0 & 0 & 1 & 0 \\ 0 & 0 & 0 & 1 \\ -\frac{\partial^4}{\partial x^4} & 0 & 2\frac{\partial^2}{\partial x^2} & 0 \end{pmatrix}.$$

Since

$$e^{tA} = \sum_{k=0}^{\infty} \frac{t^k A^k}{k!}$$

is well defined,

$$\begin{pmatrix} y^a(t,x) \\ \dot{y}^a(t,x) \\ \ddot{y}^a(t,x) \\ \dddot{y}^a(t,x) \end{pmatrix} = \sum_{k=0}^{\infty} \frac{t^k A^k}{k!} \begin{pmatrix} y^a(0,x) \\ \dot{y}^a(0,x) \\ \ddot{y}^a(0,x) \\ \dddot{y}^a(0,x) \end{pmatrix}$$

is a solution of the Cauchy problem at $t = 0$.

5.2 De Donder Forms

De Donder forms are $\pi_2^{3*}\Lambda + \Xi$, where

$$\Xi = p_a^i(\mathrm{d}y^a - z_j^a \mathrm{d}x^j) \wedge \left(\frac{\partial}{\partial x^i} \lrcorner \mathrm{d}_2 x\right) + p_a^{i_1 i_2}(\mathrm{d}z_{i_2}^a - z_{i_2 j}^a \mathrm{d}x^j) \wedge \left(\frac{\partial}{\partial x^{i_1}} \lrcorner \mathrm{d}_2 x\right) \tag{82}$$

is a local boundary form corresponding to $\mathrm{d}\Lambda$. For a section σ of π,

$$\Theta_{|\mathrm{range}\ j^3\sigma} = \pi_2^{3*}\Lambda_{|\mathrm{range}\ j^3\sigma} + \Xi_{|\mathrm{range}\ j^3\sigma} \tag{83}$$

$$= L\mathrm{d}_2 x + P_a^{i_1}(\mathrm{d}y^a - z_j^a \mathrm{d}x^j) \wedge \left(\frac{\partial}{\partial x^{i_1}} \lrcorner \mathrm{d}_2 x\right) + P_a^{i_1 i_2}(\mathrm{d}z_{i_2}^a - z_{i_2 j}^a \mathrm{d}x^j) \wedge \left(\frac{\partial}{\partial x^{i_1}} \lrcorner \mathrm{d}_2 x\right)$$

$$= P_a^{i_1}\mathrm{d}y^a \wedge \left(\frac{\partial}{\partial x^{i_1}} \lrcorner \mathrm{d}_2 x\right) + P_a^{i_1 i_2}\mathrm{d}z_{i_2}^a \wedge \left(\frac{\partial}{\partial x^{i_1}} \lrcorner \mathrm{d}_2 x\right) - \left(P_a^i y_{,i}^a + P_a^{ij} y_{,ij} - L\right)\mathrm{d}_2 x,$$

where the functions $P_a^{i_1}$ and $P_a^{i_1 i_2}$ satisfy the equations

$$\frac{\partial L}{\partial z_{ij}^a} - P_a^{(i_1 i_2)} = 0, \tag{84}$$

$$\frac{\partial L}{\partial z_i^a} - P_a^i - P_{a,i_1}^{i_1 i} = 0,$$

that follow from Eq. (20).

Since

$$\mathrm{d}\Lambda = \frac{\partial L}{\partial z_{ij}^a}\mathrm{d}z_{ij}^a \wedge \mathrm{d}_2 x + \frac{\partial L}{\partial z_i^a}\mathrm{d}z_i^a \wedge \mathrm{d}_2 x + \frac{\partial L}{\partial y^a}\mathrm{d}y^a \wedge \mathrm{d}_2 x = 2g_{ab}g^{ij}g^{kl}z_{kl}^b\,\mathrm{d}z_{ij}^a \wedge \mathrm{d}_2 x,$$

it follows that

$$\frac{\partial L}{\partial z_{ij}^a} = 2g_{ab}g^{ij}g^{kl}y_{,kl}^b,$$

$$\frac{\partial L}{\partial z_i^a} = 0.$$

Hence, the symmetric solution is

$$P_a^{i_1 i_2} = \Phi_a^{i_1 i_2} = 2g_{ab}g^{i_1 i_2}g^{kl}y_{,kl}^b,$$

$$P_{a,i_1}^{i_1 i_2} = \Phi_{a,i_1}^{i_1 i_2} = 2g_{ab}g^{i_1 i_2}g^{kl}y_{,kli_1}^b,$$

$$P_a^i = \Phi_a^i - P_{a,i_1}^{i_1 i} = -2g_{ab}g^{i_1 i}g^{kl}y_{,kli_1}^b.$$

In this case

$$
\begin{aligned}
\Xi_{|\text{range } j^3\sigma} &= P_a^i(\mathrm{d}y^a - z_j^a \mathrm{d}x^j) \wedge \left(\frac{\partial}{\partial x^i} \lrcorner \mathrm{d}_2 x\right) \\
&+ P_a^{i_1 i_2}(\mathrm{d}z_{i_2}^a - z_{i_2 j}^a \mathrm{d}x^j) \wedge \left(\frac{\partial}{\partial x^{i_1}} \lrcorner \mathrm{d}_2 x\right) \\
&= -2 g_{ab} g^{i_1 i_2} g^{kl} y_{,kli_1}^b (\mathrm{d}y^a - z_j^a \mathrm{d}x^j) \wedge \left(\frac{\partial}{\partial x^{i_2}} \lrcorner \mathrm{d}_2 x\right) \\
&+ 2 g_{ab} g^{i_1 i_2} g^{kl} y_{,kl}^b (\mathrm{d}z_{i_2}^a - z_{i_2 j}^a \mathrm{d}x^j) \wedge \left(\frac{\partial}{\partial x^{i_1}} \lrcorner \mathrm{d}_2 x\right),
\end{aligned}
$$

and

$$
\begin{aligned}
\Theta_{|\text{range } j^3\sigma} &= \pi_2^{3*} \Lambda_{|\text{range } j^3\sigma} + \Xi_{|\text{range } j^3\sigma} \qquad (85) \\
&= L d_2 x + P_a^{i_1 i_2}(\mathrm{d}y^a - z_j^a \mathrm{d}x^j) \wedge \left(\frac{\partial}{\partial x^i} \lrcorner \mathrm{d}_2 x\right) \\
&+ P_a^{i_1 i_2}(\mathrm{d}z_{i_2}^a - z_{i_2 j}^a \mathrm{d}x^j) \wedge \left(\frac{\partial}{\partial x^{i_1}} \lrcorner \mathrm{d}_2 x\right) \\
&= g_{ab} g^{ij} g^{kl} z_{ij}^a z_{kl}^b \mathrm{d}_2 x - 2 g_{ab} g^{i_1 i_2} g^{kl} y_{,kli_1}^b (\mathrm{d}y^a - z_j^a \mathrm{d}x^j) \wedge \left(\frac{\partial}{\partial x^{i_2}} \lrcorner \mathrm{d}_2 x\right) \\
&+ 2 g_{ab} g^{i_1 i_2} g^{kl} y_{,kl}^b (\mathrm{d}z_{i_2}^a - z_{i_2 j}^a \mathrm{d}x^j) \wedge \left(\frac{\partial}{\partial x^{i_1}} \lrcorner \mathrm{d}_2 x\right).
\end{aligned}
$$

A non-symmetric solution of Eq. (84) is

$$
\begin{aligned}
P_a^{\prime i_1 i_2} &= P_a^{i_1 i_2} + Q_a^{i_1 i_2} = 2 g_{ab} g^{i_1 i_2} g^{kl} y_{,kl}^b + Q_a^{i_1 i_2}, \qquad (86) \\
P_{a,i_1}^{\prime i_1 i_2} &= P_{a,i_1}^{i_1 i_2} + Q_{a,i_1}^{i_1 i_2} = 2 g_{ab} g^{i_1 i_2} g^{kl} y_{,kli_1}^b + Q_{a,i_1}^{i_1 i_2}, \\
P_a^{\prime i_2} &= \Phi_a^{i_2} - P_{a,i_1}^{\prime i_1 i_2} = -2 g_{ab} g^{i_1 i_2} g^{kl} y_{,kli_1}^b - Q_{a,i_1}^{i_1 i_2},
\end{aligned}
$$

where $Q_a^{i_1 i_2}$ is skew symmetric in i_1 and i_2,

$$
Q_{a,i_1}^{i_1 i_2} = -Q_{a,i_1}^{i_2 i_1}. \qquad (87)
$$

5.3 Symmetries

A vector field $Y = Y^i \frac{\partial}{\partial x^i} + Y^a \frac{\partial}{\partial y^a}$ is a symmetry if

$$
\mathcal{L}_{Y^2}(L\mathrm{d}_2 x) = 0, \qquad (88)
$$

where $Y^2 = Y^i\frac{\partial}{\partial x^i} + Y^a\frac{\partial}{\partial y^a} + Y_i^a\frac{\partial}{\partial z_i^a} + Y_{ij}^a\frac{\partial}{\partial z_{ij}^a}$ of Y is the prolongation of Y to $J^2(M, N)$ and

$$\begin{aligned} Y_i^a &= Y^a_{,b}z_i^b - z_i^a Y^i_{,j} + Y^a_{,i}, \\ Y_{ij}^a &= z^b_{(j}Y^a_{i),b} - z^a_{k(i}Y^k_{,j)} + Y^a_{(i,j)}, \end{aligned} \tag{89}$$

see Eqs. (105) and (106) in the Appendix. The Lorentz metric $g_{ij}\mathrm{d}x^i\mathrm{d}x^j = (\mathrm{d}t)^2 - (\mathrm{d}x)^2$ occurring in the Lagrangian has Killing vector $Y_T = \frac{\partial}{\partial x^1}$, $Y_S = \frac{\partial}{\partial x^2}$ corresponding to the time and the space translations, and the infinitesimal Lorentz transformation $Y_L = x^2\frac{\partial}{\partial x^1} + x^1\frac{\partial}{\partial x^2}$.

We discuss here conservation of energy corresponding to the time translations Y_T. Equation (89) show that the jet components of Y_T^2 and Y_T^3 vanish, so that $Y_T^2 = \frac{\partial}{\partial t}$ and $Y_T^3 = \frac{\partial}{\partial t}$. Hence,

$$\begin{aligned} \mathcal{L}_{Y_T^2}(L\mathrm{d}_2x) &= \mathcal{L}_{Y_T^2}\left(g_{ab}g^{ij}g^{kl}z_{ij}^a z_{kl}^b \mathrm{d}_2x\right) \\ &= Y_T^2 \lrcorner (2g_{ab}g^{ij}g^{kl}z_{kl}^b)\mathrm{d}z_{ij}^a \wedge \mathrm{d}_2x + \mathrm{d}\left[\left(g_{ab}g^{ij}g^{kl}z_{ij}^a z_{kl}^b\right) Y_T^{i_1}\frac{\partial}{\partial x^{i_i}} \lrcorner \mathrm{d}_2x\right] \\ &= (2g_{ab}g^{ij}g^{kl}z_{kl}^b)Y_{Tij}^a\mathrm{d}_2x + (2g_{ab}g^{ij}g^{kl}z_{kl}^b)Y_T^{i_1}\mathrm{d}z_{ij}^a \wedge \left(\frac{\partial}{\partial x^{i_i}} \lrcorner \mathrm{d}_2x\right) + \\ &\quad + \left[\left(g_{ab}g^{ij}g^{kl}z_{ij}^a z_{kl}^b\right)\right]\mathrm{d}Y_T^{i_1} \wedge \left(\frac{\partial}{\partial x^{i_i}} \lrcorner \mathrm{d}_2x\right) \\ &= 0. \end{aligned}$$

Since

$$\left(Y_T^3 \lrcorner \mathrm{d}x^j\right) = \delta_1^j,$$

$$\frac{\partial}{\partial x^{i_1}} \lrcorner \mathrm{d}_2x = \frac{\partial}{\partial x^{i_1}} \lrcorner \mathrm{d}x^1 \wedge \mathrm{d}x^2 = \delta_{i_1}^1\mathrm{d}x^2 - \delta_{i_1}^2\mathrm{d}x^1,$$

and

$$\left(Y_T^3 \lrcorner \left(\frac{\partial}{\partial x^{i_1}} \lrcorner \mathrm{d}_2x\right)\right) = \left(Y_T^2 \lrcorner \left(\delta_{i_1}^1\mathrm{d}x^2 - \delta_{i_1}^2\mathrm{d}x^1\right)\right) = -\delta_{i_1}^2,$$

equation (83) yields

$$j^3\sigma^*\left(Y_T^3 \lrcorner \Theta\right) = j^3\sigma^*\left[-P_a^{i_1}dy^a \wedge \left(Y_T^2 \lrcorner \left(\frac{\partial}{\partial x^{i_1}} \lrcorner d_2x\right)\right)\right]$$

$$+j^3\sigma^*\left[P_a^{i_1i_2}dz^a_{i\,2}\wedge\left(Y_T^2\lrcorner\left(\frac{\partial}{\partial x^{i_1}}\lrcorner d_2x\right)\right)\right]$$
$$-\left(P_a^i y^a_{,i}+P_a^{ij}y^a_{,ij}-L\right)\left(Y_T^2\lrcorner d_2x\right)$$
$$=P_a^2y^a_{,j}dx^j+P_a^{2i_2}y^a_{,i_2j}dx^j-\left(P_a^iy^a_{,i}+P_a^{ij}y^a_{,ij}-L\right)dx^2.$$

For an open relatively compact manifold K with boundary $\partial K=\Sigma-\Sigma'$,

$$\int_\Sigma j^3\sigma^*\left(Y_T^2\lrcorner\Theta\right)=\int_\Sigma\left[P_a^2y^a_{,j}\mathrm{d}x^j+P_a^{2i_2}y^a_{,i_2j}\mathrm{d}x^j-\left(P_a^iy^a_{,i}+P_a^{ij}y^a_{,ij}-L\right)\mathrm{d}x^2\right].$$

Let us decompose P_a^{ij} into its symmetric and antisymmetric parts in the upper indices

$$P_a^{ij}=P_a^{(ij)}+P_a^{[ij]}.\tag{90}$$

Then,

$$P_a^i=\frac{\partial L}{\partial z_i^a}-P^{i_1i}_{a,i_1}=\frac{\partial L}{\partial z_i^a}-P^{(i_1i)}_{a,i_1}-P^{[i_1i]}_{a,i_1},\tag{91}$$

so that

$$\int_\Sigma j^3\sigma^*\left(Y_T^2\lrcorner\Theta\right)=\tag{92}$$
$$=\int_\Sigma\left[P_a^2y^a_{,j}\mathrm{d}x^j+P_a^{2i_2}y^a_{,i_2j}\mathrm{d}x^j-\left(P_a^iy^a_{,i}+P_a^{ij}y^a_{,ij}-L\right)\mathrm{d}x^2\right]$$
$$=\int_\Sigma\left[\left(\frac{\partial L}{\partial z_2^a}-P^{(i_12)}_{a,i_1}-P^{[i_12]}_{a,i_1}\right)y^a_{,j}\mathrm{d}x^j+\left(P_a^{(2i_2)}+P_a^{[2i_2]}\right)y^a_{,i_2j}\mathrm{d}x^j\right]$$
$$-\int_\Sigma\left(\left(\frac{\partial L}{\partial z_i^a}-P^{(i_1i)}_{a,i_1}-P^{[i_1i]}_{a,i_1}\right)y^a_{,i}+P_a^{ij}y^a_{,ij}-L\right)\mathrm{d}x^2$$
$$=\int_\Sigma\left[\left(\frac{\partial L}{\partial z_2^a}-P^{(i_12)}_{a,i_1}\right)y^a_{,j}\mathrm{d}x^j+\left(P_a^{(2i_2)}\right)y^a_{,i_2j}\mathrm{d}x^j-\right]+$$
$$-\int_\Sigma\left(\left(\frac{\partial L}{\partial z_2^a}-P^{(i_12)}_{a,i_1}\right)y^a_{,i}+P_a^{ij}y^a_{,ij}-L\right)\mathrm{d}x^2+$$
$$+\int_\Sigma\left[-P^{[i_12]}_{a,i_1}y^a_{,j}\mathrm{d}x^j+P_a^{[2i_2]}y^a_{,i_2j}\mathrm{d}x^j+P^{[i_1i]}_{a,i_1}y^a_{,i}\mathrm{d}x^2\right].$$

The last integral in Eq. (92) involves only the odd terms $P_a^{[ij]}$. It can be rewritten as follows

$$\int_\Sigma \left[-P^{[i_1 2]}_{a,i_1} y^a_{,j} \mathrm{d}x^j + P^{[2i_2]}_a y^a_{,i_2 j} \mathrm{d}x^j + P^{[i_1 i]}_{a,i_1} y^a_{,i} \mathrm{d}x^2\right] = \tag{93}$$

$$= \int_\Sigma \left[\left(-P^{[12]}_{a,1} - P^{[22]}_{a,2}\right) y^a_{,j} \mathrm{d}x^j + \left(P^{[21]}_a y^a_{,1j} + P^{[22]}_a y^a_{,2j}\right) \mathrm{d}x^j\right]$$

$$+ \int_\Sigma \left(P^{[11]}_{a,1} y^a_{,1} + P^{[12]}_{a,1} y^a_{,2} + P^{[21]}_{a,2} y^a_1 + P^{[22]}_{a,2} y^a_{,2}\right) \mathrm{d}x^2$$

$$= \int_\Sigma \left[-P^{[12]}_{a,1}\left(y^a_{,1}\mathrm{d}x^1 + y^a_{,2}\mathrm{d}x^2\right) + P^{[21]}_a y^a_{,1j}\mathrm{d}x^j + \left(P^{[12]}_{a,1} y^a_{,2} + P^{[21]}_{a,2} y^a_1\right)\mathrm{d}x^2\right]$$

$$= \int_\Sigma \left[-P^{[12]}_{a,1} y^a_{,1}\mathrm{d}x^1 + P^{[21]}_a y^a_{,1j}\mathrm{d}x^j + P^{[21]}_{a,2} y^a_1 \mathrm{d}x^2\right]$$

$$= \int_\Sigma \left[P^{[21]}_a y^a_{,1j}\mathrm{d}x^j - P^{[12]}_{a,1} y^a_{,1}\mathrm{d}x^1 + P^{[21]}_{a,2} y^a_1 \mathrm{d}x^2\right]$$

$$= \int_\Sigma \left[\mathrm{d}\left(P^{[21]}_a y^a_{,1}\right) - P^{[21]}_{a,j} y^a_{,1}\mathrm{d}x^j - P^{[12]}_{a,1} y^a_{,1}\mathrm{d}x^1 + P^{[21]}_{a,2} y^a_1 \mathrm{d}x^2\right]$$

$$= \int_\Sigma \left[\mathrm{d}\left(P^{[21]}_a y^a_{,1}\right) - P^{[21]}_{a,1} y^a_{,1}\mathrm{d}x^1 - P^{[21]}_{a,2} y^a_{,1}\mathrm{d}x^2 - P^{[12]}_{a,1} y^a_{,1}\mathrm{d}x^1 + P^{[21]}_{a,2} y^a_1 \mathrm{d}x^2\right]$$

$$= \int_\Sigma \left[\mathrm{d}\left(P^{[21]}_a y^a_{,1}\right) - \left(P^{[21]}_{a,1} y^a_{,1}\mathrm{d}x^1 + P^{[12]}_{a,1} y^a_{,1}\mathrm{d}x^1\right) - P^{[21]}_{a,2} y^a_{,1}\mathrm{d}x^2 + P^{[21]}_{a,2} y^a_1 \mathrm{d}x^2\right]$$

$$= \int_\Sigma \mathrm{d}\left(P^{[21]}_a y^a_{,1}\right).$$

Since, we consider an evolution equation with non-compact Cauchy surfaces, replace K by a slice

$$S = \{(x^1, x^2) \in \mathbb{R}^2 \mid 0 < x^1 < t\}$$

with boundary

$$\partial S = \Sigma_t - \Sigma_0 = \{(t, x_2) \in \mathbb{R}^2\} - \{(0, x_2) \in \mathbb{R}^2\}.$$

We assume that the fields $y^a(x_1, x_2)$ vanish sufficiently fast as $x_2 \to \pm\infty$, so that integrals over K, Σ_t, and Σ_0 converge and permit integration by parts. For $\Sigma = \Sigma_t$, Eqs. (92) and (93) yield

$$\int_{\Sigma_t} j^3\sigma^*\left(Y^2_T \lrcorner \Theta\right) = \int_{\Sigma_t}\left[\left(\frac{\partial L}{\partial z^a_2} - P^{(i_1 2)}_{a,i_1}\right) y^a_{,j}\mathrm{d}x^j + \left(P^{(2i_2)}_a\right) y^a_{,i_2 j}\mathrm{d}x^j\right]$$

$$- \int_{\Sigma_t}\left(\left(\frac{\partial L}{\partial z^a_2} - P^{(i_1 2)}_{a,i_1}\right) y^a_{,i} + P^{ij}_a y_{,ij} - L\right)\mathrm{d}x^2 + \int_{\Sigma_t} \mathrm{d}\left(P^{[21]}_a y^a_{,1}\right)$$

$$
\begin{aligned}
&= \int_{\Sigma_t} \left[\left(\frac{\partial L}{\partial z_2^a} - P_{a,i_1}^{(i_1 2)} \right) y_{,j}^a \mathrm{d}x^j + \left(P_a^{(2 i_2)} \right) y_{,i_2 j}^a \mathrm{d}x^j \right] + \\
&\quad - \int_{\Sigma_t} \left(\left(\frac{\partial L}{\partial z_2^a} - P_{a,i_1}^{(i_1 2)} \right) y_{,i}^a + P_a^{ij} y_{,ij} - L \right) \mathrm{d}x^2 + \\
&\quad + \lim_{x_2 \to \infty} \left(P_a^{[21]} y_{,1}^a \right)(t, x_2) - \lim_{x_2 \to -\infty} \left(P_a^{[21]} y_{,1}^a \right)(t, x_2) \\
&= \int_{\Sigma_t} \left[\left(\frac{\partial L}{\partial z_2^a} - P_{a,i_1}^{(i_1 2)} \right) y_{,j}^a \mathrm{d}x^j + \left(P_a^{(2 i_2)} \right) y_{,i_2 j}^a \mathrm{d}x^j \right] \\
&\quad - \int_{\Sigma_t} \left(\left(\frac{\partial L}{\partial z_2^a} - P_{a,i_1}^{(i_1 2)} \right) y_{,i}^a + P_a^{ij} y_{,ij} - L \right) \mathrm{d}x^2
\end{aligned}
$$

because our asymptotic conditions require that $\lim_{x_2 \to \infty} \left(P_a^{[21]} y_{,1}^a \right)(t, x_2) = 0$ and $\lim_{x_2 \to -\infty} \left(P_a^{[21]} y_{,1}^a \right)(t, x_2) = 0$. Hence the potential non-uniqueness of constants of motion is taken care of by the appropriate choice of boundary conditions.

Appendix

Jets

Let $\pi : N \to M$ be a locally trivial fibration. A local section σ of π is a smooth map $\sigma : M \to N$, defined on an open subset U of M, such that $\pi \circ \sigma(x) = x$ for every $x \in U$. If $U = M$, we say that $\sigma : M \to N$ is a global section of π. In the following, we say $\sigma : M \to N$ is a section of π if σ is either local or global section.

Suppose that $m = \dim M$ and $m + n = \dim N$. We use local coordinates (x^i) on M, where $i = 1, \ldots, m$, and (x^i, y^a) on N, where $a = 1, \ldots, n$. The local coordinate description of a section $\sigma : M \to N$ is given by $y^a = \sigma^a(x^1, \ldots, x^m)$ for $a = 1, \ldots n$.

For each $x \in M$ and k $= 1, 2, \ldots$, sections σ and $\check{\sigma}$ of π are k-equivalent at x if $\sigma(x) = \check{\sigma}(x)$ and, in local coordinates,

$$
\sigma_{,i_1 \ldots i_l}^a(x) = \check{\sigma}_{,i_1 \ldots i_l}^a(x), \tag{94}
$$

where

$$
\sigma_{,i_1 \ldots i_l}^a(x) = \frac{\partial^l \sigma^a}{\partial x^{i_1} \ldots \partial x^{i_l}}(x^1(x), \ldots, x^m(x)), \tag{95}
$$

for all $l = 1, \ldots,$ k.

The k-equivalence class at x of a section σ is called the k-jet of σ at x and denoted $j^{\mathrm{k}}\sigma(x)$. The space of k-equivalence classes at x of all section σ is denoted $J_x^{\mathrm{k}}(M, N)$ and

$$J^{\mathrm{k}}(M, N) = \bigcup_{x \in M} J_x^{\mathrm{k}}(M, N)$$

is called the space of k-jets of sections of π. In terms of local coordinates, $j^{\mathrm{k}}\sigma(x)$ has coordinates $(x^i, y^a, z_{i_1}^a, \ldots, z_{i_1 \ldots i_k}^a)$, where

$$z_{i_1 \ldots i_l}^a = \sigma_{,i_1 \ldots i_l}^a(x),$$

for $l = 1, \ldots, k$, $i^1, \ldots, i^l = 1, \ldots, m$, and $a = 1, \ldots, n$. Since partial derivatives of a smooth function commute, the variables $z_{i_1 \ldots i_l}^a$ cannot be considered as independent coordinates. In the case when it matters, we use an independent collection

$$\{z_{i_1 \ldots i_l}^a \mid a = 1, \ldots, n, \text{ and } 1 \leq i_1 \leq i_2 \leq \ldots \leq i_l\}; \tag{96}$$

see Eq. (7). However, in general, we use symmetry of variables $z_{i_1 \ldots i_l}^a$ in the indices $i_1, \ldots, i_l$.

There are several maps defined on $J^{\mathrm{k}}(M, N)$:
the source map

$$\pi^{\mathrm{k}} : J^{\mathrm{k}}(M, N) \to M : j^{\mathrm{k}}\sigma(x) \mapsto x,$$

the target map

$$\pi_0^{\mathrm{k}} : J^{\mathrm{k}}(M, N) \to N : j^{\mathrm{k}}\sigma(x) \mapsto \sigma(x),$$

the (k, l)-*forgetful*

$$\pi_{\mathrm{l}}^{\mathrm{k}} : J^{\mathrm{k}}(M, N) \to J^{\mathrm{l}}(M, N) : j^{\mathrm{k}}\sigma(x) \mapsto j^{\mathrm{l}}\sigma(x) \text{ for } \mathrm{k} > \mathrm{l} > 0.$$

Each of these maps defines a fibre bundle structure in $J^{\mathrm{k}}(M, N)$. For this reason, $J^{\mathrm{k}}(M, N)$ is also called the k-jet bundle of sections of π.

Let $\sigma : M \to N$ be a section of $\pi : N \to M$. We denote the k-*jet extension* of σ by

$$j^{\mathrm{k}}\sigma : M \to J^{\mathrm{k}}(M, N) : x \mapsto j^{\mathrm{k}}\sigma(x).$$

For every integer $\mathrm{k} > 0$,

$$\pi_0^{\mathrm{k}} \circ j^{\mathrm{k}}\sigma = \sigma. \tag{97}$$

Similarly, for each k > 1 > 0,

$$\pi_1^{\mathrm{k}} \circ j^{\mathrm{k}}\sigma = j^{1}\sigma.$$

A section $\rho : M \to J^{\mathrm{k}}(M, N)$ of the source map $\pi^{\mathrm{k}} : J^{\mathrm{k}}(M, N) \to M$ is called *holonomic* if there exists a section σ of π such that

$$\rho = j^{\mathrm{k}}\sigma.$$

It follows from Eq. (97) that ρ is holonomic if and only if

$$\rho = j^{\mathrm{k}}(\pi_0^{\mathrm{k}} \circ \rho).$$

Each local chart M gives rise to *local contact forms* on $J^k(M, N)$ given by

$$\vartheta^a = \mathrm{d}y^a - \sum_{i=1}^{m} z_i^a \mathrm{d}x^i, \quad \vartheta_i^a = \mathrm{d}z_i^a - \sum_{j=1}^{m} z_{ij}^a \mathrm{d}x^j, \quad \ldots.. \tag{98}$$

$$\vartheta^a_{i_1 \ldots i_{k-1}} = \mathrm{d}z^a_{i_1 \ldots i_{k-1}} - \sum_{i_k=1}^{m} z^a_{i_1 \ldots i_{k-1} i_k} \mathrm{d}x^{i_k}.$$

A section $\rho : M \to J^{\mathrm{k}}(M, N)$ of the source map π^{k} is holonomic if the tangent space of its range is annihilated by the contact forms $\vartheta^a_{i_1 \ldots i_l}$ for all $l = 0, \ldots, k-1$ and all indices $i_1, \ldots, i_l = 1, \ldots, m$ and any collection of coordinate charts covering M.

Prolongations

Let Y be a vector field on N which projects to a vector field Y^0 on M. In other words, Y is π-related to a vector field Y^0, that is

$$T\pi \circ Y = Y^0 \circ \pi. \tag{99}$$

This implies that $\pi : N \to M$ intertwines the actions of local one-parameter local groups e^{tY} and e^{tY^0} generated by Y and Y^0, respectively.

$$\begin{array}{ccc} & \mathrm{e}^{tY} & \\ N & \rightarrow & N \\ \pi \downarrow & & \downarrow \pi \\ M & \rightarrow & M \\ & \mathrm{e}^{tY^0} & \end{array} \tag{100}$$

Hence, for every section σ of π,

$$\mathrm{e}^{tY*}\sigma = \mathrm{e}^{tY} \circ \sigma \circ \mathrm{e}^{-tY^0} \tag{101}$$

is a local section of π. For every integer k, the map $\sigma \mapsto \mathrm{e}^{tY*}\sigma$ induces a local one-parameter local group

$$\mathrm{e}^{tY^{\mathrm{k}}} : J^{\mathrm{k}}(M,N) \to J^{\mathrm{k}}(M,N) : j^{\mathrm{k}}\sigma(x) \mapsto [j^{\mathrm{k}}(\mathrm{e}^{tY*}\sigma)](\mathrm{e}^{tY^0}x) \tag{102}$$

of diffeomorphisms of $J^{\mathrm{k}}(M,N)$ to itself, generated by a vector field Y^{k} on $J^{\mathrm{k}}(M,N)$, called the prolongation of Y to $J^{\mathrm{k}}(M,N)$. In other words,

$$\mathrm{e}^{tY^{\mathrm{k}}} \circ j^{\mathrm{k}}\sigma = j^{\mathrm{k}}(\mathrm{e}^{tY*}\sigma). \tag{103}$$

For every $0 < \mathrm{l} < \mathrm{k}$, Y^{k} is $\pi^{\mathrm{k}}_{\mathrm{l}}$-related to Y^{l},

$$T\pi^{\mathrm{k}}_{\mathrm{l}} \circ Y^{\mathrm{k}} = Y^{\mathrm{l}} \circ \pi^{\mathrm{k}}_{\mathrm{l}}, \tag{104}$$

where $\pi^{\mathrm{k}}_{\mathrm{l}} : J^{\mathrm{k}}(M,N) \to J^{\mathrm{l}}(M,N)$ is the forgetful map.

Following reference [13], we show how to find the prolongation Y^{k} of a vector field $Y = Y^i(x)\frac{\partial}{\partial x^i} + Y^a(x,y)\frac{\partial}{\partial y^a}$ on N that is π-related to $Y^0 = Y^i(x)\frac{\partial}{\partial x^i}$ on M using the condition that, for every local contact form ϑ on $J^{\mathrm{k}}(M,N)$, the Lie derivative $\mathcal{L}_{Y^{\mathrm{k}}}\vartheta$ of ϑ with respect to Y^{k} is a linear combination of local contact forms. Let

$$Y^{\mathrm{k}} = Y^i\frac{\partial}{\partial x^i} + Y^a\frac{\partial}{\partial y^a} + Y^a_i\frac{\partial}{\partial z^a_i} + \ldots + Y^a_{i_1 i_2 \ldots i_k}\frac{\partial}{\partial z^a_{i_1 i_2 \ldots i_k}}$$

be the prolongation of Y to $J^{\mathrm{k}}(M,N)$. Then,

$$\begin{aligned}
&\mathcal{L}_{Y^{\mathrm{k}}}[\mathrm{d}y^a - z^a_i \mathrm{d}x^i] \\
&= Y^{\mathrm{k}} \lrcorner \left(\mathrm{d}[\mathrm{d}y^a - z^a_i\mathrm{d}x^i]\right) + \mathrm{d}\left(Y^{\mathrm{k}} \lrcorner [\mathrm{d}y^a - z^a_i\mathrm{d}x^i]\right) \\
&= -Y^{\mathrm{k}} \lrcorner \left(\mathrm{d}z^a_i \wedge \mathrm{d}x^i\right) + \mathrm{d}\left(Y^a - z^a_i Y^i\right) \\
&= -Y^a_i \mathrm{d}x^i + Y^i \mathrm{d}z^a_i + Y^a_{,b}\mathrm{d}y^b + Y^a_{,i}\mathrm{d}x^i - Y^i\mathrm{d}z^a_i - z^a_i Y^i_{,j}\mathrm{d}x^j \\
&= Y^a_{,b}(\mathrm{d}y^b - z^b_i\mathrm{d}x^i) + Y^a_{,b}z^b_i\mathrm{d}x^i - Y^a_i\mathrm{d}x^i + Y^a_{,i}\mathrm{d}x^i - z^a_i Y^i_{,j}\mathrm{d}x^j \\
&= Y^a_{,b}(\mathrm{d}y^b - z^b_i\mathrm{d}x^i) + \left[\left(Y^a_{,b}z^b_i - z^a_j Y^j_{,i} + Y^a_{,i}\right) - Y^a_i\right]\mathrm{d}x^i,
\end{aligned}$$

which implies that

$$Y_i^a = Y^a_{,b} z_i^b - z_j^a Y^j_{,i} + Y^a_{,i}. \tag{105}$$

Similarly,

$$\begin{aligned}
&\mathcal{L}_{Y^{\mathrm{k}}}[\mathrm{d}z_i^a - z_{ij}^a \mathrm{d}x^j] \\
&= Y^{\mathrm{k}} \lrcorner \left(\mathrm{d}[\mathrm{d}z_i^a - z_{ij}^a \mathrm{d}x^j]\right) + \mathrm{d}\left(Y^k \lrcorner [\mathrm{d}z_i^a - z_{ij}^a \mathrm{d}x^j]\right) \\
&= -Y^{\mathrm{k}} \lrcorner \left(\mathrm{d}z_{ij}^a \wedge \mathrm{d}x^j\right) + \mathrm{d}\left(Y_i^a - z_{ij}^a Y^i\right) \\
&= -Y_{ij}^a \mathrm{d}x^j + Y^j \mathrm{d}z_{ij}^a + Y^a_{i,b} \mathrm{d}y^b + Y^a_{i,j} \mathrm{d}x^j - Y^i \mathrm{d}z_{ij}^a - z_{ij}^a Y^i_{,k} \mathrm{d}x^k \\
&= Y^a_{i,b}\left(\mathrm{d}y^b - z_j^b \mathrm{d}x^j\right) + Y^a_{i,b} z_j^b \mathrm{d}x^j - Y_{ij}^a \mathrm{d}x^j + +Y^a_{i,j} \mathrm{d}x^j - z_{ij}^a Y^i_{,k} \mathrm{d}x^k \\
&= Y^a_{i,b}\left(\mathrm{d}y^b - z_j^b \mathrm{d}x^j\right) + \left[\left(Y^a_{i,b} z_j^b - z_{ki}^a Y^k_{,j} + Y^a_{i,j}\right) - Y_{ij}^a\right] \mathrm{d}x^j,
\end{aligned}$$

so that, for $i \leq j$,

$$Y_{ij}^a = Y^a_{i,b} z_j^b - z_{ki}^a Y^k_{,i} + Y^a_{i,j}.$$

Symmetrizing, we get

$$Y_{ij}^a = z^b_{(j} Y^a_{i),b} - z^a_{k(i} Y^k_{,j)} + Y^a_{(i,j)}. \tag{106}$$

In general,

$$\begin{aligned}
&\mathcal{L}_{Y^{\mathrm{k}}}[\mathrm{d}z^a_{i_1 \dots i_l} - z^a_{i_1 \dots i_l j} \mathrm{d}x^j] = \\
&= Y^{\mathrm{k}} \lrcorner \left(\mathrm{d}[\mathrm{d}z^a_{i_1 \dots i_l} - z^a_{i_1 \dots i_l j} \mathrm{d}x^j]\right) + \mathrm{d}\left(Y^{\mathrm{k}} \lrcorner\right) \\
&= -Y^{\mathrm{k}} \lrcorner \left(\mathrm{d}z^a_{i_1 \dots i_l j} \wedge \mathrm{d}x^j\right) + \mathrm{d}\left(Y^a_{i_1 \dots i_l} - z^a_{i_1 \dots i_l j} Y^j\right) \\
&= -Y^a_{i_1 \dots i_l j} \mathrm{d}x^j + Y^j \mathrm{d}z^a_{i_1 \dots i_l j} + Y^a_{i_1 \dots i_l, b} \mathrm{d}y^b + Y^a_{i_1 \dots i_l, j} \mathrm{d}x^j \\
&\quad -Y^j \mathrm{d}z^a_{i_1 \dots i_l j} - z^a_{i_1 \dots i_l j} Y^j_{,k} \mathrm{d}x^k \\
&= -Y^a_{i_1 \dots i_l j} \mathrm{d}x^j + Y^a_{i_1 \dots i_l, b} \mathrm{d}y^b + Y^a_{i_1 \dots i_l, j} \mathrm{d}x^j - z^a_{i_1 \dots i_l j} Y^j_{,k} \mathrm{d}x^k \\
&= Y^a_{i_1 \dots i_l, b}\left(\mathrm{d}y^b - z_j^b \mathrm{d}x^j\right) + Y^a_{i_1 \dots i_l, b} z_j^b \mathrm{d}x^j - Y^a_{i_1 \dots i_l j} \mathrm{d}x^j \\
&\quad +Y^a_{i_1 \dots i_l, j} \mathrm{d}x^j - z^a_{i_1 \dots i_l j} Y^j_{,k} \mathrm{d}x^k \\
&= Y^a_{i_1 \dots i_l, b}\left(\mathrm{d}y^b - z_j^b \mathrm{d}x^j\right) + \left[\left(Y^a_{i_1 \dots i_l, b} z_j^b - z^a_{i_1 \dots i_l k} Y^k_{,j} + Y^a_{i_1 \dots i_l, j}\right) - Y^a_{i_1 \dots i_l, j}\right] \mathrm{d}x^j.
\end{aligned}$$

Therefore

$$Y^a_{i_1 i_2 \ldots i_l j} = z^b_{(j} Y^a_{i_1 \ldots i_l),b} - z^a_{k(i_1 i_2 \ldots i_l} Y^k_{,j)} + Y^a_{(i_1 \ldots i_l, j)}. \tag{107}$$

Acknowledgements The authors are grateful to BIRS for sponsoring the Banff Workshop on "Material Evolution", June 11–18, 2017, which led to this collaboration. The first author (Jędrzej Śniatycki) greatly appreciates constructive discussions with Mark Gotay and Oğul Esen.

References

1. Aldaya, V., and de Azcárraga, J.A., "Variational principles on rth order jets of fibre bundles in field theory", *Journal of Mathematical Physics*, **19** (1978), pp. 1869–1875.
2. Binz, E., Śniatycki, J., and Fischer, H., Geometry of Classical Fields, Elsevier Science Publishers, New York, 1988. Reprinted by Dover Publications, Mineola N.Y., 2006.
3. Campos, C.M., *Geometric Methods in Classical Field Theory and Continuous Media,* Thesis, Departamento de Mathemáticas, Facultad de Ciencias, Universidad Autónoma de Madrid, 2010.
4. Cartan, É., "Les espaces métriques fondés sur la notion d'aire", *Actualités Scientifique et Industrielles,* no 72 (1933). Reprinted by Hermann, Paris, 1971.
5. De Donder, Th., "Théorie invariantive du calcul des variations", *Bull, Acad. de Belg.* 1929, chap. 1. [this reference appears in Cartan [4]].
6. De Donder, Th., "Théorie invariantive du calcul des variations" (nouvelle édit.), Gauthier Villars, Paris, 1935.
7. Gotay, M., "A Multisymplectic Framework for Classical Field Theory and Calculus of Variations", *Differential Geometry and its Applications*, **1** (1991) pp. 375–390.
8. Grabowska, K., and Vitagliano, L., Tulczyjew triples in higher derivative field theory, arXiv:1406.6503v2 [math-ph] 20 Feb 2015.
9. Kijowski, J., and Szczyrba, W., "Multisymplectic manifolds and a geometric construction of the Poisson bracket in the classical field theory", in *Géométrie symplectique et physique mathématique.* J.-M. Souriau (ed.), Publications CNRS No 237, Paris, 1975, pp. 347–379.
10. Kijowski, J., and Tulczyjew, W.M., *A Symplectic Framework for Field Theories*, Lecture Notes in Physics, vol. 107, Springer, Berlin, 1979.
11. Kupferman, R., Olami, E. and Segev R., "Stress theory for classical fields", *Mathematics and Mechanics of Solids,* https://doi.org/10.1177/1081286517723697, 2017.
12. Lepage, Th., "Sur les champs géodésiques du calcul des variations, I et II", *Académie royale de Belgique, Bulletins de la classe de sciences*, 5e série, **22** (1936) pp. 716–729 and 1034–1046.
13. Olver, P., *Equivalence, Invariance and Symmetry,* Cambridge University Press, Cambridge, UK, 1995.
14. M. Ostrogradski, "Memoires sur les equations differentielles relatives au probleme des isoperimetres", *Mem. Ac. St. Petersbourg,* VI, **4** (1850) 385.
15. R. Segev, "Geometric analysis of hyper-stresses", *International Journal of Engineering Science,* **120** (2017) pp. 100–118.
16. Śniatycki, J., "On the geometric structure of classical field theory in Lagrangian formulation", *Proc. Camb. Phil. Soc.*, **68** (1970), pp. 475–484.
17. Śniatycki, J., *Differential Geometry of Singular Spaces and Reduction of Symmetry*, Cambridge University Press, Cambridge, UK, 2013.
18. Śniatycki, J. and Esen, O. "De Donder form for second order gravity", arXiv:1902.07616v2 [math-phys], 2019.

Part II
Defects, Uniformity and Homogeneity

Regular and Singular Dislocations

Marcelo Epstein and Reuven Segev

Contents

Abstract The theory of continuous distributions of dislocations and other material defects, when formulated in terms of differential forms, is shown to comprise also the discrete, or singular, counterpart, in which defects are concentrated on lower dimensional regions, such as surfaces, lines, and points. The mathematical tool involved in this natural transition is the theory of de Rham currents, which plays in regard to differential forms the same role as the theory of Schwartz distributions plays with respect to ordinary functions. After a review of the main mathematical aspects, the theory is illustrated with a profusion of examples and applications.

1 Introduction

The theory of material defects has its origins not in the works of engineers but rather in the prodigious curiosity of the great Italian mathematician Vito Volterra (1860–1940), for whom the theory of elasticity was just one of the many possible

M. Epstein
Department of Mechanical and Manufacturing Engineering, University of Calgary, Calgary, AB, Canada
e-mail: mepstein@ucalgary.ca

R. Segev (✉)
Department of Mechanical Engineering, Ben-Gurion University of the Negev, Beer-Sheva, Israel
e-mail: rsegev@bgu.ac.il

R. Segev, M. Epstein (eds.), *Geometric Continuum Mechanics*, Advances in Mechanics and Mathematics 43, https://doi.org/10.1007/978-3-030-42683-5_5

applications of mathematical analysis, a field to which he made many pioneering contributions. In the year 1907, Volterra published a celebrated article [23], whose contents he amplified and further expanded in a series of lectures delivered in the USA in 1909 and published in 1912 [24]. The topic of these articles revolved around the equilibrium, in the absence of external forces, of multiply connected elastic bodies, which, by a process of cutting and welding, sustain non-vanishing stresses. In a now famous illustration, reproduced in texts of material science to our very day, Volterra identifies 6 possible independent processes of this kind, which he named *distortions*. Volterra's original illustration[1] is in fact a photograph of an actual realization of each of these distortions.

In 1899, a few years before Volterra's publications, the French mathematician Élie Cartan (1869–1951), acknowledging his debt to Pfaff and to Grassmann, launched a new era in the calculus of differential forms [2] and the theory of differential systems. A separate mathematical development, which formalizes the idea of 'singularity functions', informally used already at that time in many engineering and physics applications, is the work of the French mathematician Laurent Schwartz on the theory of distributions, published in a short article in 1945 [20]. The Swiss mathematician Georges de Rham (1903–1990), combined the method of Schwartz with the calculus of differential forms to produce his original theory of currents, elegantly summarized in a book he published in 1955 [4].

Although these are by no means the only mathematical tools relevant to our discussion, the chain of transmission outlined above conveys a general picture of the flow of some of the main ideas that may be necessary to invoke for the modelling of material defects. The situation is quite different when it comes to summarizing the rich engineering and applied mechanics tradition that ensued from Volterra's work. Whereas mathematics seems to have a way of incorporating old knowledge into new in a matter of one or two generations, the accelerated progress made in various areas of theoretical and experimental physics, liquid crystals and soft matter, materials science, metallurgy, numerical and computational techniques, various branches of engineering, industrial demand, and support from granting agencies, among other factors, has made the communication between various schools of research, even those interested in the same phenomena, not only very slow but, in some cases, also next to impossible, as if between people speaking different languages trying to converse in a noisy environment.

A short but comprehensive historical account of dislocation theory up to 1985 is given in [11], ranging from the early studies on isolated defects to the differential geometric theories of continuous distributions of dislocations. Historically, of course, the continuum theories grew out of their discrete counterparts, starting in the 1950s, although a rigorous passage to the continuum limit was not claimed, but only used in a heuristic way. A typical example of this kind of inspiration is the relation between the Burgers vector (that is, the lack of closure of an atomic lattice circuit enclosing a defect) and the torsion of the distant parallelism induced by a Cartan frame field. Our objective in this chapter is to demonstrate that the two extremes

[1] Available at https://archive.org/details/lecturesdelivere00clarrich, page 43.

represented, on the one hand, by the discrete models and, on the other hand, by the smooth theories, can be reconciled and encompassed under the embrace of a single mathematical apparatus, namely, the theory of de Rham currents. The passage from the smooth model to various singular versions is achieved naturally when, having formulated the smooth theory in terms of differential forms, the quantities and the governing equations are reinterpreted in the weak sense afforded by the theory of currents.

Section 2 reviews the classical setting in terms of frame fields representing an underlying geometric structure, and, having recast the theory in terms of dual co-frames, prepares the stage for the use of exterior differential calculus. The presentation is free of any metric connotations, such as available in a Cartesian space. In the case when a single differential 1-form ω is specified on an n-dimensional manifold $\mathscr{M}$, the physical counterpart can be interpreted as the prescription of a field of stacks of oriented hyperplanes, such as is the case in type-A smectics. The defectivity of this structure is measured by the lack of closure of the exterior differential of ω. Consequences of Stokes' theorem of integral calculus are explored, in anticipation of stronger results in the weak formulation, which is introduced in Sect. 3. This section contains the main results of the chapter. A brief review of the concept of current is presented, according to which the weak counterpart of a p-form is recognized as an $(n-p)$-current. Similarly, the counterpart of the exterior derivative operator of forms is the boundary operator on currents, and the lack of closure of the former is reflected in the non-vanishing of the boundary of the latter. Classical isolated edge and screw dislocations are shown to emerge naturally in this context. Examples of coherent and non-coherent interfaces are presented and discussed. When $(n-1)$-forms, or 1-currents, are used as the point of departure for the analysis, we obtain a filamented structure, such as is encountered in nematic liquid crystals and many biological tissues. Section 4 suggests how the continuous theory can in principle be reconstructed from the discrete one by means of a passage to the limit as the number of isolated dislocations in a fixed volume element grows without bound. Finally, Sect. 5 provides a framework for the description of the movement of defects in terms of time-dependent principal-bundle morphisms. Although the presentation is by no means elementary or self-contained, an effort has been made to deemphasize the formal aspects of the theory, a fact that is reflected in a lighter than usual style that makes as few demands as possible from the reader.

2 Regular Lattices

2.1 *Frame Fields*

The most natural, though not the only, continuous extension of the picture of an orderly crystalline array of atoms consists of defining a field of *frames* over an open connected region $\mathscr{R}$ of $\mathbb{R}^3$. Somewhat more technically, we may say that a *continuous crystalline array* is a (smooth) section σ of the linear frame bundle $F\mathscr{R}$.

In other words, to each point $X \in \mathscr{R}$, the map $\sigma : \mathscr{R} \to F\mathscr{R}$ assigns three linearly independent vectors $\mathbf{v}_\alpha \in T_X\mathscr{R}$ $(\alpha = 1, 2, 3)$ in the tangent space $T_X\mathscr{R}$. Within this restricted framework, the theory of continuous distributions of dislocations attempts to define and quantify measures of densities of defects. It should also consider whether other interpretations of the discrete picture are amenable to extension to the continuous realm.

In the discrete picture, it is customary to define the *Burgers vector* as the lack of closure of an approximately plane and quadrilateral atomic circuit with the same number of atomic spaces on its opposite sides. The non-vanishing of the Burgers vector is an indication of the presence of a lattice defect within the quadrilateral. In the continuous picture, we have three independent vector fields $\mathbf{e}_\alpha$, each of which gives rise to a *flow* $\phi_\alpha(s^\alpha)$, where s^α is the natural parameter, defined up to an additive constant, on the integral curve of the field $\mathbf{e}_\alpha$. The analogue of the lack of closure of a Burgers circuit is the lack of commutativity of two of these flows. Indeed, only if the flows ϕ_α and ϕ_β commute, we arrive at the same final point Y when, starting from point X, we first advance by an amount Δs^α along the integral line of $\mathbf{e}_\alpha$ and then by an amount Δs^β along the integral line of $\mathbf{e}_\beta$, or repeat this procedure in the opposite order.

The infinitesimal version of the procedure just described gives rise to the notion of a new vector field $[\mathbf{u}, \mathbf{v}]$, namely, the *Lie bracket* between the two vector fields, $\mathbf{u}$ and $\mathbf{v}$. In terms of components in a coordinate system X^I $(I = 1, 2, 3)$, the Lie bracket is expressed as

$$[\mathbf{u}, \mathbf{v}]^I = u^J \frac{\partial v^I}{\partial X^J} - v^J \frac{\partial u^I}{\partial X^J}, \tag{1}$$

where the summation convention for diagonally repeated indices is implied.

The vector

$$\mathbf{c}_{\alpha\beta} = [\mathbf{e}_\alpha, \mathbf{e}_\beta] = c^I_{\alpha\beta} \left.\frac{\partial}{\partial X^I}\right|_X = c^\gamma_{\alpha\beta}\, \mathbf{e}_\gamma(X) \tag{2}$$

will be called the *local Burgers vector* at X associated with the families α and β. It represents the defects gathered along the 'infinitesimal circuit' generated by $\mathbf{e}_\alpha$ and $\mathbf{e}_\beta$. If $\mathbf{c}_{\alpha\beta}$ at a point X happens to be a linear combination of $\mathbf{e}_\alpha$ and $\mathbf{e}_\beta$ alone, we say that at X we have a *pure edge dislocation* density associated with the families α and β. This condition can be written as

$$c^\gamma_{\alpha\beta} = 0 \qquad \text{for } \gamma \neq \alpha \text{ and } \gamma \neq \beta. \tag{3}$$

Otherwise, that is, if there is a non-vanishing component on the third base vector (not α or β), we have a mixture of edge and *screw dislocations*.

Given any two vectors, $\mathbf{u} = u^\rho \mathbf{e}_\rho$ and $\mathbf{v} = v^\rho \mathbf{e}_\rho$, we define the Burgers vector associated with their corresponding infinitesimal circuit as

$$\mathbf{c}_{u,v} = c^\gamma_{\alpha\beta} u^\alpha v^\beta\, \mathbf{e}_\gamma. \tag{4}$$

The components $c^\gamma_{\alpha\beta} = -c^\gamma_{\beta\alpha}$ are called the *structure constants* of the frame field at X. When the structure constants vanish identically throughout $\mathscr{R}$, the frame field is called *holonomic*. In that case, and only in that case, there exists a coordinate system on $\mathscr{R}$ such that its natural basis coincides with the frame field at each point $X \in \mathscr{R}$. Otherwise, the frame field is said to be *anholonomic*. From the physical point of view, holonomicity corresponds to a defect-free continuous crystalline array.

Let $\chi : \mathscr{R} \to \mathscr{R}'$ be a diffeomorphism from $\mathscr{R}$ to another region $\mathscr{R}'$. The pushforward of the crystal frame $\mathbf{e}_\alpha$ at X is a frame $\mathbf{e}'_\alpha$ at $X' = \chi(X)$. The collection of structure constants vanishes at X' if, and only if, it vanishes at X, as can be verified by a direct calculation. From the physical standpoint, this remark means that it is not possible to remove defects by a mere deformation. This fact also shows that our initial choice of $\mathscr{R}$ as a region in $\mathbb{R}^3$ might as well be replaced with any sub-body of a body manifold. The metric properties of $\mathbb{R}^3$ need not be invoked to detect the presence of material defects. Given these considerations, we see no reason to limit the theory either by the metric properties or by the dimensionality of our ordinary space. We will, therefore, proceed with a formulation based on an underlying n-dimensional manifold $\mathscr{M}$.

2.2 Material Parallelism

A somewhat different way to look at the continuous version of a discrete lattice is to regard it as a *parallelism* on $\mathscr{M}$. Starting from a given frame field, as in the previous section, two vectors, $\mathbf{v}$ at X, and $\mathbf{w}$ at Y, are said to be *materially parallel* if their components in their respective local bases, $\mathbf{e}_\alpha(X)$ and $\mathbf{e}_\alpha(Y)$, are correspondingly the same for each $\alpha = 1, \ldots, n$. If, given a vector field, this condition is satisfied for every pair of points, we say that the vector field is materially parallel or *materially constant*. Clearly, the frame field of departure consists of three materially constant vector fields. The notion of material parallelism is more general than that of a continuous lattice in two senses, namely,

- The parallelism is preserved under any lattice transformation of the form

$$\mathbf{f}_\alpha(X) = A^\rho_\alpha \, \mathbf{e}_\rho(X), \tag{5}$$

  where A^ρ_α are constants. Thus, we are not beholden to a particular frame field, but rather to a more general geometric notion.
- The concept of material parallelism can be induced directly from a given constitutive law via the notion of *material isomorphism* [15, 25]. The importance of this remark can be appreciated both from a practical and from an epistemological point of view. It means that a theory of dislocations can, at least in principle, be built upon a continuum scaffolding alone, without any need to resort to a lower, more fundamental, level of discourse.[2] In particular, the incorporation

[2]On this point, it is pertinent to mention the works of the philosopher of science Mario Bunge [1]. Bunge, though, warns us about the dangers of 'explicatio obscurum per obscurius'.

of continuous symmetry groups becomes available as an additional degree of freedom of the theory, which will not be considered herein.

A parallelism on a manifold can be construed in terms of a *linear connection* with vanishing curvature. Starting from a materially parallel frame field $\mathbf{e}_\alpha$ ($\alpha = 1, \ldots, n$), expressed in terms of components in a coordinate system X^I ($I = 1, \ldots, n$) as

$$\mathbf{e}_\alpha = e^I_\alpha \frac{\partial}{\partial X^I}, \tag{6}$$

the *Christoffel symbols* of this connection, expressed in the coordinate system, are obtained as

$$\Gamma^I_{JK} = -e^I_{\alpha,K}\, e^\alpha_J. \tag{7}$$

In this equation, we use commas to indicate partial derivatives with respect to the coordinates. Moreover, we have denoted by e^α_J the entries in the inverse of the matrix with entries e^J_α, that is, $e^J_\alpha\, e^\alpha_K = \delta^J_K$.

We remark that the vanishing of the curvature is guaranteed automatically by the independence of path implicit in the very definition. The *torsion* of this connection can be expressed in terms of the *torsion tensor* $\mathbf{T}$, whose components in the coordinate system are

$$T^I_{JK} = \Gamma^I_{JK} - \Gamma^I_{KJ}. \tag{8}$$

We can verify that

$$\mathbf{T} = T^I_{JK} \frac{\partial}{\partial X^I} \otimes dX^J \otimes dX^K = -c^\gamma_{\alpha\beta}\, \mathbf{e}_\gamma \otimes \mathbf{e}^\alpha \otimes \mathbf{e}^\beta, \tag{9}$$

where $\mathbf{e}^\alpha$ represents the point-wise algebraic dual basis of $\mathbf{e}_\alpha$. In other words, the field of structure constants of the anholonomic frame field $\mathbf{e}_\alpha$ is nothing but the anholonomic component expression of the torsion tensor of the associated material connection.

It is interesting to point out that the torsion tensor gives rise to a *trace 1-form* $\omega = \omega_I\, dX^I$ defined as

$$\omega_I = T^K_{KI}. \tag{10}$$

Following [5], we can use this 1-form to canonically decompose the torsion tensor into the sum of a *diagonal part*

$$\tilde{T}^I_{JK} = \frac{1}{n-1}\left(\delta^I_J\, \omega_K - \delta^I_K\, \omega_J\right), \tag{11}$$

and a *traceless part*

$$\hat{T}^I_{JK} = T^I_{JK} - \tilde{T}^I_{JK}. \tag{12}$$

It is not difficult to verify that, if we go back to the case $n = 3$, the vanishing of the traceless part implies that, for every pair of vectors **u**, **v**, the Burgers vector $\mathbf{c}_{u,v}$ corresponds to a pure edge dislocation. Thus, the traceless part is an indication of the presence of screw dislocations at a given point.

2.3 The Dual View

To each frame $\mathbf{e}_\alpha$ at a point $X \in \mathscr{M}$, we can uniquely assign its dual *co-frame* $\mathbf{e}^\alpha$, which is a basis of the cotangent space $T^*_X\mathscr{M}$. As linear operators on $T_X\mathscr{M}$, the action of the covectors of the co-frame on tangent vectors is completely defined by the algebraic condition

$$\langle \mathbf{e}^\alpha, \mathbf{e}_\beta \rangle = \delta^\alpha_\beta. \tag{13}$$

Since a frame field is, by definition, a (smooth) section of the frame bundle $F\mathscr{M}$, each of the three base covectors $\mathbf{e}^\alpha$ is a (smooth) section of the cotangent bundle $T^*\mathscr{M}$, namely, a differential 1-form. From the algebraic standpoint, it does not appear that working with co-frames could provide further insight into the description of distributed defects. Nevertheless, one can foresee at the very least three possible features of the dual approach that might bear fruit and shed new light on our topic, to wit

- (i) Broadly speaking, it can be said that, if a frame represents lines joining atoms with their lattice neighbours, a co-frame can be regarded as a collection of atomic planes, a point of view not foreign to crystallographic science.
- (ii) In contradistinction to vector fields, differential forms offer a fully fledged *exterior calculus* that elegantly summarizes and generalizes to arbitrary manifolds the machinery of classical vector calculus in Euclidean space.
- (iii) The theory of integration on manifolds is intimately connected with the theory of differential forms. Important facts pertaining to global properties of a defective structure can, therefore, be obtained in a natural way with the dual approach.
- (iv) Finally, and most strikingly, the calculus of differential forms can be extended to a weak formulation, known as the *theory of currents*, much in the way that the theory of distributions generalizes that of ordinary functions. This circumstance opens the door to a unified treatment of regular and singular dislocations under the overarching safeguard of a common mathematical apparatus.

To address point (i) above, we may start by considering a covector **W** at a point $X \in T\mathscr{M}$. The collection of vectors that annihilate **W**, namely,

$$\mathscr{H}_0 = \{\mathbf{v} \in T_X\mathscr{R} \mid \langle \mathbf{W}, \mathbf{v} \rangle = 0\}, \tag{14}$$

is clearly a hyperplane (an $(n-1)$-dimensional subspace of $T_X\mathscr{M}$). For each value $s \in \mathbb{R}$, we can also define the collection

$$\mathscr{H}_s = \{\mathbf{v} \in T_X\mathscr{M} \mid \langle \mathbf{W}, \mathbf{v} \rangle = s\}. \tag{15}$$

Any two vectors in $\mathscr{H}_s$ differ by an element in $\mathscr{H}_0$, whence it follows that $\mathscr{H}_s$ is an affine hyperplane parallel to $\mathscr{H}_0$. Letting s vary, we obtain a stack of oriented hyperplanes, the positive orientation pointing towards increasing s. The relative 'density' of this stack can be intuited by letting s attain only integer values. If we multiply **W** by, say, 2, we obtain a stack made of the same kind of hyperplanes, but with double the density. If we multiply by -1, we obtain the same stack, but with reversed orientation. Visualizing a vector $\mathbf{u} \in T_X\mathscr{R}$ as an arrow, and a covector $\mathbf{W} \in T^*_X\mathscr{R}$ as a stack of hyperplanes endowed with a density, the evaluation $\langle \mathbf{W}, \mathbf{u} \rangle$ of **W** on **u** can be visualized as the 'number of stack hyperplanes' pierced by the arrow. This intuitive idea is illustrated in Fig. 1.

As far as the co-frame field $\mathbf{e}^\alpha$ is concerned, we can regard it as the smooth assignation at each point of $\mathscr{M}$ of n mutually intersecting infinitesimal stacks of atomic hyperplanes. We may now focus attention on one of these families, say $\mathbf{e}^\beta$, and ask ourselves the question: do these stacks at different points of each neighbourhood fit well together? The first notable feature of this question is that it can be meaningfully asked. This is not the case if we were to isolate one of the n vector fields, say $\mathbf{e}_\beta$, since, according to the fundamental theorem of the theory of ordinary differential equations, any vector field has local integral curves. The vectors of one family always fit well together! When we were looking for defects in

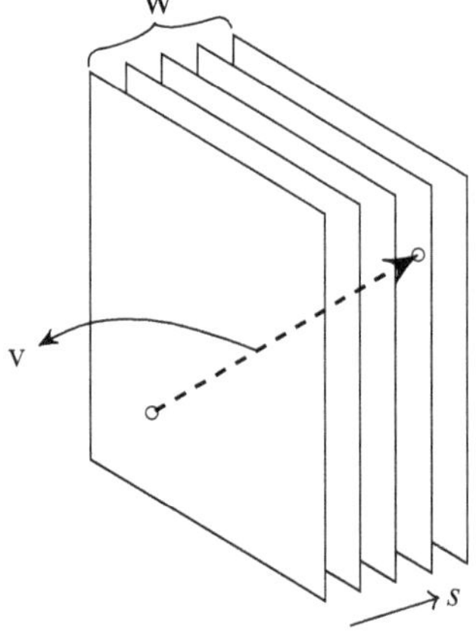

Fig. 1 A covector **W** as a stack of oriented hyperplanes. The evaluation $\langle \mathbf{W}, \mathbf{v} \rangle$ if the number of hyperplanes pierced. This 'number' is negative if the vector points in the direction of decreasing s

the crystal frame approach, therefore, we needed to look at least at 2 such families. The second feature of the question just posed is that it can be answered in any one of two respects. The first, and stricter, criterion consists of finding a family of hypersurfaces such that they match the hyperplanes at each point both in tangency and in density. The less severe way to look at the question of fitness is to look for a family of hypersurfaces that at each point are tangent to the hyperplane $\mathscr{H}_0$, ignoring the issue of stack density.

To answer the question according to the first criterion, we make use of point (ii) above, and invoke *Poincaré's lemma*. It asserts that a differential 1-form ω is *locally exact* if, and only if, it is *closed*, namely, its exterior derivative $d\omega$ vanishes identically on a simply connected open neighbourhood. By exact, we mean that there exists a scalar-valued function ϕ on this neighbourhood such that $\omega = d\phi$. In other words, local exactness means that the surfaces ϕ = constant do precisely the job of coinciding locally with the stacks defined by ω. For our context, we can call ω a *layering form*. It induces a field of hyperplane stacks. This field is *defect-free* if $d\omega = 0$, namely, if the layering form is closed. Otherwise, the 2-form $D = d\omega$, called the *dislocation form*, is a local measure of the defect density.

Identifying the layering form with $\mathbf{e}^\alpha = e^\alpha_I \, dX^I$, we conclude that the condition for it being defect-free can be stated as

$$d\mathbf{e}^\alpha = e^\alpha_{I,J} \, dX^J \wedge dX^I = 0. \tag{16}$$

Invoking Eqs. (13) and (2), this condition can be rewritten as

$$c^\alpha_{\beta\gamma} \, \mathbf{e}^\beta \wedge \mathbf{e}^\gamma = 0. \tag{17}$$

Thus, we recover, with little effort, the same characterization of a defective crystalline structure in terms of the structure constants of the frame-field formulation.

Having found the conditions ensuring the fitness of the layering structure in respect of both tangency and density, we turn to the less restrictive criterion, according to which we only require the hyperplanes $\mathscr{H}_0(X)$ to derive from local hypersurfaces ($(n-1)$-dimensional integral manifolds). This criterion can be interpreted in terms of the *involutivity*, or lack thereof, of the distribution $\mathscr{H}_0(X)$, for which the *theorem of Frobenius* provides the answer.

Recall that a (geometric) *distribution* on an n-dimensional manifold consists of a smooth assignation of an r-dimensional subspace of $T_x\mathscr{M}$ at each point $x \in \mathscr{M}$. In our case, the layering 1-form ω induces an $(n-1)$-dimensional distribution on $\mathscr{M}$ consisting of the hyperplanes $\mathscr{H}_0(X)$. A distribution is *involutive* if the Lie bracket of every pair of vector fields in the distribution is also in the distribution. Thus, employing the terminology introduced in Sect. 2.1, the distribution associated with two vector fields, $\mathbf{u}$ and $\mathbf{v}$, is involutive if, and only if, it contains only pure edge dislocations.

An r-dimensional distribution is *completely integrable* if around each point $x \in \mathscr{M}$ we can find a coordinate chart x^i $(i = 1, \ldots, n)$ such that the r-dimensional submanifolds with equation $x^j = k^j$ $(j = r+1, \ldots, n)$, where each k^j is a constant, are everywhere tangent to the distribution. The theorem of Frobenius asserts that every involutive distribution is completely integrable. In terms of the physical picture, we may say that if two vector fields give rise to no screw dislocations, we can find (local) integral surfaces to which the vectors are tangent.

The theorem of Frobenius can alternatively be expressed in the language of differential forms. For the particular case of the $(n-1)$-dimensional distribution generated by a 1-form ω, involutivity is equivalent to the condition

$$\omega \wedge d\omega = 0. \tag{18}$$

If we set $\omega = \mathbf{e}^\alpha$, we recover Eq. (3).

Example 1 (Actualization in the realm of liquid crystals) One of the attractive features of the dual approach is its ability to deal with defects in phases other than the crystalline solid. As stated by Chandrasekhar in his classical treatise [3], the 'term *liquid crystal* signifies a state of aggregation that is intermediate between the crystalline solid and the amorphous liquid'. Liquid crystals, therefore, offer a rich variety of examples for the application of the dual approach. Quoting again from [3],

> Smectic liquid crystals have stratified structures but a variety of molecular arrangements are possible within each stratification. In smectic A the molecules are upright in each layer ... The interlayer attractions are weak as compared with the lateral forces between molecules and in consequence the layers are able to slide over one another relatively easily.[3]

In addition to this obvious physical instance to which the above dislocation modelling can be applied, we also mention a potential application to *discotic liquid crystals*, made of disc-shaped molecules.[4] In all these cases, the physical reality imposes the consideration of local stacks of planes, rather than that of an ordered molecular lattice. □

Remark 1 In [16], Nye uses a tetrahedron argument, similar to the one used to relate the stress tensor to the traction vector on a surface element with unit normal **n**, to prove that there exists a second-order tensor controlling the local Burgers vectors associated with the lack of closure of infinitesimal circuits on all possible area elements at a point. This reasoning is hardly necessary, since Equation (2) above can be regarded as a tensor equation doing precisely what Nye's 'state of dislocation' tensor accomplishes. The vector-valued 2-form with components $c^\alpha_{\beta\gamma}$ acts on a 2-vector $\mathbf{u} \wedge \mathbf{v}$ to produce the (Burgers) vector $c^\alpha_{\beta\gamma} u^\beta v^\gamma$. More traditionally, when working in $\mathbb{R}^3$ with all its metric structure, we can define the tensor $\mathbf{C}$ associated with the structure constants $c^\alpha_{\beta\gamma}$ as

$$C^{\alpha\rho} = \epsilon^{\rho\beta\gamma} c^\alpha_{\beta\gamma},$$

[3]See [3], p. 6.

[4]Ibid., p. 8.

where ϵ denotes the alternating symbol. Since, in the case of $\mathbb{R}^3$, the 2-vector $2\,\mathbf{u}\wedge\mathbf{v}$ corresponds to an element of area $\mathbf{n}\,dA = \mathbf{u}\times\mathbf{v}$, where $\times$ stands for the ordinary cross product, Nye's result follows suit. Indeed,

$$\begin{aligned} C^{\alpha\rho} n_\rho dA &= \epsilon^{\rho\beta\gamma} c^{\alpha}_{\beta\gamma} n_\rho dA \\ &= c^{\alpha}_{\beta\gamma}\epsilon^{\rho\beta\gamma}\epsilon_{\rho\sigma\tau}u^\sigma v^\tau \\ &= (\delta^{\beta}_{\sigma}\delta^{\gamma}_{\tau} - \delta^{\beta}_{\tau}\delta^{\gamma}_{\sigma})c^{\alpha}_{\beta\gamma}u^\sigma v^\tau \\ &= (c^{\alpha}_{\sigma\tau} - c^{\alpha}_{\tau\sigma})u^\sigma v^\tau \\ &= 2c^{\alpha}_{\sigma\tau}u^\sigma v^\tau . \end{aligned}$$

□

2.4 Integral Perspective

As anticipated in point (iii) of Sect. 2.3, there are still some natural consequences that will emerge from the theory of integration of differential forms. A central result of this theory is Stokes' theorem. It establishes that

$$\int_{\partial\mathscr{M}} \omega = \int_{\mathscr{M}} d\omega. \tag{19}$$

In this equation, $\mathscr{M}$ is an oriented n-dimensional manifold-with-boundary, ∂ is the boundary operator, and ω is a compactly supported $(n-1)$-form in $\mathscr{M}$. The boundary $\partial\mathscr{M}$ is assumed to have been consistently oriented. More importantly, the boundary $\partial\mathscr{M}$, as an $(n-1)$-dimensional manifold, is not necessarily the same as the boundary of $\mathscr{M}$ as a topological space (typically, a subspace with the subset topology). Bearing in mind this distinction, the boundary of a manifold-with-boundary satisfies the identity

$$\partial(\partial\mathscr{M}) = 0. \tag{20}$$

The duality of this identity with respect to the fundamental identity of the exterior differential operator, namely, for any differential form ρ,

$$d(d\rho) = 0, \tag{21}$$

is striking. Their consistency is mediated by Stokes' theorem, as can be gathered choosing $\omega = d\rho$ in (20).

At this point, going back for a moment to the case $n = 3$, for a given layering 1-form ω, we will be able to define a scalar dislocation measure associated with any circuit γ in $\mathscr{M}$. Its meaning is intended to convey the idea of the net amount of

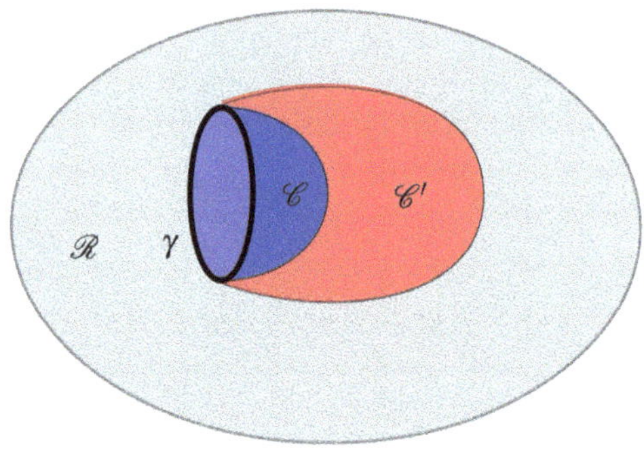

Fig. 2 A Burgers circuit γ and two associated surfaces $\mathscr{C}$ and $\mathscr{C}'$

dislocations embraced, as it were, by the circuit. Let $\mathscr{C}$ be an oriented 2-dimensional submanifold with boundary (a surface) such that $\gamma = \partial\mathscr{C}$. We may think of γ as a *Burgers' circuit*. The surface $\mathscr{C}$ encounters defects (smoothly distributed mismatches in slope and/or density between neighbouring stacks). The net amount of defects encountered is given by the integral of the defect density $D = d\omega$ over $\mathscr{C}$. This net amount may vanish even if there are defects on $\mathscr{C}$ which end up cancelling out mutually. According to Stokes' Theorem, we must have

$$\int_{\mathscr{C}} D = \int_{\gamma} \omega. \tag{22}$$

Consider a different surface $\mathscr{C}'$ with *the same* boundary curve γ, as shown in Fig. 2. We conclude from Eq. (22) that the net amount of defects is the same as before. In other words, the net amount of defects is the same on all surfaces sharing a common boundary. Thus, a Burgers circuit γ can be said to embrace a fixed net amount of defects. This generalization of the concept of Burgers circuit (and, in this case, the associated Burgers *scalar*) is a direct elementary consequence of the geometrical setting, rather than of a clever physical insight.

Consider, furthermore, the integration of the defect density $D = d\omega$ over the boundary $\partial\mathscr{S}$ of an n-dimensional submanifold with boundary $\mathscr{S}$ in $\mathscr{M}$. Applying Stokes' theorem, we obtain

$$\int_{\partial\mathscr{S}} D = \int_{\partial\partial\mathscr{S}} \omega = 0. \tag{23}$$

This result can be interpreted physically as the fact that there can be no isolated sources of defects. It was obtained by exploiting the vanishing of the boundary of a boundary, that is, $\partial^2 = 0$. It could also have been independently obtained from $d^2 = 0$. Indeed, recalling that, by its very definition, D is an exact form, we could have claimed that

$$\int_{\partial\mathscr{S}} D = \int_{\mathscr{S}} dD = \int_{\mathscr{S}} d(d\omega) = 0. \tag{24}$$

Assume, finally, that the defect density vanishes identically everywhere except within a very thin, wire-like, domain traversing the body. Enclosing this domain with a slightly thicker tubular neighbourhood, and focusing attention on the portion comprised between two cross sections, we conclude that the net defect density in any cross section of this tube is constant. We have thus essentially recovered Frank's rule [10] for line dislocations. A similar result can be obtained for branching dislocation lines. We again emphasize that, in this approach, the physical results emerge naturally from the geometric setting.[5]

Remark 2 **Cartesian meditation**: With the luxury of a 3-dimensional Euclidean background, 1-forms can be construed as ordinary vectors. Starting with the frame $\mathbf{e}_\alpha$ $(\alpha = 1, 2, 3)$, with Cartesian components forming the rows of a matrix $[A]$ with entries $e_{\alpha i}$, the vector version of the dual basis $\mathbf{e}^\alpha$ $(\alpha = 1, 2, 3)$ is given by the columns of the inverse matrix $[A]^{-1}$ with entries $e^{-1}_{i\alpha}$. We look now for the meaning of the exterior differential of $\mathbf{e}^\alpha$, with α fixed. We notice that $d\mathbf{e}^\alpha$ is a 2-form, whose components are entries of a skew-symmetric matrix, namely, $f_{\alpha ij} = (e^{-1}_{i\alpha,j} - e^{-1}_{j\alpha,i})/2$. The metric structure, however, allows us to view this 2-form as a vector with components $f_{\alpha k} = \varepsilon_{kij} f_{\alpha ij}$, where ε_{ijk} is the Cartesian alternating symbol. We conclude that, in the Cartesian world, the vector field representing the differential 2-form $d\mathbf{e}^\alpha$ is precisely the curl of the vector with Cartesian components $e^{-1}_{i\alpha}$. Stokes' theorem, used to derive the elegant results above, is replaced now by its Euclidean version, which states that the flux of the curl of a vector field over a surface equals the circulation of this field over the bounding curve. In short, when thinking in a Cartesian way, there is a loss of generality and of elegance, but not of truth, as long as the context is appropriate. These Cartesian gymnastics, however, can hardly survive the treatment of singular dislocations in terms of currents.

3 Singular Lattices

Although, for merely heuristic purposes, we have commented on the extension of the concept of a discrete crystal lattice into the continuous realm, this vaguely described extension has not played any role in our treatment of continuous distributions of defects. We are now concerned with the opposite paradigm, formulated as the question: Can one obtain the discrete picture out of the stand-alone continuous one, and, if so, how? To answer this question, we will make use of the theory of currents, as developed by Georges de Rham [4], whose general lineaments are briefly reviewed below.

[5]We revisit these ideas in greater detail in Sects. 3.4 and 3.5.

3.1 De Rham Currents

On an n-dimensional oriented manifold $\mathscr{M}$, we consider the collection $\Lambda_c^p(\mathscr{M})$ of all C^∞ differential p-forms with compact support in $\mathscr{M}$. It is clear that $\Lambda_c^p(\mathscr{M})$ is an infinite-dimensional vector space.[6]

Definition 1 A *p-current* on $\mathscr{M}$ is a continuous linear functional on $\Lambda_c^p(\mathscr{M})$.

As de Rham explains in [4], he interprets the notion of continuity following Laurent Schwartz's definition of a distribution, of which a current is a generalization.[7] De Rham, in fact, affirms that 'a current can be considered as a differential form for which the coefficients are distributions'.[8]

Since de Rham states[9] that the concept of a current is 'a notion so general that it includes as special cases both differential forms and chains', it seems appropriate at this point to consider a few examples to confirm and expand this assertion.

Example 2 Let $\mathbf{v} \in T_X\mathscr{M}$ be a vector at $X \in \mathscr{M}$. We define the 1-current T_v associated with this vector as the operator

$$T_v[\phi] = \langle \phi(X), \mathbf{v} \rangle \qquad \forall \phi \in \Lambda_c^1(\mathscr{M}). \tag{25}$$

Regarding $\mathbf{v}$ as the value of some vector field on $\mathscr{M}$ at a point X, we may say, with an understandable abuse of language, that the 1-current $T_{v(X)}$ is equal to the vector $\mathbf{v}(X)$. In the same spirit, we can associate with a p-vector at X a corresponding p-current. The case $p = 0$ delivers Dirac's delta with an intensity determined by the 0-vector. □

Example 3 Let ω be a differential p-form, not necessarily with compact support, on an n-dimensional manifold-with-boundary $\mathscr{M}$. We define the associated $(n-p)$-current T_ω as

$$T_\omega[\phi] = \int_{\mathscr{M}} \omega \wedge \phi \qquad \forall \phi \in \Lambda_c^{n-p}(\mathscr{M}). \tag{26}$$

---

[6] By virtue of the point-wise vector-space character of $\Lambda^p(T_X^*\mathscr{M})$, with $X \in \mathscr{M}$.

[7] In his momentous article [20], Schwartz describes the continuity of a distribution $T(\cdot)$ as follows: 'Si une suite de fonctions ϕ_i, ont leurs noyaux contenus dans un compact fixe et si elles convergent uniformément vers 0, ainsi que chacune de leurs dérivées, alors les $T(\phi_i)$ convergent vers 0'. In other words, the requirement placed on the continuity of the functional T is stronger than the mere uniform convergence of the functions, since it entails that each of the sequences of derivatives must also converge uniformly to zero. When generalizing this idea to manifolds, Schwartz and de Rham demand that the supports must all be contained in a single compact set within the domain of a chart. The derivatives are taken with respect to the chart coordinates.

[8] See [4], p. 1.

[9] Ibid.

To justify the heuristic identification of T_ω with ω itself, we remark the obvious fact that a p-covector based on an n-dimensional vector space is completely defined if we know the value of its exterior product with every $(n-p)$-covector. From this observation, we can reason like Schwartz does in [20] in respect of the identification of a function with its associated distribution. It should be pointed out, however, that this identification of a p-form with its corresponding $(n-p)$-current is not to be taken at face value. While we can define the exterior product of two differential forms, a definition of the tensor product of two currents is not trivially available. □

Example 4 Let $\mathscr{S}$ be an oriented immersed p-dimensional submanifold (with boundary) of the n-dimensional manifold $\mathscr{M}$. Let, moreover, $\mathscr{S}$ be such that the restriction to $\mathscr{S}$ of every compactly supported p-form in $\mathscr{M}$ has compact support in $\mathscr{S}$. This is the case, for example, when $\mathscr{S}$ itself is compact. We define the associated p-form $T_{\mathscr{S}}$ as

$$T_{\mathscr{S}}[\phi] = \int_{\mathscr{S}} \phi \qquad \forall \phi \in \Lambda_c^p(\mathscr{M}). \tag{27}$$

On the right-hand side of this equation, we interpret ϕ as the restriction $\phi_{|\mathscr{S}}$. □

The continuity condition, involving the uniform convergence of the sequences of each component of the p-forms, and each of its partial derivatives in a chart, is crucial for the calculus of currents. As compared with, say, the uniform convergence of the sequences of components alone, we conclude that in the former case fewer sequences converge. It is this feature that allows us to enlarge the family of continuous functionals, that is, of currents. Expressed somewhat differently, we may say that currents can be more irregular than measures, which are thus a particular case.

Example 5 As in Example 2, consider a vector $\mathbf{v}$ at a point $X \in \mathscr{M}$. We can define the 0-current D_v by

$$D_v[\phi] = \mathbf{v}(\phi) \qquad \forall \phi \in \Lambda_c^0(\mathscr{M}). \tag{28}$$

In other words, this current assigns to a smooth function with compact support in $\mathscr{M}$ its directional derivative in the direction of the vector $\mathbf{v}$ at X. The subtle point in this example is that we need to make sure this linear operator is also continuous. If our criterion of continuity had been based on the uniform convergence of the functions alone (without including its derivatives), we could have chosen (in, say $\mathbb{R}^2$, with coordinates x, y) the sequence of functions

$$\phi_n(x) = B(x, y)\,\frac{1}{n} sin(nx) \qquad n = 1, 2, \ldots, \tag{29}$$

where $B(x, y)$ is a bump function around the origin, at which point its first derivatives vanish. If $\mathbf{v} = \frac{\partial}{\partial x}\big|_{0,0}$, although the sequence ϕ_n converges uniformly to zero, the sequence of derivatives $\mathbf{v}(\phi_n)$ does not. □

Example 6 **The principal-value 0-current**: A 0-current on $\mathbb{R}$ is a (Schwartz) distribution. Our purpose in this example is to show yet another way in which singularities or discontinuities can be subsumed under a distribution or, more generally, under a current. Let $f : \mathbb{R} \to \mathbb{R}$ be a function, possibly with a pole at $x = 0$. Recall that the *Cauchy principal value* of the integral of f over an interval $(-a, a)$ is defined as

$$\text{pv} \int_{-a}^{a} f \, dx = \lim_{\varepsilon \to 0^+} \left(\int_{-a}^{-\varepsilon} f(x) \, dx + \int_{\varepsilon}^{a} f(x) \, dx \right). \tag{30}$$

In particular, for the function $f = x^{-1}$, the principal value vanishes. Following Schwartz [20], we define the *principal-value distribution* associated with x^{-1} as the functional P given by

$$P[\phi] = \text{pv} \int_{-a}^{a} \frac{\phi}{x} \, dx, \tag{31}$$

for all C^∞-functions ϕ with compact support in $(-a, a)$. The function x^{-1} being odd, only the odd part of ϕ survives the integration and, therefore,

$$P[\phi] = \int_{0}^{a} \frac{\phi(x) - \phi(-x)}{x} \, dx. \tag{32}$$

□

The collection of all p-currents, that is, of all continuous linear operators on $\Lambda_c^p(\mathscr{M})$, is, by definition, precisely the *topological dual* $\Lambda_c^p(\mathscr{M})'$. A linear combination of two p-currents, T_1 and T_2, is the p-current $a_1 T_1 + a_2 T_2$ defined as

$$(a_1 T_1 + a_2 T_2)[\phi] = T_1[a_1 \phi] + T_2[a_2 \phi] \qquad \forall \phi_1, \phi_2 \in \Lambda_c^p(\mathscr{M}), \qquad \forall a_1, a_2 \in \mathbb{R}. \tag{33}$$

The *boundary* of a p-current T is the $(p-1)$-current ∂T defined as

$$\partial T[\phi] = T[d\phi] \qquad \forall \phi \in \Lambda_c^{p-1}(\mathscr{M}). \tag{34}$$

The operator symbol ∂ appears to have been abused. That this may not quite be the case can be gathered from an attempt at iterating the operator, namely,

$$\partial^2 T[\phi] = \partial \partial T[\phi] = \partial T[d\phi] = T[d^2 \phi] = 0 \qquad \forall \phi \in \Lambda_c^{p-2}(\mathscr{M}). \tag{35}$$

The kinship between the various uses of the operators d and ∂ is further emphasized by working out the boundary of the current associated with a submanifold $\mathscr{S}$, as given in Example 4. We obtain

$$\partial T_{\mathscr{S}}[\phi] = T_{\mathscr{S}}[d\phi] = \int\limits_{\mathscr{S}} d\phi = \int\limits_{\partial\mathscr{S}} \phi = T_{\partial\mathscr{S}}[\phi]. \tag{36}$$

Thus, Stokes' theorem, as used above, mediates between the various uses of the symbols and furnishes an elegant formulaic consistency, to wit,

$$\partial T_{\mathscr{S}} = T_{\partial\mathscr{S}}. \tag{37}$$

In words, this equation states that the boundary of a current of a manifold is equal to the current of its boundary.

The relation between the operators d and ∂ is further confirmed by calculating, in a manifold $\mathscr{M}$ with vanishing boundary, the current associated with the exterior derivative of a p-form ω, which yields, for each $\phi \in \Lambda_c^{n-p-1}(\mathscr{M})$,

$$\begin{aligned} T_{d\omega}[\phi] &= \int\limits_{\mathscr{M}} d\omega \wedge \phi = \int\limits_{\mathscr{M}} \big(d(\omega \wedge \phi) - (-1)^p \omega \wedge d\phi\big) \\ &= \int\limits_{\mathscr{M}} -(-1)^p \omega \wedge d\phi = (-1)^{p+1}\, \partial T_\omega[\phi], \end{aligned} \tag{38}$$

where we have invoked Stokes' theorem. Thus, except possibly for sign, the current associated with the exterior derivative of a form is equal to the boundary of the current associated with the form itself.

Example 7 The boundary of the 1-current of Example 2 is equal to the 0-current of Example 5. Indeed, for each function (0-form) ω with compact support in $\mathscr{M}$, we have

$$\partial T_v[\omega] = T_v[d\omega] = \langle d\omega, \mathbf{v}\rangle = \mathbf{v}(\omega). \tag{39}$$

□

A p-current T is *equal to zero*, written as $T = 0$, in an open set $U \subset \mathscr{M}$ if $T[\phi] = 0$ for all p-forms ϕ with compact support in U. Denoting by $\mathscr{U}$ the maximal open set[10] in which $T = 0$, the *support* of the current T is defined as

$$\operatorname{supp} T = \mathscr{M} \setminus \mathscr{U}. \tag{40}$$

The support of a current is not necessarily compact.

[10]The existence of this maximal set is proved as a theorem in [4], p. 35.

3.2 Singular Layerings

As introduced in [6–8], a *singular layering* is obtained by specifying an $(n-1)$-current T on $\mathscr{M}$. As a particular case when $T = T_\omega$ for some 1-form ω, we recover a regular layering form as described in Sect. 2.3. In the physical picture, therefore, the specification of a layering current T entails a possible loss of smoothness in the assignation of the stacks of lattice planes, and/or an infinite density of the stacks along some singular surface.

Example 8 Let a body manifold $\mathscr{B}$ be identified with the open unit cube $(-1, 1)^3 \subset \mathbb{R}^3$, and let $\mathscr{S}$ be the intersection of $\mathscr{B}$ with the (oriented) closed half-plane $\mathscr{H} = \{(x, y, z) \in \mathbb{R}^3 \mid y = 0, z \geq 0\}$, as depicted in Fig. 3. The 2-current $T_\mathscr{S}$ can be regarded as a 1-form with components given by Schwartz distributions. Specifically, $T = H(z)\ \delta(y)\ dy$, where H is the Heaviside step function, and δ is Dirac's delta distribution. Seen in this light, we have obtained a layering that leaves all points of $\mathscr{B}$ without any layering, except that at each point of $\mathscr{S}$ we have a high (infinite, if you will) concentration of layers parallel to the plane x, z. The 2-current $T_{dy} - T_\mathscr{S}$ will combine this singular layering with a regular one. This combination can be recognized as the classical textbook edge dislocation, in which we 'subtract a half-plane of atoms' from the lattice. □

In Sect. 2.3 we introduced the dislocation form associated with a given layering 1-form ω as the 2-form $D = d\omega$. Its identical vanishing ensures an everywhere defect-free structure. By virtue of the result embodied in Eq. (38), we define the *dislocation current* associated with a layering $(n-1)$-current T as the $(n-2)$-current $S = \partial T$. The vanishing of S is expressed mathematically by saying that the layering current T is *closed*. By analogy with the regular case, we will declare a singular layering T to be defect-free if it is closed.[11]

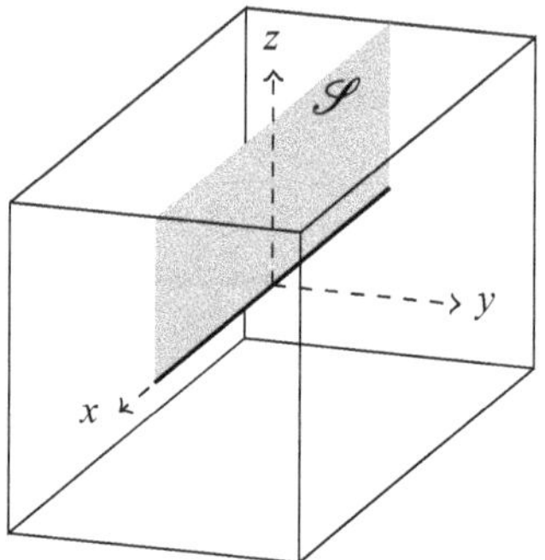

Fig. 3 A singular layering as an edge dislocation

[11]The justification for this statement can be found in a theorem [4] stating that every closed current is homologous to a differential form.

Example 9 The dislocation current of Example 8 is the current $S = \partial T_{\mathscr{S}} = T_{\partial \mathscr{S}}$. The support of this current is $\partial \mathscr{S}$ itself, which in our case is the open interval $(-1, 1)$ along the x-axis. Thus, just as asserted in material science textbooks, the defect is concentrated along a *dislocation line*.

3.3 A Screw Dislocation

3.3.1 Two-Dimensional Prelude

The screw dislocation, as conceived by Volterra [23, 24], is an essentially three-dimensional phenomenon. Volterra describes it by considering a doubly connected domain (a thick cylinder) which, when cut with a plane containing the axis and then rejoined after a relative axial displacement of both exposed faces, gives rise to the screw. Although this procedure has no equivalent in a two-dimensional setting, the mathematical difficulty of the screw dislocation resides in the passage from a multiply connected domain to a simply connected one, a passage that can be best illustrated in $\mathbb{R}^2$.

Consider, therefore, the 1-form $\omega = d\theta$, which is well-defined and closed (but not exact) in the punctured plane $\mathscr{R} = \mathbb{R}^2 \setminus \{O\}$, where θ is the standard angular coordinate of the polar system (r, θ), and O is the origin of $\mathbb{R}^2$. We remark that $\mathscr{R}$ is diffeomorphic to $\mathscr{C} = \mathbb{R}^+ \times S^1$, where S^1 is the unit circle. This diffeomorphism is precisely established by the polar coordinate system in the punctured plane and the corresponding axial and circumferential coordinates in $\mathscr{C}$.

The 1-current $T_{d\theta}$ on $\mathscr{R}$ is, as usual, defined by the prescription

$$T_{d\theta}[\phi] = \int_{\mathscr{R}} d\theta \wedge \phi, \tag{41}$$

for all 1-forms ϕ with compact support in $\mathscr{R}$. We want to extend this definition by considering all 1-forms ϕ *with compact support in* $\mathbb{R}^2$. In other words, we want to define a current on the whole of $\mathbb{R}^2$ while using an integration on $\mathscr{R}$. To this end, consider the effect of removing the open ball B_ϵ with centre at O and radius ϵ. The set $\mathbb{R}^2 \setminus B_\epsilon$ is diffeomorphic to the product $\mathscr{C}_\epsilon = [\epsilon, \infty) \times S^1$. On both $\mathscr{R}_\epsilon$ and $\mathscr{C}_\epsilon$ the integral on the right-hand side of (41) is well-defined for any 1-form with compact support thereat. To extend the domain of integration, however, the latter alternative avoids the singularity of the polar coordinate system.

Indeed, let $\phi = \phi_x\, dx + \phi_y\, dy$ be a 1-form with compact support in $\mathbb{R}^2$. Its restriction to $\mathscr{R}_\epsilon$ and, therefore, to $\mathscr{C}_\epsilon$, via the natural pullback provided by the inverse of the natural diffeomorphism described above, is the 1-form $\phi = \phi_x\, dx + \phi_y\, dy = \phi_r\, dr + \phi_\theta\, d\theta$, where $\phi_r = \phi_x\, cos\theta + \phi_y\, sin\theta$. On each line $\theta =$ constant the limit $\lim\limits_{\epsilon \to 0^+} \phi_r$ exists and is finite. Therefore, the integral

$$\int_{\mathscr{C}_0} d\theta \wedge \phi = -\int_0^{2\pi}\int_0^{\infty} \phi_r \, dr \, d\theta \tag{42}$$

is well-defined and finite. We declare this to be the value of $T_{d\theta}$ over a 1-form ϕ with compact support in $\mathbb{R}^2$.

The boundary 0-current $\partial T_{d\theta}$ can be evaluated over any function f with compact support in $\mathbb{R}^2$ as

$$\partial T_{d\theta}[f] = T_{d\theta}[df] = -\int_0^{2\pi}\int_0^{\infty} \frac{df}{dr} \, dr \, d\theta = 2\pi \; f(0). \tag{43}$$

In other words, the dislocation current associated with the layering $d\theta$ is proportional to the Dirac delta at the origin.

Remark 3 A possible physical interpretation of this situation can be gathered by imagining that an edge dislocation (subtraction of one atomic row) has been 'smeared' along a circumference, as suggested in the Fig. 4. Equivalently, we may say that we have the limit of a circumferential incoherent interface, of the type described below in Sect. 3.6, as the extent of the interface shrinks to zero. □

3.3.2 The Screw

Let $\mathscr{R} = \mathbb{R}^3 \setminus \mathscr{V}$, where $\mathscr{V}$ denotes the z-axis in the natural Cartesian coordinate system (x, y, z) of $\mathbb{R}^3$. In $\mathscr{R}$, with the standard cylindrical coordinates (r, θ, z), we consider the 1-form $\omega = d\theta$. We remark that ω is closed but not exact, since θ is not a globally defined function on $\mathscr{R}$. With this proviso, we propose ω as a layering 1-form on $\mathscr{R}$. As far as this layering is concerned, there are no dislocations in $\mathscr{R}$.

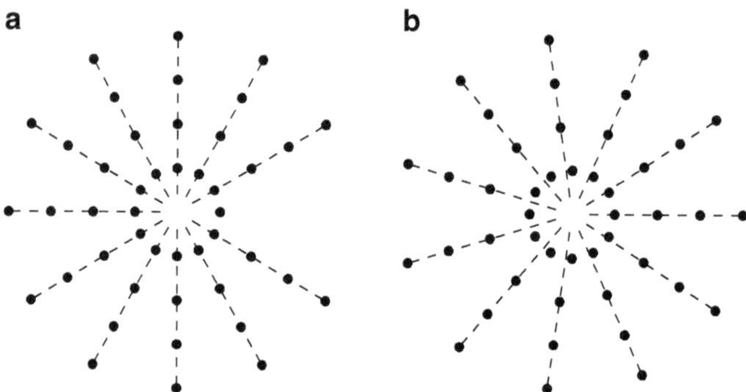

Fig. 4 Smearing (**b**) an edge dislocation (**a**) over a circle

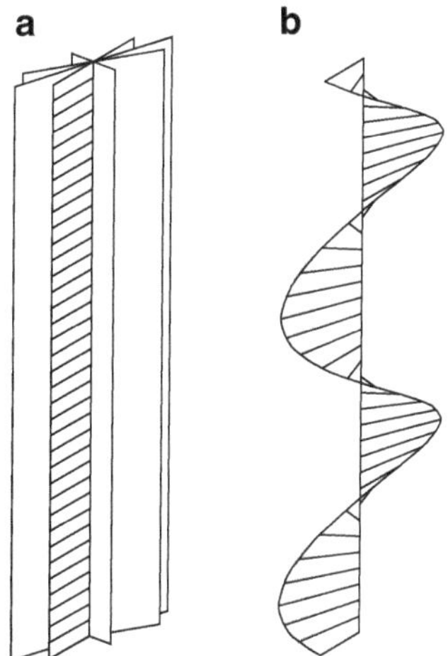

Fig. 5 A screw dislocation: the passage from each page of the open book (**a**) to the helicoid (**b**) is achieved by adding an exact 1-form to the layering current. The dislocation current is not affected

The corresponding integral manifolds look like the pages of an open book around the (excluded) spine, as shown in Fig. 5a.

We can extend this layering current from $\mathscr{R}$ to $\mathbb{R}^3$ by using, for each z, the same scheme as we used for the two-dimensional description above. The dislocation current is given by

$$D_\omega[\psi] = \partial T_\omega[\psi] = 2\pi \int_{\mathscr{V}} \psi_z = 2\pi \int_{-\infty}^{\infty} \psi_z(0, 0, z)dz, \tag{44}$$

where $\psi = \psi_x\, dx + \psi_y\, dy + \psi_z\, dz$ has compact support in $\mathbb{R}^3$. In short, this rather complicated layering 2-current has exactly the same boundary as an edge dislocation. For the sake of illustration, let us add to our layering current (the 'open book') the closed 1-form $a\ dz$, where a is a constant. The boundary of the limiting current is not affected by this addition. The integral surfaces of the modified form are no longer the pages of an open book but rather the surfaces with equation $\theta + az =$ constant, one of which is shown in Fig. 5b, thus justifying the appellation of 'screw dislocation'.

Remark 4 The transformation of a book page into the helicoidal shape could also be achieved by a diffeomorphism of $\mathscr{R}$ obtained by the coordinate transformation $\theta \mapsto \theta + a\, z$.

3.4 A Conservation Law

Suppose that we want to build a p-current in $\mathscr{M}$ whose support is a subset of a p-dimensional submanifold $\mathscr{T}$ (without boundary) such that the restriction of every compactly supported p-form ϕ in $\mathscr{M}$ has compact support in $\mathscr{T}$.[12] The 'most regular' way to achieve this is to choose a smooth function $u : \mathscr{T} \to \mathbb{R}$ and to define the corresponding p-current, denoted by $T_{u\mathscr{T}}$, as

$$T_{u\mathscr{T}}[\phi] = T_{\mathscr{T}}[u\,\phi] = \int\limits_{\mathscr{T}} u\,\phi. \tag{45}$$

We will call $u(X)$ the *intensity* of $T_{u\mathscr{T}}$ at the point $X \in \mathscr{T}$.

Consider the particular case when $T_{u\mathscr{T}}$ is closed, that is, $\partial T_{u\mathscr{T}} = 0$. Then

$$\begin{aligned}\partial T_{u\mathscr{T}}[\psi] = T_{u\mathscr{T}}[d\psi] &= \int\limits_{\mathscr{T}} u\,d\psi \\ &= \int\limits_{\mathscr{T}} (d(u\psi) - du \wedge \psi) = -\int\limits_{\mathscr{T}} du \wedge \psi = 0.\end{aligned} \tag{46}$$

Since this equation is satisfied identically for all $(p-1)$-forms ψ with compact support, we conclude that $u =$ constant. This result, which is a particular case of the *constancy theorem* of geometric measure theory,[13] means, in our terminology, that the intensity of a closed p-current (of the type considered), whose support is a subset of a p-dimensional submanifold, is necessarily constant.

Example 10 To provide an intuitive motivation for the idea of intensity of a current of the kind discussed above, consider the cylindrical domain $\mathscr{U}$ defined by $y^2+z^2 \leq \varepsilon^2 \ll 1$ comprised within the open cube $\mathscr{B} = (-1, 1)^3 \subset \mathbb{R}^3$. The axis of this cylinder is, trivially, an embedded 1-dimensional submanifold $\mathscr{T}$ of $\mathscr{B}$. Let $\omega = f(x, y, z)\,dy \wedge dz$ be a smooth 2-form with support in $\mathscr{U}$. Its associated 1-current is denoted by T_ω. For any 1-form $\phi = a(x, y, z)dx + b(x, y, z)dy + c(x, y, z)dz$ with compact support in $\mathscr{B}$, we obtain

$$T_\omega[\phi] = \int\limits_{\mathscr{B}} \omega \wedge \phi = \int\limits_{\mathscr{B}} f\,a\,dy \wedge dz \wedge dx = \int\limits_{-1}^{1}\int\limits_{-\varepsilon}^{\varepsilon}\int\limits_{-\varepsilon}^{\varepsilon} f\,a\,dy\,dz\,dx = \int\limits_{-1}^{1} \tilde{u}(x)\,dx, \tag{47}$$

where we have set

$$\tilde{u}(x) = \int\limits_{-\varepsilon}^{\varepsilon}\int\limits_{-\varepsilon}^{\varepsilon} f\,a\,dy\,dz. \tag{48}$$

[12] A good example in a body $\mathscr{B}$ is a loop, or a curve whose ends are not in $\mathscr{B}$.

[13] See [9], p. 357.

The tilde is meant to remind us of the dependence on ε. By the mean-value theorem, for each x we can find a point P within the circle of radius ε such that $\tilde{u}(x) = a(x, y_P, z_P) \int_{-\varepsilon}^{\varepsilon} \int_{-\varepsilon}^{\varepsilon} f \quad dy\,dz$. Choosing ε sufficiently small while modifying the function f so that for each x the integral $\int_{-\varepsilon}^{\varepsilon} \int_{-\varepsilon}^{\varepsilon} f \quad dy\,dz$ is kept at a fixed value $u(x)$, we can approximate to any degree of accuracy the 1-current $T_{u\,\mathscr{T}}$. If the 2-form ω is closed, $u(x)$ is constant, as can be concluded by applying Stokes' theorem to arbitrary closed balls in $\mathscr{B}$. □

Applying these ideas to a layering 2-current T in a 3-dimensional body, we conclude that, if the support of its dislocation current $S = \partial T$ is a 1-dimensional manifold $\mathscr{T}$, the strength of the dislocation is constant. This fact is known as *Frank's first rule* [10]. Any circuit enclosing $\mathscr{T}$ cannot be removed from it. It embraces a constant amount of dislocations, just as a Burgers circuit.

Example 11 **Dislocation shedding**: Let $\mathscr{T} \subset \mathscr{M}$ be a p-dimensional submanifold with boundary such that the restriction of every compactly supported p-form ϕ in $\mathscr{M}$ has compact support in $\mathscr{T}$.[14] Rewriting Eq. (46) without the assumption that the current $T_{u\,\mathscr{T}}$ is closed, we obtain the general result

$$\partial T_{u\,\mathscr{T}}[\psi] = \int_{\partial\mathscr{T}} u\,\psi - \int_{\mathscr{T}} du \wedge \psi. \tag{49}$$

If $\partial T_{u\,\mathscr{T}}$ is supported in $\partial\mathscr{T}$, the first term on the right-hand side of Eq. (49) vanishes for every $(p-1)$-form whose (compact) support is disjoint with $\partial\mathscr{T}$. It follows that in this case u must be constant on $\mathscr{T}$. Consequently, if u is not constant on $\mathscr{T}$, we must conclude that the support of $\partial T_{u\,\mathscr{T}}$ contains points not belonging to the boundary $\partial\mathscr{T}$. A dramatic application of this result is provided by the data of Example 8, when we assume that the measured dislocation intensity along $\partial\mathscr{S}$ is not constant. We immediately deduce that there are additional dislocations continuously distributed on $\mathscr{S}$. We call this mechanism *dislocation shedding*. □

3.5 Branching

Another application of these mathematical ideas, whereby the physical meaning emerges naturally from the geometrical context, is the content of *Frank's second rule* of branching dislocation lines. In Frank's own words, 'if the Burgers vectors of all dislocation lines meeting at a node are defined by right-handed circuits, when

[14]For example, $\mathscr{T}$ could be a compact submanifold of $\mathscr{M}$, or a submanifold such as $\mathscr{S}$ in Example 8.

looking outwards from the node, the sum of these Burgers vectors is zero. This corresponds to a vectorial version of Kirchhoff's law'.[15]

Consider, therefore, a node $P \in \mathscr{M}$ at which N non-intersecting dislocation lines $\mathscr{S}_i$ $(i = 1, \ldots, N)$ converge. For simplicity, we assume the manifold $\mathscr{M}$ to be of dimension 3. As a 1-dimensional submanifold with boundary, each of these lines is diffeomorphic to the real interval $[0, 1)$, with 0 corresponding to P. We assume, moreover, that each of these lines coincides with its closure within $\mathscr{M}$,[16] which is a manifold without boundary (such as a connected open subset of $\mathbb{R}^3$).

Each of the lines $\mathscr{S}_i$ is assumed to be part of the support of a dislocation of constant intensity u_i. We associate with the union $\mathscr{S} = \bigcup_i \mathscr{S}_i$ the dislocation current $D = \sum_i u_i T_{\mathscr{S}_i}$. If this is indeed a dislocation current, it must be closed, and we obtain, therefore, for every 0-form f with compact support,

$$0 = \partial D[f] = \sum_i u_i \int_{\mathscr{S}_i} df = f(P) \sum_i u_i. \tag{50}$$

By the arbitrariness of f, we obtain Frank's second rule (in a scalar context) as

$$\sum_i u_i = 0. \tag{51}$$

Example 12 An example that enhances the understanding of this conservation law can be gathered if we start from a body $\mathscr{M}$ identified with an open cylinder $\mathscr{M}$, as shown in Fig. 6, with equation $r < 1$ in a cylindrical coordinate system (r, θ, z).[17] We identify the branching point P with the coordinate origin. Consider the intersections $\mathscr{T}_i$ $(i = 1, 2, 3)$ of this cylinder with each of the oriented quarter planes with equations $r \geq 0$, $\theta = 2\pi i/3$, and $z \leq 0$. For constant u_i, the layering current $T_{u_i \mathscr{T}_i}$ has a boundary whose support is the bracket-like set made from the union of the lower z-semiaxis and the segment $\mathscr{S}_i$ given by $0 \leq r < 1, \theta = 2\pi i/3$, $z = 0$. If, as stated by Frank's second rule, the 3 horizontal segments constitute the support of the boundary of the resulting current $T = \sum_i u_i T_{u_i \mathscr{T}_i}$, we must require the vanishing of the boundary current ∂T on the open lower z-semiaxis. This is possible only if $\sum_i u_i = 0$. □

[15] See [10], p. 813. The allusion to Kirchhoff's law of electrical circuits, where the algebraic sum of electrical currents at a node is zero, is very pertinent.

[16] Intuitively speaking, each line comes out, as it were, from $\mathscr{M}$. This condition guarantees that the restriction to $\mathscr{S}_i$ of each compactly supported form in $\mathscr{M}$ is also compactly supported in $\mathscr{S}_i$.

[17] Although the cylindrical coordinate system is not well-defined on the z-axis, this fact is not of significance for this example.

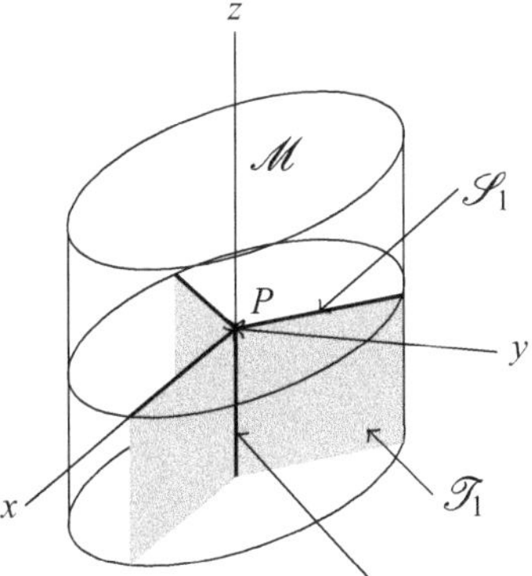

Fig. 6 Frank's second rule for branching dislocation lines

3.6 Interfaces

Since the pioneering work of Read and Shockley [18], it has been recognized that dislocation models of grain boundaries between crystals can provide accurate predictions of experimental measurements, such as the energy associated with such interfaces. Our interest here is only to demonstrate how the description of various geometric arrangements associated with grain boundaries can be encompassed under the general framework of de Rham currents.

Within a framework suggestively reminiscent of the use of Hadamard's lemma for the establishment of geometric compatibility conditions on a wave front,[18] we work in a simply connected n-dimensional oriented manifold $\mathscr{M}$ which is divided into two disjoint open consistently oriented submanifolds, $\mathscr{M}^+$ and $\mathscr{M}^-$, by an embedded oriented $(n-1)$-dimensional manifold (a hypersurface) $\mathscr{T}$, as shown in Fig. 7. Let ω^+ and ω^- denote two closed[19] smooth layering 1-forms defined, respectively, on $\mathscr{M}^+$ and $\mathscr{M}^-$. Unless ω^+ and ω^- happen to be the restrictions of one and the same smooth 1-form ω on $\mathscr{M}$, we are in the presence of a *grain boundary* $\mathscr{T}$.

The 1-forms ω^+ and ω^- can be extended to the closures $\bar{\mathscr{M}}^+ = \mathscr{M}^+ \cup \mathscr{T}$ and $\bar{\mathscr{M}}^- = \mathscr{M}^- \cup \mathscr{T}$, which are submanifolds with boundary of $\mathscr{M}$. We denote by T_{ω^+} and T_{ω^-} the $(n-1)$-currents associated, respectively, with these extensions of ω^+ and ω^-. We define the total layering current

$$T = T_{\omega^+} + T_{\omega^-}, \tag{52}$$

[18] See, e.g., [22], p. 492.

[19] The assumption that these two 1-forms are closed is made explicitly to concentrate on the role of the discontinuity hypersurface $\mathscr{T}$, thus ignoring explicitly the possibility of existence of smooth dislocations.

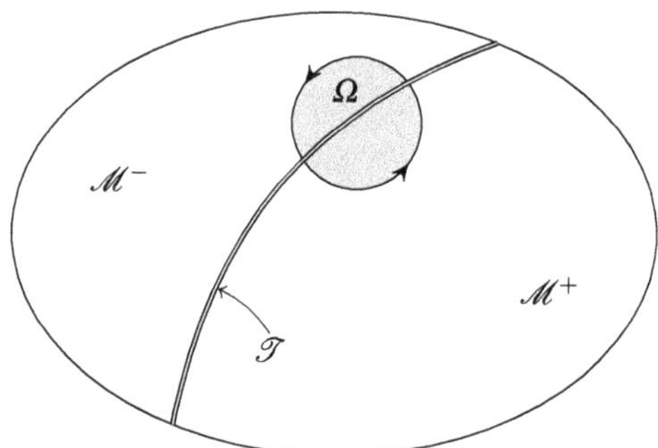

Fig. 7 An interface

and notice that, if ω^+ and ω^- are restrictions of a single 1-form ω on $\mathscr{M}$, that is, if there is no grain boundary, the boundary ∂T of the total layering current vanishes.

The converse is not true. In other words, it is possible to have an absence of dislocations consistent with a lack of smoothness of the layering across a hypersurface of discontinuity. This should correspond precisely to the freedom afforded by Hadamard's lemma! To make matters explicit, let us set $\partial T = 0$. Recalling that we have assumed the forms ω^+ and ω^- to be closed, any $(n-2)$-form ϕ with compact support Ω intersecting the hypersurface $\mathscr{T}$ gives rise to the evaluation

$$0 = \partial T[\phi] = \partial T_{\omega^+}[\phi] + \partial T_{\omega^-}[\phi] = -\int\limits_{\Omega\cap\bar{\mathscr{M}}^+} d(\omega^+ \wedge \phi) - \int\limits_{\Omega\cap\bar{\mathscr{M}}^-} d(\omega^- \wedge \phi)$$

$$= -\int\limits_{\Omega\cap\bar{\mathscr{T}}} [\![\omega]\!] \wedge \phi\Big|_{\mathscr{T}}, \tag{53}$$

where, on $\mathscr{T}$,

$$[\![\omega]\!] = \omega^+ - \omega^-. \tag{54}$$

We have purposely indicated in Eq. (53) that the arbitrary forms $[\![\omega]\!]$ and ϕ involved in the last term, via Stokes' theorem, are to be restricted to $\mathscr{T}$. Consequently, if the discontinuity hypersurface $\mathscr{T}$ is represented locally as $\Psi = \text{constant}$, where Ψ is some smooth function $\Psi : \mathscr{M} \to \mathbb{R}$, we conclude that, when

$$[\![\omega]\!] \propto d\Psi, \tag{55}$$

the dislocation vanishes altogether. If a Riemannian structure were to be introduced on $\mathscr{M}$, this result would be expressed by saying that an arbitrary jump in the direction perpendicular to the discontinuity surface can be tolerated. This is precisely the same as the geometric compatibility condition derived classically from Hadamard's lemma applied to a surface of discontinuity of a field. For this reason,

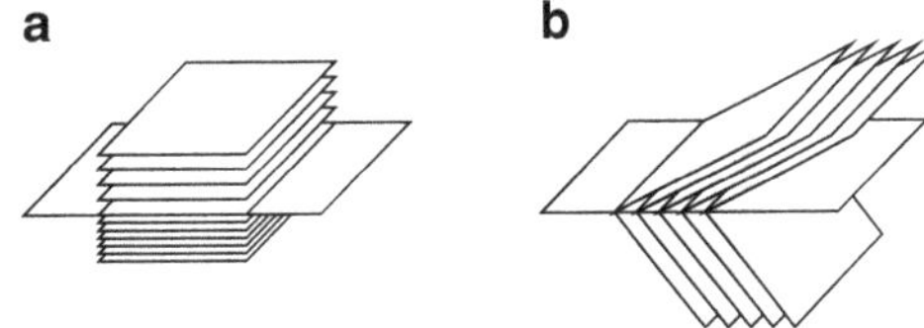

Fig. 8 Hadamard (coherent) interfaces

we may call this tolerable lack of smoothness a *Hadamard interface*, known in material science as a *coherent interface*. Such an interface involves no dislocations. We remark that our definitions are strictly local.

A particular case of a Hadamard interface is obtained when $\omega^+ \propto d\Psi$ at the discontinuity hypersurface. Since we have assumed Eq. (55) to apply, we conclude that also $\omega^- \propto d\Psi$. In other words, the local hyperplane stacks on both sides of $\mathscr{T}$ are parallel to the tangent hyperplane, but undergo a jump in density, as schematically shown in Fig. 8a. The more general case of a coherent interface is shown in Fig. 8b. Intuitively, the stacks intersect $\mathscr{T}$ along the same family of oriented lines (that is, hyperplanes of dimension $n - 2$).

Remark 5 **Orientation and coherence**: In physical presentations of examples of coherent and non-coherent interfaces, it is often implicitly assumed that the lattice planes do not carry a specific orientation. Thus, Fig. 8b, which clearly shows the coincidence of the intersections of the lattice planes, is sufficient to explain the notion of coherence of the interface. Our mathematical model, however, contains an extra degree of freedom, namely, the signature of the layering form arising from the assumed orientation of the stacks. Apart from the intrinsic mathematical interest of the fact that a differential form is clearly different from its negative, it should be clear that interfaces involving chiral molecules[20] will require consideration of the orientation of the lattice planes and, consequently, of the induced orientation of their intersections with the interface. □

If Eq. (55) is not satisfied at a point $P \in \mathscr{T}$, we say that at that point there is a *non-coherent interface*. In a 3-dimensional context, a further classification can be established on the basis of the exterior product $\sigma = \omega^+ \wedge \omega^- \wedge d\Psi$. When $\sigma = 0$, the dislocation consists of a mere mismatch in the density of the stacks restricted to $\mathscr{T}$, which are otherwise parallel, as shown in Fig. 9a. When $\sigma \neq 0$ the stacks are mismatched also in rotational terms, as shown in Fig. 9b.

Example 13 In $\mathbb{R}^3$, let the surface of discontinuity $\mathscr{T}$ be the plane $z = 0$, and let ω^+ and ω^- be closed 1-forms defined on the closures of the upper and lower half spaces, respectively. The coherence condition (55) requires that, at $z = 0$,

$$\omega_x^+ = \omega_x^- \qquad \omega_y^+ = \omega_y^-, \tag{56}$$

[20]A *chiral molecule* can exist in two isomeric varieties, known as *enantiomers*. They are mutual mirror images.

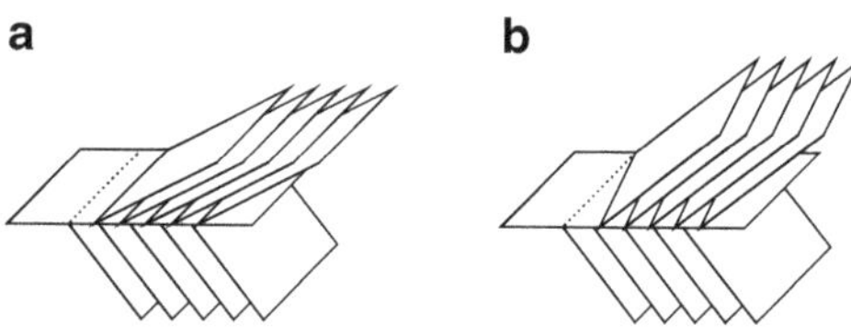

Fig. 9 Non-coherent interfaces

while the components ω_z^+ and ω_z^- can be arbitrary. We remark that the representation shown in Fig. 8b would remain unaltered if we had supposed that, at $z = 0$, $\omega_x^+ = -\omega_x^-$ and $\omega_y^+ = \omega_y^-$. In that case, however, the boundary of the total current would have been given by

$$\partial T[\phi] = -\iint\limits_{\{z=0\}} 2(\omega_x \phi_y - \omega_y \phi_x)\, dx\, dy, \tag{57}$$

for all 1-forms $\phi = \phi_x dx + \phi_y dy + \phi_z dz$ with compact support in $\mathbb{R}^3$. In other words, the interface would have been non-coherent. As indicated in Remark 5, this could very well be the case with chiral crystals if the molecules at both sides of the (apparently coherent) interface do not have the same handedness. □

3.7 A Volterra Disclination

Although not using the terms 'dislocation' and 'disclination', Volterra [23, 24] described three displacement-induced, and three rotation-induced defects, which were later so named. In the case of a *wedge disclination*, in Volterra's description, a wedge-like segment (such as when cutting a cake) is removed from a thick-walled cylinder, which is then repaired by joining the two exposed faces. From this perspective, it follows that this kind of defect can be described as an interface. Indeed, the process of repair just described brings two surfaces into contact, resulting in a potential coherent or non-coherent boundary. A different, certainly cleverer but essentially equivalent, point of view[21] consists of regarding the wedge removal as a superposition of an infinite number of edge dislocations, whose dislocation lines span a plane through the cylinder axis. We will briefly describe and analyze these two constructions below. Similar ideas can be applied to the other two types of Volterra disclinations.

[21] See, e.g. [19].

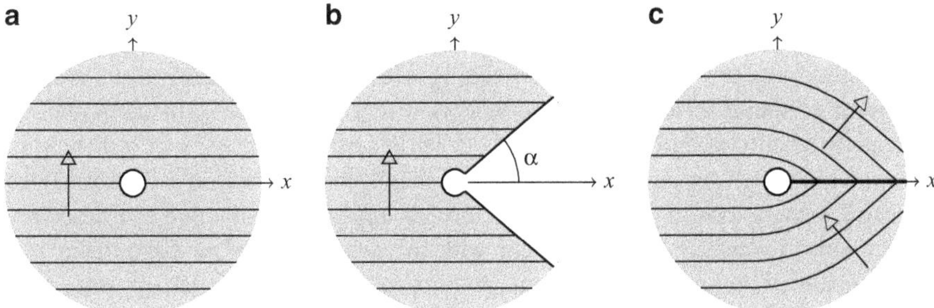

Fig. 10 Volterra's cylinder (**a**) is deprived of a wedge (**b**) and then repaired (**c**) by a diffeomorphism, thus generating an interface. The orientation of the integral surfaces is indicated with hollow-headed arrows

3.7.1 Wedge Disclination as Interface

Although not strictly necessary, we will assume that a small open tube of radius $\epsilon \ll 1$ has been removed from the open cylinder of radius 1 and axis z in $\mathbb{R}^3$. The typical x, y section of the remaining manifold with boundary $\mathscr{R}$ is shown in Fig. 10a, in which some of the integral surfaces $y =$ constant of the closed layering 1-form $\omega = dy$ have been shown. Following Volterra's scheme, we now remove a wedge with an semi-aperture angle α and we are left with a new domain $\mathscr{R}'$, as shown in Fig. 10b, without affecting the integral lines. The repaired state is obtained by means of a diffeomorphism between the interior of this now simply connected manifold and the manifold obtained from the original cylinder $\mathscr{R}$ minus the positive half-plane $\mathscr{H}$ with equations $y = 0, x > 0$. We can assume (if so desired) that the angle α between the integral surfaces and the surface of the cut is preserved, and that the cut surface remains rigid. Since diffeomorphisms preserve the closed character of differential forms, the only source of dislocation arises from the discontinuity surface whose trace is indicated by a thick line in Fig. 10c.

By the assumed conditions of the diffeomorphism used, we obtain the following values for the restriction of the layering form on either side of the glued joint

$$\omega_x^+ = -\omega_x^- = sin\alpha \qquad \omega_y^+ = \omega_y^- = cos\alpha \qquad \omega_z^+ = \omega_z^- = 0. \tag{58}$$

The dislocation current evaluated on a 1-form $\phi = \phi_x dx + \phi_y dy + \phi_z dz$ with compact support in the cylinder is given by

$$\partial T_\omega[\phi] = -\int_{\mathscr{H}} [\![\omega]\!] \wedge \phi = -2\, sin\alpha \int_{-\infty}^{\infty} \int_0^1 \phi_z \, dx\, dz. \tag{59}$$

3.7.2 Wedge Disclination as Superposition of Edge Dislocations

The method just presented, while in keeping with the physical motivation, is not in the spirit of the general geometric conception that we advocate. In particular, the operation of cutting and welding needs to be described by means of a diffeomorphism which itself becomes a component of the problem. In fact, a different, no less intuitive, way to look at the physical picture, as appealing as the one just described, is available. Indeed, if we are content with the description of an edge dislocation as the result of removing a half-plane from an otherwise regular lattice, and, as we have demonstrated in Example 8, if this operation is represented mathematically by a singular layering associated with the removed submanifold, then the removal of a wedge is nothing but the cumulative effect of the removal of additional parallel half-planes starting at progressively advancing locations. This operation is pictorially represented in Fig. 11.

Let k denote the constant strength of each of the edge dislocations, where a negative k describes a removal. We obtain a 1-parameter family $\mathscr{H}_\xi$ of (oriented) half-planes, where ξ is the x-coordinate of their respective edge. The coordinate system is the same as in Fig. 10. The total 2-current evaluated on a 2-form ψ with compact support in $\mathbb{R}^3$ is

$$T[\psi] = \int_0^\infty T_{\mathscr{H}_\xi}[\psi]\, d\xi = k \int_0^\infty \left(\int_{\mathscr{H}_\xi} \psi \right) d\xi. \tag{60}$$

Its boundary, evaluated on a 1-form ϕ with compact support in $\mathbb{R}^3$, is

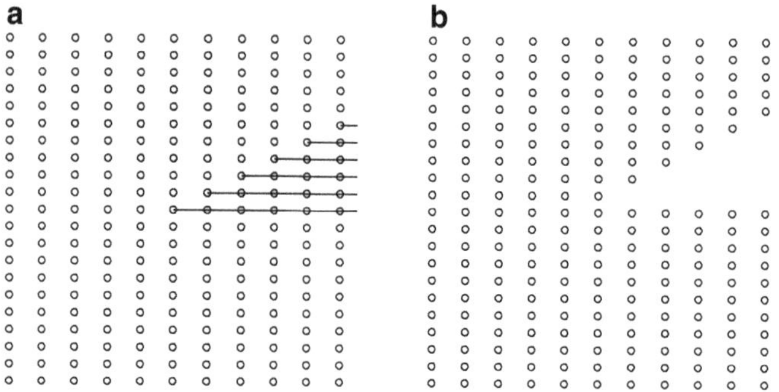

Fig. 11 Regular lattice (**a**) with half-planes (perpendicular to drawing) slated for removal as edge dislocations marked with lines. Lattice with half-planes removed (**b**), showing a disclination. In the limit of vanishing lattice unit, all these half-planes lie on one and the same plane

$$\partial T[\phi] = T[d\phi] = k \int_0^\infty \left(\int_{\mathscr{H}_\xi} d\phi \right) d\xi = k \int_0^\infty \left(\int_{\partial \mathscr{H}_\xi} \phi \right) d\xi$$

$$= k \int_0^\infty \left(\int_{-\infty}^\infty \phi_z(\xi, 0, z) dz \right) d\xi. \qquad (61)$$

With $k = -2\, sin\alpha$, the same result (59) is obtained, except that here we worked on $\mathbb{R}^3$ rather than on a cylinder of radius 1.

3.8 Disengagements or Distriations

Of the immense geometric and analytic apparatus provided by differential forms and de Rham currents, we have so far utilized a small part only. In fact, we have concentrated on 1-forms or, correspondingly, $(n - 1)$-currents, mainly because, as suggested in Sect. 2.3, a covector can be regarded as an oriented stack of planes with a certain stack density. A differential 1-form ω, accordingly, can be regarded as the geometrical representation of a field of such stacks as encountered, for example, in smectic liquid crystals or in lattice planes of an atomic crystalline array.

In the smooth case, we identified a *dislocation* with the lack of closure of the layering 1-form ω. Correspondingly, in the singular case, the presence of dislocations is identified with the non-vanishing of the boundary of the layering $(n - 1)$-current T. One can expect that in an n-dimensional manifold $\mathscr{M}$, if consideration is given to forms of arbitrary order $0 \leq r \leq n$, these r-forms, and their $(n - r)$-current counterparts, may be useful to represent other smooth or singular internal structures and their defectiveness.

In this section, we will explore the case $r = n - 1$, a choice that can be justified for various reasons. One of these is that every $(n - 1)$-form is *decomposable*, namely, expressible as a monomial, that is, a wedge product of $n - 1$ 1-forms, a fact that facilitates the physical and geometrical interpretation. A second reason is that in dimension $n = 3$, having already considered 1-forms, and observing that 0-forms and n-forms are of scarce practical interest, we are only left with 2-forms.[22] Finally, and most importantly, $(n-1)$-forms and 1-currents turn out to be associated with imperfections known in various applied fields, such as *Frank disclinations* in nematic liquid crystals.

[22] It is remarkable that $n = 3$ is the maximum dimension for which all forms are automatically decomposable. This can be used as a somewhat banal argument for our space to be 3-dimensional, but nor more banal than the acoustic argument according to which wave fronts propagate sharply only in odd dimensional spaces. In a 2-dimensional world, we would not be able to communicate by sharp signals.

3.8.1 Affine Subspaces, and Decomposable Multivectors and Multicovectors

In an n-dimensional vector space U, an r-vector $\mathbf{V}$ is *simple* or *decomposable* if it can be expressed as

$$\mathbf{V} = \mathbf{v}_1 \wedge \mathbf{v}_2 \wedge \ldots \wedge \mathbf{v}_r, \tag{62}$$

where each $\mathbf{v}_i$ is a vector in U. The r-vector $\mathbf{V}$ vanishes if, and only if, these vectors are linearly dependent. Therefore, each non-zero $\mathbf{V}$ spans an r-dimensional subspace, $S_V \subset U$. Vice versa, one can show that each r-dimensional subspace of U is determined uniquely, up to a non-zero multiplicative constant, by a decomposable r-vector. Specifically, the subspace S_V is given by

$$S_V = \{\mathbf{v} \in U \mid \mathbf{V} \wedge \mathbf{v} = 0\}. \tag{63}$$

Similar considerations apply to the exterior algebra of the dual space U^*, thus giving rise to the notion of decomposable multicovectors.[23] As a vector space in its own right, the collection of r-covectors is the dual vector space of the collection of r-vectors, whether decomposable or not. Since r-vectors and r-covectors can be understood, respectively, as completely skew-symmetric contravariant and covariant tensors of order r, the linear action of r-covectors on r-vectors can be regarded as the full contraction of the corresponding tensors (perhaps with a factorial multiplier).

For the particular case $r = 1$, we suggested[24] in Sect. 2.3 that a covector ω can be conceived as a stack of oriented hyperplanes (affine $(r-1)$-dimensional subspaces of U) with a certain stack density. The action of $\omega \in U^*$ on $\mathbf{v} \in U$ can be pictorially described as the 'number' of hyperplanes cut by $\mathbf{v}$. We would like to elicit a similar intuitive picture for arbitrary r.

Given a non-zero decomposable r-covector $\Omega = \omega_1 \wedge \omega_2 \wedge \ldots \wedge \omega_r$ we obtain r stacks of hyperplanes, one for each covector ω_i. For $r = n - 1$, the mutual intersections of these hyperplanes constitute an $(n-1)$-parameter family of equally oriented parallel lines, endowed with a density. We will call such a family a *fascicle*, a terminology borrowed from the anatomy of striated muscles. For any given decomposable $(n-1)$-vector $\mathbf{V}$, the linear action of Ω on $\mathbf{V}$ can be interpreted as the number of lines intersected by the oriented r-dimensional parallelepiped associated with $\mathbf{V}$. This parallelepiped lies on S_V and can be also called its associated r-volume element. These ideas are schematically illustrated in Fig. 12.

[23]For an illuminating presentation of these ideas, see [21].

[24]Following [14].

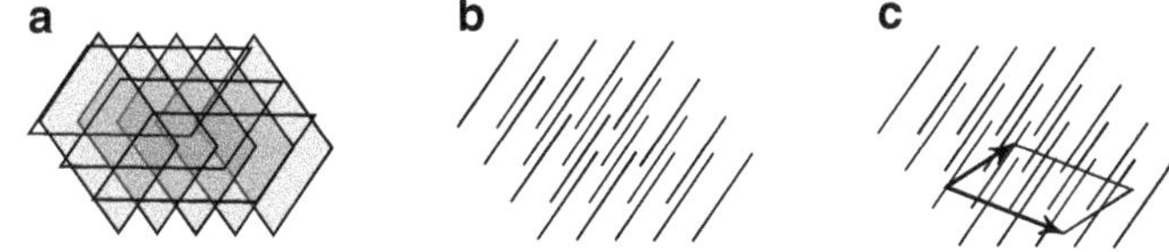

Fig. 12 Two stacks (**a**) intersect to generate a fascicle (**b**), whose evaluation over a 2-vector is obtained by counting the intersections of the fascicle with the corresponding parallelogram (**c**)

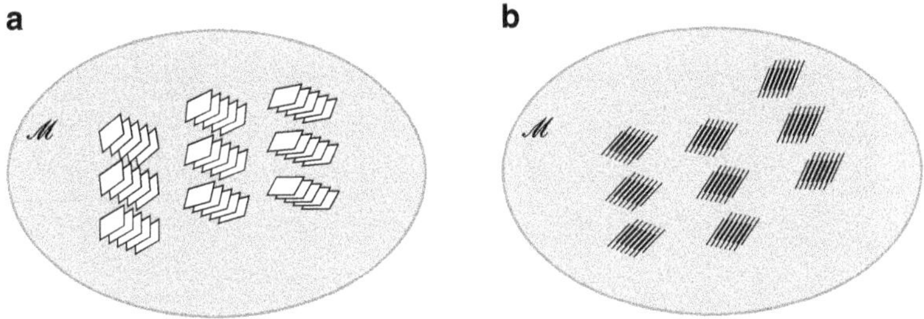

Fig. 13 Layering 1-form (**a**), and filamenting $(n - 1)$-form (**b**)

3.8.2 The Smooth Case

An $(n-1)$-form on an n-dimensional manifold $\mathscr{M}$ will be called a *filamenting form*, or *threading form*. It smoothly assigns to each point of $\mathscr{M}$ a fascicle. Figure 13 shows a schematic comparison with the case of a layering 1-form.

Although the dimension of the spaces of 1-covectors and $(n - 1)$-covectors are the same (that is, n), when turning to analysis there is a fundamental asymmetry in the treatment. It arises from the fact that the exterior derivative of an r-form is an $(r + 1)$-form, which introduces an obvious bias in the determination of the exactness of the differential form. Another fundamental difference, which pertains to the treatment of material defects, is that, as discussed in Sect. 2.3, in the case of a layering form, a less strict notion of defectiveness can be introduced by requiring merely that the geometric 2-dimensional distribution be involutive. This condition guarantees that the slopes of neighbouring stacks fit properly, in the sense that they can be derived from the tangent spaces of an integral surface. The stronger condition of the vanishing of the exterior derivative guarantees also that the stack densities are in harmony with each other.

In the case of the filamenting $(n - 1)$-form, however, since the underlying distribution is 1-dimensional, the theorem of existence and uniqueness of solutions of systems of ordinary differential equations, ensures the involutivity of the distribution. In other words, integral curves always exist, at least locally. The defectiveness, therefore, can only manifest itself in the lack of fitting of the densities of neighbouring fascicles, which is measured by the lack of closure of the exterior derivative of the filamenting form. A smoothly defective filamented structure is

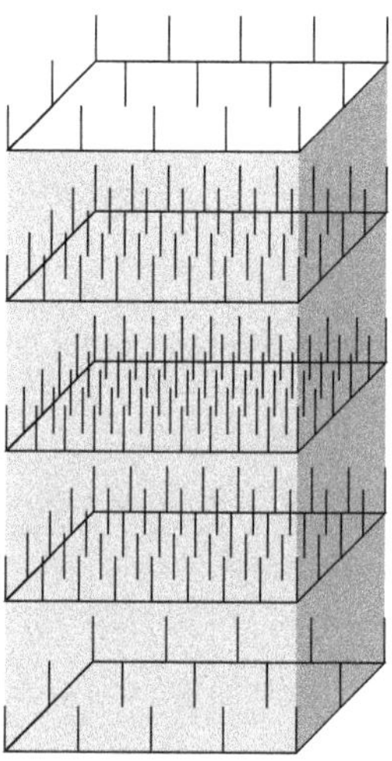

Fig. 14 A smooth disengagement progressively tearing apart the striated structure

said to represent a *continuous distribution of disengagements or distriations*.[25] We remark, finally, that in the case of a 2-dimensional manifold, there is no distinction between layering and filamenting forms.

Example 14 Let $\mathscr{M}$ be an open rectangular prism, as shown in Fig. 14, and let the filamenting 2-form be given by

$$\Omega = \left(1 + e^{-z^2}\right) dx \wedge dy, \tag{64}$$

where x, y, z are the standard coordinates of $\mathbb{R}^3$. In Fig. 14 the z-axis is drawn vertically and pointing upwards, while the origin is at the centroid of the prism. Any vertical line is an integral curve of the 1-dimensional distribution associated with Ω, but there is a vertical mismatch in the density of the fascicles. Specifically, this mismatch is measured by the *disengagement 3-form*

$$D = d\Omega = -2ze^{-z^2} dx \wedge dy \wedge dz. \tag{65}$$

□

3.8.3 The Singular Case

The singular counterpart of a smooth filamentous structure can be represented by a filamenting 1-current T. Its boundary ∂T is the disengagement 0-current, whose non-vanishing is a measure of defectiveness. In the smooth case, we remarked that,

[25]The neologism 'distriation' is meant to suggest the disruption of the striated structure implied by the filaments in compatible fascicles.

since the associated 1-dimensional distribution is always involutive, the only cause of defectiveness is a smooth mismatch of the fascicle density. In the singular case, quite apart from the obvious concentration of such mismatch along a submanifold of $\mathcal{M}$, we may also have a violation of the Lipschitz condition necessary to guarantee the existence of local continuous integral curves.

Example 15 **An edge distriation**: Let $\mathcal{S}$ be a 1-dimensional submanifold with boundary of $\mathcal{M}$. We define its associated 1-current as

$$T_{\mathcal{S}}[\phi] = \int_{\mathcal{S}} \phi, \tag{66}$$

for all 1-forms ϕ with compact support in $\mathcal{M}$. The corresponding distriation 0-current is

$$D[f] = \partial T_{\mathcal{S}}[f] = T_{\mathcal{S}}[df] = \int_{\mathcal{S}} df = \int_{\partial \mathcal{S}} f, \tag{67}$$

for all functions f with compact support in $\mathcal{M}$. The boundary $\partial \mathcal{S}$ may be empty (such as when $\mathcal{S}$ is a loop, or when its terminal points are not in $\mathcal{M}$), or it may consist of 1 or 2 points within $\mathcal{M}$. In the case of two terminal points $p, q \in \mathcal{M}$, we obtain

$$D[f] = f(q) - f(p), \tag{68}$$

and the distriation can be interpreted as residing in two points with opposite polarity. If only one point (p, say) lies within $\mathcal{M}$, we have a case of a single point distriation, or *edge distriation*. It may be regarded as a sudden tear of a strand, and it is the distriation analogue of an edge dislocation. □

Example 16 **A Volterra conical disclination**: We have already pointed out that, in the case of a 2-dimensional manifold $\mathcal{M}$, there is no distinction between dislocations and distriations. In particular, the construction of a Volterra disclination in Sect. 3.7.2 can be interpreted as either a distribution of edge dislocations or of edge disclinations of the type discussed in Example 15. If, however, $\mathcal{M}$ is 3-dimensional, a continuous accumulation of edge distriations will give rise to new kinds of defects, namely, those generated by a one- or two-parameter family of edge distriations. The two-parameter case is easier to grasp, since it can be described as a *conical disclination* of the Volterra type, schematically illustrated in Fig. 15.

Just as in the case with the wedge disclination, in this formulation, based on a continuous superposition of 1-currents, all the filaments removed lie on the same line (the cone axis, which is the positive x-semiaxis), and only their terminal points vary. The total dislocation 0-current is given by

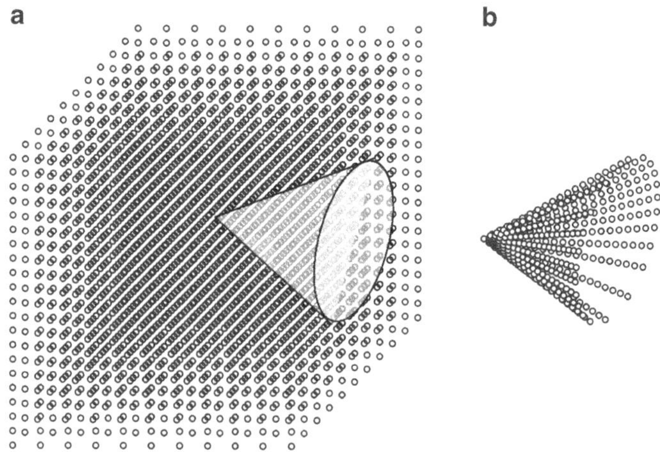

Fig. 15 A conical disclination generated by the removal of all filaments within the conical volume shown in (**a**). The total dislocation current is obtained by adding up (integrating) the values of a scalar function f over the terminal points indicated in (**b**)

$$D[f] = 2\pi k \int_0^\infty f(x, 0, 0)\, dx, \tag{69}$$

where k is the common strength of the distriations, and f is a scalar function with compact support in $\mathscr{M}$. The Volterra process for creating a conical disclination would consist in actually indenting a material with a conical indenter and then bringing the generators of the cone into coincidence with its axis. The question as to whether or not this kind of process would require too much energy to be of common occurrence is beyond the scope of our conceptual analysis. □

4 From Discrete to Continuous Dislocations

The question we want to address[26] is the following: Can a continuous distribution of defects be rigorously obtained as the limit of a sequence of singular dislocations? As an example, consider a body $\mathscr{B}$ identified with an open rectangle, as shown in Fig. 16, aligned with the natural coordinate axes of $\mathbb{R}^2$. At a given stage of the limiting process, we subdivide the rectangle into $M \times N$ small identical bricks of area $h \times k$. Each brick contains the terminal point of a manifold $\mathscr{S}_{ij}$ consisting of a

[26] A more rigorous limiting process is presented in another chapter of this volume authored by Kupferman and Olami [13].

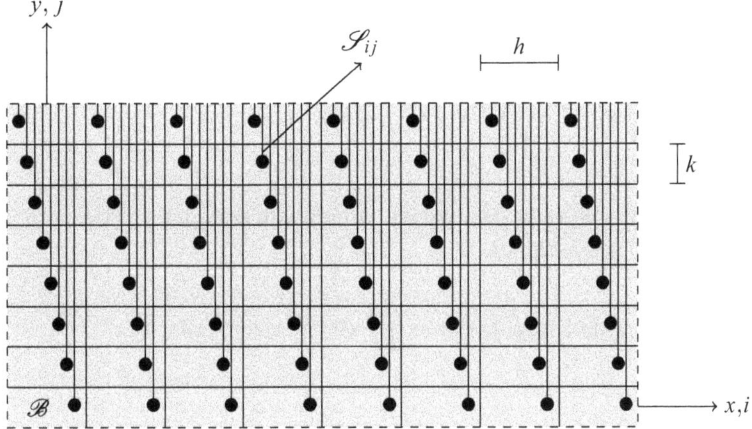

Fig. 16 A multiple singular layering

vertical semi-open segment coming from above, as suggested in the figure. The end point of $\mathscr{S}_{ij}$ is located at the centroid of the box i, j, namely, it has the coordinates

$$(x_i, y_j) = ((i-1)h, (j-1)k). \tag{70}$$

Let K_{ij} denote the constant intensity of the edge dislocation associated with $\mathscr{S}_{ij}$. At this stage of the approximation, the total layering current T is obtained as the sum

$$T = \sum_{i=1}^{M} \sum_{j=1}^{N} K_{ij} T_{\mathscr{S}_{ij}}, \tag{71}$$

and the total dislocation current ∂T acts on an arbitrary function (0-form) f with compact support in $\mathscr{B}$ according to

$$\partial T[f] = \sum_{i=1}^{M} \sum_{j=1}^{N} K_{ij} \partial T_{\mathscr{S}_{ij}}[f] = \sum_{i=1}^{M} \sum_{j=1}^{N} K_{ij} f(x_i, y_j). \tag{72}$$

We remark that we would obtain exactly the same result if we assume the submanifolds to be horizontal and coming from the right.

Let $\omega = \omega_1 dx + \omega_2 dy$ be the putative 1-form we are trying to approximate. We want to adjust the values of K_{ij} in such a way that the limit of this dislocation current, as $M, N \to \infty$ is, in some sense, the dislocation 2-form $D = d\omega$. To this end, we choose the strengths K_{ij} to be

$$K_{ij} = (\omega_{1,2} - \omega_{2,1})hk, \tag{73}$$

where the derivatives are evaluated at the centre of the corresponding rectangle. With this choice, indeed, we verify that

$$\lim_{M,N\to\infty} \sum_{i=1}^{M} \sum_{j=1}^{N} K_{ij} f(x_i, y_j) = \int_{\mathscr{B}} d\omega \wedge f = T_{d\omega} f. \tag{74}$$

Let $\phi = \phi_1 dx + \phi_2 dy$ be an arbitrary 1-form with compact support in $\mathscr{B}$. Recall that T_ω acts on ϕ according to

$$T_\omega[\phi] = \int_{\mathscr{B}} \omega \wedge \phi = \iint_{\mathscr{B}} (\omega_1 \phi_2 - \omega_2 \phi_1)\, dx\, dy. \tag{75}$$

On the other hand, the approximation (71) yields

$$T[\phi] = \sum_{i=1}^{M} \sum_{j=1}^{N} K_{ij} T_{\mathscr{S}_{ij}}[\phi] = \sum_{i=1}^{M} \sum_{j=1}^{N} K_{ij} \int_{y_j}^{N-0.5k} \phi\, dy \approx \sum_{i=1}^{M} \sum_{j=1}^{N} K_{ij} \sum_{l=j}^{N} \phi_2(x_i, y_l) k. \tag{76}$$

We collect, for each fixed i, the coefficients of each successive $\phi_2(x_i, y_j)$ and rearrange the sums to obtain

$$T[\phi] \approx \sum_{i=1}^{M} \sum_{j=1}^{N} \phi_2(x_i, y_j) \sum_{l=1}^{j} K_{il} k. \tag{77}$$

We can see that the only hope for this series to converge to a definite 1-form is to assume that $\omega_2 = 0$. We may also assume that ω_1 vanishes identically at $y = 0$. Under these conditions, taking account of (73), we have in the limit

$$\lim_{k\to 0} \sum_{l=1}^{j} K_{il} k = hk\, \omega_1(x, y). \tag{78}$$

Finally, we obtain from Eq. (77) the Riemann integral

$$\lim_{h,k\to 0} T[\phi] = \iint_{\mathscr{B}} (\omega_1 \phi_2)\, dx\, dy = \omega[\phi]. \tag{79}$$

In conclusion, when the boundary ∂T of the current T of a denumerably infinite collection of manifolds $\mathscr{S}_{ij}$ tends to a definite continuous dislocation 2-form D, the current T tends to a definite layering 1-form ω whose exterior derivative is precisely D. There is a physical feel to the fact that our particular choice of $\mathscr{S}_{ij}$ as vertical lines corresponds to a vanishing vertical component of ω.

5 The Movement of Dislocations

5.1 Introduction

Dislocations move and, in so doing, they are responsible for important phenomena such as metal plasticity and observable effects of glide and climb of faults in liquid crystals.[27] In this section, we provide a geometrical framework that may serve to describe these phenomena against a differential geometric background.

Although not usually expressed in these terms, the gist of plasticity theory lies in the time evolution of the frame bundle of the body manifold. A similar mental picture applies to various other theories, such as biological growth and remodelling, in which the material properties do not change in time, but the material undergoes a process of re-accommodation and/or growth which results generally in the development of residual stresses. This process can, therefore, be described as a point-wise time-dependent change of reference frame for the tangent space. In more mundane terms, the constitutive response at a point remains unchanged, except that the strain argument is measured with respect to a variable reference frame. From this description, we retain just the kinematic component, since we are not addressing the cause-effect paradigm involved in constitutive laws, but only distilling the underlying geometrical apparatus.

5.2 Frame Bundle Automorphisms

We consider the collection $\mathscr{A}(\mathscr{M})$ of smooth bundle automorphisms of the frame bundle $F\mathscr{M}$ of a fixed n-dimensional manifold $\mathscr{M}$.[28] We denote the bundle projection by π, and we recall that a *frame bundle automorphism* can be represented as the commutative diagram

$$\begin{array}{ccc} F\mathscr{M} & \xrightarrow{\ \Phi\ } & F\mathscr{M} \\ \Big\downarrow{\scriptstyle\pi} & & \Big\downarrow{\scriptstyle\pi} \\ \mathscr{M} & \xrightarrow{\ \varphi\ } & \mathscr{M} \end{array} \tag{80}$$

In this diagram, whose commutativity implies fibre preservation, the diffeomorphism φ is naturally implied in Φ. Moreover, Φ commutes with the right action

[27] See, e.g., [12].

[28] For simplicity of the exposition, we will assume that $F\mathscr{M}$ is globally trivializable.

of the structure group, which, in the case of the frame bundle, is the general linear group $GL(n, \mathbb{R})$.

Remark 6 There is a one-to-one correspondence between frame bundle automorphisms and automorphisms of any of its associated (tensor) bundles. Consider the tangent bundle $T\mathscr{M}$. An automorphism Φ of $F\mathscr{M}$ gives us, for each point $X \in \mathscr{M}$, a map $\Phi_X : F_X\mathscr{M} \to F_{\varphi(X)}\mathscr{M}$ satisfying $\Phi_X(fg) = \Phi_X(f)\, g$, for every frame $f \in F_X\mathscr{M}$ and for every $g \in GL(n, \mathbb{R})$, where the right action is indicated by simple apposition. Let $\mathbf{v} \in T_X\mathscr{M}$ be a vector at X. We declare its image at $\varphi(X)$ to be the vector whose components in the frame $\Phi_X(f)$ are the same as the corresponding components of $\mathbf{v}$ in the frame f. This correspondence is independent of the frame f chosen, due to the indicated commutativity with the right action just described. Vice versa, given an automorphism of the tangent bundle $T\mathscr{M}$, it maps every base vector of the frame $f \in F_X\mathscr{M}$ to a base vector at the image point. Again, it is easy to show that this assignment commutes with the right action of the structure group. A similar argument can be used for the cotangent bundle, in which case the assignment is obtained by the inverse of the transpose map used for the tangent bundle. □

Given a diffeomorphism $\varphi : \mathscr{M} \to \mathscr{M}$, we can construct its associated *lifted frame bundle automorphism*, $\hat{\varphi} : F\mathscr{M} \to F\mathscr{M}$, by using the tangent map φ_* to induce the fibre-wise maps $\Phi(X)$ for each $X \in \mathscr{M}$. We will denote the collection of all these induced bundle automorphisms by $\mathscr{A}_*(\mathscr{M})$.

Of particular interest too is the subcollection $\mathscr{A}_0(\mathscr{M})$ based upon the identity map $\varphi = id_{\mathscr{M}}$. Every element of $\mathscr{A}_0(\mathscr{M})$ can be regarded as a smooth section of $\mathscr{M} \times GL(n, \mathbb{R})$,[29] so that $\mathscr{A}_0(\mathscr{M})$ is essentially identifiable with the space of sections $\Lambda(F\mathbb{R}^n)$, which has the natural structure of a Banach manifold [17]. An automorphism $\Phi \in \mathscr{A}_0(\mathscr{M})$ is *holonomic* if it coincides, as a section of $\mathscr{M} \times GL(n, \mathbb{R})$, with the natural basis of a coordinate system in $\mathscr{M}$.

Every frame bundle automorphism Φ can be trivially and uniquely expressed as the composition of an element of $\mathscr{A}_*(\mathscr{M})$ with an element of $\mathscr{A}_0(\mathscr{M})$, in either order, namely,

$$\Phi = \left(\Phi \circ \hat{\varphi}^{-1}\right) \circ \hat{\varphi} = \hat{\varphi} \circ \left(\hat{\varphi}^{-1} \circ \Phi\right). \tag{81}$$

Although trivial, this decomposition will turn out to demarcate an important physical distinction between processes of material convection and processes of material evolution.

[29] More precisely, in a given trivialization of the bundle, at each point $X \in \mathscr{M}$, $\Phi(X)$ amounts to a *left* translation of the fibre $\pi^{-1}(X) = GL(n, \mathbb{R})$ by some element of $GL(n, \mathbb{R})$.

5.3 Material Convection and Material Evolution

A process of *material convection* in the time interval $[0, T]$ is defined as a smooth curve $\rho : [0, T] \to \mathscr{A}_*(\mathscr{M})$. We assume that $\rho(0)$ is the identity bundle automorphism. Recalling that the elements of $\mathscr{A}_*(\mathscr{M})$ are lifts of material diffeomorphisms, these processes serve to describe phenomena of migration of quantities defined as sections of any of the bundles associated with the frame bundle, namely, all tensor bundles. The convection, therefore, takes place as a result of an imagined material flow. By construction, a defective structure, represented by a non-vanishing dislocation current, will remain defective during the convection process. Processes of material convection were carefully studied in [8].

A *material evolution* in the time interval $[0, T]$ is defined as a smooth curve $\gamma : [0, T] \to \mathscr{A}_0(\mathscr{M})$. We assume that $\gamma(0)$ is the identity bundle automorphism. A material evolution is said to be *holonomic* if $\gamma(t)$ is holonomic for every $t \in [0, T]$. In processes of material evolution, quantities defined as sections of any of the associated bundles change their local values smoothly as time goes on. There is no convection, but rather a local rearrangement of the tangent spaces. If the process is non-holonomic, these tangent spaces do not fit well together, as it were. We note that, in our theory, the defects are not associated with this lack of compatibility, but rather with the lack of integrability of certain forms or currents. Thus, a non-holonomic evolution process may as well end up leading, at some time t, to the integrability of an evolving structure that was initially non-integrable.

Let, for example, $\omega = \omega_i \, dx^i$ be a 1-form on $\mathscr{M}$, and let $\gamma : [0, T] \to \mathscr{A}_0(\mathscr{M})$ be a material evolution. In a given trivialization,[30] this evolution is expressed by means of a point- and time-dependent group element $G^i_j = G^i_j(X, t)$, with $G^i_j(X, 0) = \delta^i_j$. The corresponding change of the components of ω will then be given by

$$\omega_i(X, t) = \left(G(X, t)^{-1}\right)^j_i \, \omega_j(X, 0). \tag{82}$$

The support of ω remains unchanged. Consequently, at least in principle, as continuous linear operators on the space of compactly supported forms ψ, currents can be transformed according to

$$T(t)[\psi] = T(t)[\psi(\cdot, 0)] = T(0)[\psi(., t)]. \tag{83}$$

Remark 7 The transformation of currents just outlined turns out to also be consistent with the boundary operator, so that both layering currents and their associated dislocation currents are transformed consistently. The proof is an immediate consequence of the fact that Eq. (83) can be applied to exact forms $\psi = d\alpha$. □

[30] This is the coordinate expression of the map defined in Remark 6.

5.4 Evolution Laws

It is common practice (in constitutive based theories, for instance) to postulate an evolution law as a first order system of PDEs, in our case for the variables $G^i_j(X, t)$. Since we have no other constitutive context, the only natural driving force behind the evolution must be the dislocation form or current itself. Thus, a typical evolution law will read something like

$$G^{-1}\frac{dG}{dt} = g(D), \tag{84}$$

where (for the smooth case) D is the local and present value of the dislocation form. This evolution law is akin to the so-called self-driven evolution laws. Spatial derivatives of D may also be included to simulate evolution laws driven by some measure of 'torsion' or of 'curvature' of the dislocation. Examples can be produced, but at this stage the theory itself is on shaky ground, so we should wait.

References

1. Bunge M A (1998), *Philosophy of Science*, Transaction Publishers, New Brunswick, New Jersey.
2. Cartan É (1899), Sur certaines expressions différentielles et le problème de Pfaff, *Annales Scientifiques de l'École Normale Supŕieure*, Série 3, **16**, 239–332
3. Chandrasekhar S (1992), *Liquid crystals*, 2nd. edition, Cambridge University Press.
4. de Rham G (1955), *Variétés différentiables*, Hermann. English version: *Differentiable manifolds*, Springer, 1984.
5. Elżanowski M and Preston S (2007), A model of the self-driven evolution of a defective continuum, *Mathematics and Mechanics of Solids* **12**/4, 450–465.
6. Epstein M and Segev R (2014), Geometric aspects of singular dislocations, *Mathematics and Mechanics of Solids* **19**/4, 337–349.
7. Epstein M and Segev R (2014), Geometric theory of smooth and singular defects, *International Journal of Nonlinear Mechanics* **66**, 105–110.
8. Epstein M and Segev R (2015), On the geometry and kinematics of smoothly distributed and singular defects, in *Differential Geometry and Continuum Mechanics*, Ed. by G-Q G Chen, M Grinfeld and R J Knops, Springer, 203–234.
9. Federer H (1969), *Geometric Measure Theory*, Springer-Verlag.
10. Frank F C (1951), Crystal dislocations - Elementary concepts and definitions, *Phil. Mag.* **42**, 809–819.
11. Hirth J P (1985), A brief history of dislocation theory, *Metallurgical Transactions A* **16A**, 2085–2090.
12. Kleman M and Lavrentovich O D (2003), *Soft matter physics: an introduction*, Springer-Verlag, New York.
13. Kupferman R and Olami E (2020), Homogenization of edge-dislocations as a weak limit of de-Rham currents, In: Segev R and Epstein M (eds), *Geometric Continuum Mechanics*. Advances in Mechanics and Mathematics, Springer, New York. https://doi.org/10.1007/978-3-030-42683-5_6.
14. Misner W, Thorne K S and Wheeler J A (1973), *Gravitation*, Freeman, San Francisco.

15. Noll W (1967), Materially Uniform Bodies with Inhomogeneities, *Archive for Rational Mechanics and Analysis* **27**, 1–32.
16. Nye J F (1953), Some geometrical relations in dislocated crystals, *Acta Metallurgica* **1**, 153–162.
17. Palais R S (1970), Banach manifolds of fiber bundle sections, *Actes du Congrès International des Mathématiciens*, **2**, 243–249.
18. Read W T and Shockley W (1950), Dislocation models of crystal grain boundaries, *Physical Review* **78**/3, 275–289.
19. Romanov A E and Kolesnikova A L (2009), Application of disclination concept to solid structures, *Progress in Material Science* **54**, 740–769.
20. Schwartz L (1945), Généralisation de la notion de fonction, de dérivation, de transformation de Fourier et applications mathématiques et physiques, *Annales de l'Université de Grenoble* **21**, 57–74.
21. Sternberg S (1983), *Lectures on Differential Geometry*, 2nd edition, Chelsea Publishing, New York.
22. Truesdell C and Toupin R A (1960), *The Classical Field Theories*, in *Encyclopedia of Physics*, ed. by S. Fl'ugge, Springer-Verlag, Berlin.
23. Volterra V (1907), Sur l'équilibre des corps élastiques multiplement connexes, *Annales scientifiques de l'École Normale Supérieure* **24**, 401–517.
24. Volterra V (1912), *Lectures delivered at the twentieth anniversary of the foundation of Clark University*, published by Clark University.
25. Wang C-C (1967), On the Geometric Structure of Simple Bodies, a Mathematical Foundation for the Theory of Continuous Distributions of Dislocations, *Archive for Rational Mechanics and Analysis* **27**, 33–94.

Homogenization of Edge-Dislocations as a Weak Limit of de-Rham Currents

Raz Kupferman and Elihu Olami

Contents

Abstract We consider the geometric homogenization of edge-dislocations as their number tends to infinity. The material structure is represented by 1-forms and their singular counterparts, de-Rham currents. Isolated dislocations are represented by closed 1-forms with singularities concentrated on submanifolds of co-dimension one (the defect locus), whereas a continuous distribution of dislocations is represented by smooth, non-closed 1-forms. We prove that every smooth distribution of dislocations is a limit, in the sense of weak convergence of currents, of increasingly dense and properly scaled isolated edge-dislocations. We also define a notion of singular torsion current (associated with isolated dislocations), and prove that the torsion currents converge, in the homogenization limit, to the smooth torsion field which is the continuum measure of the dislocation density.

1 Introduction

Models of Dislocations The study of material defects, and notably dislocations, is a central theme in material science. The modeling of solid bodies, with or without defects, often follows a paradigm in which the elemental object is that of a **body**

R. Kupferman (✉) · E. Olami
Institute of Mathematics, The Hebrew University, Jerusalem, Israel
e-mail: raz@math.huji.ac.il

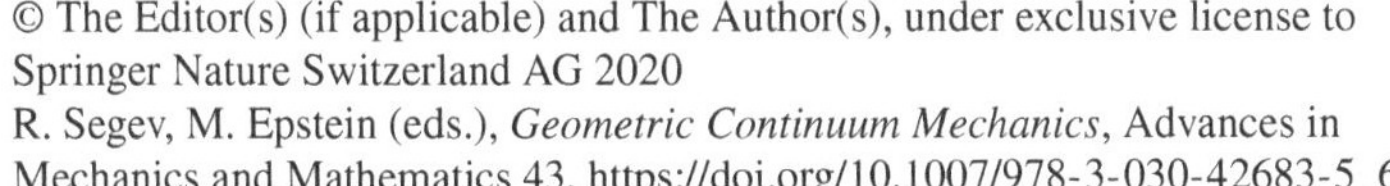

R. Segev, M. Epstein (eds.), *Geometric Continuum Mechanics*, Advances in Mechanics and Mathematics 43, https://doi.org/10.1007/978-3-030-42683-5_6

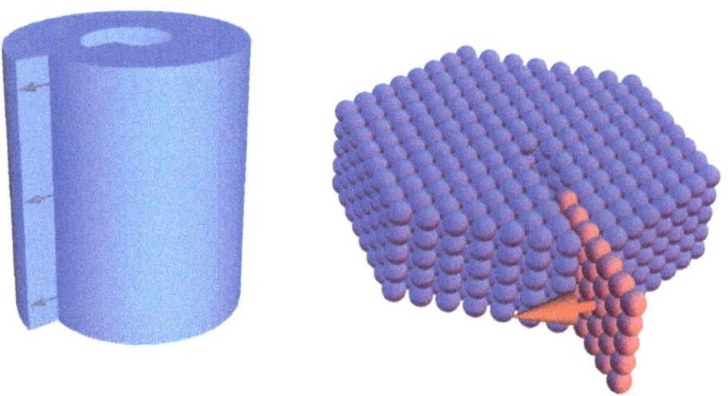

Fig. 1 Left: An edge-dislocation generated by a cut-and-weld protocol in a continuum setting. Right: An edge-dislocation generated by removing a half-plane in a lattice

manifold: solid bodies are modeled as geometric objects—manifolds—and their internal structure is represented by additional structures such as a frame field, a metric or an affine connection. The mechanical properties of the body enter through a **constitutive relation**, whose structure is correlated with the geometric structure of the body.

There have been two distinct approaches to the modeling of body manifolds with dislocations:

1. **Isolated dislocations**: One starts with a defect-free body, which is either modeled as a subset of Euclidean space or as a perfect lattice.[1] Defects are introduced by Volterra cut-and-weld protocols [1] (see Fig. 1) resulting in a locally flat manifold with singularities. The singularities are identified as the defect loci and the presence of dislocations is detected by measuring a non-trivial circulation, known as the **Burgers vector**, along closed paths encircling the defect loci.
2. **Distributed dislocations**: The body is modeled as a smooth manifold endowed with a flat (curvature-free) affine connection. The density of the dislocations is identified with the torsion tensor of the affine connection [2–5]. If, in addition, one adds a basis of the tangent space at one point, then the affine connection induces a smooth frame field, which is the kinematic model, for example, in [6]. In later literature [7], the continuum model is that of a Weitzenböck manifold, which is a smooth manifold endowed with a Riemannian metric and a metrically consistent, curvature-free affine connection. Note that a frame field induces an intrinsic metric and a material connection, so that all three descriptions are essentially identical.

[1]A perfect lattice may be related to a smooth Euclidean structure by assigning lengths and angles to inter-particle bonds and letting the lattice size tend to infinity with the inter-particle bonds scaled appropriately.

Homogenization A longstanding problem has been to rigorously justify, in the spirit of homogenization theories, the continuum model of distributed dislocations as a dense limit of properly scaled isolated dislocations. In particular, one would like to understand how torsion, which is the continuum measure of the dislocation density, emerges in the homogenization limit.

In order to obtain a body manifold endowed with a smooth geometric structure as a limit of body manifolds endowed with localized defects, we must first cast these two seemingly distinct models into the same framework. One possible approach is to "remove" small neighborhoods of the isolated dislocations. Thus, bodies with isolated and smoothly distributed dislocations are both modeled as smooth Weitzenböck manifolds, where in the former case, the bodies are multiply connected; see [8, 9] for a homogenization of defects using this approach.

Another possible approach is to account for the localized defects using singular geometric fields, which is the approach used in this work. As described above, the internal structure of a d-dimensional body $\mathcal{M}$ can be modeled by a frame field, $\{e_i\}_{i=1}^d$, or equivalently, by its dual coframe, which is a set of d 1-forms, $\{\vartheta^i\}_{i=1}^d$. Every smooth 1-form $\omega \in \Omega^1(\mathcal{M})$ induces a distribution for the tangent bundle

$$\ker(\omega) \subset T\mathcal{M}, \qquad \ker(\omega)_p = \ker(\omega_p) \leq T_pM.$$

Under certain integrability conditions [10, Chap. 19], $\ker(\omega)$ induces a foliation (or layering) of $\mathcal{M}$ as a union of *Bravais hypersurfaces*, which are tangent to $\ker(\omega)$ at every point. These surfaces represent the infinitesimal atomic/molecular layers composing the body. Henceforth, we call a 1-form inducing a foliation a **layering form**.

Bodies with localized defects are modeled using singular layering forms, which are represented by the distributional counterpart of differential forms—de-Rham **currents**. As pointed out by Epstein and Segev [11], even a single layering form may detect the presence of defects. Following [11], we define:

A body with dislocations is a d-dimensional manifold $\mathcal{M}$ endowed with a possibly singular layering form ω on $\mathcal{M}$, viewed as a de-Rham $(d-1)$-current,

$$T_\omega : \Omega_c^{d-1}(\mathcal{M}) \to \mathbb{R}, \qquad T_\omega(\eta) = \int_{\mathcal{M}} \omega \wedge \eta, \tag{1}$$

where $\Omega_c^k(\mathcal{M})$ is the space of smooth, compactly supported k-forms on $\mathcal{M}$.

The defect density associated with ω is represented by the **boundary current** ∂T_ω, which is defined in the next section.

A layering form ω models a density of Bravais surfaces. Given a vector $v_p \in T_p\mathcal{M}$, $\omega(v_p)$ is interpreted as the signed number of Bravais planes intersecting v_p. For a closed curve $C \subset \mathcal{M}$, the **Burgers scalar**

$$\oint_C \omega$$

is interpreted as the signed number of Bravais hyperplanes intersecting C. In particular if the Burgers scalar along C is non-vanishing, there is a discrepancy in the layering structure, that is, a defect. By Stokes' theorem, the defect density may be identified with the exterior derivative $d\omega$.

Since a defect-free structure is represented by a closed layering form, isolated dislocations are represented by layering forms ω that are closed everywhere except in a set $\Gamma \subset \mathcal{M}$, which we identify as the locus of the dislocations. Moreover, the existence of non-trivial Burgers scalars around Γ implies that ω must be singular at Γ.

To conclude, both isolated and smoothly distributed dislocations are represented by de-Rham currents; in the smooth case, the currents are induced by smooth layering forms and in the isolated case, by closed forms with singularities. We may now state our main homogenization theorem (see Theorem 6.1 below) in terms of convergence of currents:

> Let $\mathcal{M}$ be a compact, orientable two-dimensional surface, possibly with boundary. Let $\omega \in \Omega^1(\mathcal{M})$ be a (generally non-closed) layering form on $\mathcal{M}$. Then, there exist sequences ω_n and Γ_n such that
>
> 1. Γ_n is a finite disjoint union of segments in $\mathcal{M}$ and is bounded away from $\partial\mathcal{M}$.
> 2. ω_n are closed C^1-bounded layering forms on $\mathcal{M} \setminus \Gamma_n$.
> 3. ω_n converge to ω in the sense of currents. That is, $T_{\omega_n} \to T_\omega$ as $n \to \infty$.

We prove this homogenization theorem in three main steps:

Step I: A Single Dislocation Given a (generally non-closed) layering form β on the unit square $\mathcal{M} = [0, 1]^2$, we construct in Sect. 3 a closed layering form ν on $\mathcal{M} \setminus \Gamma$, where Γ is a segment. The layering form ν has the same circulation around $\partial\mathcal{M}$ as β. The layering form ν induces a 1-current T_ν on $\mathcal{M}$; its boundary is a 0-current supported on Γ. Thus, we may view the layering form ν as representing a singular edge-dislocation, whose locus is Γ, and whose intensity is equal to the integrated intensity of the layering form β.

Step II: Homogenization for the Square In Sect. 4, we prove that every (possibly non-closed) layering form $\beta \in \Omega^1(\mathcal{M})$ can be approximated by a sequence of closed layering forms ν_n, representing an n-by-n array of edge-dislocations ($\mathcal{M}$ is still the unit square). We construct ν_n by gluing together properly rescaled versions of the form ν constructed in Sect. 3. We then prove that T_{ν_n} converges as $n \to \infty$ to the 1-current T_β.

Step III: The General Case In Sect. 6, we prove a homogenization theorem for a general compact and orientable surface $\mathcal{M}$. We show that for every layering form $\beta \in \Omega^1(\mathcal{M})$, there exists a sequence ν_n of closed layering forms supported everywhere except for a lower-dimensional submanifold, such that T_{ν_n} converges to T_β. The proof relies on a classical classification theorem for two-dimensional manifolds, along with gluing techniques for 1-forms (presented in the appendix). The homogenization problem is thus reduced to two elemental building blocks: the closed disk and a "pair of pants" for which homogenization follows from the homogenization theorem for the square.

Singular Torsion In Sect. 5, we generalize the analysis to the case where $\mathcal{M}$ is a d dimensional manifold equipped with a full lattice structure, that is, a frame field $\{e_i\}_{i=1}^d$ or equivalently, the dual coframe of d layering forms $\{\vartheta^i\}_{i=1}^d$. A frame–coframe pair induces a path-independent parallel transport $\Pi_p^q : T_p\mathcal{M} \to T_q\mathcal{M}$ between every two points $p, q \in \mathcal{M}$. The corresponding material connection ∇ is flat but may be non-symmetric; the torsion tensor is given by

$$\tau = e_i \otimes d\vartheta^i,$$

and it is non-zero if the layering forms ϑ^i are not closed. In the case of isolated dislocations, the torsion is identically zero in the smooth set and not defined on the singular set. Note that the above expression for τ cannot be interpreted as a de-Rham current on $\mathcal{M}$ (it behaves like a product of a Heaviside function and a delta function).

Using the distant parallelism induced by the frame field (defined also for isolated dislocations) we define a notion of singular torsion for a singular frame as a vector valued de-Rham current. We show how the singular torsion generalizes the notion of a smooth torsion field and proves a homogenization theorem for the torsion tensor; if a sequence of coframes $\{\vartheta_n^i\}$ converges in the sense of currents to a smooth coframe $\{\vartheta^i\}$, then the corresponding singular torsions converge to the smooth torsion associated with the limit.

There are several differences between the present work and the earlier work in [8, 9, 12, 13]: In the earlier work, the loci of the dislocations were "removed,"yielding a geometric convergence of smooth multiply connected manifolds to a smooth simply connected limit. Furthermore, the mode of convergence was a strong L^p-convergence of frame fields, which is stronger than the weak convergence of currents; a stronger convergence is particularly important for obtaining a convergence of the associated mechanical models. On the other hand, the current approach is more physical, as it accounts explicitly for the singular region; also, our notion of singular dislocations chimes in with the classical case of cone singularities, i.e., disclinations. Finally, the emergence of torsion in the continuous case no longer occurs "out of the blue," but is shown to be a bona fide limit of singular torsion fields.

Three points should be emphasized: (1) This work focuses on the geometry of bodies with dislocations. There exists a wealth of literature addressing the mechanics of dislocations, which we do not mention here. (2) A body manifold is our elemental object of consideration, and it should not be confused with a (deformed) configuration, which is an embedding of that manifold in the ambient space. Since the body manifold and the deformed configuration are diffeomorphic, the same defect structure would be observed in the deformed configuration. (3) In our model, the locus of a dislocation is a submanifold of co-dimension one, whereas it is often described in the literature as a submanifold of co-dimension two, e.g., a point in $2d$. Geometrically, a dislocation is a curvature dipole, or a pair of disclinations of opposite signs (e.g., a 5–7 pair in a hexagonal lattice). Since the Frank vector of a positive disclination is bounded by 2π, one cannot obtain a non-zero point dislocation as a limit of disclination dipoles, as in the case of electrostatics.

This paper is organized as follows. In Sect. 2 we review the definition of de-Rham currents on manifolds, which are the kinematic variables of our model. Section 3 is devoted to the first step of our homogenization proof—the construction of a layering form representing a single dislocation. The second step—the homogenization construction for the square—is conducted in Sect. 4. We then consider in Sect. 5 the notion of singular torsion and its homogenization. Finally, we extend in Sect. 6 the homogenization proof to general compact orientable surfaces.

2 De-Rham Currents

We start by reviewing the definition of de-Rham currents on manifolds. For a full introduction, see the classical monographs of Federer [14] or de-Rham [15]; see [16, 17] for more recent reviews.

Let $\mathcal{M}$ be a smooth, compact, orientable d-dimensional manifold with boundary. For every $1 \leq k \leq d$, let $\Omega^k(\mathcal{M})$ denote the space of smooth k-forms on $\mathcal{M}$ and let

$$\Omega_c^k(\mathcal{M}) = \left\{\omega \in \Omega^k(\mathcal{M}) \ : \ \operatorname{supp}(\omega) \Subset \mathcal{M}\right\}$$

denote the $C^\infty(\mathcal{M})$-module of smooth k-forms compactly supported in $\mathcal{M}$. Choose a Riemannian metric g on $\mathcal{M}$, and define for every compact $K \Subset \mathcal{M}$ a family of seminorms $\phi^k_{K,j} : \Omega^k_c(\mathcal{M}) \to \mathbb{R}^+$ by

$$\phi^k_{K,j}(\omega) = \sup_{0 \leq i \leq j} \|D^i \omega\|_K,$$

where $D^i\omega : \mathcal{M} \to \operatorname{Hom}(\otimes^i T\mathcal{M}, \Lambda^k T^*\mathcal{M})$ is the i-th differential of ω (not to be confused with the exterior derivative), and

$$\|D^i \omega\|_K = \sup_{p \in K} \|(D^i\omega)_p\|,$$

where $\|\cdot\|$ is the norm on $\operatorname{Hom}(\otimes^i T\mathcal{M}, \Lambda^k T^*\mathcal{M})$ induced by the metric g. Since $\mathcal{M}$ is compact, a different choice of g gives equivalent seminorms. As a result, it makes sense to say that a k-form is C^j-bounded without reference to any particular metric (recall that in a topological vector space a set is bounded if every open neighborhood of zero can be inflated to include that set).

The seminorms $\{\phi^k_{K,j}\}_{j=1}^\infty$ turn

$$\Omega^k_K(\mathcal{M}) = \{\omega \in \Omega^k_c(\mathcal{M}) \ : \ \operatorname{supp}(\omega) \subset K\}$$

into a Fréchet space, that is, a locally convex topological vector space which is complete with respect to a translationally invariant metric [18, p. 9]. Endow $\Omega_c^k(\mathcal{M})$ with the finest topology for which the inclusion maps

$$\Omega_K^k(\mathcal{M}) \hookrightarrow \Omega_c^k(\mathcal{M})$$

are continuous for all compact $K \Subset \mathcal{M}$. A sequence $\omega_n \in \Omega_c^k(\mathcal{M})$ converges in this topology to 0 if and only if there exists a compact set $K \Subset \mathcal{M}$ such that $\operatorname{supp}(\omega_n) \subset K$ for all n large enough, and $\omega_n \to 0$ in the $\Omega_K^k(\mathcal{M})$ topology.

Definition 2.1 (de-Rham Current) A de-Rham k-current is a continuous linear functional on $\Omega_c^k(\mathcal{M})$. The vector space of de-Rham k-currents is denoted by $\mathscr{D}_k(\mathcal{M})$.

A linear functional $T : \Omega_c^k(\mathcal{M}) \to \mathbb{R}$ is a k-current if and only if there exists for every $K \Subset \mathcal{M}$ an $N_K \in \mathbb{N}$ and a constant $C_K > 0$, such that for every $\omega \in \Omega_K^k(\mathcal{M})$,

$$|T(\omega)| \leq C_K\, \phi_{K,N_K}^k(\omega).$$

(See, e.g., [18, Th. 6.8] in the context of distributions in $\mathbb{R}^d$.) We endow $\mathscr{D}_k(\mathcal{M})$ with the weak-star topology: a sequence of k-currents T_n converges to a k-current T if

$$\lim_{n\to\infty} T_n(\omega) = T(\omega)$$

for every $\omega \in \Omega_c^k(\mathcal{M})$. The **support** of a k-current $T \in \mathscr{D}_k(\mathcal{M})$ is defined by $\operatorname{supp}(T) = \mathcal{M} \setminus A(T)$, where $A(T)$ is the annihilation set of T, i.e., the union of all open subsets $U \subset \mathcal{M}$ for which $T(\alpha) = 0$ whenever $\operatorname{supp}(\alpha) \subset U$.

Example 2.1 Every locally integrable k-form β on Ω defines a $(d-k)$-current $T_\beta \in \mathscr{D}_{d-k}(\mathcal{M})$ by

$$T_\beta(\alpha) = \int_{\mathcal{M}} \beta \wedge \alpha, \qquad \alpha \in \Omega_c^{d-k}(\mathcal{M}).$$

In other words, currents may be viewed as generalized differential forms.

Example 2.2 Let $S \subset \mathcal{M}$ be a k-dimensional oriented submanifold. Then, S induces a k-current $[S] \in \mathscr{D}_k(\mathcal{M})$ given by

$$[S](\alpha) = \int_S \alpha, \qquad \alpha \in \Omega_c^k(\mathcal{M}).$$

In other words, currents also generalize the concept of a submanifold.

Definition 2.2 (Boundary of Current) The **boundary operator** of a k-current is a map $\partial : \mathscr{D}_k(\mathcal{M}) \to \mathscr{D}_{k-1}(\mathcal{M})$, defined by

$$\partial T(\alpha) = T(d\alpha), \qquad \alpha \in \Omega_c^{k-1}(\mathcal{M}).$$

Since $d^2 = 0$, it immediately follows by duality that $\partial^2 = 0$. Moreover, it follows from integration by parts and Stokes' theorem that

$$\partial T_\beta = (-1)^{k-1} T_{d\beta}$$

for every smooth k-form β.

3 Layering Form for an Edge-Dislocation

Let V be a vector space. A covector $\omega \in V^*$ induces a family of hyperplanes (Bravais planes),

$$H_t = \{v \in V \,:\, \omega(v) = t\}, \qquad t \in \mathbb{R}$$

foliating V (i.e., forming a disjoint cover of V). The action of ω on a vector $v \in V$ can be interpreted as the "number of hyperplanes intersected by v." In a smooth manifold $\mathcal{M}$, the role of the covector is played by a 1-form foliating $\mathcal{M}$: given a 1-form ν and an oriented curve $C \subset \mathcal{M}$, the integral

$$\int_C \nu$$

can be interpreted as the (signed) "number" of ν-hyperplanes intersected by C.

Definition 3.1 (Layering Form) Let $\mathcal{M}$ be a smooth manifold. A 1-form $\nu \in \Omega^1(\mathcal{M})$ is called a layering form if it foliates $\mathcal{M}$. That is, if $\mathcal{M}$ is the disjoint union of smooth hypersurfaces—leaves—such that the tangent bundle of each leaf coincides with the kernel of ν.

A sufficient and necessary condition for a 1-form ν to induce a smooth layering structure is that locally

$$d\nu = \alpha \wedge \nu$$

for some $(d-1)$-form α [10, Chap. 19]. In particular, for a simply connected two-dimensional manifold, every non-vanishing 1-form induces a smooth layering structure.

If ν is a closed layering form, $d\nu = 0$, it follows from Stokes' theorem that for every simple, oriented, closed curve $C \subset \mathcal{M}$, the "number" of hyperplanes intersected by C vanishes,

$$\int_C \nu = \int_{\Sigma_C} d\nu = 0, \tag{2}$$

where $\Sigma_C \subset \mathcal{M}$ is any two-dimensional submanifold of $\mathcal{M}$ bounded by C. In other words, there are no "extra" layers, and the layering structure is defect-free. In view of (2), we may interpret $d\nu$ as a **defect density** associated with the layering form ν.

Definition 3.2 (Continuously Distributed Dislocations) Let $\mathcal{M}$ be a smooth simply connected manifold. A smooth layering form is said to represent a continuous distribution of dislocations if there exists a closed curve C, such that

$$\int_C \nu \neq 0.$$

The quantity on the left-hand side is called the **Burgers scalar**, or the circulation of the layering form ν around the loop C. We consider Burgers scalars, rather than Burgers vectors, since we account for only one layering form. When representing the structure by a coframe, one obtains d Burgers scalars, which are the components in the local frame of the Burgers vector.

Clearly, ν represents a continuous distribution of dislocations if and only if it is non-closed.

Definition 3.3 (Singular Dislocation) Let $\mathcal{M}$ be a smooth manifold and let $\Gamma \subset \mathcal{M}$ be a hypersurface. A layering form ν on $\mathcal{M} \setminus \Gamma$ is said to represent a dislocation concentrated on Γ, if ν is closed and there exists a closed curve $C \in \mathcal{M} \setminus \Gamma$, such that

$$\int_C \nu \neq 0.$$

Suppose that $\nu \in \Omega^1(\mathcal{M} \setminus \Gamma)$ represents a dislocation concentrated on Γ. Since ν is closed, its Burgers scalar vanishes for every contractible loop. Therefore, $\mathcal{M} \setminus \Gamma$ is necessarily not simply connected, i.e., the removal of the dislocation locus Γ changes the topology of the manifold. Let C be a loop in $\mathcal{M}$ encircling Γ (Fig. 2), such that

$$\int_C \nu \neq 0.$$

Since ν is closed on $\mathcal{M} \setminus \Gamma$, the circulation remains unchanged under homotopic variations of C, and in particular, as C shrinks to Γ. Hence, ν is necessarily singular at Γ.

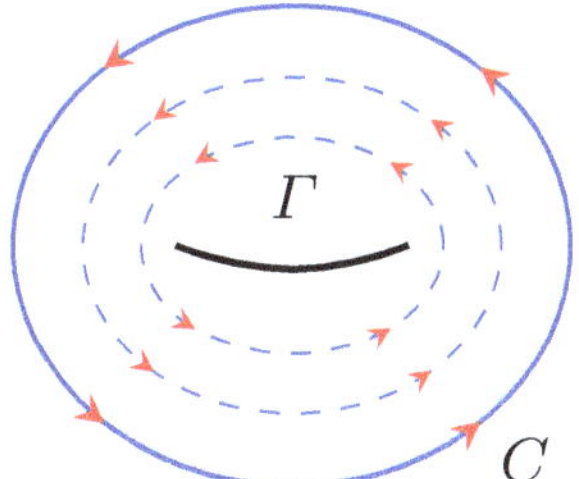

Fig. 2 A body endowed with a layering form ν with a singular dislocation located on a hypersurface Γ. The circulation of ν is homotopic-invariant for loops encircling the locus of the dislocation

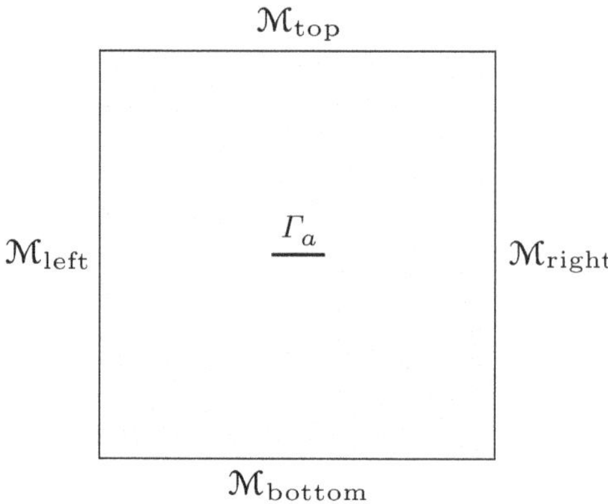

Fig. 3 The topological rectangle $\mathcal{M}$ and the locus Γ_a of the dislocation

We next consider a two-dimensional manifold $\mathcal{M}$ endowed with a non-closed smooth layering form β (representing a continuous distribution of dislocations). We construct a layering form ν representing a singular dislocation concentrated on a curve $\Gamma \subset \mathcal{M}$, which approximates β is a sense made precise. In a sense, this construction concentrates the "defectiveness" of β onto the submanifold Γ. This construction will be used in the next section to prove the homogenization theorem.

Consider then a topological rectangle, i.e., a manifold that can be parametrized as follows:

$$\mathcal{M} = [0, 1]^2 = \{(x, y) \, : \, 0 \le x, y \le 1\}.$$

We denote the left, right, top, and bottom edges of $\mathcal{M}$ by $\mathcal{M}_{\text{left}}$, $\mathcal{M}_{\text{right}}$, $\mathcal{M}_{\text{top}}$, and $\mathcal{M}_{\text{bottom}}$, respectively. The locus of the singular dislocation will be the closed parametric segment

$$\Gamma_a = [1/2 - a/2, 1/2 + a/2] \times \{1/2\} \subset \mathcal{M}, \tag{3}$$

where $0 < a < 1$ is a parameter (see Fig. 3).

Proposition 3.1 *Let $\beta \in \Omega^1(\mathcal{M})$ be a nowhere-vanishing (generally non-closed) layering form. Then, there exists a continuously differentiable layering form ν_a on $\mathcal{M} \setminus \Gamma_a$ satisfying the following properties:*

(a) ν_a is C^1-bounded (see definition in Sect. 2).
(b) ν_a is closed.
(c) ν_a coincides with β on $\mathcal{M}_{left}$ and $\mathcal{M}_{right}$.
(d) ν_a has the same circulation as β around $\partial\mathcal{M}$,

$$\int_{\partial\mathcal{M}} \nu_a = \int_{\partial\mathcal{M}} \beta.$$

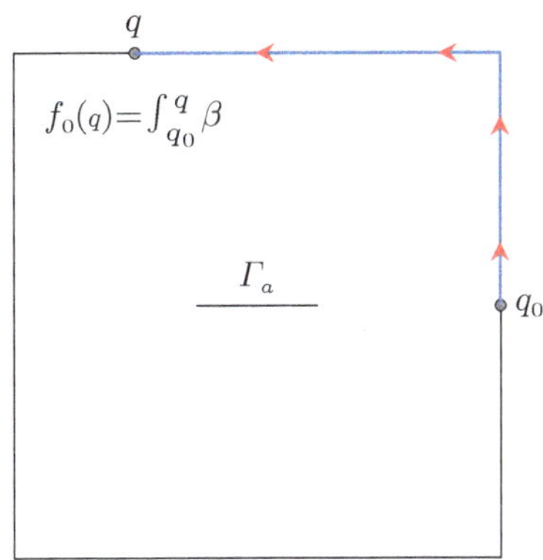

Fig. 4 The first stage in the construction of f: the 1-form β is integrated along $\partial\mathcal{M}$

(e) The horizontal components of v_a *and* β *coincide,*

$$v_a(\partial_x) = \beta(\partial_x),$$

whenever $|x - 1/2| > a/2$.

Proof We construct v_a as the (continuous) differential of a discontinuous function f. First, define $f_0 : \partial\mathcal{M} \to \mathbb{R}$ by fixing $q_0 = (1, 1/2)$ and letting

$$f_0(q) = \int_{q_0}^{q} \beta,$$

where the integration from q_0 to q is counter-clockwise along $\partial\mathcal{M}$. If the circulation of β around $\partial\mathcal{M}$ is non-zero, then f_0 is discontinuous at q_0. However, its differential is well-defined and smooth at q_0 as it coincides with the tangential component of β (see Fig. 4).

Next, consider the vertical strip of width a,

$$\mathcal{M}_a = \{(x, y) \in [0, 1]^2 \, : \, |x - 1/2| < a/2\},$$

and define $\bar{f} : \mathcal{M} \setminus \mathcal{M}_a \to \mathbb{R}$ by integrating β horizontally, from the boundaries inward,

$$\bar{f}(x, y) = \begin{cases} f_0(0, y) + \int_{[(0,y),(x,y)]} \beta, & x < 1/2 - a/2 \\ f_0(1, y) + \int_{[(1,y),(x,y)]} \beta, & x > 1/2 + a/2 \end{cases}$$

(see Fig. 5).

It remains to define f on $\mathcal{M}_a/\Gamma_a$. Denote by $p_L, p_R : \mathcal{M} \to \mathbb{R}$ the second-order Taylor expansions of $\bar{f}$ about $x_L = 1/2 - a/2$ and $x_R = 1/2 + a/2$ along the x-direction, i.e.,

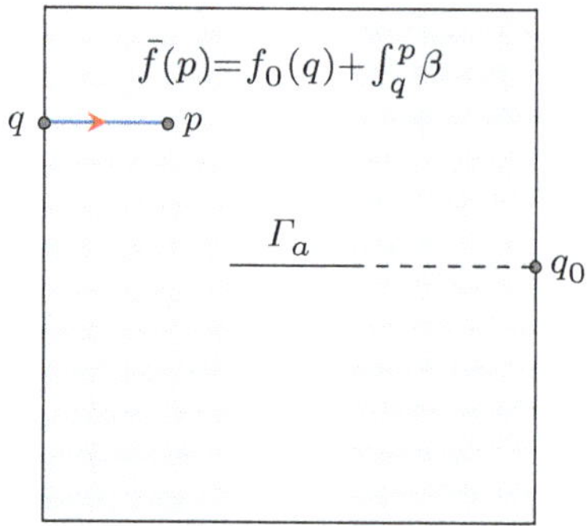

Fig. 5 The second stage in the construction of f: $\bar{f}$ is defined on the set $|x-1/2|>a/2$ by integrating the horizontal component of β from the nearest vertical boundary point. The dashed segment connecting Γ_a to q_0 is the discontinuity line of f

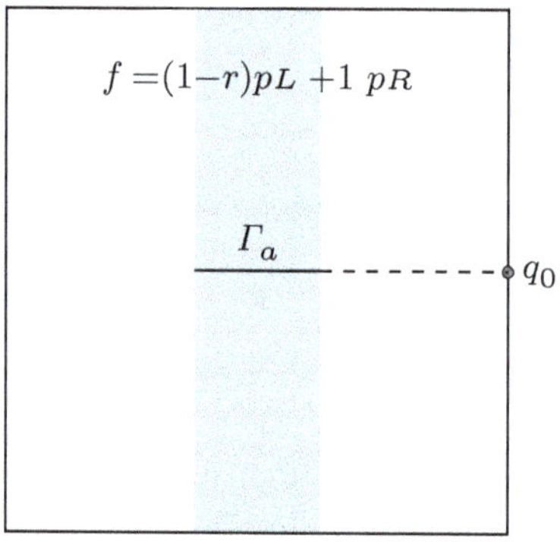

Fig. 6 The third stage in the construction of f: f is extended from $\bar{f}$ to the set $|x-1/2|\leq a/2$ by interpolation

$$p_L(x,y)=\bar{f}(x_L,y)+\frac{\partial \bar{f}}{\partial x}(x_L,y)(x-x_L)+\frac{1}{2}\frac{\partial^2 \bar{f}}{\partial x^2}(x_L,y)(x-x_L)^2$$

$$p_R(x,y)=\bar{f}(x_R,y)+\frac{\partial \bar{f}}{\partial x}(x_R,y)(x-x_R)+\frac{1}{2}\frac{\partial^2 \bar{f}}{\partial x^2}(x_R,y)(x-x_R)^2.$$

Let $r\in C^\infty(\mathbb{R})$ be a monotonically increasing function satisfying,

$$r(t)=0 \quad \forall t\leq -1/2 \qquad \text{and} \qquad r(t)=1 \quad \forall t\geq 1/2.$$

We extend $\bar{f}$ to $\mathcal{M}\setminus\Gamma_a$ by interpolating between p_L and p_R, using the smooth "connecting" function r (see Fig. 6),

$$f(x,y)=\begin{cases}\bar{f}(x,y) & |x-1/2|\geq a/2\\ (1-r(\frac{x-1/2}{a}))p_L(x,y)+r(\frac{x-1/2}{a})p_R(x,y) & |x-1/2|<a/2.\end{cases} \tag{4}$$

We obtain $\nu_a=df$ by differentiating (4). For $x>a/2+1/2$, an explicit calculation yields

$$df_{(x,y)}=\beta_1(x,y)\,dx+\left(\beta_2(1,y)+\int_1^x \frac{\partial \beta_1}{\partial y}(x',y)dx'\right)dy, \tag{5}$$

where β_1 and β_2 are the components of β,

$$\beta = \beta_1\, dx + \beta_2\, dy.$$

Similarly, for $x < 1/2 - a/2$,

$$df_{(x,y)} = \beta_1(x, y)\, dx + \left(\beta_2(0, y) + \int_0^x \frac{\partial \beta_1}{\partial y}(x', y) dx'\right) dy. \tag{6}$$

While f has a discontinuity along the segment $[1/2 + a, 1] \times \{1/2\}$, its one-sided derivatives along this segment are continuous, as they are expressed in terms of the smooth layering form β. Moreover,

$$df|_{\mathcal{M}_{\text{left}}} = \beta|_{\mathcal{M}_{\text{left}}} \qquad \text{and} \qquad df|_{\mathcal{M}_{\text{right}}} = \beta|_{\mathcal{M}_{\text{right}}},$$

proving Property (c). Likewise, for $|x - 1/2| \geq a/2$,

$$df(\partial_x) = \beta_1 = \beta(\partial_x),$$

proving Property (e).

For $(x, y) \in \mathcal{M}_a$,

$$\begin{aligned} df_{(x,y)} &= \frac{1}{a} r'\left(\tfrac{x-1/2}{a}\right)(p_R(x, y) - p_L(x, y))dx \\ &+ \left[\left(1 - r\left(\tfrac{x-1/2}{a}\right)\right)\frac{\partial p_L}{\partial x}(x, y) + r\left(\tfrac{x-1/2}{a}\right)\frac{\partial p_R}{\partial x}(x, y)\right] dx \\ &+ \left[\left(1 - r\left(\tfrac{x-1/2}{a}\right)\right)\frac{\partial p_L}{\partial y}(x, y) + r\left(\tfrac{x-1/2}{a}\right)\frac{\partial p_R}{\partial y}(x, y)\right] dy. \end{aligned} \tag{7}$$

The layering form df is continuous at $x = 1/2 \pm a/2$. For example,

$$\begin{aligned} \lim_{x \nearrow 1/2+a/2} df_{(x,y)} &= \frac{\partial p_R}{\partial x}(1/2 + a/2, y)\, dx + \frac{\partial p_R}{\partial y}(1/2 + a/2, y)\, dy \\ &= \frac{\partial \bar{f}}{\partial x}(1/2 + a/2, y)\, dx + \frac{\partial \bar{f}}{\partial y}(1/2 + a/2, y)\, dy \\ &= d\bar{f}(1/2 + a/2, y). \end{aligned}$$

A second differentiation shows that v_a is continuously differentiable at $x = 1/2 \pm a/2$. This together with (7) proves Property (a) and consequently also Property (b).

It remains to prove Property (d). From our construction of f_0 on $\partial\mathcal{M}$,

$$\begin{aligned}\int_{\partial\mathcal{M}} df &= \lim_{\varepsilon\to 0}\left(f(1,1/2-\varepsilon)-f(1,1/2+\varepsilon)\right)\\ &= \lim_{\varepsilon\to 0}\left(f_0(1,1/2-\varepsilon)-f_0(1,1/2+\varepsilon)\right)\\ &= \int_{\partial\mathcal{M}}\beta,\end{aligned}$$

which concludes the proof. □

Regardless of the particular construction of ν_a, since ν_a is closed in $\mathcal{M}\setminus\Gamma_a$, it follows that

$$\oint_C \nu_a = 0$$

along every contractible loop C in $\mathcal{M}\setminus\Gamma_a$. Let g be a metric on $\mathcal{M}$, and denote by Γ_a^ε, $\varepsilon>0$, a family of ε-tubular neighborhoods of Γ_a. By Stokes' law, for every small enough $\varepsilon>0$,

$$0=\int_{\mathcal{M}\setminus\Gamma_a^\varepsilon} d\nu_a = \int_{\partial\mathcal{M}}\nu_a - \int_{\partial\Gamma_a^\varepsilon}\nu_a.$$

Since ν_a has the same circulation as β along $\partial\mathcal{M}$,

$$\int_{\partial\Gamma_a^\varepsilon}\nu_a = \int_{\partial\mathcal{M}}\beta.$$

Letting $\varepsilon\to 0$, we obtain

$$\int_{\Gamma_a}[\nu_a] = \int_{\partial\mathcal{M}}\beta, \tag{8}$$

where $[\nu_a]$ is the discontinuity jump of ν_a along Γ_a, whose sign is determined by the orientation of $\mathcal{M}$ (hence of Γ_a^ε) and Γ_a. Note that the one-sided limits of ν_a at Γ_a exist since ν_a is C^1-bounded. Moreover, since $\mathcal{M}$ is compact, the limit leading to Identity (8) does not depend on the choice of the metric g. We conclude that if β (hence, ν_a) has non-vanishing circulation along $\partial\mathcal{M}$, then $[\nu_a]\neq 0$, that is, ν_a is discontinuous along Γ_a.

Remark 3.1 The singular set Γ_a of ν_a is uncountable. Generally, if $\mathcal{M}$ is a compact two-dimensional manifold with or without boundary, Γ is a submanifold of $\mathcal{M}$, and ν is a C^0-bounded closed 1-form on $\mathcal{M}\setminus\Gamma$, such that there exists a closed curve C for which

$$\oint_C \nu \neq 0,$$

then Γ cannot be a finite set. Suppose, by contradiction that $\Gamma = \{p_1, p_2, \ldots, p_k\}$ is finite, and assume without loss of generality that all the points in Γ are enclosed by the curve C. Assuming as above a metric g, setting $\Gamma^\varepsilon = \cup_i B_\varepsilon(p_i)$, and performing the same calculation,

$$\sum_{i=1}^{k} \oint_{\partial B_\varepsilon(p_i)} \nu = -\oint_C \nu.$$

If ν is bounded, then the left-hand side vanishes as $\varepsilon \to 0$, yielding a contradiction. The physical interpretation of this observation is that in our setting *there is no such thing as an edge-dislocation supported at a point* (or on a line in three dimensions).

The 1-form ν_a (which is only defined on $\mathcal{M} \setminus \Gamma_a$) induces a 1-current on $\mathcal{M}$,

$$T_{\nu_a}(\alpha) = \int_{\mathcal{M}} \nu_a \wedge \alpha \qquad \alpha \in \Omega^1_c(\mathcal{M}).$$

Its boundary is the 0-current,

$$\partial T_{\nu_a}(f) = T_{\nu_a}(df) = \int_{\mathcal{M}} \nu_a \wedge df \qquad f \in C^\infty_c(\mathcal{M}).$$

Integrating by parts on $\mathcal{M} \setminus \Gamma_a^\epsilon$ and taking $\epsilon \to 0$ (as above), we obtain

$$\partial T_{\nu_a}(f) = \int_{\Gamma_a} f[\nu_a],$$

where for $|x - 1/2| < a/2$,

$$\begin{aligned}
(x) &= \lim_{\varepsilon \to 0} (df(x, 1/2 + \varepsilon) - df(x, 1/2 - \varepsilon)) \\
&= \frac{1}{a} r'\left(\tfrac{x-1/2}{a}\right) \lim_{\varepsilon \to 0} (p_R(x, 1/2 + \varepsilon) - p_R(x, 1/2 - \varepsilon)) \\
&= \frac{1}{a} r'\left(\frac{x - 1/2}{a}\right) \int_{\partial \mathcal{M}} \beta,
\end{aligned}$$

where we substituted (7), used the facts that p_L is continuous at $y = 1/2$ and that the discontinuity of p_R at $y = 1/2$ equals the circulation of β.

To conclude, ν_a represents a layering form on $\mathcal{M}$ having an edge-dislocation concentrated on the hypersurface Γ_a. The locus of the dislocation is revealed by the boundary of the differential current induced by ν_a. Note that $\mathcal{M} \setminus \Gamma_a$ is defect-free only to the extent detectable by ν_a. Generally, $\mathcal{M} \setminus \Gamma_a$ may contain defects detected by other layering forms.

4 Homogenization of Distributed Edge-Dislocations

In this section we show how a non-closed layering from (representing continuously distributed dislocations) can be approximated, in the sense of currents, by an n-by-n array of singular edge-dislocations, each of magnitude of order $1/n^2$. We construct the approximation by "gluing" properly rescaled copies of the layering form ν_a constructed in Proposition 3.1.

For $(x_0, y_0) \in \mathbb{R}^2$, denote by $\tau_{(x_0,y_0)} : \mathbb{R}^2 \to \mathbb{R}^2$ the translation operator

$$\tau_{(x_0,y_0)}(x, y) = (x + x_0, y + y_0).$$

Likewise, for $\lambda > 0$, denote by $S_\lambda : \mathbb{R}^2 \to \mathbb{R}^2$ the scaling operator

$$S_\lambda(x, y) = (\lambda x, \lambda y).$$

Let $n \in \mathbb{N}$ be given; for every $0 \le k, j < n$, let

$$\mathcal{M}_{n;kj} = S_{1/n} \circ \tau_{(k,j)}(\mathcal{M})$$

be translated and rescaled copies of $\mathcal{M}$, forming an n-by-n tiling of $\mathcal{M}$. By construction,

$$\iota_{n;kj} = S_{1/n} \circ \tau_{(k,j)} : \mathcal{M} \to \mathcal{M}_{n;kj} \tag{9}$$

are diffeomorphisms (see Fig. 7). Similarly, let

$$\Gamma_{n;kj} = \iota_{n;kj}(\Gamma_{a/n})$$

be segments of lengths a/n^2 located at the centers of each square. Finally, denote by

$$\Gamma_n = \bigcup_{k,j=0}^{n-1} \Gamma_{n;kj}$$

the union of those segments.

Let $\beta \in \Omega^1(\mathcal{M})$ be a layering form. Let

$$\beta_{n;kj} = (\iota_{n;kj})^\star \beta|_{\mathcal{M}_{n;kj}} \in \Omega^1(\mathcal{M}), \tag{10}$$

be the pullback[2] of β (restricted to $\mathcal{M}_{n;kj}$) to $\mathcal{M}$ and let $\mu_{n;kj} \in \Omega^1(\mathcal{M} \setminus \Gamma_{a/n})$ be the singular layering form defined in Proposition 3.1, with $\beta_{n;kj}$ playing the role of

[2]For a smooth map $f : \mathcal{M} \to \mathcal{N}$ between two manifolds and a k-form $\beta \in \Omega^k(\mathcal{N})$, we denote by $f^*\beta \in \Omega^k(\mathcal{M})$ its pullback,

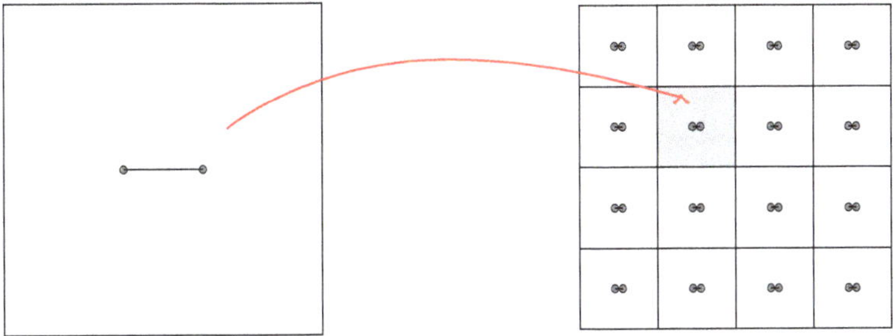

Fig. 7 The diffeomorphism $\iota_{n;kj}$ for $n = 4$, $k = 1$, and $j = 2$

β and the parameter a is scaled by a factor of $1/n$. We approximate β by a sequence of singular layering forms,

$$\nu_n \in \Omega^1(\mathcal{M} \setminus \Gamma_n),$$

by pushing forward $\mu_{n;kj}$ into $\mathcal{M}_{n;kj}$,

$$\nu_n|_{\mathcal{M}_{n;kj}} = (\iota_{n;kj})_\star \mu_{n;kj}. \tag{11}$$

Proposition 4.1 *Equation* (11) *for* $0 \le k, j < n$ *defines a layering form* ν_n *on* $\mathcal{M} \setminus \Gamma_n$, *satisfying*

(a) ν_n *is* C^1*-bounded.*
(b) ν_n *is closed.*
(c) ν_n *has the same circulation as* β *in each sub-domain: for every* $0 \le k, j \le n-1$,

$$\int_{\partial\mathcal{M}_{n;kj}} \nu_n = \int_{\partial\mathcal{M}_{n;kj}} \beta.$$

(d) ν_n *coincides with* β *on the vertical segments* $L_k = \{k/n\} \times [0, 1]$ *for* $0 \le k \le n$.

Proof We first show that ν_n is well-defined and satisfies Property (a). Since the $\mu_{n;kj}$ are smooth and C^1-bounded, ν_n is smooth and C^1-bounded in the interior of each $\mathcal{M}_{n;kj} \setminus \Gamma_{n;kj}$. It remains to prove that it is continuously differentiable on the "skeleton" $\cup_{k,j} \partial\mathcal{M}_{n;kj}$. Note that

$$\partial\mathcal{M}_{n;kj} = \iota_{n;kj}(\mathcal{M}_{\text{left}}) \cup \iota_{n;kj}(\mathcal{M}_{\text{right}}) \cup \iota_{n;kj}(\mathcal{M}_{\text{top}}) \cup \iota_{n;kj}(\mathcal{M}_{\text{bottom}}).$$

$$(f^*\beta)_p(v_1, \ldots, v_k) = \beta_{f(p)}(df_p(v_1), \ldots, df_p(v_k)).$$

If f is a diffeomorphism, then k-forms can also be pushed forward.

By (10), since the diffeomorphism $\iota_{n;kj}$ is a combination of a translation and a scaling,

$$\beta_{n;kj}(\partial_x) = \frac{1}{n}\beta(\partial_x) \circ \iota_{n;kj} \qquad \text{and} \qquad \beta_{n;kj}(\partial_y) = \frac{1}{n}\beta(\partial_y) \circ \iota_{n;kj},$$

which are equalities between functions on $\mathcal{M}$. In particular, since $\iota_{n;k\,j+1}(x, 0) = \iota_{n;kj}(x, 1)$ and $\iota_{n;k\,j-1}(x, 1) = \iota_{n;kj}(x, 0)$, it follows that for every $x, y \in [0, 1]$, and $v \in \{\partial_x, \partial_y\}$

$$\beta_{n;k\,j+1}(v)(x, 0) = \beta_{n;kj}(v)(x, 1)$$
$$\beta_{n;k+1\,j}(v)(0, y) = \beta_{n;kj}(v)(1, y).$$

By the same argument, for $w \in \{\partial_x, \partial_y\}$

$$\mathcal{L}_w\beta_{n;k\,j+1}(v)(x, 0) = \mathcal{L}_w\beta_{n;kj}(v)(x, 1)$$
$$\mathcal{L}_w\beta_{n;k+1\,j}(v)(0, y) = \mathcal{L}_w\beta_{n;kj}(v)(1, y),$$

where $\mathcal{L}_w$ is the Lie derivative along w. By (5), (6), and (7), the construction of $\mu_{n;kj}$ only depends on $\beta_{n;kj}$ (and the smooth function r). Moreover, $\mu_{n;kj}$ and its derivative on every side of $\partial\mathcal{M}$ depend only on $\beta_{n;kj}$ and its derivatives on that side. As a result, for every $x, y \in [0, 1]$, and $v, w = \{\partial_x, \partial_y\}$,

$$\mu_{n;k\,j+1}(v)(x, 0) = \mu_{n;kj}(v)(x, 1)$$
$$\mu_{n;k+1\,j}(v)(0, y) = \mu_{n;kj}(v)(1, y)$$
$$\mathcal{L}_w\mu^n_{k\,j+1}(v)(x, 0) = \mathcal{L}_w\mu_{n;kj}(v)(x, 1)$$
$$\mathcal{L}_w\mu^n_{k+1\,j}(v)(0, y) = \mathcal{L}_w\mu_{n;kj}(v)(1, y).$$

Since the relation between $\mu_{n;kj}$ and ν_n is once again a pullback under a combination of scaling and translation, we obtain that ν_n is continuously differentiable along the skeleton.

We proceed to prove Property (d): by Property (c) of Proposition 3.1:

$$\begin{aligned}
\nu_n|_{\iota_{n;kj}(\mathcal{M}_{\text{left}})} &= (\iota_{n;kj})_\star\mu_{n;kj}|_{\iota_{n;kj}(\mathcal{M}_{\text{left}})} \\
&= (\iota_{n;kj})_\star\beta_{n;kj}|_{\iota_{n;kj}(\mathcal{M}_{\text{left}})} \\
&= (\iota_{n;kj})_\star(\iota_{n;kj})^\star\beta|_{\iota_{n;kj}(\mathcal{M}_{\text{left}})} \\
&= \beta|_{\iota_{n;kj}(\mathcal{M}_{\text{left}})},
\end{aligned}$$

i.e., ν_n coincides with β on the vertical components of the skeleton.

Property (b) is immediate as $\mu_{n;kj}$ are closed and closedness is invariant under pullback. Finally, Property (c) follows from Property (d) in Proposition 3.1: using the change of variables formula and the fact that $\mu_{n;kj}$ and $\beta_{n;kj}$ have the same circulation along $\partial\mathcal{M}$,

$$\int_{\partial\mathcal{M}_{n;kj}} \nu_n = \int_{\iota_{n;kj}(\partial\mathcal{M})} ((\iota_{n;kj})^{-1})^{\star}\mu_{n;kj}$$
$$= \int_{\partial\mathcal{M}} \mu_{n;kj} = \int_{\partial\mathcal{M}} \beta_{n;kj} = \int_{\partial\mathcal{M}_{n;kj}} \beta.$$

□

As in the case of a single dislocation, we define for each n the 1-current induced by ν_n:

$$T_{\nu_n}(\alpha) = \int_{\mathcal{M}} \nu_n \wedge \alpha, \qquad \alpha \in \Omega_c^1(\mathcal{M}).$$

Its boundary ∂T_{ν_n} is a 0-current given by

$$\partial T_{\nu_n}(f) = \sum_{k,j=1}^{n-1} \int_{\Gamma_{n;kj}} f[\nu_n]_{\Gamma_{n;kj}}, \qquad f \in C_c^{\infty}(\mathcal{M}),$$

where $[\nu_n]_{\Gamma_{n;kj}}$ is the discontinuity jump of ν_n along $\Gamma_{n;kj}$, given by

$$[\nu_n]_{\Gamma_{n;kj}}(x, (j+1/2)/n) = \frac{n}{a} r'\left(\frac{nx - k - 1/2}{a}\right) \int_{\partial\mathcal{M}_{n;kj}} \beta.$$

We view ν_n as a layering form on $\mathcal{M}$ having n^2 edge-dislocations concentrated on Γ_n. The loci of the dislocations are revealed by the boundary of the differential current induced by ν_n. Once again, $\mathcal{M}\setminus\Gamma_n$ is defect-free only to the extent detectable by ν_n.

Theorem 4.1 (Homogenization) *The sequence ν_n of layering forms converges to β in the sense of currents: for every $\alpha \in \Omega_c^1(\mathcal{M})$,*

$$\lim_{n\to\infty} \int_{\mathcal{M}} \nu_n \wedge \alpha = \int_{\mathcal{M}} \beta \wedge \alpha,$$

or equivalently,

$$\lim_{n\to\infty} T_{\nu_n - \beta}(\alpha) = \lim_{n\to\infty} \int_{\mathcal{M}} (\nu_n - \beta) \wedge \alpha = 0. \tag{12}$$

Proof Choose any metric on $\mathcal{M}$; for concreteness we take the Euclidean metric associated with the parametrization.

If $\beta = \beta_1\, dx + \beta_2\, dy$, then

$$\|\beta_{(x,y)}\|^2 = \beta_1^2(x, y) + \beta_2^2(x, y).$$

For every $\alpha \in \Omega_c^1(\mathcal{M})$,

$$
\begin{aligned}
T_{\nu_n-\beta}(\alpha) &= \sum_{k,j=0}^{n-1} \int_{\mathcal{M}_{n;kj}} (\nu_n - \beta) \wedge \alpha \\
&= \sum_{k,j=0}^{n-1} \int_{\iota_{n;kj}(\mathcal{M})} ((\iota_{n;kj})^{-1})^\star (\mu_{n;kj} - \beta_{n;kj}) \wedge \alpha \\
&= \sum_{k,j=0}^{n-1} \int_{\mathcal{M}} (\mu_{n;kj} - \beta_{n;kj}) \wedge (\iota_{n;kj})^\star \alpha,
\end{aligned}
\tag{13}
$$

where the second equality follows from the definitions of ν_n and $\beta_{n;kj}$, and the third equality follows from the change of variables formula. Fix $0 \le k, j \le n-1$. Since $\iota_{n;kj}$ involves a contraction by a factor of n,

$$
\left\| (\iota_{n;kj})^\star \alpha|_{\mathcal{M}_{n;kj}} \right\|_\infty \le \frac{1}{n} \|\alpha\|_\infty.
$$

It follows that

$$
\begin{aligned}
\left| \int_{\mathcal{M}} (\mu_{n;kj} - \beta_{n;kj}) \wedge (\iota_{n;kj})^\star \alpha \right| &\le \frac{1}{n} \|\alpha\|_\infty \sup_{\|\xi\|_\infty = 1} \left| \int_{\mathcal{M}} (\mu_{n;kj} - \beta_{n;kj}) \wedge \xi \right| \\
&\le \frac{1}{n} \|\alpha\|_\infty \int_{\mathcal{M}} |\mu_{n;kj} - \beta_{n;kj}|\, dx \wedge dy.
\end{aligned}
$$

Combining with (13),

$$
\left| T_{\nu_n-\beta}(\alpha) \right| \le n \,\|\alpha\|_\infty \sup_{0 \le k, j < n} \int_{\mathcal{M}} |\mu_{n;kj} - \beta_{n;kj}|\, dx \wedge dy.
$$

Next, writing $\beta_{n;kj}$ explicitly,

$$
(\beta_{n;kj})_{(x,y)} = \frac{1}{n}\beta_1\left(\frac{x+k}{n}, \frac{y+j}{n}\right) dx + \frac{1}{n}\beta_2\left(\frac{x+k}{n}, \frac{y+j}{n}\right) dy.
$$

By (6), for $x < 1/2 - a/2n$,

$$
\begin{aligned}
(\mu_{n;kj})_{(x,y)} &= \frac{1}{n}\beta_1\left(\frac{x+k}{n}, \frac{y+j}{n}\right) dx \\
&\quad + \left(\frac{1}{n}\beta_2\left(\frac{k}{n}, \frac{y+j}{n}\right) + \int_0^x \frac{1}{n^2} \frac{\partial \beta_1}{\partial y} \left(\frac{x'+k}{n}, \frac{y+j}{n}\right) dx'\right) dy,
\end{aligned}
$$

so that

$$
\begin{aligned}
n\,|\mu_{n;kj} - \beta_{n;kj}|(x,y) &\le \left|\beta_2\left(\frac{x+k}{n},\frac{y+j}{n}\right) - \beta_2\left(\frac{k}{n},\frac{y+j}{n}\right)\right| \\
&\quad + \frac{1}{n}\int_0^x \left|\frac{\partial\beta_1}{\partial y}\left(\frac{x'+k}{n},\frac{y+j}{n}\right)\right| dx' \\
&\le \frac{1}{n}\left(\left\|\frac{\partial\beta_2}{\partial x}\right\|_\infty + \left\|\frac{\partial\beta_1}{\partial y}\right\|_\infty\right).
\end{aligned}
$$

The same bound is obtained for $x > 1/2 + a/2n$. Finally, for $|x - 1/2| < a/2n$, using (7), and noting that p_L and p_R are $O(1/n)$, we obtain that

$$
n\,|\mu_{n;kj} - \beta_{n;kj}|(x,y) \le \frac{C}{a}\|r'(x)\|_\infty,
$$

where $C > 0$ is some constant. Putting it all together,

$$
\begin{aligned}
|T_{\nu_n-\beta}(\alpha)| &\le n\,\|\alpha\|_\infty \sup_{0\le k,j<n} \int_{\mathcal{M}\setminus\mathcal{M}_{a/n}} |\mu_{n;kj} - \beta_{n;kj}|\, dx\wedge dy \\
&\quad + n\,\|\alpha\|_\infty \sup_{0\le k,j<n} \int_{\mathcal{M}_{a/n}} |\mu_{n;kj} - \beta_{n;kj}|\, dx\wedge dy \\
&\le \frac{\|\alpha\|_\infty}{n}\left(\left\|\frac{\partial\beta_2}{\partial x}\right\|_\infty + \left\|\frac{\partial\beta_1}{\partial y}\right\|_\infty + \tilde{C}\,\|r'(x)\|_\infty\right),
\end{aligned}
$$

where in the estimation of the third term we used the fact that the volume of $\mathcal{M}_{a/n}$ is $O(1/n)$. Letting $n\to\infty$ we obtain the desired result. □

5 Singular Torsion and Its Homogenization

Thus far, we analyzed a lattice structure through a single layering form, representing a single family of Bravais surfaces. In d dimensions, a lattice structure is fully determined by a set of d linearly independent layering forms, i.e., by a coframe $\{\vartheta^i\}$. Denote by $\{e_i\}$ the frame field dual to $\{\vartheta^i\}$.

A frame–coframe structure induces a path-independent parallel transport,

$$
\Pi_p^q : T_p\mathcal{M} \to T_q\mathcal{M} \qquad \text{given by} \qquad \Pi_p^q = e_i|_q \otimes \vartheta^i|_p. \tag{14}
$$

The latter induces a connection ∇ having trivial holonomy, which locally implies zero curvature. By construction, the frame field $\{e_i\}$ and its dual $\{\vartheta^i\}$ are ∇-parallel sections,

$$
\nabla e_i = 0 \qquad \text{and} \qquad \nabla\vartheta^i = 0.
$$

The torsion tensor associated with ∇ is a $T\mathcal{M}$-valued 2-form τ, given by

$$\tau(e_i, e_j) = \nabla_{e_i} e_j - \nabla_{e_j} e_i - [e_i, e_j] = [e_j, e_i].$$

Since for every $1 \le i, j, k \le d$,

$$\begin{aligned} d\vartheta^i(e_j, e_k) &= e_j(\vartheta^i(e_k)) - e_k(\vartheta^i(e_j)) - \vartheta^i([e_j, e_k]) \\ &= \vartheta^i([e_k, e_j]) \\ &= \vartheta^i(\tau(e_j, e_k)), \end{aligned}$$

we conclude that $d\vartheta^i = \vartheta^i \circ \tau$, or equivalently,

$$\tau = e_i \otimes d\vartheta^i, \tag{15}$$

where we adopt henceforth Einstein's summation rule, whereby repeated upper and lower indexes imply a summation. In particular, torsion vanishes if and only if $d\vartheta^i = 0$ for all $1 \le i \le d$, or equivalently, if $[e_i, e_j] = 0$ for all $1 \le i, j \le d$.

The question we are addressing henceforth is in what sense is the smooth torsion τ given by (15) a limit of singular torsions associated with singular dislocations. For example, let $\mathcal{M}$, β, and ν_n be defined as in the previous section, and suppose that

$$\vartheta_n^1 = \nu_n \qquad \text{and} \qquad \vartheta_n^2 = dx$$

is a sequence of coframe fields (namely, ν_n are dx are linearly independent everywhere in $\mathcal{M}$). By the analysis of the previous section (and trivially for ϑ^2),

$$\lim_{n\to\infty} T_{\vartheta_n^1} = T_\beta \qquad \text{and} \qquad \lim_{n\to\infty} T_{\vartheta_n^2} = T_{dx},$$

i.e.,

$$\lim_{n\to\infty} \{\vartheta_n^1, \vartheta_n^2\} = \{\beta, dx\}$$

in the sense of weak convergence of currents.

Since the coframe field $\{\vartheta_n^1, \vartheta_n^2\}$ consists of closed layering forms, the induced torsion on $\mathcal{M} \setminus \Gamma_n$ vanishes identically for every n,

$$\tau_n = e_i^n \otimes d\vartheta_n^i = 0,$$

which, if $d\beta \neq 0$, does not converge to the torsion

$$\tau = \frac{1}{\beta_2} \partial_y \otimes d\beta$$

associated with the limiting coframe field in any classical sense (we used here the fact that the frame dual to $\{\beta, dx\}$ is $\{\partial_y/\beta_2, \partial_x - \beta_1/\beta_2\, \partial_y\}$).

The question is how to cast a weak convergence of torsion in the framework of de-Rham currents. Torsion is a tangent bundle-valued 2-form. While it is possible to define currents associated with tangent bundle-valued forms, see, e.g., [19], this approach does not seem applicable here. A simple heuristic argument shows that if we try to interpret torsion as a distribution for a discontinuous coframe field, we obtain the product of a discontinuous section e_i and the derivative $d\vartheta^i$ of a discontinuous section (loosely speaking, a product of a Heaviside function and a delta function), which is not well-defined.

A hint toward a correct interpretation of singular torsion is obtained by considering Burgers circuits: Let C be a simple, oriented, regular closed curve in $\mathcal{M}$. The Burgers vector associated with the curve C is a parallel vector field B [20], whose value at a reference point p is given by

$$B_p = \oint_C \Pi^p_\gamma(d\gamma),$$

where Π^p is the parallel-transport to p, which by (14) is given by

$$\Pi^p = e_i|_p \otimes \vartheta^i,$$

and γ is a parametrization for C. Interpreting Π^p as a $T_p\mathcal{M}$-valued 1-form, we rewrite the Burgers vector B_p in a more succinct form,

$$B_p = \oint_C \Pi^p.$$

Applying Stokes' theorem,

$$B_p = \int_\Sigma d\Pi^p,$$

where $\partial\Sigma = C$. Hence,

$$B_p = e_i|_p \int_\Sigma d\vartheta^i.$$

Thus, having chosen a reference point p, the Burgers vector for a loop C is an integral over the area enclosed by this loop of a Burgers vector density

$$e_i|_p \otimes d\vartheta^i,$$

which is a $T_p\mathcal{M}$-valued 2-form; it is nothing but the torsion τ, whose output, once acting on a bivector, is parallel-transported to the reference point p. We henceforth denote by

$$\tau_p = \Pi^p \circ \tau = e_i|_p \otimes d\vartheta^i$$

the torsion transported to p. The notion of singular torsion may now be easily defined as the distributional counterpart of τ_p by replacing $d\vartheta^i$ with the boundary current ∂T_{ϑ^i}. However, we first need to define the notion of a singular frame. Rather than choosing the most general possible framework, we adopt a possibly restrictive but yet sufficiently rich and physically motivated approach:

Definition 5.1 Let $\mathcal{M}$ be a compact d-dimensional manifold. A collection $\{\vartheta^i\}_{i=1}^d$ of 1-forms is called a **singular coframe** for $\mathcal{M}$ if for every $1 \le i \le d$, there exists a compact $(d-1)$-dimensional submanifold $\Gamma_i \subset \mathcal{M}$, such that

1. Each ϑ^i is a C^1-bounded 1-form on $\mathcal{M} \setminus \Gamma_i$.
2. $\{\vartheta^i_p\}$ is a basis for $T^*_p\mathcal{M}$ for every $p \in \mathcal{M} \setminus \Gamma$ where $\Gamma = \cup_i \Gamma_i$.
3. $\mathcal{M} \setminus \Gamma$ is path connected and $\partial\mathcal{M} \cap \Gamma = \emptyset$.

A **closed singular coframe** is a singular coframe $\{\vartheta^i\}$ satisfying $d\vartheta^i = 0$ on $\mathcal{M} \setminus \Gamma_i$ for every $1 \le i \le d$.

Recall that if a layering form $\omega \in \Omega^1(\mathcal{M})$ is closed, its induced layering structure (foliation) is defect-free. A closed singular coframe therefore corresponds to isolated defects which are concentrated on a set of measure zero.

Definition 5.2 Let $\{\vartheta^i\}$ be a singular coframe field on $\mathcal{M}$ and let $p \in \mathcal{M} \setminus \Gamma$ be an arbitrary reference point. The **torsion current** is a $T_p\mathcal{M}$-valued $(d-2)$-current given by

$$\mathcal{T} = e_i|_p\, \partial T_{\vartheta^i}.$$

For a smooth coframe $\{\vartheta^i\}$, the torsion current is given by

$$\mathcal{T}(\alpha) = e_i|_p\, \partial T_{\vartheta^i}(\alpha) = e_i|_p T_{d\vartheta^i}(\alpha) = T_{\tau_p}(\alpha), \qquad \alpha \in \Omega_c^{d-2}(\mathcal{M}). \tag{16}$$

In other words, in the smooth case, the torsion current $\mathcal{T}$ is the $T_p\mathcal{M}$-valued $(d-2)$-current induced by the smooth $T_p\mathcal{M}$-valued 2-form τ_p.

In the case of a closed singular coframe (isolated defects), the singular torsion is supported on the singular hypersurfaces $\{\Gamma_i\}$ and is given explicitly by

$$\mathcal{T}[p](\eta) = \sum_{i=1}^{d} \left(\int_{\Gamma_i} [\vartheta^i]_{\Gamma_i} \wedge \eta \right) e_i(p), \tag{17}$$

where $[\vartheta^i]_{\Gamma_i}$ is the discontinuity jump of ϑ^i along Γ_i and $\eta \in \Omega_c^{d-2}(\mathcal{M})$. For a general (non-closed) singular frame $\{\vartheta^i\}$, the torsion current naturally decomposes into a smooth component as in (16) and a singular component as in (17).

We have thus obtained the following corollary:

Corollary 5.1 (Homogenization of Torsion) *Let $\{\vartheta^i_n\}$ be a sequence of (possibly) singular coframes and $p \in \mathcal{M}$ a reference point, satisfying:*

orientable, connected surface $\mathcal{M}$ with boundary is diffeomorphic to either S^2 or T_n, with k holes, namely,

$$\mathcal{M} = S^2 \setminus \amalg_{i=1}^k U_i \qquad \text{or} \qquad \mathcal{M} = T_n \setminus \amalg_{i=1}^k U_i,$$

where U_i are disjoint open sets diffeomorphic to a disc; see, e.g., [21] for a proof using Morse theory. Moreover, each of those surfaces can be constructed by gluing together a finite number of two building blocks: a closed disc and a "pair-of-pants."

To prove a homogenization for compact, orientable surfaces with or without boundary we adopt the following strategy: We first prove the homogenization property for the two above-mentioned building blocks. Then, using a gluing lemma (Lemma A.2), we deduce the homogenization property for S^2 and T_n with k holes. We finally obtain the general case by combining Lemma 6.1 and the classification theorem of surfaces.

We start by constructing a layering form containing a single dislocation on the unit disk

$$D = \left\{(x, y) \in \mathbb{R}^2 : x^2 + y^2 \leq 1\right\}.$$

Lemma 6.2 *Let $\omega \in \Omega^1(D)$ and let $\Gamma_a = [-a, a] \times \{0\}$, where $0 < a < 1/2$. Then there exists a closed, C^1-bounded 1-form $\omega_a \in \Omega^1(D \setminus \Gamma_a)$ satisfying*

(a) $\int_{\partial D} \omega = \int_{\partial D} \omega_a$.
(b) $\omega|_{\partial D} = \omega_a|_{\partial D}$.
(c) $\left(\mathcal{L}_{\partial_r} \omega_a\right)|_{\partial D} = \left(\mathcal{L}_{\partial_r} \omega\right)|_{\partial D}$.

Proof *(Sketch)* The proof follows the same lines as the proof of Proposition 3.1. We construct ω_a as the differential of a discontinuous function $f : D \to \mathbb{R}$. First fix $q_0 = (0, 1) \in \partial D$ and define $f_0 : \partial D \to \mathbb{R}$ by

$$f_0(q) = \int_{q_0}^{q} \omega,$$

where the integration is counter-clockwise along ∂D. As in the case of a square, f_0 is discontinuous at q_0 but its differential is well-defined. Next, define $f : D \setminus \Gamma_a \to \mathbb{R}$ as follows: for every $q = (q_1, q_2) \in \partial D$, let l_q be the segment connecting q to $(aq_1, 0) \in \Gamma_a$ (see Fig. 8). Then every $p \in D$ lies on a unique segment l_q, and we may define

$$f(p) = f_0(q) + \int_{[q,p]} \omega, \quad p \in l_q.$$

A straightforward computation as in the proof of Proposition 3.1 shows that $\omega_a = df$ satisfies the desired properties. □

We next prove the homogenization property for the closed disk.

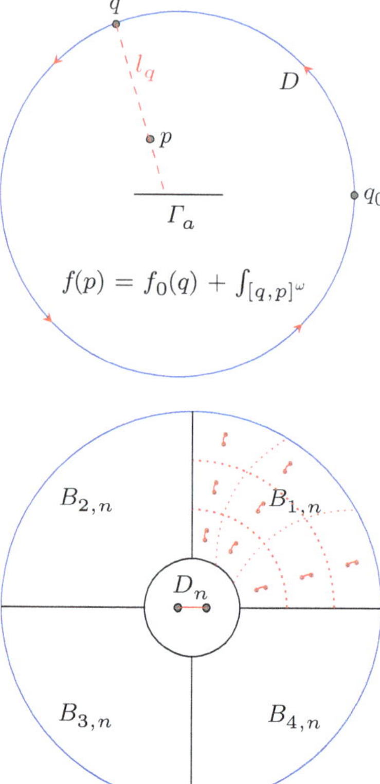

Fig. 8 Construction of a layering form on a disk containing a single dislocation

Fig. 9 The disk D is decomposed into four sectors (diffeomorphic to a square) and a small disk. A layering form containing an "array of dislocations" is constructed in each sector, and glued together to obtain ω_n

Lemma 6.3 *The homogenization property holds for the disk*

$$D = \left\{(x, y) \in \mathbb{R}^2 \,:\, x^2 + y^2 \le 1\right\}.$$

Proof Let $\omega \in \Omega^1(D)$. For every $n \in \mathbb{N}$, let $D_n = \frac{1}{n}D$ and let $B_{i,n}$, $1 \le i \le 4$, be the sectors given by

$$B_{i,n} = \{(r\,\cos\theta, r\,\sin\theta) \,:\, 1/n \le r \le 1, \quad i\pi/4 \le \theta \le (i+1)\pi/4\}\,.$$

Then $B_{i,n} \simeq [0,1]^2$ and $D \simeq D_n \cup_{i=1}^4 B_{i,n}$ (see Fig. 9). Let $\phi : B_{1,n} \to [0,1]^2$ be a diffeomorphism which preserves the left/right and upper/lower edges/arcs. Then its rotations $\phi_i = \phi \circ R_{(i-1)\pi/4} : B_{i,n} \to [0,1]^2$ $(i = 2, 3, 4)$ are diffeomorphisms as well. Using Proposition 4.1, we may construct singular closed layering forms $\omega_{i,n}$ on $B_{i,n}$ which combine together into a singular layering form $\tilde{\omega}_n$ on $D \setminus D_n$, whose singularity set of $\tilde{\omega}_n$ is a union of segments and it coincides with ω on $\partial D \setminus \frac{1}{n}D$. Finally, by Lemma 6.2, we may complete $\tilde{\omega}_n$ into a singular layering form ω_n on D. That

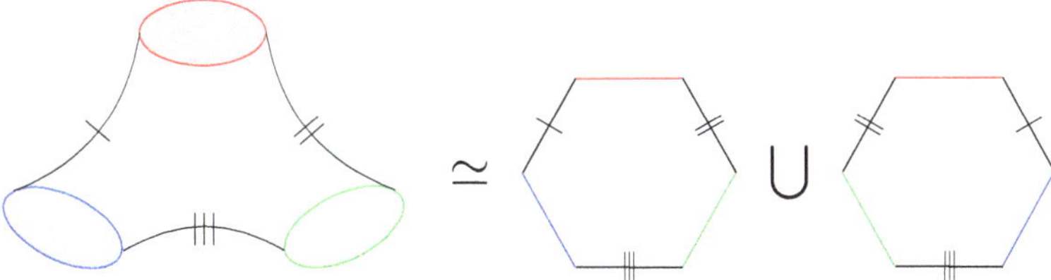

Fig. 10 A pair-of-pants. It can be obtained by gluing hexagons along three pairs of edges. The remaining pairs of (colored) edges are glued at their ends thus forming three boundary circles

$$T_{\omega_n} \to T_\omega$$

follows from Theorem 4.1 (applied separately for each sector). □

We next prove the homogenization property for a pair-of-pants which is diffeomorphic to the three-holed sphere (see Fig. 10).

Lemma 6.4 *The homogenization property holds for a pair-of-pants.*

Proof First, note that the hexagon, denoted by $\mathcal{O}$, also satisfies the homogenization property as well as the gluing conditions as in Lemma 6.3. The proof is almost identical to the proof for the disk (taking 6 rather than 4 sectors). Let $\mathcal{M}$ be a pair-of-pants. It can be obtained by identifying three pairs of edges of two hexagons $\mathcal{O}_1$ and $\mathcal{O}_2$ (see Fig. 10). Hence, a layering form $\omega \in \Omega^1(\mathcal{M})$ induces layering forms $\tilde{\omega}_1$ and $\tilde{\omega}_2$ on $\mathcal{O}_{1,2}$ satisfying (trivially) the gluing conditions of Lemma A.2. Since the homogenization property holds for each hexagon, there exist approximating sequences $\tilde{\omega}_{i,n}$ for $\tilde{\omega}_i$ $(i = 1, 2)$ which satisfy the gluing conditions and therefore form together an approximating sequence ω_n for ω. □

We next prove the following gluing argument.

Proposition 6.1 *Suppose that the homogenization property holds for compact orientable surfaces $\mathcal{M}_1$ and $\mathcal{M}_2$. Let $A_i \subset \partial\mathcal{M}_i$ be connected components of the boundaries and $h : A_1 \to A_2$ a diffeomorphism. Finally, let $\iota_{A_i} : [0, 1) \times A_i \to \mathcal{M}_i$ be collar neighborhoods for A_i. Then the glued manifold*

$$\mathcal{M} = \mathcal{M}_1 \amalg_h \mathcal{M}_2$$

satisfies the homogenization property (see the appendix for the definition of collar neighborhoods and gluing constructions).

Proof *(Sketch)* Let $\omega \in \Omega^1(\mathcal{M})$. Then ω induces layering forms $\omega^i \in \Omega^1(\mathcal{M}_i)$ satisfying the gluing conditions, so that the restriction of ω and its first derivatives to $\mathcal{M}_i$ coincides with those of ω_i. Apply the homogenization property to obtain sequences of closed singular layering forms $\omega_{i,n}$, so that $T_{\omega_{i,n}}$ converges weakly to T_{ω_i}. We may choose the $\omega_{i,n}$ such that their values and their Lie derivatives coincide with those of ω_i at A_i. By the gluing Lemma A.2, $\omega_{1,n}$ and $\omega_{2,n}$ induce a closed and singular C^1-bounded layering form ω_n on $\mathcal{M} = \mathcal{M}_1 \amalg_h \mathcal{M}_2$. It follows directly from the construction that the sequence ω_n satisfies the required properties. □

Remark 6.2 We are mostly interested in the case where $\mathcal{M}_1$ and $\mathcal{M}_2$ are submanifolds of $\mathcal{M}$ with a common boundary (circle) component $A = A_1 = A_2 \simeq S^1$. In such a case, one can take a collar neighborhood induced by a vector field $X \in \Gamma(T\mathcal{M})$ which is transversal to A. Taking, $h = Id_A : A_1 \to A_2$ one obtains $\mathcal{M} \simeq \mathcal{M}_1 \amalg_h \mathcal{M}_2$. In other words, it is not necessary in this case to specify the collar neighborhoods and the boundary identifications, and the conditions for the gluing lemma to apply are satisfied automatically.

By applying Proposition 6.1 inductively we may finally prove Theorem 6.1:

Proof *(Of Theorem 6.1)* Let $\mathcal{M}$ be a compact, orientable surface (possibly with boundary). By the classification of surfaces, we may decompose $\mathcal{M}$ into a finite number of pairs of pants and disks (glued along circles). By Lemmas 6.3 and 6.4, the homogenization property holds for the disk and for the pair-of-pants. Hence, given $\omega \in \Omega^1(\mathcal{M})$, we may inductively apply Proposition 6.1 (on larger and larger components of $\mathcal{M}$) to obtain the desired sequence ω_n. The convergence $T_{\omega_n} \to T_\omega$ follows immediately from the construction and the compactness of $\mathcal{M}$. □

Acknowledgements We would like to thank Marcelo Epstein, Pavel Giterman, Cy Maor, Reuven Segev, and Amitay Yuval for many helpful discussions. This research was partially funded by the Israel Science Foundation (Grant No. 1035/17), and by a grant from the Ministry of Science, Technology and Space, Israel and the Russian Foundation for Basic Research, the Russian Federation.

Appendix: Gluing Constructions

The homogenization procedure presented in Sects. 4 and 6 relies on gluing diffeomorphic copies of single isolated dislocations and their structure forms. To this end, we review some basic definitions and facts, following [10, Chap. 9] and prove a gluing lemma for 1-forms.

Let $\mathcal{M}$ be a smooth manifold with boundary. A neighborhood of $\partial\mathcal{M}$ is called a **collar neighborhood** if it is the image of a smooth embedding $\iota : [0,1) \times \partial\mathcal{M} \hookrightarrow \mathcal{M}$ sending (identically) $\{0\} \times \partial\mathcal{M}$ to $\partial\mathcal{M}$. It follows from the theory of flows that every smooth manifold with boundary admits a collar neighborhood; see [10, Theorem 9.25].

Let $\mathcal{M}_1$ and $\mathcal{M}_2$ be smooth manifolds with boundary of the same dimension, and let $A \subset \partial\mathcal{M}_1$, and $B \subset \partial\mathcal{M}_2$ be nonempty connected (possibly closed) submanifolds. Suppose that $h : B \to A$ is a diffeomorphism. h defines an equivalence relation on the disjoint union $\mathcal{M}_1 \amalg \mathcal{M}_2$ whereby $p \sim_h q$ if and only if $p = h(q)$. Let

$$\mathcal{M}_1 \amalg_h \mathcal{M}_2 := \{[p]_h \mid p \in \mathcal{M}_1 \amalg \mathcal{M}_2\},$$

where $[p]_h$ is the $\sim_h$-equivalence class of p. Then, $\mathcal{M}_1 \amalg_h \mathcal{M}_2$ is a topological manifold (possibly with boundary and corners); it admits a smooth structure such that the natural embeddings

$$\mathcal{M}_1 \hookrightarrow \mathcal{M}_1 \amalg_h \mathcal{M}_2, \qquad \mathcal{M}_2 \hookrightarrow \mathcal{M}_1 \amalg_h \mathcal{M}_2,$$

are smooth, $[\mathcal{M}_1]_h \cup [\mathcal{M}_2]_h = \mathcal{M}_1 \amalg_h \mathcal{M}_2$ and $[\mathcal{M}_1]_h \cap [\mathcal{M}_2]_h = [A]_h = [B]_h$. We will denote by

$$\pi : \mathcal{M}_1 \amalg \mathcal{M}_2 \to \mathcal{M}_1 \amalg_h \mathcal{M}_2$$

the projection map sending every point $p \in \mathcal{M}_1 \amalg \mathcal{M}_2$ to its equivalence class $[p]_h \in \mathcal{M}_1 \amalg_h \mathcal{M}_2$.

The construction of the smooth structure relies on gluing collar neighborhoods of A and B along h. In particular the smooth structure depends on the chosen collar neighborhoods; see [10, Theorem 9.29] for details.

Let $\iota_A : [0, 1) \times A \to \mathcal{M}_1$ and $\iota_B : [0, 1) \times B \to \mathcal{M}_2$ be collar neighborhoods for A and B; define also the inclusions $\eta_A : A \hookrightarrow \mathcal{M}_1$ and $\eta_B : B \hookrightarrow \mathcal{M}_1$ by $\eta_A(p) = \iota_A(0, p)$ and $\eta_B(p) = \iota_B(0, p)$; see diagram below.

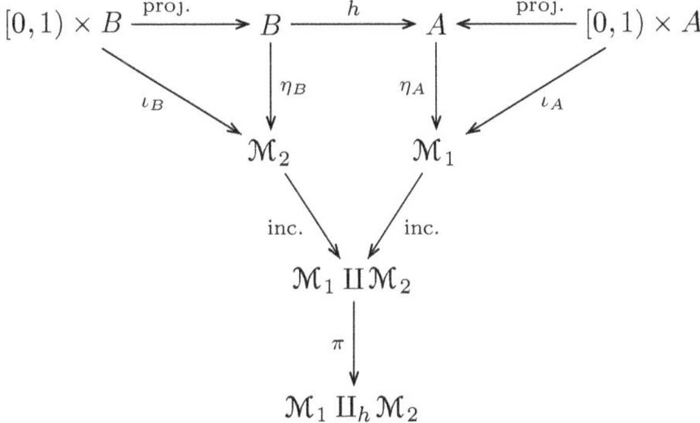

For later use, we note that

$$\pi \circ \eta_B = \pi \circ \eta_A \circ h,$$

hence, differentiating, for $p \in B$,

$$d\pi_{\eta_B(p)} \circ (d\eta_B)_p = d\pi_{\eta_A(h(p))} \circ (d\eta_A)_{h(p)} \circ dh_p. \tag{18}$$

The collar neighborhoods define a decomposition of $T\mathcal{M}_1$ and $T\mathcal{M}_2$ at A and B: for example,

$$T\mathcal{M}_1|_{\eta(A)} = T\mathcal{M}_1^{\parallel} \oplus T\mathcal{M}_1^{\perp},$$

where

$$T\mathcal{M}_1^{\|} = (\eta_A)_\star TA,$$

and

$$T\mathcal{M}_1^{\perp} = \operatorname{span}(n_A),$$

where

$$n_A = (\iota_A)_\star(\partial_t)|_{A\times\{0\}} \tag{19}$$

is a vector field normal to $T\mathcal{M}_1^{\|}$ with respect to the collar neighborhood ι_A. Similar definitions apply for the tangent bundle of $\mathcal{M}_2$ at B.

We turn to characterize tangent vectors on the quotient space $\mathcal{M}_1 \amalg_h \mathcal{M}_2$. Suppose first that $p \in \mathcal{M}_1 \amalg \mathcal{M}_2 \setminus (A \cup B)$. Then, π is a local diffeomorphism in a neighborhood of p, hence $d\pi_p$ is a linear isomorphism. In other words, tangent vectors at $[p]_h$ can be identified with tangent vectors at p.

In contrast, let $p \in B$, i.e.,

$$\pi^{-1}(\pi(p)) = \{h(p), p\},$$

and let $v \in T_{\pi(p)}(\mathcal{M}_1 \amalg_h \mathcal{M}_2)$. Then, $d\pi^{-1}(v) = \{v_1, v_2\}$, where $v_1 \in T_{h(p)}\mathcal{M}_1$ and $v_2 \in T_p\mathcal{M}_2$. Each of the two vectors can be written in the form

$$v_1 = (\eta_A)_\star(v_1^{\|}) + v_1^{\perp}\, n_A \qquad \text{and} \qquad v_2 = (\eta_B)_\star(v_2^{\|}) + v_2^{\perp}\, n_B,$$

where $v_1^{\|} \in TA$, $v_2^{\|} \in TB$ and $v_1^{\perp}, v_2^{\perp} \in \mathbb{R}$.

We state without a proof:

Lemma A.1 *The following relations hold:*

$$v_1^{\|} = h_\star(v_2^{\|}), \tag{20}$$

and

$$v_1^{\perp} = -v_2^{\perp}. \tag{21}$$

Moreover,

$$\pi_\star(n_A) = -\pi_\star(n_B). \tag{22}$$

Our next goal is to glue together 1-forms along $A \subset \partial\mathcal{M}_1$ and $B \subset \partial\mathcal{M}_2$:

Lemma A.2 (Gluing of Forms) *Let $\omega_1 \in \Omega^1(\mathcal{M}_1)$ and $\omega_2 \in \Omega^1(\mathcal{M}_2)$ satisfy the following conditions:*

(i) Equality of tangential component:

$$h^\star(\eta_A^\star \omega_1) = \eta_B^\star \omega_2 \tag{23}$$

(this is an equality of 1-forms on B).

(ii) Matching of normal component:

$$\omega_1(n_A) \circ h = -\omega_2(n_B) \tag{24}$$

(this is an equality of functions on B).

(iii) Matching of normal derivative:

$$(\mathcal{L}_{n_A}\omega_1(n_A)) \circ h = -\mathcal{L}_{n_B}\omega_2(n_B),$$

and

$$h^\star\left(\eta_A^\star(\mathcal{L}_{n_A}\omega_1)\right) = -\eta_B^\star(\mathcal{L}_{n_B}\omega_2),$$

where $\mathcal{L}$ is the Lie derivative and n_A and n_B are extended to neighborhoods of $A \subset \mathcal{M}_1$ and $B \subset \mathcal{M}_2$ via (19).

Then, there exists a 1-form ω on $\mathcal{M}_1 \amalg_h \mathcal{M}_2$ which is C^1 with respect to the smooth structure induced by ι_A and ι_B, such that the restrictions of ω to $\mathcal{M}_1$ and $\mathcal{M}_2$ coincide with ω_1 and ω_2.

Proof Let $\omega_1 \amalg \omega_2 \in \Omega^1(\mathcal{M}_1 \amalg \mathcal{M}_2)$ be the induced form on the disjoint union. We first show that Conditions (i) and (ii) imply that $\omega_1 \amalg \omega_2$ projects to a well-defined 1-form ω on $\mathcal{M}_1 \amalg_h \mathcal{M}_2$.

Consider first $p \in \mathcal{M}_1 \amalg \mathcal{M}_2 \setminus (A \cup B)$, and let $v \in T_{\pi(p)}(\mathcal{M}_1 \amalg_h \mathcal{M}_2)$. Since $\pi^{-1}(\pi(p)) = \{p\}$ and $d\pi_p$ is an isomorphism, we may define

$$\omega_{\pi(p)}(v) = (\omega_1 \amalg \omega_2)_p(d\pi^{-1}(v)).$$

Next, let $p \in B$ and let $v \in T_{\pi(p)}(\mathcal{M}_1 \amalg_h \mathcal{M}_2)$. Now $\pi^{-1}(\pi(p)) = \{h(p), p\}$ and $d\pi^{-1}(v) = \{v_1, v_2\}$, where $v_1 \in T_{h(p)}\mathcal{M}_1$ and $v_2 \in T_p\mathcal{M}_2$. In order to define $\omega_{\pi(p)}(v)$ unambiguously, it suffices to show that $\omega_1(v_1) = \omega_2(v_2)$.

Write as above

$$v_1 = d\eta_A(v_1^{\parallel}) + v_1^{\perp}\,(n_A)_{h(p)}$$

$$v_2 = d\eta_B(v_2^{\parallel}) + v_2^{\perp}\,(n_B)_p,$$

where $v_1^{\|} \in T_{h(p)}A$ and $v_2^{\|} \in T_pB$. Then,

$$\omega_1(d\eta_A(v_1^{\|})) = \eta_A^\star\omega_1(v_1^{\|}) \overset{(20)}{=} \eta_A^\star\omega_1(dh(v_2^{\|})) = h^\star\eta_A^\star\omega_1(v_2^{\|}) \overset{(23)}{=} \eta_B^\star\omega_2(v_2^{\|}) = \omega_2(d\eta_B(v_2^{\|})),$$

and

$$\omega_1(v_1^\perp(n_A)_{h(p)}) = v_1^\perp\,\omega_1(n_A)_{h(p)} \overset{(21)}{=} -v_2^\perp\,\omega_1(n_A)_{h(p)} \overset{(24)}{=} v_2^\perp\,\omega_2(n_B)_p = \omega_2(v_2^\perp n_B)_p.$$

We have thus proved that ω is well-defined. It remains to show that ω (or equivalently $\Phi_\star\omega$) is continuously differentiable. For $(t,p) \in (-1,1)\times A$ and $\alpha\partial_t \oplus v \in T((-1,1)\times A) \simeq T_t(-1,1)\oplus T_pA$,

$$(\Phi_\star\omega)\,|_{(t,p)}(\alpha\partial_t \oplus v) = \begin{cases} \omega_1|_{\iota_A(-t,p)}(-\alpha n_A + v) & t<0 \\ \omega_2|_{\iota_B(t,p)}(\alpha n_B + dh(v)) & t\geq 0. \end{cases}$$

Condition (i) then implies that the tangential (to A) derivatives of $\Phi_\star\omega$ are continuous and Condition (iii) shows (by a similar calculation) that it is continuously differentiable in the "t" direction (one-sided limits coincide). This completes the proof. □

References

1. V. Volterra, Ann. Sci. Ecole Norm. Sup. Paris 1907 **24**, 401 (1907)
2. J. Nye, Acta Met. **1**, 153 (1953)
3. K. Kondo, in *Memoirs of the Unifying Study of the Basic Problems in Engineering Science by Means of Geometry*, vol. 1, ed. by K. Kondo (1955), pp. 5–17
4. B. Bilby, R. Bullough, E. Smith, Proc. Roy. Soc. A **231**, 263 (1955)
5. E. Kröner, in *Les Houches Summer School Proceedings*, ed. by R. Balian, M. Kleman, J.P. Poirier (North-Holland, Amsterdam, 1981)
6. C. Davini, Arch. Rat. Mech. Anal. **96**, 295 (1986)
7. A. Yavari, A. Goriely, Arch. Rational Mech. Anal. **205**, 59 (2012)
8. R. Kupferman, C. Maor, J. Geom. Mech. **7**, 361 (2015)
9. R. Kupferman, C. Maor, Proc. Roy. Soc. Edinburgh **146A**, 741 (2016)
10. J. Lee, *Introduction to smooth manifolds*, second edition edn. (Springer, 2012)
11. M. Epstein, R. Segev, Int. J. Non-Linear Mech. **66**, 105 (2014)
12. R. Kupferman, C. Maor, Calc. Var. PDEs **55**, 40 (2016)
13. M. Epstein, R. Kupferman, C. Maor, Limit of distributed dislocations in geometric and constitutive paradigms (2019). ArXiv:1902.02410
14. H. Federer, *Geometric Measure Theory* (Springer-Verlag, 1969)
15. G. de Rham, *Differentiable Manifolds* (Springer, 1984)

16. F. Lin, X. Yang, *Geometric Measure Theory: An Introduction*. Advanced Mathematics (Science Press, 2002)
17. S. Krantz, H. Parks, *Geometric Integration Theory* (Springer, 2008)
18. W. Rudin, *Functional Analysis*, 2nd edn. (McGraw-Hill, 1991)
19. R. Segev, L. Falach, Disc. Cont. Dyn. Sys. B **17**, 687 (2012)
20. R. Kupferman, M. Moshe, J. Solomon, Arch. Rat. Mech. Anal **216**, 1009 (2015)
21. M. Hirsch, *Differential Topology* (Springer, 1997)

A Kinematics of Defects in Solid Crystals

Marek Z. Elżanowski and Gareth P. Parry

Contents

Abstract We review various mathematical constructions relevant to the kinematical model of defective crystals that Davini proposed in 1986. Partly, the motivation for this is the need to place quantities that are useful in phenomenological theories of inelastic behaviour (which are many and rather varied) in a general mathematical framework. Partly, too, simple assumptions regarding defective crystal symmetries are inadequate, so a re-evaluation of those assumptions is necessary. Also, as motivation, we take the current effort in continuum mechanics to rationalize the connection between continuum and discrete models of materials, and so review results which elucidate the rigorous connection between continuous and discrete structures in the context of Davini's model.

M. Z. Elżanowski (✉)
Fariborz Maseeh Department of Mathematics and Statistics, Portland State University, Portland, OR, USA
e-mail: elzanowskim@pdx.edu

G. P. Parry
School of Mathematical Sciences, University of Nottingham, Nottingham, UK
e-mail: gareth.parry1@nottingham.ac.uk

R. Segev, M. Epstein (eds.), *Geometric Continuum Mechanics*, Advances in Mechanics and Mathematics 43, https://doi.org/10.1007/978-3-030-42683-5_7

1 Introduction

We review various mathematical constructions relevant to the kinematical model of defective crystals that Davini [9] proposed in 1986. Partly, the motivation for this is the need to place quantities that are useful in phenomenological theories of inelastic behaviour (which are many and rather varied) in a general mathematical framework. Partly, too, simple assumptions regarding defective crystal symmetries are inadequate [7, 17], so a re-evaluation of those assumptions is necessary. Also, as motivation, we take the current effort in continuum mechanics to rationalize the connection between continuum and discrete models of materials, and so review results which elucidate the rigorous connection between continuous and discrete structures in the context of Davini's model.

In [9] a crystal state Σ, in $\mathbb{R}^3$, is given by the prescription of three smooth linearly independent 'lattice vector' fields $\boldsymbol{l}_1(\cdot), \boldsymbol{l}_2(\cdot), \boldsymbol{l}_3(\cdot)$ defined at each point of some region $M \subseteq \mathbb{R}^3$. The lattice vector fields vary on a scale which is finer than that commonly associated with continuum mechanics, but which is coarser than the interatomic length scale. So one may imagine that the fields are obtained by some averaging procedure from a given (discrete) atomic structure, and that by virtue of the separation of scales the lattice vector fields are not necessarily embedded in the macroscopic deformations of continuum mechanics. Thus the lattice vector fields carry geometrical information regarding the inelastic behaviour of the material.

A further important point, regarding [9], is that the 'current' state of the material, which is the prescribed crystal state Σ, is supposed to determine all geometric constitutive variables, c.f., [5, 28]. So, for example, any strain energy density is supposed to depend only on (point values of) the lattice vector fields and their derivatives. A priori, then, no 'reference configuration', or 'intermediate configuration', figures in this model—the details of the current geometry are all that is required to determine the constitutive variables. However, particular distributions of lattice vector fields may, of course, arise (for example) by elastic deformation of some 'simple' reference crystal state—our formalism is powerful enough to recognize particular cases when this is so. In fact it is a central question to determine whether or not two prescribed crystal states are such that one may be obtained from the other by elastic deformation—the answer to this question requires a mathematical perspective which emphasizes the importance of quantities which generalize the notions of Burgers vectors, dislocation density tensor, and so forth, prominent in theories of continuous distributions of dislocations [4, 25], and in engineering plasticity theories [26, 27]. It turns out to be necessary to consider quantities that are of higher order in the derivatives of the lattice vector fields than the Burgers vectors, dislocation density tensor, etc., in order to answer the question posed.

It is a global question, whether or not two prescribed crystal states are related by elastic deformation, since the two regions over which the respective lattice vector fields are defined must map, one to the other, under the deformation. The corresponding local question, whether or not the lattice vector fields in a neighbourhood of some point in a given crystal state are related by elastic deformation to the lattice vector fields in some neighbourhood of a prescribed point in the other crystal state, is easier to address. One progresses, in addressing the local question, by constructing objects (functions dependent on the lattice vector fields and their derivatives of arbitrary finite order) which are unchanged by elastic deformation—primary amongst these objects is the dislocation density tensor function (denoted **ddt** for brevity), which is a (first order) measure of the non-commutativity of pairs of lattice vector fields. In fact there is an infinite number of such objects, though a certain finite number of them is sufficient for the purposes of this chapter. This allows us to generate a certain finite number of conditions, necessary in order that two crystal states be 'locally elastically related' (see (4) below), by requiring that these objects are unchanged by elastic deformation. Naturally, one enquires also if these necessary conditions are sufficient for the stated purpose, and we elaborate on this in the sequel. Note that one might call these objects '**plastic strain variables**', by analogy with terminology in the continuum mechanics literature, for they quantify some aspects of inelastic behaviour—if these objects are different, at two points in two crystal states, the respective sets of lattice vectors in neighbourhoods of the two points are not obtained, one from the other, by any elastic deformation. However, we prefer to call these objects '**scalar invariants**' (c.f., Sect. 2.2). Just as the ddt measures the non-commutativity of pairs of lattice vector fields, the higher order scalar invariants are higher order measures of non-commutativity (e.g., relating to the non-commutativity of a lattice vector field and the Lie bracket of a pair of vector fields, see (41), (43), Sect. 4.4).

The non-commutativity of pairs of vector fields is characteristic of defective crystals, in this context, so it is not surprising that there is a duality between descriptions and classifications of defective crystals in terms of the ddt and its higher order generalizations, and descriptions in terms of Lie brackets and their higher order generalizations. The geometrical interpretation of higher order generalizations of the ddt is not well developed (it hardly exists), by way of contrast with the vast literature relating to Lie groups and algebras. In this chapter we develop the two descriptions almost side by side, to begin with, emphasizing their inter-relatedness. Then we exploit existing results available in the Lie group literature to address geometrical questions of interest from the point of view of continuum mechanics, for example, we identify a point of departure which leads to rigorous elastic-plastic decompositions of certain changes of state (such decompositions of changes of crystal states are derived results, based on the geometry of the vector fields, rather than constitutive hypotheses), we show that certain discrete structures are naturally associated with particular crystal states and consider how the symmetries of the discrete structures relate to those of the crystal states, and we describe how certain invariant measures of curvature and torsion may be calculated from given crystal states.

In each section, we summarize, introduce concepts, and give definitions, to begin with, then (for the most part) we choose to develop the material by focusing on what seems to be simplest non-trivial instance of the defective crystal state, where results are less intricate than in other, more general, cases—it is the case where a certain Lie group is nilpotent. (The reader may refer to [1, 32, 33], for example, for discussion of more general cases). In later sections we move to descriptions of crystal states in terms of modern differential geometry, and finally we indicate very briefly how the work may be extended and improved.

Note that we do not give explicit derivations of results in many cases, as this is a review, but refer the reader to the relevant sources.

2 Crystal States and Scalar Invariants

In this section we introduce the lattice vector fields that, together with the region $M \subseteq \mathbb{R}^3$, define the crystal state Σ. We define what is meant by an elastic deformation and also what is meant by stating that two crystal states are locally elastically related. The dual lattice vector fields are introduced, and this allows us to define the dislocation density tensor (ddt). We define scalar (elastic) invariants and note that each component of the ddt is one such. Also we make the observation that there is an infinite number of scalar elastic invariants, in general, for example each directional derivative of the ddt is such an invariant, c.f., [35]. We argue that, in the case $M \subseteq \mathbb{R}^3$, there can be at most three independent scalar invariants and that however many such invariants there are, they occur amongst the directional derivatives of the ddt of order less than or equal to two. Finally we introduce the classifying manifold [35] associated with a crystal state and quote a result which gives a set of conditions sufficient that two crystal states be locally elastically related to one another.

2.1 Crystal States

Let $\mathbf{l}$ denote a set of three linearly independent smooth vector fields $\boldsymbol{l}_1(\cdot), \boldsymbol{l}_2(\cdot), \boldsymbol{l}_3(\cdot)$, defined on a manifold M of dimension 3 in $\mathbb{R}^3$, so that $l_i : M \to TM, i = 1, 2, 3$, where TM denotes the tangent space of M. We shall call these the **lattice** vector fields. The corresponding **crystal state** Σ is defined by

$$\Sigma \equiv \{\mathbf{l}; M\}. \tag{1}$$

An **elastic deformation** of Σ is a smooth invertible mapping $\boldsymbol{u} : M \to \boldsymbol{u}(M) \subseteq \mathbb{R}^3$, with smooth inverse, such that the set of fields $\mathbf{l} = \{\boldsymbol{l}_i(\cdot); i = 1, 2, 3\}$ is transformed to $\tilde{\mathbf{l}} = \{\tilde{\boldsymbol{l}}_i(\cdot); i = 1, 2, 3\}$ defined on $\boldsymbol{u}(M)$, where

$$\tilde{\boldsymbol{l}}_i\,(\boldsymbol{u}(\boldsymbol{x})) \equiv \nabla \boldsymbol{u}(\boldsymbol{x})\boldsymbol{l}_i(\boldsymbol{x}),\ i = 1, 2, 3,\ \boldsymbol{x} \in M. \tag{2}$$

The crystal state

$$\tilde{\Sigma} \equiv \{\tilde{\mathbf{l}}; \tilde{M}\}. \tag{3}$$

is said to be elastically related to Σ when (2) holds and $\tilde{M} = \boldsymbol{u}(M)$, and vice versa via $\boldsymbol{u}^{-1}(\cdot)$.

Also, it is convenient to introduce the notion of local elastic relatedness: states Σ, $\tilde{\Sigma}$ are **locally elastically related** to one another if for each $\boldsymbol{x}_0 \in M$ there exists a diffeomorphism $\boldsymbol{u}_{\boldsymbol{x}_0}$ defined on a neighbourhood $N_{\boldsymbol{x}_0}$ of $\boldsymbol{x}_0$ in M, with $\boldsymbol{u}_{\boldsymbol{x}_0}(N_{\boldsymbol{x}_0}) \subseteq \tilde{M}$ such that

$$\tilde{\boldsymbol{l}}_i(\boldsymbol{u}_{\boldsymbol{x}_0}(\boldsymbol{x})) = \nabla \boldsymbol{u}_{\boldsymbol{x}_0}(\boldsymbol{x})\boldsymbol{l}_i(\boldsymbol{x}), \quad i = 1, 2, 3, \quad \boldsymbol{x} \in N_{\boldsymbol{x}_0}, \boldsymbol{x}_0 \in M, \tag{4}$$

and vice versa. Clearly, if two crystal states are elastically related to each other, then they are locally elastically related to each other—however the converse proposition is false.

2.2 Scalar Invariants

The archetypical scalar invariant is the dislocation density tensor field S (ddt). It is defined in dimension 3 in terms of the duals of the lattice vector fields: subject to the choice of a local chart and using the Euclidean structure of $\mathbb{R}^3$, the lattice $\mathbf{l}(x)$, $x \in M$, induces a dual frame (**dual lattice**) $\mathbf{d} = \{\boldsymbol{d}_i(\cdot); i = 1, 2, 3\}$ such that $d_i(x) \cdot l_j(x) = \delta_{ij}, i, j = 1, \ldots, n, x \in M$, where δ_{ij} denotes the usual Kronecker delta. The components of the ddt S are denoted S_{ij} and defined by the equations

$$n(\boldsymbol{x})S_{ij}(\boldsymbol{x}) = \nabla \wedge d_i(\boldsymbol{x}) \cdot d_j(\boldsymbol{x}), i, j = 1, 2, 3, \quad \boldsymbol{x} \in M, \tag{5}$$

where $n(\boldsymbol{x})$ is the lattice volume element ($n(\boldsymbol{x})$ is the determinant of the dual lattice at $\boldsymbol{x}$).

One notes that the ddt depends on the values of the dual lattice vector fields and their first derivatives at any point of M. Via an analogue of (5) one can calculate the value of the ddt corresponding to the lattice vector fields defined in (2) above, denoted $\widetilde{S_{ij}}$, and it turns out that

$$\widetilde{S}_{ij}(\boldsymbol{u}(x)) = S_{ij}(x), \quad i, j = 1, \ldots, \quad \boldsymbol{x} \in M. \tag{6}$$

Objects which are unchanged by elastic deformation in the sense of (analogues of) (6) are called **scalar (elastic) invariants**—to be precise, let $\Delta^{(r)}$ consist of the fields of gradients of lattice vector fields $\boldsymbol{l}_i(\cdot)$ of order r, and let $\tilde{\Delta}^{(r)}$ derive, similarly, from lattice vector fields $\tilde{\boldsymbol{l}}_i(\cdot)$. Then $f : \Delta^{(r)} \to \mathbb{R}$ is a scalar (elastic) invariant if whenever (2) holds, $f(\Delta^{*(r)}(\boldsymbol{u}(\boldsymbol{x})) = f(\Delta^{(r)}(\boldsymbol{x})), \boldsymbol{x} \in M$.

There is an infinite number of such objects, depending on gradients of the lattice vector fields of arbitrary order (see Davini and Parry [10, 11], Olver [35], Parry and Silhavy [41], for more information). For example, one can show that $\boldsymbol{l}_i \cdot \nabla S_{jk}$ is also a scalar elastic invariant, for each $i, j, k = 1, 2, \ldots$, ($\boldsymbol{l}_i \cdot \nabla S_{jk}$ is called a first order directional derivative of S_{jk}) and that successive directional derivatives of the dislocation density tensor are also unchanged under a diffeomorphism of M. We shall call the set of first order directional derivatives $l_i \cdot \nabla S_{jk}$, the scalar invariants of 'first order', with the analogous nomenclature for the sets of higher order directional derivatives.

Note that in dimension 3 at most three of these scalar invariants can be independent non-constant functions, since three independent functions parameterize a local chart. Now one can view (6) as a necessary condition on crystal states Σ, $\tilde{\Sigma}$ that those two states be elastically related—the condition is that there must exist a diffeomorphism $\boldsymbol{u} : M \to \tilde{M}$ such that (6) holds, and there is a corresponding necessary condition for each independent scalar invariant. However, conditions such as (6) are not particularly useful as they stand, because one cannot immediately verify, given states Σ, $\tilde{\Sigma}$, whether or not there exists a smooth invertible mapping $\boldsymbol{u}$ such that (6) holds—a useful reformulation is given in the next subsection.

Consider the case where there are precisely three independent scalar invariants (in dimension 3)—they must occur amongst the directional derivatives of the ddt of order ≤ 2: for if the first such invariant is some component of the ddt, and if no other component of the ddt is independent of the first then a second independent invariant must be found amongst the first order directional derivatives of the ddt, and so on.

In the general case (still in dimension 3) suppose that three independent scalar invariants occur amongst the first k directional derivatives of the ddt, where $k \leq 2$. These invariants parameterize a local chart, and so the scalar invariants of order $k+1$ may be expressed as functions of the three independent invariants. Given those functions, it is straightforward to find, inductively, the form of the function which expresses any invariant of arbitrary finite order in terms of the three independent invariants, see [40]. The case where there are fewer than three independent scalar invariants may be discussed analogously.

2.3 Classifying Manifold

Let us introduce $\mathfrak{F}$, the set of (fields of) directional derivatives of S of order ≤ 3:

$$\mathfrak{F} = \{(S(\cdot), (\boldsymbol{l}_a \cdot \nabla S)(\cdot), (\boldsymbol{l}_b \cdot \nabla(\boldsymbol{l}_a \cdot \nabla S))(\cdot), (\boldsymbol{l}_c \cdot \nabla(\boldsymbol{l}_b \cdot \nabla(\boldsymbol{l}_a \cdot \nabla S)))(\cdot)); \\ a, b, c = 1, 2, 3\}. \tag{7}$$

Then $\mathfrak{F}$ is a **functional basis** of all scalar elastic invariants deriving from the lattice vector fields in the sense that all scalar elastic invariants (of all orders) can be calculated if $\mathfrak{F}$ is known in M. For by the discussion in the previous section, the independent scalar invariants (≤ 3 in number) must occur as directional derivatives of order ≤ 2. The third order directional derivatives can be expressed as functions of the independent invariants, and these functions determine all the higher order invariants. (Let f denote any one of the third order scalar directional derivatives in (7), and let I denote the set of independent invariants, then f is defined on $\{I(\boldsymbol{x}), \boldsymbol{x} \in M\}$, via (7). The higher order directional derivatives are determined in terms of the derivatives of f, assuming sufficient regularity.)

Now let

$$\mathrm{CM}_{\Sigma} = \{g(x); x \in M, g \in \mathfrak{F}\} \tag{8}$$

be the **classifying manifold** associated with Σ (Olver [35]), and define $\mathrm{CM}_{\tilde{\Sigma}}$ via (2) similarly.

The following result addresses the question—if two crystal states are given, that is, if two distributions of vector fields (defined over different regions of $\mathbb{R}^3$) are given, then how does one decide if there is a macroscopic (elastic) deformation, mapping one region to the other, which also maps one set of fields to the other?

Suppose that $\mathrm{CM}_{\Sigma} = \mathrm{CM}_{\tilde{\Sigma}}$ and that $g(x_0) = \tilde{g}(\tilde{x}_0)$ where g and $\tilde{g}$ are corresponding elements of $\mathfrak{F}$ and $\tilde{\mathfrak{F}}$, for some $x_0 \in M, \tilde{x}_0 \in \tilde{M}$ in particular. Then Σ and $\tilde{\Sigma}$ are locally elastically related to each other, and (4) holds with $u_{x_0}(x_0) = \tilde{x}_0$.

The result gives a sufficient condition that two crystal states be **locally** elastically related, by doing so it identifies a particular set of scalar invariant functions, $\mathfrak{F}$, and shows that it is the **range** of those functions, $\{\mathfrak{F}(x); x \in M\}$, which provides the relevant condition. So $\mathfrak{F}$ gives a quite general set of 'plastic strain variables', in continuum mechanics terminology, in this context.

3 Burgers Vectors, Invariant Integrals, and Neutrally Related States

Here we define the Burgers vectors and Burgers integrals. The Burgers integrals are particular 'invariant integrals' (they are objects invariant under elastic deformation more general than the scalar invariants of the previous section)—we define these integrals and note that there is a basis of invariant integrals, in a certain sense. The basis of invariant integrals **strictly includes** the Burgers integrals. We continue to address the question of how to decide, if two crystal states are given, whether or not there is a macroscopic (elastic) deformation, mapping one region to the other, which also maps one set of lattice vector fields to the other. The integral invariants must match, in the two states, if such an (global) elastic deformation exists—and this gives us a set of corresponding necessary conditions. These necessary conditions can be reformulated in terms of the existence, or not, of a non-trivial solution of a set of partial differential equations: when two such states are not elastically related, the partial differential equations have a non-trivial solution, and in that case we say that the states are **neutrally related**. It is a fact that neutrally related exist, so these necessary conditions are not sufficient conditions.

One can interpret neutrally related states in terms of the 'slip', or rearrangement, of the lattice vector fields, and decompose the changes of state (rigorously) in terms of elastic deformation and slip, but we do not do so in this chapter, see [10, 11]. Slip has an ad hoc status in many phenomenological theories of inelastic behaviour, as a kinematical or constitutive assumption (likewise for elastic-plastic decomposition)—but here these concepts arise naturally from analysis subordinate to the construction of invariants relating to the crystal states. Finally, in this section, we show that neutrally related states are locally elastically related, and for completeness we give constraints on the ddt, and the set of first order directional derivatives of the ddt, which guarantee that those quantities derive from a set of lattice vector fields in the prescribed manner.

3.1 Burgers Vectors and Invariant Integrals

The **Burgers integral**, defined by $\int_{\mathscr{C}} \boldsymbol{d}_i \cdot \boldsymbol{dx}$, where $\mathscr{C}$ is a circuit in M, is a prototypical **invariant integral**. Proof of this fact requires discussion of the transformation properties, induced by (2), of the various derivatives of the lattice vector fields. Thus the Burgers vector is unchanged if the dual lattice vector fields are transformed in a manner compatible with (2), and the circuit $\mathscr{C}$ is mapped to $\boldsymbol{u}(\mathscr{C})$.

More generally, one searches for circuit, closed surface, or volume integrals with integrands depending on the lattice vector fields and their derivatives of any order (succinctly, 'differential functions') which are functionals independent of elastic deformation, see Davini, Parry and Šilhavý [10, 11, 41].

Thus if $F_{\Omega}(\{\boldsymbol{l}_i(\cdot)\})$ is any such functional, and fields $\{\tilde{\boldsymbol{l}}_i(\cdot)\}$ are defined by (2), then $F_{\tilde{M}}(\{\tilde{\boldsymbol{l}}_i(\cdot)\}) = F_M(\{\boldsymbol{l}_i(\cdot)\})$, with $\tilde{M} = \boldsymbol{u}(M)$. For example, since $\det(\{\tilde{\boldsymbol{l}}_a\}) = \det(\nabla \boldsymbol{u}) \det(\{\boldsymbol{l}_a\})$ (where $\det(\cdot)$ denotes the determinant) then, with the definition $n = \boldsymbol{d}_1 \cdot \boldsymbol{d}_2 \wedge \boldsymbol{d}_3$ one obtains that $\int_V n dV = \int_{\tilde{V} \equiv \boldsymbol{u}(V)} \tilde{n} d\tilde{V}$ is an invariant integral, $V \subseteq M$. In fact, it is straightforward to show that

$$\int_{\mathscr{C}} \boldsymbol{d}_a \cdot \boldsymbol{dx}, \int_V n\, dV, \int_V Sn\, dV, \text{ are invariant integrals,} \tag{9}$$

where $\mathscr{C}$ is a circuit, by recalling the definition of the dual fields $\{(\boldsymbol{d}_a(\cdot))\}$ and the dislocation density S, and by calculating relevant transformation properties. Note that integrals over closed surfaces can be expressed, alternatively, as volume integrals.

Now it is important to note that if ν is any scalar (i.e., real-valued differential function) invariant, then

$$\{(\boldsymbol{l}_i \cdot \nabla \nu)(\cdot)\} \text{ is a set of scalar invariant fields.} \tag{10}$$

Recall that if S has at least one non-constant component, then there is in general an infinite number of scalar invariants, obtained from that particular component of

S by successive directional differentiation. So there is an infinite number of integral invariants too, because (for example)

$$\int_{\mathscr{C}} \nu \boldsymbol{d}_i \cdot d\boldsymbol{x} \text{ is an integral invariant if } \nu \text{ is any scalar invariant.} \tag{11}$$

It turns out that there is a basis of integral invariants in the sense that if densities corresponding to the basis integral invariants are given, as fields in M, then the densities of all integral invariants can be determined. The basis integral invariants are

$$\int_{\mathscr{C}} \nu \boldsymbol{d}_a \cdot d\boldsymbol{x}, \int_V \nu n \, dV : \quad \nu \in \{1, S, \{(\boldsymbol{l}_a \cdot \nabla S)\}\}. \tag{12}$$

3.2 Neutral Related States

The basis integral invariants match, in the obvious sense, in elastically related crystal states, so they provide necessary conditions that two crystal states, Σ and $\tilde{\Sigma}$, be elastically related. Now suppose that Σ and $\tilde{\Sigma}$ are not necessarily elastically related, but that the basis integral invariants nevertheless match in the two states. Then the Burgers integrals match, for example, in the sense that there exists a smooth invertible mapping $\boldsymbol{u} : M \to \boldsymbol{u}(M) = \tilde{M}$ such that $\int_{\mathscr{C}} \boldsymbol{d}_i \cdot d\boldsymbol{x} = \int_{\boldsymbol{u}(\mathscr{C})} \tilde{\boldsymbol{d}}_i \cdot d\boldsymbol{x}$. We can simplify this condition a little: we map $\tilde{\Sigma}$ elastically, via the inverse of that mapping $\boldsymbol{u}$, to a state $\Sigma' = \left\{\left\{\boldsymbol{l}'_i(\cdot)\right\} ; M\right\}$. The integral invariants match in states Σ and $\tilde{\Sigma}$ if and only if they match in states Σ and Σ'. So $\int_{\mathscr{C}} \boldsymbol{d}_i \cdot d\boldsymbol{x} = \int_{\mathscr{C}} \boldsymbol{d}'_i \cdot d\boldsymbol{x}$, for example (the point being that the circuits are identical on the left and right sides of this relation), and by noting that the circuit $\mathscr{C}$ is arbitrary one deduces that $\nabla \wedge \boldsymbol{d}_i = \nabla \wedge \boldsymbol{d}'_i$, via Stokes' theorem. It is usual to call the quantities $\nabla \wedge \boldsymbol{d}_i, i = 1, 2, 3$, the **Burgers Vectors**, in this context.

Applying the simplification of the last paragraph to the basis of integral invariants one obtains the relations

$$\nabla \wedge \boldsymbol{d}_i = \nabla \wedge \boldsymbol{d}'_i, n = n', \nu = \nu', \nabla \nu \wedge \boldsymbol{d}_i = \nabla \nu \wedge \boldsymbol{d}'_i, \ \nu \in \{S, \{(\boldsymbol{l}_i \cdot \nabla S)\}\}, \tag{13}$$

which may be regarded as partial differential equations to determine the fields $\left\{\boldsymbol{l}'_i(\cdot)\right\}$, given $\{\boldsymbol{l}_i(\cdot)\}$. Of course if Σ and $\tilde{\Sigma}$ are elastically related, then Σ and Σ' are identical, hence if (13) has a non-trivial solution for $\left\{\boldsymbol{l}'_i(\cdot)\right\}$, given $\{\boldsymbol{l}_i(\cdot)\}$ (i.e., $\left\{\boldsymbol{l}'_i(\cdot)\right\} \neq \{\boldsymbol{l}_i(\cdot)\}$), then Σ and $\tilde{\Sigma}$ are not elastically related. If (13) has a non-trivial solution for $\left\{\boldsymbol{l}'_i(\cdot)\right\}$, given $\{\boldsymbol{l}_i(\cdot)\}$, we say that Σ and Σ' are **neutrally related states**, and we also say that Σ **allows neutral deformations**.

A simple example of the non-uniqueness of solutions of (13) is given in [11]. The non-uniqueness of solutions of (13) shows that the necessary conditions derived from the invariant integrals are not sufficient to determine whether or not two crystal states are elastically related (indeed (13) provides just local conditions)—there are crystal states which are not elastically related to one another such that all the invariant integrals match in the two states, for some choice of the mapping $\boldsymbol{u}$. One can, however, interpret neutrally related states in terms of 'slip', or rearrangement, of vector fields, but we do not do so here, see [10, 11] for details.

One implication of (13) is very important for subsequent discussion. Suppose that (13) has a solution $\{\boldsymbol{l}_i'(\cdot)\} \neq \{\boldsymbol{l}_i(\cdot)\}$ and that ν is a non-constant component of S. Then by differentiating $(13)_3$ ((13) holds for all points in M), $\nabla\nu = \nabla\nu'$, and so from $(13)_4$, $\nabla\nu \wedge (\boldsymbol{d}_i - \boldsymbol{d}_i') = \boldsymbol{0}$, $a = 1, 2, 3$. If ϑ is any of $\boldsymbol{l}_i \cdot \nabla S, i = 1, 2, 3$, we have from $(13)_3$ that $\nabla\vartheta \wedge (\boldsymbol{d}_i - \boldsymbol{d}_i') = \boldsymbol{0}$, $i = 1, 2, 3$. Since $\boldsymbol{d}_i(\cdot) \neq \boldsymbol{d}_i'(\cdot)$ for some $i = 1, 2, 3$, it follows that $\vartheta = \vartheta(\nu)$ (in loose notation), when (13) has non-unique solutions and S has a non-constant component. Furthermore, since $\boldsymbol{l}_i \cdot \nabla\vartheta = (\boldsymbol{l}_i \cdot \nabla\nu)\frac{d\vartheta}{d\nu}$, it follows that all second order directional derivatives of S are functions of ν, and by induction all directional derivatives of S have the same property. Also, let ϑ' be any of $\boldsymbol{l}_i' \cdot \nabla S', i = 1, 2, 3$, then one shows quite readily that $\vartheta'(\nu') = \vartheta'(\nu) = \vartheta(\nu)$, and that all higher order directional derivatives have the analogous property.

Therefore the classifying manifolds of states Σ and Σ' are identical, $\text{CM}_\Sigma = \text{CM}_{\Sigma'}$, so that neutrally related states are locally elastically related, by the result highlighted in Sect. 2.3.

Note that if we consider only crystal states Σ where (13) has non-unique solutions for Σ', then

either $S =$ constant, or all elastic invariants are functions of one non-constant component of S (denoted ν). (14)

Also note that there are constraints on the functions S and its first order directional derivatives, see [40]:

(a) From (5), $n\, S_{ab}\boldsymbol{l}_b = \nabla \wedge \boldsymbol{d}_a$ and so one has $\nabla \cdot (nS_{ab}\boldsymbol{l}_b) = 0,\ a = 1, 2, 3$. This gives

$$S_{ai}(\varepsilon_{ijk}S_{jk}) + \boldsymbol{l}_k \cdot \nabla S_{ak} = 0, \quad a = 1, 2, 3. \tag{15}$$

In the case that S is constant, (15) reduces to

$$S_{ai}(\varepsilon_{ijk}S_{jk}) = 0, \quad a = 1, 2, 3, \tag{16}$$

which is a representation of the Jacobi identity (from the theory of Lie algebras);

(b) From $\nabla \nu = (\boldsymbol{l_a} \cdot \nabla \nu)\boldsymbol{d}_a$ one has $\nabla \wedge ((\boldsymbol{l_a} \cdot \nabla \nu)\boldsymbol{d}_a) = \boldsymbol{0}$, and this gives

$$S_{ab} F_a = \varepsilon_{bac} F'_a F_c, \quad b = 1, 2, 3, \tag{17}$$

where we introduce the notation $F_a \equiv \boldsymbol{l_a} \cdot \nabla \nu$, $a = 1, 2, 3$, and $F'_a \equiv \frac{dF_a}{d\nu}$, $a = 1, 2, 3$.

In the case that S is constant, (17) is trivially satisfied.

(c) In the case that Σ allows neutral deformations, a short calculation based on the first three equations in (13) gives, in addition, that

$$S_{ab} F_b = 0, \quad a = 1, 2, 3. \tag{18}$$

When (15), (17), and (18) hold, it can be shown that there exists a crystal state, with the corresponding functions S and its first order directional derivatives the same as those in Σ, such that the lattice vector fields themselves depend only on the chosen non-constant component of S, denoted ν.

Finally in this section, according to [8, 16], (15) and (17) together are sufficient for the local integrability of

$$n S_{ab} \boldsymbol{l}_b = \nabla \wedge \boldsymbol{d}_a, \quad \nabla \nu = F_a \boldsymbol{d}_a, \quad a = 1, 2, 3, \tag{19}$$

for the fields $\{\boldsymbol{l}_a(\cdot)\}$, given S, F_a as functions of ν. Therefore (15) and (17) are the only constraints on the functions S, F_a, in general.

4 Lie Groups

The theory of Lie groups is intimately related to the classification of distributions of vector fields modulo diffeomorphism, and so also related to the 'elastic-plastic' decompositions of continuum mechanics (because if a change of state has a non-trivial plastic part, corresponding vector fields are inequivalent modulo diffeomorphism). Here we review aspects of the theory of Lie groups and algebras which are necessary in order to understand this relationship, and explore the duality between descriptions/classifications of crystal states in terms of the dislocation density functions, and in terms of Lie groups/Lie algebras.

For example, in the case where the ddt is constant in a given crystal state, it turns out that one can regard the elements of the domain M, which are the 'material points' of continuum mechanics, as elements of a Lie group. Note also that, if the ddt is constant in M, the corresponding lattice vector fields satisfy a self-similarity condition, and that the function which encapsulates the self-similarity can be regarded as a Lie group composition (multiplication) function [45]. Vector fields which have this type of self-similarity are called right invariant fields, in the

Lie group literature—the self-similarity of the fields is a statement that the lattice vector fields 'fit together' in a certain way, in a defective crystal with constant ddt.

In the first two subsections below, we expand on the items of the previous paragraph, then we recall the construction of the Lie algebra corresponding to a given Lie group, and introduce the structure constants of the algebra and the exponential mapping from the algebra to the group. Also we recall the notions of isomorphic Lie groups and algebras.

4.1 Constant Dislocation Density Tensor

Let us consider crystal states where the lattice vector fields $\{\boldsymbol{l}_i(\cdot)\}$ are defined in $M \equiv \mathbb{R}^3$, and are such that the dislocation density tensor S is constant in $\mathbb{R}^3$. The condition that S is constant is an integrability condition [45], which guarantees that for given lattice vector fields, the partial differential system

$$\boldsymbol{l}_i(\boldsymbol{\psi}(\boldsymbol{x}, \boldsymbol{y})) = \nabla_1 \boldsymbol{\psi}(\boldsymbol{x}, \boldsymbol{y}) \boldsymbol{l}_i(\boldsymbol{x}), \qquad i = 1, 2, 3, \tag{20}$$

where $\nabla_1 \boldsymbol{\psi}(\cdot, \cdot)$ denotes the gradient of $\boldsymbol{\psi}$ with respect to its first argument, has a solution for the function $\boldsymbol{\psi}$. **Note that (20) expresses a self-similarity of the lattice vector fields** $\{\boldsymbol{l}_i(\cdot)\}$ in M. Moreover, the function $\boldsymbol{\psi} : \mathbb{R}^3 \times \mathbb{R}^3 \to \mathbb{R}^3$ can be taken to satisfy the properties required for it to be a Lie group composition function, i.e.,

$$\begin{aligned} &\boldsymbol{\psi}(\mathbf{0}, \boldsymbol{x}) = \boldsymbol{\psi}(\boldsymbol{x}, \mathbf{0}) = \boldsymbol{x}, \ \boldsymbol{\psi}(\boldsymbol{x}, \boldsymbol{x}^{-1}) = \boldsymbol{\psi}(\boldsymbol{x}^{-1}, \boldsymbol{x}) = \mathbf{0}, \\ &\boldsymbol{\psi}(\boldsymbol{\psi}(\boldsymbol{x}, \boldsymbol{y}), \boldsymbol{z}) = \boldsymbol{\psi}(\boldsymbol{x}, \boldsymbol{\psi}(\boldsymbol{y}, \boldsymbol{z})), \end{aligned}$$

where $\mathbf{0}$ is regarded as the group identity element and $\boldsymbol{x}^{-1}$ is the unique inverse of the element $\boldsymbol{x}$ [37, 45]. So the points of M can be represented as Lie group elements. Given an appropriate value of the dislocation density tensor S one can specify a corresponding Lie group G by constructing fields $\boldsymbol{l}_i(\cdot)$, $i = 1, 2, 3$ such that the dual fields satisfy (5) in Sect. 2.2 and then solving (20) for the group composition function $\boldsymbol{\psi}$. Note that when (20) holds, the vector field $\{\boldsymbol{l}_i(\cdot)\}$ is said to be **right invariant** with respect to the composition function ψ. So when the lattice vector fields $\{\boldsymbol{l}_i(\cdot)\}$ are such that S is constant in $M \equiv \mathbb{R}^3$, they represent right invariant fields on a certain Lie group.

It is standard that each Lie group $G = (\mathbb{R}^3, \boldsymbol{\psi})$ has a corresponding Lie algebra $\mathfrak{g}$ consisting of the vector space $\mathbb{R}^3$, in this case, with Lie bracket operation $[\cdot, \cdot] : \mathbb{R}^3 \times \mathbb{R}^3 \to \mathbb{R}^3$ defined by

$$[\boldsymbol{x}, \boldsymbol{y}] = C_{ijk} x_j y_k \boldsymbol{e}_i, \qquad \boldsymbol{x}, \boldsymbol{y} \in \mathbb{R}^3, \tag{21}$$

where $\{e_1, e_2, e_3\}$ is a basis of $\mathbb{R}^3$, $x = x_j e_j$, $y = y_j e_j$, and where the structure constants C_{ijk}, with respect to this choice of basis, are given by

$$C_{ijk} = \frac{\partial^2 \psi_i}{\partial x_j \partial y_k}(\mathbf{0}, \mathbf{0}) - \frac{\partial^2 \psi_i}{\partial x_k \partial y_j}(\mathbf{0}, \mathbf{0}), \tag{22}$$

where $\boldsymbol{\psi}(x, y) = \psi_i(x, y)e_i$. The structure constants with respect to the basis $\{e_1, e_2, e_3\}$ are directly related to the components of the dislocation density tensor, see Elzanowski and Parry [12].

Let $l_i(\cdot)$, $i = 1, 2, 3$, satisfy (20) and let ν_1, ν_2, ν_3 be given real numbers. Define the integral curve through x_0 of the field $\nu_i l_i(\cdot)$ to be the solution $\{x(t) : t \in \mathbb{R}\}$ of the ordinary differential equation $\dot{x}(t) = \nu_i l_i(x(t))$, $x(0) = x_0$. Note that $\boldsymbol{\nu} := \nu_i l_i(\mathbf{0})$ determines the field $\nu_i l_i(\cdot)$ by (20). (By virtue of this remark one can think of the Lie algebra of G either as the vector space $\mathbb{R}^3$, with Lie bracket given by (21), or as the vector space of right invariant vector fields on G, with the Lie bracket of vector fields defined by (41), Sect. 4.4 below. Note that the Lie bracket of right invariant fields is right invariant.)

One defines the mapping $\exp(\boldsymbol{\nu}) : G \to G$, and the group element $e^{(\boldsymbol{\nu})}$, by

$$\exp(\boldsymbol{\nu})(x_0) = x(1), \ e^{(\boldsymbol{\nu})} = \exp(\boldsymbol{\nu})(\mathbf{0}). \tag{23}$$

It is a fact that

$$\exp(\boldsymbol{\nu})(x) = \boldsymbol{\psi}(e^{(\boldsymbol{\nu})}, x), \tag{24}$$

which states that the flow along the integral curves of the lattice vector fields corresponds to group multiplication by the group element $e^{(\boldsymbol{\nu})}$.

4.2 Isomorphic Lie Groups and Algebras

Recall the following facts and definitions from the theory of Lie groups and algebras, see for example [2, 3, 18, 19, 21, 23, 24, 45, 46].

Let $\mathfrak{g}$ and $\mathfrak{h}$ be Lie algebras with Lie brackets $[\cdot, \cdot]_{\mathfrak{g}}$, $[\cdot, \cdot]_{\mathfrak{h}}$, respectively. (In the context of this chapter, both brackets $[\cdot, \cdot]_{\mathfrak{g}}$, $[\cdot, \cdot]_{\mathfrak{h}}$ map $\mathbb{R}^3 \times \mathbb{R}^3 \to \mathbb{R}^3$). A linear transformation $L : \mathfrak{g} \to \mathfrak{h}$ which satisfies

$$[Lx, Ly]_{\mathfrak{h}} = L[x, y]_{\mathfrak{g}}, \quad x, y \in \mathfrak{g}, \tag{25}$$

is called a **Lie algebra homomorphism**. If $C^{\mathfrak{g}}_{ijk}$, $C^{\mathfrak{h}}_{ijk}$ are the structure constants for $\mathfrak{g}$, $\mathfrak{h}$, respectively, then (8) implies

$$C^{\mathfrak{h}}_{ijk} L_{jp} L_{kq} = L_{ir} C^{\mathfrak{g}}_{rpq}, \tag{26}$$

where $Le_i = L_{ji} e_j$, $i, j = 1, 2, 3$.

Let G and H be Lie groups with group multiplication functions ψ_G, ψ_H, respectively. A smooth mapping $\phi : G \to H$ which satisfies

$$\psi_H(\phi(x), \phi(y)) = \phi(\psi_G(x, y)), \quad x, y \in G \tag{27}$$

is called a **Lie group homomorphism**.

If $\mathfrak{g}$ is the Lie algebra of G, and $\mathfrak{h}$ is the Lie algebra of H, and $\phi : G \to H$ is a Lie group homomorphism, then $\nabla\phi(\mathbf{0}) \equiv L$ is a Lie algebra homomorphism. Conversely if L satisfies (25), then there exists a Lie group homomorphism ϕ such that $\nabla\phi(\mathbf{0}) = L$. Also,

$$\phi\left(e^{\nu}\right) = e^{(\nabla\phi(\mathbf{0})\nu)}, \quad \nu \in \mathfrak{g} \equiv \mathbb{R}^3, \tag{28}$$

where ϕ satisfies (27), where the exponential on the left hand side of (28) is the exponential which maps $\mathfrak{g}$ to G, and that on the right hand side maps $\mathfrak{h}$ to H. Relation (28) allows one to calculate the Lie group homomorphisms explicitly if the Lie algebra homomorphisms are found by solving (26). $\phi(\cdot)$ (resp. L) is called a **Lie group (resp. algebra) isomorphism** if it (resp. L) is invertible. An isomorphism $\phi : G \to G$ (resp. $L : \mathfrak{g} \to \mathfrak{g}$) is called an automorphism ($\phi(\cdot)$ and $\phi^{-1}(\cdot)$ have to be smooth).

The above facts are useful when it comes to constructing vector fields with given scalar invariants—since such vector fields are determined only modulo diffeomorphism (i.e., modulo elastic deformation), one can choose the diffeomorphisms to simplify calculations in many cases.

4.3 *Campbell–Baker–Hausdorff Formula, Canonical Group J*

Here we introduce the particular Lie group which we have selected to illustrate ideas and methods—it is a three dimensional **nilpotent**[1] Lie group which we refer to as the **canonical group**. The group elements are represented by elements of $\mathbb{R}^3$, and the group composition is given explicitly, similarly the Lie algebra is defined on

[1]Let G be a three dimensional Lie group, with commutator $(x, y) \equiv x^{-1}y^{-1}xy$. Let $G \equiv G_0$ and define $G_1 \equiv (G, G_0)$, the group generated by elements of the form (x, y), $x \in G$, $y \in G_0$. Define $G_k \equiv (G, G_{k-1})$ inductively, $k \geq 1$. G is called nilpotent if and only if G_k is the trivial group $\{\mathbf{0}\}$ for sufficiently large k. For three dimensional nilpotent groups, $G \equiv G_0 \supseteq G_1 \supseteq G_2 = \{\mathbf{e}\}$, where $\mathbf{e}$ is a temporary notation for the group identity $\mathbf{0}$.

Let $\mathfrak{g}$ be the Lie algebra corresponding to a Lie group G, with Lie bracket $[x, y]$, $x, y \in \mathfrak{g}$. Let $\mathfrak{g} \equiv \mathfrak{g}_0$ and define $\mathfrak{g}_1 \equiv [\mathfrak{g}, \mathfrak{g}_0]$, the subspace generated by elements of the form $[x, y]$, $x \in \mathfrak{g}$, $y \in \mathfrak{g}_0$. Define $\mathfrak{g}_k \equiv [\mathfrak{g}, \mathfrak{g}_{k-1}]$ inductively, $k \geq 1$. $\mathfrak{g}$ is called nilpotent if and only if $\mathfrak{g}_k$ is the trivial subspace $\{\mathbf{0}\}$ for sufficiently large k. For three dimensional nilpotent algebras, $\mathfrak{g} \equiv \mathfrak{g}_0 \supseteq \mathfrak{g}_1 \supseteq \mathfrak{g}_2 = \{\mathbf{0}\}$.

A Lie group is nilpotent if and only if the corresponding Lie algebra is nilpotent (Gorbatsevich, Onishchik, Vinberg [19]).

$\mathbb{R}^3$, with corresponding Lie bracket—it is a special feature of this group that one may identify Lie group and algebra elements, that is, the exponential map from the group to the algebra is the identity mapping. This special feature simplifies many calculations, e.g., it is a fact that the commutator (see footnote) of two group elements may be identified with the Lie bracket of the corresponding algebra element. We show that the automorphisms (symmetries) of the canonical group are linear mappings (homogeneous deformations, in continuum mechanics terms), by virtue of this special feature.

Let $\boldsymbol{\psi}$ be the composition function for a three dimensional Lie Group G, and let $e^{(\cdot)} : \mathfrak{g} \equiv \mathbb{R}^3 \to G$ be the exponential function. The Campbell–Baker–Hausdorff (CBH) formula gives an explicit expression for the quantity $\boldsymbol{c}$ in the relation,

$$e^{(\boldsymbol{c})} = e^{(\boldsymbol{a})} e^{(\boldsymbol{b})}, \quad \boldsymbol{a}, \boldsymbol{b} \in \mathbb{R}^3. \tag{29}$$

One finds the full formula in Varadarajan [50]. We give the formula as it applies to three dimensional nilpotent Lie groups G, it is

$$\boldsymbol{c} = \boldsymbol{a} + \boldsymbol{b} + \frac{1}{2}\left[\boldsymbol{a}, \boldsymbol{b}\right]. \tag{30}$$

The simple form of (30) exposes the following fact—the expression on the right hand side of (30) depends only on the Lie bracket $[\cdot, \cdot]$, i.e., it only depends on the Lie algebra $\mathfrak{g}$; it does not depend on the choice of group G in the isomorphism class of groups which have the Lie algebra determined by the given bracket operation.

Now put

$$\boldsymbol{c} = \boldsymbol{\psi}'(\boldsymbol{a}, \boldsymbol{b}) \tag{31}$$

and note that $\boldsymbol{\psi}'$ has the properties required in order to regard it as a group composition function on the vector space associated with the given Lie algebra ($\mathbb{R}^3$ in this case). We call this group the **canonical group** J associated with the given structure constants, i.e., with the Lie algebra given implicitly in (30) via the choice of Lie bracket.

Let us choose the structure constants with respect to a basis $\{\boldsymbol{e}_1, \boldsymbol{e}_2, \boldsymbol{e}_3\}$ of $\mathbb{R}^3$ to have the form

$$C_{ijk} = \lambda \varepsilon_{rjk} p_i p_r, \tag{32}$$

in terms of real parameters λ, p_i, $i = 1, 2, 3$. (See [37] for discussion of this choice). One may check that the corresponding Lie algebra is nilpotent. Let $\boldsymbol{p}$ denote the vector in $\mathbb{R}^3$ with components p_i, $i = 1, 2, 3$ with respect to the chosen basis. Then, the composition function in the group J (dropping the prime in $\boldsymbol{\psi}'$) is given by

$$\boldsymbol{\psi}(\boldsymbol{x}, \boldsymbol{y}) = \boldsymbol{x} + \boldsymbol{y} + \frac{1}{2}\lambda \boldsymbol{p}(\boldsymbol{p} \cdot \boldsymbol{x} \wedge \boldsymbol{y}), \tag{33}$$

and one can show that the three vector fields $\boldsymbol{\ell}_i(\boldsymbol{x}) = \nabla_1 \boldsymbol{\psi}(\boldsymbol{0}, \boldsymbol{x})\boldsymbol{e}_i = \boldsymbol{e}_i + \frac{1}{2}\lambda \boldsymbol{p}(\boldsymbol{x} \wedge \boldsymbol{p} \cdot \boldsymbol{e}_i)$, (so $\boldsymbol{e}_i = \boldsymbol{\ell}_i(\boldsymbol{0})$ in particular), $i = 1, 2, 3$, provide a basis of the set of right invariant vector fields. Also from $\dot{\boldsymbol{x}} = v_a \boldsymbol{\ell}_a(\boldsymbol{x}) = \boldsymbol{v} + \frac{1}{2}\lambda \boldsymbol{p}(\boldsymbol{x} \wedge \boldsymbol{p} \cdot \boldsymbol{v})$, one obtains

$$\exp(\boldsymbol{v}t)(\boldsymbol{x}) = \boldsymbol{x} + \boldsymbol{v}t + \frac{1}{2}\lambda \boldsymbol{p}(\boldsymbol{x} \wedge \boldsymbol{p} \cdot \boldsymbol{v}t). \tag{34}$$

Therefore, $\exp(\boldsymbol{v}t)(\boldsymbol{0}) = \boldsymbol{v}t$, (which implies that the one-parameter groups in J are straight lines), and the corresponding exponential mapping is

$$e^{(\boldsymbol{x})} = \exp(\boldsymbol{x})(\boldsymbol{0}) = \boldsymbol{x}, \quad \boldsymbol{x} \in \mathbb{R}^3. \tag{35}$$

Remarkably, **Lie group and Lie algebra elements may be identified**, via (35), in this particular case.

Also we have

$$(\boldsymbol{x}, \boldsymbol{y}) = [\boldsymbol{x}, \boldsymbol{y}], \quad \boldsymbol{x}, \boldsymbol{y} \in \mathbb{R}^3, \tag{36}$$

so that the group commutator, $(\boldsymbol{x}, \boldsymbol{y}) \equiv \boldsymbol{x}^{-1}\boldsymbol{y}^{-1}\boldsymbol{x}\boldsymbol{y}$, (which represents the 'finite' Burgers vector obtained by successive flow along the right invariant fields defined by $\boldsymbol{y}, \boldsymbol{x}, \boldsymbol{y}^{-1}, \boldsymbol{x}^{-1}$, given (24) above) may be calculated via the Lie bracket of algebra elements;

Now by a slight extension of (28), if a linear transformation L is a Lie algebra automorphism (i.e., an isomorphism from the algebra to itself), there is a Lie group automorphism $\boldsymbol{\phi} : J \to J$ such that $\nabla\boldsymbol{\phi}(\boldsymbol{0}) = L$. Then $\boldsymbol{\phi}(\boldsymbol{e}^{(\boldsymbol{x})}) = e^{(\nabla\boldsymbol{\phi}(0)\boldsymbol{x})}$ gives, noting that the exponentials on both sides of this relation satisfy (35), that

$$\boldsymbol{\phi}(\boldsymbol{x}) = \nabla\boldsymbol{\phi}(\boldsymbol{0})\boldsymbol{x}, \quad \boldsymbol{x} \in \mathfrak{g} \equiv J. \tag{37}$$

Relation (37) shows that the **automorphisms of J are linear mappings ('homogeneous deformations' in continuum mechanical terms)**, and this fact helps a great deal when one comes to calculate the symmetries of discrete subgroups of J later on.

Let $\boldsymbol{j}$ denote the Lie algebra of J. For later use we calculate the Lie algebra automorphisms of $\boldsymbol{j}$ in the case that $\lambda = 1$, $p_1 = p_2 = 0$, $p_3 = 1$, so that $\boldsymbol{p} = \boldsymbol{e}_3$, referring to (32) above. One finds that

$$L_{13} = L_{23} = 0, L_{33} = L_{11}L_{22} - L_{12}L_{21} \neq 0, \tag{38}$$

where $L\boldsymbol{x} = L(x_i e_i) = x_i(L\boldsymbol{e}_i) \equiv x_i(L_{ji}\boldsymbol{e}_j) = (L_{ji}x_i)\boldsymbol{e}_j$. Also for later use, in the same particular case of (32), we show immediately below that there are precisely two inequivalent one-dimensional subgroups of J, modulo automorphisms of J: the one-dimensional subgroups of J have the form

$$H_{\boldsymbol{v}} = \{t\boldsymbol{v} : \boldsymbol{v} \in \boldsymbol{j}, t \in \mathbb{R}\}, \tag{39}$$

by the comment which precedes (35). Let $\boldsymbol{\phi} : J \to J$ be an automorphism of J, with $L \equiv \nabla\boldsymbol{\phi}(0)$. Then

$$\boldsymbol{\phi}(H_{\boldsymbol{v}}) = H_{L\boldsymbol{v}}. \tag{40}$$

We have that:

1. $H_{\boldsymbol{h}}$ is equivalent to $H_{\boldsymbol{e}_3}$ modulo automorphism provided that $L\boldsymbol{e}_3 = \boldsymbol{h}$ for some L satisfying (38). So $\boldsymbol{h} = L_{33}\boldsymbol{e}_3$ and $\boldsymbol{h}$ must be parallel to $\boldsymbol{e}_3$.
2. $H_{\boldsymbol{h}}$ is equivalent to $H_{\boldsymbol{e}_1}$ provided $L\boldsymbol{e}_1 = \boldsymbol{h}$. This requires that, and is satisfied if, $\boldsymbol{h}$ is any vector not parallel to $\boldsymbol{e}_3$.

4.4 Higher Dimensional Lie Groups

A further connection between the ddt and Lie group ideas occurs at higher order than that above, in cases where the ddt is not constant in M. We note to begin that when the ddt is constant, the corresponding right invariant lattice vector fields provide a basis of a Lie algebra of vector fields with Lie bracket defined by (41) below, so that the dimension of the algebra equals the number of lattice vector fields. When the ddt is not constant, it may be that the 'lattice vector fields and certain of their (iterated) Lie brackets' provide a basis of a Lie algebra (with the same choice of Lie bracket), and in that case the dimension of the algebra is strictly greater than the number of lattice vector fields. According to [36], subject to some technical conditions, there is then a corresponding Lie group with Lie algebra isomorphic to that consisting of the lattice vector fields and certain of their (iterated) Lie brackets, which has a dimension equal to the number of 'lattice vector fields and certain of their (iterated) Lie brackets'. So this group has a dimension strictly greater than that of M, and according to [36] there is a group action which, for any choice of group element, maps M to M. We expand on these remarks below, and focus on a particular case which will be useful later in Theorem 1.

Let us use the following sign convention for the Lie bracket of pairs of vector fields $\boldsymbol{v}(\cdot)$, $\boldsymbol{w}(\cdot)$, thus

$$[\boldsymbol{v}, \boldsymbol{w}](\cdot) \equiv \{(\boldsymbol{w} \cdot \nabla)\boldsymbol{v} - (\boldsymbol{v} \cdot \nabla)\boldsymbol{w}\}(\cdot). \tag{41}$$

Denote the Lie bracket of lattice vector fields $\boldsymbol{l}_1(\cdot)$, $\boldsymbol{l}_2(\cdot)$ by $\boldsymbol{L}_3(\cdot)$, thus

$$\boldsymbol{L}_3(\cdot) = [\boldsymbol{l}_1, \boldsymbol{l}_2](\cdot), \tag{42}$$

and introduce $\boldsymbol{L}_1(\cdot)$, $\boldsymbol{L}_2(\cdot)$, analogously. Then one may compute, as in [40, 41], that in general

$$\boldsymbol{L}_b = S_{ab}\boldsymbol{l}_a, \; [\boldsymbol{L}_b, \boldsymbol{l}_c] = S_{ac}[\boldsymbol{l}_a, \boldsymbol{l}_b] + (\boldsymbol{l}_b \cdot \nabla S_{ac})\boldsymbol{l}_a, \tag{43}$$

and so, in particular, relate the Burgers vectors with the set of Lie brackets via the first of (43).

We shall refer to terms such as $[\boldsymbol{l}_1, \boldsymbol{l}_2]$ as an 'iterated', or 'nested', Lie bracket of second order, to terms such as $[\boldsymbol{L}_3, \boldsymbol{l}_c] = [[\boldsymbol{l}_1, \boldsymbol{l}_2], \boldsymbol{l}_c]$ as Lie brackets of third order, etc. (and to $\boldsymbol{l}_a$ as a Lie bracket of first order, for convenience). The second of (43) gives the third order Lie brackets in terms of S and its first order directional derivatives, using the first of (43), and there are analogous higher order versions of these relations which we do not exhibit explicitly. **Each such relation expresses some 'nested' Lie bracket linearly in terms of lower order nested Lie brackets, with coefficients that are directional derivatives of** S.

We have noted in Sect. 4.1 that if S is constant in $M \equiv \mathbb{R}^3$, then (20) may be solved for the set of lattice vector fields, and that there is an associated three dimensional Lie group structure in that case. Now one can rephrase the assumption that S is constant, via the first of (43), as an assumption that the vector fields $\boldsymbol{L}_i(\cdot)$, $1 = 1, 2, 3$, can be expressed as constant linear combinations of the lattice vector fields. Via the second of (43), and its analogues, one sees that all nested Lie brackets of lattice vector fields are also then expressible as constant linear combinations of the lattice vector fields. Thus the lattice vector field provide a basis of a Lie algebra of vector fields, and this statement is equivalent to the fact that S is constant.

As a generalization of the last remark we shall investigate the following **assumption**: suppose that the lattice vector fields and certain of their nested Lie brackets provide a (finite) basis of a Lie algebra, so that the dimension of the algebra is strictly greater than that of M. Subject to a further technical assumption, it follows from result of Palais [36], see also [14], that there is an associated 'higher dimensional' Lie group that acts on M.

We shall study groups of this type in Sect. 7, but to keep the discussion as compact and as simple as seems possible (to us) in this section we restrict matters much further:

(a) We shall assume that M is a two-dimensional manifold, then there are two smooth linearly independent vector fields, denoted $\boldsymbol{l}_1(\cdot)$ and $\boldsymbol{l}_2(\cdot)$. In a **new notation** define the Lie bracket $\boldsymbol{l}_3(\cdot)$ of $\boldsymbol{l}_1(\cdot)$ and $\boldsymbol{l}_2(\cdot)$ by

$$\boldsymbol{l}_3(\cdot) \equiv [\boldsymbol{l}_1, \boldsymbol{l}_2](\cdot) \equiv \{(\boldsymbol{l}_2 \cdot \nabla)\boldsymbol{l}_1 - (\boldsymbol{l}_1 \cdot \nabla)\boldsymbol{l}_2\}(\cdot). \tag{44}$$

(b) We consider a particular example in the case where $\boldsymbol{l}_1(\cdot), \boldsymbol{l}_2(\cdot)$ and $\boldsymbol{l}_3(\cdot) \equiv [\boldsymbol{l}_1, \boldsymbol{l}_2](\cdot)$ provide a basis for all vector fields generated by taking successive Lie brackets of $\boldsymbol{l}_1(\cdot), \boldsymbol{l}_2(\cdot)$. In fact we assume that all Lie brackets of order ≥ 3 are zero. So the corresponding (nilpotent) Lie algebra has dimension 3, as does the associated Lie group. Thus

$$[\boldsymbol{l}_1, \boldsymbol{l}_2] \equiv \boldsymbol{l}_3, [\boldsymbol{l}_2, \boldsymbol{l}_3] = \boldsymbol{0}, [\boldsymbol{l}_3, \boldsymbol{l}_1] = \boldsymbol{0}. \tag{45}$$

The Lie group corresponding to (45) is called the Heisenberg group.

The following result is a particular case of Theorem 1.57 in Olver [34], it allows one to associate a three dimensional Lie group with lattice vector fields $\boldsymbol{l}_1(\cdot), \boldsymbol{l}_2(\cdot), \boldsymbol{l}_3(\cdot) \equiv [\boldsymbol{l}_1, \boldsymbol{l}_2](\cdot)$ defined on M satisfying (45).

Theorem 1 *Suppose that lattice vector fields* $\boldsymbol{l}_1(\cdot), \boldsymbol{l}_2(\cdot), \boldsymbol{l}_3(\cdot) \equiv [\boldsymbol{l}_1, \boldsymbol{l}_2](\cdot)$ *defined on M are given, such that, c.f., (45),*

$$[\boldsymbol{l}_i, \boldsymbol{l}_j] = C_{kij}\boldsymbol{l}_k, \quad i, j, k = 1, 2, 3, \tag{46}$$

where the structure constants C_{ijk} *are zero except that* $C_{312} = -C_{321} = 1$. *Then there exists a Lie group G, a corresponding Lie algebra* $\mathfrak{g}$ *with the same structure constants relative to some basis* $\boldsymbol{e}_1, \boldsymbol{e}_2, \boldsymbol{e}_3$ *of* $\mathfrak{g}$, *and a local group action* $\boldsymbol{\lambda} : G \times M \to M$ *such that*

$$\nabla_1 \boldsymbol{\lambda}(\boldsymbol{e}, \boldsymbol{x})\boldsymbol{e}_i = \boldsymbol{l}_i(\boldsymbol{x}), \quad \boldsymbol{x} \in M, \tag{47}$$

where $\boldsymbol{e}$ *is the identity element of G,* $\nabla_1 \boldsymbol{\lambda}(\cdot, \boldsymbol{x})$ *is the gradient of* $\boldsymbol{\lambda}$ *with respect to its first argument, and*

$$\boldsymbol{\lambda}(\boldsymbol{e}, \boldsymbol{x}) = \boldsymbol{x}, \boldsymbol{\lambda}(\boldsymbol{g}_1, \boldsymbol{\lambda}(\boldsymbol{g}_2, \boldsymbol{x})) = \boldsymbol{\lambda}(\boldsymbol{\psi}(\boldsymbol{g}_1, \boldsymbol{g}_2), \boldsymbol{x}), \boldsymbol{g}_1, \boldsymbol{g}_2 \in G, \boldsymbol{x} \in M, \tag{48}$$

where $\boldsymbol{\psi} : G \times G \to G$ *is the composition function in G.*

In this general situation, the lattice vector fields no longer represent right invariant fields on the Lie group, nor is it true that $\boldsymbol{e}_1$ equals $\boldsymbol{l}_1(\boldsymbol{0})$, for example ($\boldsymbol{e}_1$ lies in the Lie algebra, $\boldsymbol{l}_1(\boldsymbol{0})$ is in the tangent space at $\boldsymbol{0} \in M$)—one might contrast this with the situation in Sect. 4.1. It is the group action $\boldsymbol{\lambda}$ that links the three dimensional group structure to the flow defined by the two lattice vector fields. This leads to the brief discussion of isotropy groups and homogeneous spaces immediately below (we revisit these topics in Sect. 7).

4.5 Homogeneous Spaces

Here we introduce the notion of a homogeneous space and give some associated definitions—this is done briefly in this subsection, in more detail in Sect. 7. One thereby represents points of the manifold M in a manner which exploits the group structure deriving from the assumption in the previous subsection. The group action above fixes points of M, for some non-trivial set of group elements (in the case that the dimension of the group is greater than that of M), and this leads to the definition of the isotropy group corresponding to the group action. In fact, given a transitive group action $\boldsymbol{\lambda}$, any isotropy group is a closed subgroup of G, and given any closed subgroup of G one can construct a corresponding group action. We also define projection and section mappings, and express the group action in terms of these functions, to facilitate later computations.

The **isotropy group** of the left action $\boldsymbol{\lambda}\colon G \times M \to M$ is defined by

$$G_m = \{\boldsymbol{g} \in G\colon \boldsymbol{\lambda}(\boldsymbol{g}, m) = m\}, \tag{49}$$

for any $m \in M$. We suppose that the action is transitive, so

$$\boldsymbol{\lambda}(G, m) = M, \tag{50}$$

for any $m \in M$. Then (G, M) is called a **homogeneous space**.

Let H be a subgroup of G and define the left coset space G/H by

$$G/H = \{\boldsymbol{k}H; \boldsymbol{k} \in G\}. \tag{51}$$

(Two juxtaposed group elements represent group composition, i.e., if $\boldsymbol{a}, \boldsymbol{b} \in G$, $\boldsymbol{ab} \equiv \psi(\boldsymbol{a}, \boldsymbol{b})$, so $\boldsymbol{k}H \equiv \{\boldsymbol{\psi}(\boldsymbol{k}, \boldsymbol{h}); \boldsymbol{h} \in H\}$, in particular).

Komrakov [29], gives the following results:

- If H is a closed subgroup of G, then G/H can be given the structure of a manifold, with $\boldsymbol{\lambda}\colon G \times G/H \to G/H$ defined by $\boldsymbol{\lambda}(\boldsymbol{g}, \boldsymbol{k}H) = \boldsymbol{\psi}(\boldsymbol{g}, \boldsymbol{k})H$ smooth and transitive. Then $(G, G/H)$ is a homogeneous space.
- If (G, M) is a homogeneous space, and $m \in M$, then G_m is a closed subgroup of G.

Define the **projection** mapping $\boldsymbol{\pi}\colon G \to G/H$ by $\boldsymbol{\pi}(\boldsymbol{g}) = \boldsymbol{g}H$ and choose a **section** $\boldsymbol{\sigma}\colon G/H \to G$ such that $\boldsymbol{\pi}(\boldsymbol{\sigma}(\boldsymbol{g}H)) = \boldsymbol{g}H$. (Note that it is not generally true that there exists a well defined global section, but that the existence of such a section may be verified in each case of interest below). Then $\boldsymbol{\lambda}$ can be expressed as:

$$\boldsymbol{\lambda}(\boldsymbol{g}, \boldsymbol{k}H) = \boldsymbol{\pi}(\boldsymbol{\psi}(\boldsymbol{g}, \boldsymbol{\sigma}(\boldsymbol{k}H))), \tag{52}$$

and this shows how to construct the group action from any closed subgroup. It is a fact that $\boldsymbol{\lambda}$, defined by (52), is independent of the choice of section. One can also show that

$$H = G_{\pi(\mathbf{0})}, \tag{53}$$

if $\boldsymbol{\lambda}$ is so expressed, so any closed subgroup is an isotropy group.

5 Discrete Groups

One main purpose of this chapter is to discuss symmetries of discrete sets of points associated with a defective crystal, as part of the rather general effort in continuum mechanics to correlate discrete and continuous models of materials. To begin this discussion we first of all consider the case where the defective crystal has constant ddt, so that there is a corresponding Lie group whose dimension equals that of M. (The more general case, where these dimensions are unequal, will be considered in the next section). The sets of points that are constructed turn out to represent discrete subgroups of Lie groups. In the case where S is identically zero, for example, the elements of the relevant discrete subgroup represent the points of a perfect lattice.

In this section we first outline the way that these subgroups were introduced in Cermelli and Parry [6], Parry [37–39]. Next we paraphrase Mal'cev's general perspective [31], in the context of nilpotent Lie groups, and remark on the connections between the two approaches in the context of this chapter. We also outline some of Mal'cev's results, those that relate to the topics at hand.

5.1 Construction of Discrete Group from Given Crystal State

Choose three smooth linearly independent right invariant fields $\boldsymbol{l}_1(\cdot), \boldsymbol{l}_2(\cdot), \boldsymbol{l}_3(\cdot)$ to specify the texture of a crystal with constant ddt S, as in Davini's prescription of a crystal state. Let the corresponding Lie group be denoted G. From (20) and (22) in Sect. 4 calculate the structure constants C_{ijk} of a corresponding Lie algebra. There is an isomorphism (elastic deformation) from G to the canonical group J that has the same Lie algebra, denote it by $\boldsymbol{\theta}$, $\boldsymbol{\theta} : G \to J$, and let $\boldsymbol{l}_i'(\boldsymbol{\theta}(\boldsymbol{x})) - \nabla\boldsymbol{\theta}(\boldsymbol{x})\boldsymbol{l}_i(\boldsymbol{x})$, $i = 1, 2, 3$, by analogy with (2) in Sect. 2.1. We consider the crystal state where the corresponding Lie group is J, below, and drop the prime on the lattice vector fields.

Suppose then that $\boldsymbol{l}_i(\cdot), i = 1, 2, 3$ are right invariant fields defined on J. Let $\tilde{\boldsymbol{x}} \in J$ and say that $\tilde{\boldsymbol{y}} \in J$ is a neighbour of $\tilde{\boldsymbol{x}}$ if and only if there exists an index $i \in \{1, 2, 3\}$ such that

$$\text{either } \frac{d\boldsymbol{x}}{dt}(t)=\boldsymbol{l}_i\left(\boldsymbol{x}(t)\right), \boldsymbol{x}(0)=\tilde{\boldsymbol{x}}, \boldsymbol{x}(1)=\tilde{\boldsymbol{y}}, \text{ or } \frac{d\boldsymbol{x}}{dt}(t)=\boldsymbol{l}_i\left(\boldsymbol{x}(t)\right), \boldsymbol{x}(0)=\tilde{\boldsymbol{y}}, \boldsymbol{x}(1)=\tilde{\boldsymbol{x}}. \tag{54}$$

Thus $\tilde{\boldsymbol{x}}$ and $\tilde{\boldsymbol{y}}$ are neighbours of each other if and only if the 'unit' flow along some lattice vector field maps $\tilde{\boldsymbol{x}}$ to $\tilde{\boldsymbol{y}}$ or vice versa. (This is a generalization of the 'nearest neighbour' idea for a cubic lattice). Let $D \subset J$ be a set such that $\boldsymbol{0} \in D$ and such that if $\boldsymbol{x} \in D$, then the neighbours of $\boldsymbol{x}$ are elements of D. Then since $\boldsymbol{0} \in D$, it follows that $e^{(\boldsymbol{l}_i)} \in D$, $e^{-(\boldsymbol{l}_i)} \in D$, where $\boldsymbol{l}_i \equiv \boldsymbol{l}_i(\boldsymbol{0})$, $i = 1, 2, 3$ (noting that $\left(e^{(\boldsymbol{l}_i)}\right)^{-1} = e^{-(\boldsymbol{l}_i)}$). Also if $\boldsymbol{x} \in D$, then $\boldsymbol{\alpha}\boldsymbol{x} \in D$, where $\boldsymbol{\alpha}$ is any of the six elements $e^{(\boldsymbol{l}_i)}, e^{-(\boldsymbol{l}_i)}, i = 1, 2, 3$. So D includes all elements of J which have the form

$$\boldsymbol{x} = \boldsymbol{\alpha}_1\boldsymbol{\alpha}_2 \ldots \boldsymbol{\alpha}_n, \tag{55}$$

where n is arbitrary, and each $\boldsymbol{\alpha}_i, i = 1, 2, \ldots n$ is one of $e^{(\boldsymbol{l}_i)}, e^{-(\boldsymbol{l}_i)}, i = 1, 2, 3$. Suppose that D has no other elements. Then D is a subgroup of J (with group operation corresponding to juxtaposition of expressions such as that on the right hand side of (55), recognizing that $e^{(l_i)}e^{-(l_i)}$ is the group identity). D is said to be **generated** by the three elements $e^{(l_1)}, e^{(l_2)}, e^{(l_3)}$, when (55) holds for all $\boldsymbol{x} \in D$.

Note that, generally, an element of D has many representations of the form (55). So suppose

$$\boldsymbol{x} = \boldsymbol{\alpha}_1\boldsymbol{\alpha}_2 \dots \boldsymbol{\alpha}_n = \boldsymbol{\beta}_1\boldsymbol{\beta}_2 \dots \boldsymbol{\beta}_m, \tag{56}$$

where each of $\boldsymbol{\alpha}_i, i = 1, 2 \dots n$; $\boldsymbol{\beta}_j, j = 1, 2 \dots m$, is one of the generators or the inverse of one of the generators. Thus

$$\boldsymbol{\alpha}_1\boldsymbol{\alpha}_2 \dots \boldsymbol{\alpha}_n\boldsymbol{\beta}_m^{-1}\boldsymbol{\beta}_{m-1}^{-1} \dots \boldsymbol{\beta}_1^{-1} = \mathbf{0}, \tag{57}$$

and one sees that the non-uniqueness of the representation (55) corresponds precisely to the existence of non-trivial relations (such as (57)) between the generators and their inverses.

The above construction, Cermelli and Parry [6], Parry [37, 39], is analogous to the construction of a perfect lattice (in the case $S = 0$). Mal'cev [31], on the other hand, considers discrete subgroups D of a general Lie group G, a priori, without assuming that D has a finite number of generators. Nominally, then, his position is more general than that adopted in [6, 37, 39]. However, he finds it useful to restrict attention to **uniform** discrete subgroups of G: a discrete subgroup of G is uniform if the left coset space G/D is compact—this is the generalization of the requirement, in the case $S = 0$, that $\mathbb{R}^3/L$ (which is the 'unit cell' of the lattice, with appropriate identification of boundary points) is compact. It transpires that this criterion (that the subgroup be uniform), and the restriction to three dimensional nilpotent Lie groups, together imply that D is generated by three elements. He also shows: in order that G contains a uniform discrete subgroup D, it is necessary and sufficient that the corresponding Lie algebra $\mathfrak{g}$ have rational structure constants with respect to an appropriate basis. We shall see towards the end of the next subsection that the two perspectives coincide, in the context of this chapter—so we paraphrase Mal'cev's results in Sect. 5.3 below, as they particularize to the three dimensional case.

5.2 *Analogue of Crystallographic Restriction*

In traditional crystallography, where the Euclidean motions of $\mathbb{R}^3$, say, are central, it is the so-called **crystallographic restriction** that begins the analysis and leads to the ideas of space groups and crystallographic groups, for example. This restriction is the requirement that there is a minimum separation between pairs of points (atoms) that make up the crystal, and the classification of perfect crystal structures (due to Bieberbach, Frobenius, Shoenflies, see Senechal [47]) is based on this assumption.

This basic restriction applies just as well when we study sets of points which model defective crystals, so we search for conditions on the (continuum) crystal state which guarantee that the subgroup D of the previous subsection be discrete—the discreteness of the subgroup is the analogue of the crystallographic restriction, in this context. One can make progress quite generally, but for the purposes of this chapter we shall make a more stringent assumption, which is that the elements of this subgroup have a **minimum (non-zero) separation for arbitrarily small generators** , a, b $\left(a, b \in \{e^{(\boldsymbol{l}_1)}, e^{(\boldsymbol{l}_2)}, e^{(\boldsymbol{l}_3)}\right)$.

Recall that $(a, b) \equiv a^{-1}b^{-1}ab$ denotes the commutator of group elements a, b. If a and b are small (one parameterizes the three dimensional group by three real numbers, close to the identity element, and takes the Euclidean measure of distance), then the commutator is of second order in the size of those two elements, and one can choose ε so that

$$|(a, b)| < \tfrac{1}{2}|a|, \quad \text{if} \quad |a|, |b| < \varepsilon. \tag{58}$$

Iterating this inequality, following Thurston [48], one deduces that $|k^{th}$ nested commutator of generators $| < 2^{-k}\varepsilon$. It follows, since the separation between any pair of discrete group elements cannot be arbitrarily small, that there exists a k such that the kth nested commutators of generators are zero (identity). One can show that further $k = 2$ in the case of interest here (i.e., in $\mathbb{R}^3$), Parry [37], and that the discrete group is then such that any commutator of group elements commutes with any group element. Continuous groups with this property, such as the canonical group J introduced in Sect. 4.3, are nilpotent, according to the footnote in that subsection. In this case, Cermelli and Parry [6], Parry [37, 39] showed that D is a discrete subgroup of J if and only if the structure constants with respect to $\boldsymbol{l}_1, \boldsymbol{l}_2, \boldsymbol{l}_3$ as basis are rational, and this leads to (61) below.

We shall highlight properties of J and its discrete subgroups, or groups isomorphic to them, as running examples throughout.

5.3 *Mal'cev's Results*

We present some fundamental results of Mal'cev [31], in the context of three dimensional nilpotent Lie groups and their discrete subgroups. Those results are algebraic in character, but they lead to a 'rigidity' theorem which states, in particular, that the symmetries (automorphisms) of such discrete subgroups can be uniquely extended to symmetries of a corresponding continuous (Lie) group. In continuum mechanical terms, this implies that one can take the symmetries of those discrete subgroups as material symmetries for continuum energy densities for defective crystals of this type. Of course, in the perfect crystal case, the 'crystallographic point groups' have been used as material symmetries for continuum energy densities without question, for many years (see for example Green and Adkins [20])—the relevant rigidity result is almost self-evident in that case, although one never sees it emphasized. First we outline some necessary algebra relevant to the nilpotent Lie group and its algebra. In the next subsection we consider the discrete subgroups, and give the rigidity result. The results depend on the construction of particularly judicious sets of coordinates for the continuous group, its algebra, and the discrete groups.

Let G be a connected and simply connected three dimensional nilpotent Lie group, let $\mathfrak{g}$ be the corresponding Lie algebra, and let J be the corresponding canonical group. Recall that a subspace $\mathbf{h}\subseteq\mathfrak{g}$ is an ideal if and only if $[\mathbf{h}, \mathfrak{g}]\subseteq\mathbf{h}$.

If possible, select in $\mathfrak{g}$ an ordered basis $\{\boldsymbol{g}_1, \boldsymbol{g}_2, \boldsymbol{g}_3\}$ with the following two properties: $\{a_2\boldsymbol{g}_2 + a_3\boldsymbol{g}_3; a_2, a_3 \in \mathbb{R}\} \equiv \mathfrak{g}_2$ and $\{a_3\boldsymbol{g}_3; a_3 \in \mathbb{R}\} \equiv \mathfrak{g}_3$ are ideals in $\mathfrak{g}$. Then each element of $\mathfrak{g}$ can be uniquely represented in the form

$$\boldsymbol{g} = a_1\boldsymbol{g}_1 + a_2\boldsymbol{g}_2 + a_3\boldsymbol{g}_3, \tag{59}$$

and by definition

- the numbers a_1, a_2, a_3 are the 'coordinates of the first kind' of $\boldsymbol{g}$,
- the algebra elements $\boldsymbol{g}_1, \boldsymbol{g}_2, \boldsymbol{g}_3$ are the corresponding 'system of coordinates of the first kind'.

For example, referring to (33) of Sect. 4 above, if $[\boldsymbol{x}, \boldsymbol{y}] = \lambda\boldsymbol{p}(\boldsymbol{p}\cdot\boldsymbol{x}\wedge\boldsymbol{y}),\ \lambda \in \mathbb{R}$ and $\{\boldsymbol{l}, \boldsymbol{m}, \boldsymbol{p}\}$ is a basis of $\mathbb{R}^3$, the ordered basis $\{\boldsymbol{l}, \boldsymbol{m}, \boldsymbol{p}\}$ is a system of coordinates of the first kind (because $\mathfrak{g}_3 \equiv \mathbb{R}\boldsymbol{p}$, $[\mathbb{R}\boldsymbol{p}, \boldsymbol{y}] = \boldsymbol{0}$, $\boldsymbol{y} \in \mathfrak{g}$, and $[\mathfrak{g}_2, \boldsymbol{y}] = \mathbb{R}\boldsymbol{p} \subseteq \mathfrak{g}_2,\ \boldsymbol{y} \in \mathfrak{g}$).

Next, suppose that the Lie group G has a system of one-parameter subgroups $\boldsymbol{x}_1(t), \boldsymbol{x}_2(t), \boldsymbol{x}_3(t)$ such that three conditions hold; first each element of G can be written in the form $\boldsymbol{x}_1(t_1)\boldsymbol{x}_2(t_2)\boldsymbol{x}_3(t_3)$, $t_1, t_2, t_3 \in \mathbb{R}$; also $\{\boldsymbol{x}_2(t_2)\boldsymbol{x}_3(t_3);\ t_2, t_3 \in \mathbb{R}\} \equiv G_2$ and $\{\boldsymbol{x}_3(t_3);\ t_3 \in \mathbb{R}\} \equiv G_3$ are closed invariant (normal) subgroups of G, finally $G/G_2, G_2/G_3, G_3$ are one-parameter vector groups (i.e., they are isomorphic to $\mathbb{R}$).

These conditions imply that each element of G can be written **uniquely** in the form $\boldsymbol{x}_1(t_1)\boldsymbol{x}_2(t_2)\boldsymbol{x}_3(t_3)$, for some $t_1, t_2, t_3 \in \mathbb{R}$. Then by definition

- the numbers t_1, t_2, t_3 are called the (Mal'cev) 'coordinates of the second kind' of that element,
- the subgroups $\boldsymbol{x}_1(t), \boldsymbol{x}_2(t), \boldsymbol{x}_3(t)$ are called a 'system of coordinates of the second kind'.

For example, if $G = J$ and $\boldsymbol{g}_1, \boldsymbol{g}_2, \boldsymbol{g}_3$ is a system of coordinates of the first kind, then $\boldsymbol{x}_i(t) \equiv t\boldsymbol{g}_i,\ t \in \mathbb{R},\ i = 1, 2, 3$ is a system of coordinates of the second kind, and the converse is also true. For the next lemma, recall the discussion of uniform discrete subgroups at the end of Sect. 5.1.

Lemma 1 (Mal'cev) *If a Lie group G has a system of coordinates of the second kind, denoted $\boldsymbol{x}_1(t), \boldsymbol{x}_2(t), \boldsymbol{x}_3(t)$ and if a subgroup H contains the elements $\boldsymbol{x}_1(1), \boldsymbol{x}_2(1), \boldsymbol{x}_3(1)$, then H is uniform in G.*

For example, the subgroup generated by $\boldsymbol{x}_1(1), \boldsymbol{x}_2(1), \boldsymbol{x}_3(1)$ is uniform.

5.4 Canonical Basis of Discrete Groups

Elements $\boldsymbol{d}_1, \boldsymbol{d}_2 \ldots \boldsymbol{d}_r$ of a nilpotent group D constitute a **canonical basis** of D if each element of D can be represented in the form

$$\boldsymbol{d}_1^{n_1}\boldsymbol{d}_2^{n_2}\ldots\boldsymbol{d}_r^{n_r}, \quad \text{for some } n_i \in \mathbb{Z}, \quad i = 1, 2 \ldots r, \tag{60}$$

and the following two conditions hold: $\{\boldsymbol{d}_i^{n_i}\boldsymbol{d}_{i+1}^{n_{i+1}}, \ldots \boldsymbol{d}_r^{n_r};\ n_i, n_{i+1} \ldots n_r \in \mathbb{Z}\} \equiv D_i$ is an invariant subgroup of D, $i = 1, 2 \ldots r$, and also the quotient groups D_i/D_{i+1} (where D_{r+1} is the trivial group) are infinite cyclic. In particular a nilpotent group D with a canonical basis has countably many elements.

These conditions imply that any element of D can be written **uniquely** in the form (60). Again recall the discussion of uniform discrete subgroups at the end of Sect. 5.1.

Lemma 2 (Mal'cev) *Every uniform discrete subgroup D of a connected, simply-connected nilpotent three dimensional Lie group G contains at least one canonical basis* $\boldsymbol{d}_1, \boldsymbol{d}_2, \boldsymbol{d}_3$. *Let* $\boldsymbol{d}_1(t), \boldsymbol{d}_2(t), \boldsymbol{d}_3(t)$ *be the one-parameter groups passing through* $\boldsymbol{d}_1, \boldsymbol{d}_2, \boldsymbol{d}_3$ *such that*

$$\boldsymbol{d}_i(1) = \boldsymbol{d}_i, \quad i = 1, 2, 3.$$

Then these one-parameter groups provide a system of coordinates of the second kind.

For example, if $G = J$, each uniform discrete subgroup of J has a canonical basis, and corresponding systems of coordinates of the first and second kinds (via Lemmas 1 and 2). Also, each system of coordinates of the first kind, $\boldsymbol{g}_1, \boldsymbol{g}_2, \boldsymbol{g}_3$, induces a corresponding system of coordinates of the second kind $\boldsymbol{x}_i(t) = \boldsymbol{g}_i t$, $i = 1, 2, 3$, $t \in \mathbb{R}$, and the subgroup generated by $\boldsymbol{g}_1, \boldsymbol{g}_2, \boldsymbol{g}_3$ is uniform ($\boldsymbol{x}_i(1) = \boldsymbol{g}_i$).

It seems to us that theorems of the following type should be central in discussions of the relation between continuous and discrete models of defective crystals, since the particular theorem below establishes a rigorous connection between geometric symmetries (automorphisms) of such objects.

Theorem 2 (Mal'cev) *Let D and* D^* *be uniform discrete subgroups of connected, simply-connected nilpotent Lie groups G and* G^**, respectively. Then every isomorphism between D and* D^* *can be uniquely extended to an isomorphism between G and* G^**. In particular, every automorphism of D can be extended to an automorphism of G.*

Note This theorem is proven by noting that a canonical basis $\boldsymbol{d}_1, \boldsymbol{d}_2, \boldsymbol{d}_3$ of D maps to a canonical basis $\boldsymbol{d}_1^*, \boldsymbol{d}_2^*, \boldsymbol{d}_3^*$ of D^* under the given isomorphism. Let the corresponding system of coordinates of the second kind (in D) be $\boldsymbol{d}_1(t), \boldsymbol{d}_2(t), \boldsymbol{d}_3(t)$ and (in D^*) be $\boldsymbol{d}_1^*(t), \boldsymbol{d}_2^*(t), \boldsymbol{d}_3^*(t)$. Then the unique extension of the given mapping to an isomorphism $G \to G^*$ is shown to be the map which sends $\boldsymbol{d}_1(t_1)\boldsymbol{d}_2(t_2)\boldsymbol{d}_3(t_3)$ to $\boldsymbol{d}_1^*(t_1)\boldsymbol{d}_2^*(t_2)\boldsymbol{d}_3^*(t_3)$.

5.5 *Lattice Structure of Discrete Nilpotent Groups*

It turns out that one may choose coordinates such that the discrete nilpotent Lie groups above also have a lattice structure, in the usual crystallographic sense.

Let us choose the structure constants in the canonical group J to have the form

$$C_{ijk} = \lambda \varepsilon_{rjk} p_i p_r, \tag{61}$$

in terms of real parameters λ, $p_i, i = 1, 2, 3$, as in Sect. 4.3. Then the composition function in J has the form given in (33), Sect. 4.3, and one can show that, by virtue of the analogue of the crystallographic restriction, as in Parry [37], that λ can be taken to be rational, $p_i, i = 1, 2, 3$ can be taken to be relatively prime integers.

Let D be the discrete subgroup of J which is generated by $\mathrm{e}^{(\boldsymbol{l}_1)}, \mathrm{e}^{(\boldsymbol{l}_2)}, \mathrm{e}^{(\boldsymbol{l}_3)}$. The **translation group** T of D is defined by

$$T = \{\boldsymbol{t} \in J \;:\; \text{if } \boldsymbol{d} \in D,\; \boldsymbol{d} + \boldsymbol{t} \in D\}. \tag{62}$$

(One understands by $\boldsymbol{d} + \boldsymbol{t}$ the group element in J which has components equal to the sum of the components of $\boldsymbol{d} \in D,\ \boldsymbol{t} \in J$). Also let $\lambda = s/q \in \mathbb{Q}$ where s and $q \in \mathbb{Z}$ have no common factors, define $\nu = p_1 p_2 p_3$ and recall that $\boldsymbol{p}$ denotes the vector in $\mathbb{R}^3$ with components $p_i, i = 1, 2, 3$. Now define the integer k by

$$k = \begin{cases} s & \text{if } \nu \text{ is even or if } (\nu \text{ is odd and } s \in 4\mathbb{Z}) \\ \frac{s}{2} & \text{if } \nu \text{ is odd and } s \in 2\mathbb{Z},\ s \notin 4\mathbb{Z} \\ 2s & \text{if } \nu \text{ is odd and } s \text{ is odd.} \end{cases} \tag{63}$$

Cermelli and Parry [6] show that if k is even then $T = D$ and T consists of all integer linear combinations of $\boldsymbol{l}_1, \boldsymbol{l}_2, \boldsymbol{l}_3, \lambda \boldsymbol{p}/k$. Then the points of D form a three-dimensional lattice, since an integral basis of $T = D$ may be found in terms of $\boldsymbol{l}_1, \boldsymbol{l}_2, \boldsymbol{l}_3, \lambda \boldsymbol{p}/k$. Also, if k is odd they show that T consists of all integer linear combinations of $2\boldsymbol{l}_1, 2\boldsymbol{l}_2, 2\boldsymbol{l}_3, \lambda \boldsymbol{p}/k$, and that D/T has four elements which may be written as $T, \boldsymbol{\alpha}T, \boldsymbol{\beta}T, \boldsymbol{\alpha\beta}T$ for some $\boldsymbol{\alpha}, \boldsymbol{\beta} \in D$. Thus the points of D form a 4-lattice in the sense of Pitteri and Zanzotto [43].

5.6 Symmetries of Discrete Nilpotent Groups *D*

Let D be a uniform discrete subgroup of J. Mal'cev's results give that, if $\boldsymbol{c}_1, \boldsymbol{c}_2, \boldsymbol{c}_3$ generates D, then it provides a canonical basis of D if $\boldsymbol{c}_3$ is a basis element of $D \cap (J, J)$—we focus on such a canonical basis. So

$$\boldsymbol{c}_3 = \theta \boldsymbol{p}, \tag{64}$$

for some real θ. Then since $(\boldsymbol{c}_1, \boldsymbol{c}_2) \in \mathbb{R}\boldsymbol{p}$,

$$(\boldsymbol{c}_1, \boldsymbol{c}_2) = \boldsymbol{c}_3^k, \quad \text{for some } k \in \mathbb{Z}. \tag{65}$$

(From (65), there is an evident interpretation of the distribution of points in D in terms of screw dislocations). Then from (64) above and (33) of Sect. 4

$$(\boldsymbol{c}_1, \boldsymbol{c}_3) = (\boldsymbol{c}_2, \boldsymbol{c}_3) = \boldsymbol{0}. \tag{66}$$

It turns out that conditions (65) and (66) are sufficient in order that $\{\boldsymbol{c}_1, \boldsymbol{c}_2, \boldsymbol{c}_3\}$ be a canonical basis of D.

To calculate the symmetries of D, let us discuss general ideas regarding changes of generators in a group from the point of view of Magnus, Karrass, Solitar [30], Johnson [22]. So let X be a set (which will eventually play the role of a set of generators of D), and let $F(X)$ be the free group with X as basis, which means that: $F(X)$ consists of all 'words' in the elements of X (so if, for example, $X = \{x_1, x_2, x_3\}$, the words of $F(X)$ have the form $\omega = x_1^{\alpha_1} x_2^{\beta_1} x_3^{\gamma_1} x_1^{\alpha_2} x_2^{\beta_2} x_3^{\gamma_2} \ldots x_1^{\alpha_r} x_2^{\beta_r} x_3^{\gamma_r}$ for some integer r, integers $\alpha_i, \beta_i, \gamma_i$, $i = 1, 2, \ldots r$). The group operation in $F(X)$ is juxtaposition of words, with terms of the form $x_i x_i^{-1}, x_i^{-1} x_i$ 'cancelled' in any product of words. In fact, confine attention to the case $X = \{x_1, x_2, x_3\}$. Then a 'free substitution' of $F(X)$ is a replacement of the elements x_1, x_2, x_3 by words $\omega_1(x_1, x_2, x_3)$, $\omega_2(x_1, x_2, x_3)$, $\omega_3(x_1, x_2, x_3) \in F(X)$ such that these words are also a basis of $F(X)$. (For example, one may take $\omega_1 = x_1, \omega_2 = x_2, \omega_3 = x_1 x_3$). This implies, in particular, that each x_i, $i = 1, 2, 3$, may be written as a word in $\omega_1, \omega_2, \omega_3$ (thus, in the example, $x_1 = \omega_1, x_2 = \omega_2, x_3 = \omega_1^{-1}\omega_3$), and this fact alone is sufficient that $\{\omega_1, \omega_2, \omega_3\}$ is a basis of $F(X)$. Also, a free substitution gives rise to a mapping which sends any $\omega(x_1, x_2, x_3) \in F(X)$ to $\omega(\omega_1(x_1, x_2, x_3), \omega_2(x_1, x_2, x_3), \omega_3(x_1, x_2, x_3)) \in F(X)$, and it is a fact that this mapping is an automorphism of $F(X)$.

Thus the free substitutions represent changes in the set of generators of a free group. It is important to note that such changes of generators, if applied to a given group D with generators $\{x_1, x_2, x_3\}$, do not generally provide automorphisms of D. The condition that a free substitution may be associated with an automorphism of D is provided by a lemma of Magnus, Karrass, Solitar [30], Johnson [22]:

Lemma 4 *Let* $\omega(\boldsymbol{c}_1, \boldsymbol{c}_2, \boldsymbol{c}_3) \equiv \boldsymbol{c}_1^{\alpha_1}\boldsymbol{c}_2^{\beta_1}\boldsymbol{c}_3^{\gamma_1}\boldsymbol{c}_1^{\alpha_2}\boldsymbol{c}_2^{\beta_2}\boldsymbol{c}_3^{\gamma_2}\ldots\boldsymbol{c}_1^{\alpha_r}\boldsymbol{c}_2^{\beta_r}\boldsymbol{c}_3^{\gamma_r}$, *where* $\alpha_i, \beta_i, \gamma_i$, $i = 1, 2, \ldots r$ *are integers, be any word in the generators* $\boldsymbol{c}_1, \boldsymbol{c}_2, \boldsymbol{c}_3$ *of* D *such that*

$$\omega(\boldsymbol{c}_1, \boldsymbol{c}_2, \boldsymbol{c}_3) = \boldsymbol{0}, \tag{67}$$

$\boldsymbol{0}$ *the group identity. Then a free substitution* $\boldsymbol{\phi}$ *extends to an automorphism of* D *if and only if*

$$\omega\left(\boldsymbol{\phi}(\boldsymbol{c}_1), \boldsymbol{\phi}(\boldsymbol{c}_2), \boldsymbol{\phi}(\boldsymbol{c}_3)\right) = \boldsymbol{0}, \quad \text{and} \quad \omega\left(\boldsymbol{\phi}^{-1}(\boldsymbol{c}_1), \boldsymbol{\phi}^{-1}(\boldsymbol{c}_2), \boldsymbol{\phi}^{-1}(\boldsymbol{c}_3)\right) = \boldsymbol{0} \tag{68}$$

for each such word, where $\boldsymbol{\phi}^{-1}$ *is the free substitution that maps* $\boldsymbol{\phi}(\boldsymbol{c}_i)$ *to* $\boldsymbol{c}_i$.

This lemma allows the automorphisms of D to be calculated explicitly, for all relations in D of the form $\omega(\boldsymbol{c}_1, \boldsymbol{c}_2, \boldsymbol{c}_3) = \boldsymbol{0}$ follow from (65) and (66).

One can show that the automorphisms of D include the mappings:

$$\left.\begin{array}{llll}
\bullet & \boldsymbol{c}_1 \to \boldsymbol{c}_1^{-1}, & \boldsymbol{c}_2 \to \boldsymbol{c}_2, & \boldsymbol{c}_3 \to \boldsymbol{c}_3^{-1}, \\
\bullet & \boldsymbol{c}_1 \to \boldsymbol{c}_1, & \boldsymbol{c}_2 \to \boldsymbol{c}_1\boldsymbol{c}_2, & \boldsymbol{c}_3 \to \boldsymbol{c}_3, \\
\bullet & \boldsymbol{c}_1 \to \boldsymbol{c}_2, & \boldsymbol{c}_2 \to \boldsymbol{c}_1^{-1}, & \boldsymbol{c}_3 \to \boldsymbol{c}_3, \\
\bullet & \boldsymbol{c}_1 \to \boldsymbol{c}_1\boldsymbol{c}_3, & \boldsymbol{c}_2 \to \boldsymbol{c}_2, & \boldsymbol{c}_3 \to \boldsymbol{c}_3, \\
\bullet & \boldsymbol{c}_1 \to \boldsymbol{c}_1, & \boldsymbol{c}_2 \to \boldsymbol{c}_2\boldsymbol{c}_3, & \boldsymbol{c}_3 \to \boldsymbol{c}_3.
\end{array}\right\} \tag{69}$$

In fact these particular automorphisms generate the (group of) automorphisms of D, and each mapping (69) provides a free substitution of $F(\{\boldsymbol{c}_1, \boldsymbol{c}_2, \boldsymbol{c}_3\})$. Note that $\boldsymbol{c}_3$ may only be replaced by itself or its inverse, in any automorphism, so that the basis element $\boldsymbol{c}_3$ is distinguished (as an element of the canonical basis) by the fact that it is a basis element of $D \cap (J, J)$. The automorphisms preserve $D \cap (J, J)$.

Finally **each of these automorphisms extends to an automorphism of** J, according to Theorem 2. One might contrast the mappings (69) with those corresponding to the symmetries of a perfect crystal lattice, considered as a discrete subgroup of the group $\mathbb{R}^3$, with addition as group operation: that group is commutative, so no basis element is distinguished from any other.

6 Discrete Structures

Now we adapt the above methods to encompass situations where the relevant group action $\boldsymbol{\lambda} : G \times M \to M$ is that of a Lie group G which has a higher dimension than that of M. We confine attention in this section to a three dimensional nilpotent Lie group acting on $M = \mathbb{R}^2$ to illustrate the methods, since we have set out much information relating to this particular group in previous sections. The lattice vector fields in this case are just two, rather than three, and it seems at first glance that the map deriving from unit flow along the lattice vector fields does not correspond to multiplication by a group element, so that the situation is a little more intricate than that described in Sect. 5.1—however, one only has to incorporate homogeneous space ideas in order to progress.

Suppose then that the two lattice vector fields generate a three dimensional nilpotent lattice algebras of vector fields in $\mathbb{R}^2$. It is shown in Parry and Zyskin [42] that all Lie brackets of order ≥ 3, in such lattice algebras, are zero, and further that one can construct all vector fields, $\boldsymbol{l}_1(\cdot), \boldsymbol{l}_2(\cdot)$, whose components are real analytic in Ω, which solve (c.f., (45) in Sect. 4.4), modulo local elastic deformation and change of basis. Thus we take

$$[\boldsymbol{l}_1, \boldsymbol{l}_2](\cdot) \equiv \boldsymbol{l}_3(\cdot), [\boldsymbol{l}_2, \boldsymbol{l}_3](\cdot) = \mathbf{0}, [\boldsymbol{l}_3, \boldsymbol{l}_1](\cdot) = \mathbf{0} \tag{70}$$

and may choose, as in [42],

$$\boldsymbol{l}_1(x, y) = (0, x), \quad \boldsymbol{l}_2(x, y) = (1, 0), \quad \boldsymbol{l}_3(x, y) = (0, 1). \tag{71}$$

(This is case 22 of Olver [34], with $\eta_1(x), \eta_2(x)$ a basis of solutions of $\eta''(x) = 0$).

Also we choose the composition function corresponding to the canonical group J: recalling Theorem 1, define $\boldsymbol{\psi} : \mathbb{R}^3 \times \mathbb{R}^3 \to \mathbb{R}^3$, with $\boldsymbol{\psi}(\boldsymbol{r}, \boldsymbol{s}) = \psi_i \boldsymbol{e}_i, \boldsymbol{r} = r_i \boldsymbol{e}_i, s = s_i \boldsymbol{e}_i$, by

$$(\psi_i) = \left(r_1 + s_1, r_2 + s_2, r_3 + s_3 + \frac{1}{2}(r_1 s_2 - r_2 s_1)\right). \tag{72}$$

(This is (33) of Sect. 4.3 in the case that $\lambda = 1$, $p_1 = p_2 = 0$, $p_3 = 1$, $\boldsymbol{p} = \boldsymbol{e}_3$).

Next, to fix ideas regarding the material in Sect. 4.5, we derive (71) using a homogeneous space construction. From Sect. 4.3, there are precisely two inequivalent one-dimensional subgroups of J, modulo automorphisms of J: they are $H_{\boldsymbol{e}_1}$ and $H_{\boldsymbol{e}_3}$ in the notation introduced there. We choose to calculate the group action corresponding to the subgroup $H_{\boldsymbol{e}_1}$. Recalling definitions from Sect. 4.5, the projection mapping $\boldsymbol{\pi} : J \to J/H_{\boldsymbol{e}_1}$ is given by

$$\begin{aligned} \boldsymbol{\pi}(\boldsymbol{g}) = g H_{\boldsymbol{e}_1} &= \left\{\boldsymbol{\psi}(\boldsymbol{g}, t\boldsymbol{v}_1); t \in \mathbb{R}\right\} \\ &= \left\{(g_1 + t, g_2, g_3 + \tfrac{1}{2}(-g_2 t)) : t \in \mathbb{R}\right\}. \end{aligned} \tag{73}$$

There is precisely one element of this coset with first component zero (that element with $t = -g_1$), so we may parameterize $g H_{\boldsymbol{e}_1}$ by $(g_2, g_3 + \frac{1}{2} g_1 g_2)$. Let $g_2, g_3 + \frac{1}{2} g_1 g_2$ be the two components x_1, x_2 of a point $\boldsymbol{x} = x_i \boldsymbol{\gamma}_i, \quad i = 1, 2$, of $\mathbb{R}^2$, where $\boldsymbol{\gamma}_1, \boldsymbol{\gamma}_2$ is a basis of $\mathbb{R}^2$. This choice of basis allows us to identify $J/H_{\boldsymbol{e}_1}$ with $\mathbb{R}^2$. We may also choose as section mapping $\boldsymbol{\sigma} : J/H_{\boldsymbol{e}_1} \to J$:

$$\boldsymbol{\sigma}(\boldsymbol{g} H_{\boldsymbol{e}_1}) = \boldsymbol{\sigma}\left((g_2, g_3 + \tfrac{1}{2} g_1 g_2)\right) = (0, g_2, g_3 + \tfrac{1}{2} g_1 g_2), \tag{74}$$

so

$$\sigma\big((x, y)\big) = (0, x, y), \quad (x, y) \in \mathbb{R}^2. \tag{75}$$

Now we can find the group action $\boldsymbol{\lambda} : J \times \mathbb{R}^2 \to \mathbb{R}^2$ (regarding G/H_{e_1} as $\mathbb{R}^2$) corresponding to this projection from (52):

$$\begin{aligned}\boldsymbol{\lambda}(\boldsymbol{p}, (x, y)) &= \boldsymbol{\pi}(\boldsymbol{p}\boldsymbol{\sigma}(x, y)) = \boldsymbol{\pi}(\boldsymbol{\psi}(\boldsymbol{p}, (0, x, y)) \\ &= \pi(p_1, p_2 + x, p_3 + y + \tfrac{1}{2}p, x) \\ &= (p_2 + x, p_3 + y + p_1 x + \tfrac{1}{2}p_1 p_2).\end{aligned} \tag{76}$$

The lattice vector fields deriving from this projection are, from (47) in Sect. 4.4,

$$\boldsymbol{l}_i(\boldsymbol{x}) = \nabla_1\boldsymbol{\lambda}(\boldsymbol{e}, \boldsymbol{x})\boldsymbol{e}_i = \frac{\partial \lambda r}{\partial p_i}(\boldsymbol{e}, \boldsymbol{x})\boldsymbol{\gamma}_r, \quad i = 1, 2, \ \boldsymbol{x} \in \mathbb{R}^2. \tag{77}$$

So from (77) the components of $\boldsymbol{l}_i(\cdot)$ with respect to the basis $\boldsymbol{\gamma}_1, \boldsymbol{\gamma}_2$, are $\left(\frac{\partial \lambda_r}{\partial p_i}(\boldsymbol{e}, \boldsymbol{x})\right)$ and this gives

$$\boldsymbol{l}_1(\boldsymbol{x}) = (0, x), \quad \boldsymbol{l}_2(\boldsymbol{x}) = (1, 0), \quad \boldsymbol{l}_3(x) = (0, 1). \tag{78}$$

These are the canonical forms (71) of the lattice vector fields. So we have an explicit construction of the objects whose existence is asserted in Theorem 1, namely we have a Lie group J, Lie algebra $\boldsymbol{j}$, basis $\boldsymbol{e}_1, \boldsymbol{e}_2, \boldsymbol{e}_3$ of $\boldsymbol{j}$, and group action $\boldsymbol{\lambda} : J \times \mathbb{R}^2 \to \mathbb{R}^2$ such that (48) in Sect. 4.4 holds, in this case.

6.1 Structures Obtained by Discrete Flow Along the Two Lattice Vector Fields

Given a two-dimensional crystal state Σ, with vector fields $\boldsymbol{l}_1(\cdot), \boldsymbol{l}_2(\cdot)$ generating a three-dimensional Lie algebra, we generate a set of points S_Σ in M by the following iterative process: choose a point $\boldsymbol{x}_0 \in M \subseteq \mathbb{R}^2$ as initial point, construct two points $\boldsymbol{x}(1), \boldsymbol{x}(-1)$ by solving

$$\frac{d\boldsymbol{x}}{dt}(t) = \boldsymbol{l}_1(\boldsymbol{x}(t)), \quad \boldsymbol{x}(0) = \boldsymbol{x}_0, \quad t \in \mathbb{R}, \tag{79}$$

for $\boldsymbol{x}(t)$. Obtain two further points by solving the analogue of (79) with $\boldsymbol{l}_1(\cdot)$ replaced by $\boldsymbol{l}_2(\cdot)$. Iterate this process, using the four points so obtained as initial points in turn. Continue, to obtain S_Σ. This starting point is the analogue of that in Sect. 5.1.

We assume that the vector fields $\boldsymbol{l}_1(\cdot), \boldsymbol{l}_2(\cdot)$ derive from the group action (76), so that $G = J$, and we connect flow along right invariant fields in J with flow along the lattice vector fields ('infinitesimal generators') in M, when (J, M) is a homogeneous space. The following result is surely well known (see Parry and Zyskin [42])—it shows that 'unit flow along the right invariant fields commutes with projection'. Let us define

$$\nabla_1 \boldsymbol{\lambda}(\boldsymbol{e}, \boldsymbol{x})\boldsymbol{v} \equiv \boldsymbol{l}_{\boldsymbol{v}}(\boldsymbol{x}), \quad x \in M, \boldsymbol{v} \in \boldsymbol{g}, \tag{80}$$

the lattice vector field corresponding to the Lie algebra element $\boldsymbol{v}$ (if we set $\boldsymbol{l}_i(\cdot) \equiv \boldsymbol{l}_{\boldsymbol{v}_i}(\cdot)$), also called an **infinitesimal generator of the group action**.

Theorem 3 *Let $\boldsymbol{x}(\varepsilon), \varepsilon \in \mathbb{R}, \boldsymbol{x}(\varepsilon) \in M$ be defined by*

$$\boldsymbol{x}(\varepsilon) = \boldsymbol{\lambda}(e^{\varepsilon \boldsymbol{v}}\boldsymbol{g}, \boldsymbol{x}), \quad x \in M, \boldsymbol{v} \in \boldsymbol{g}. \tag{81}$$

Then $\boldsymbol{x}(\varepsilon)$ solves

$$\begin{cases} \frac{d}{d\varepsilon}\boldsymbol{x}(\varepsilon) &= \boldsymbol{l}_{\boldsymbol{v}}(\boldsymbol{x}(\varepsilon)), \quad \varepsilon \in \mathbb{R}, \\ \boldsymbol{x}(0) &= \boldsymbol{\lambda}(\boldsymbol{g}, \boldsymbol{x}). \end{cases} \tag{82}$$

Let us use this result to calculate S_Σ. So put $\boldsymbol{g} = \boldsymbol{e}$ in Theorem 3 to obtain that $\boldsymbol{x}(t) = \boldsymbol{\lambda}(e^{t\boldsymbol{e}_1}, \boldsymbol{x}_0)$ solves (79), $x_0 \in M$. Let $\boldsymbol{\pi}, \boldsymbol{\sigma}$ be the projection and section mappings of Sect. 4.5, so $\boldsymbol{\lambda}(\boldsymbol{g}, \boldsymbol{x}) = \boldsymbol{\pi}(\boldsymbol{g}\boldsymbol{\sigma}(\boldsymbol{x}))$. Then

$$\boldsymbol{x}(t) = \boldsymbol{\lambda}(e^{t\boldsymbol{e}_1}, \boldsymbol{x}_0) = \boldsymbol{\pi}(e^{t\boldsymbol{e}_1}\boldsymbol{\sigma}(\boldsymbol{x}_0)) = \boldsymbol{\pi}(e^{t\boldsymbol{e}_1}\boldsymbol{g}_0), \tag{83}$$

if we set

$$\boldsymbol{g}_0 = \boldsymbol{\sigma}(\boldsymbol{x}_0). \tag{84}$$

Hence $\boldsymbol{x}(1) = \boldsymbol{\pi}(e^{\boldsymbol{e}_1}\boldsymbol{g}_0)$, $\boldsymbol{x}(-1) = \boldsymbol{\pi}(e^{-\boldsymbol{e}_1}\boldsymbol{g}_0)$, and the two further points obtained are $\boldsymbol{\pi}(e^{\boldsymbol{e}_2}\boldsymbol{g}_0)$, $\boldsymbol{\pi}(e^{-\boldsymbol{e}_2}\boldsymbol{g}_0)$. It follows that

$$S_\Sigma = \boldsymbol{\pi}(D\boldsymbol{g}_0), \tag{85}$$

where D is the subgroup of J generated by just two group elements $e^{\boldsymbol{e}_1}$ and $e^{\boldsymbol{e}_2}$. We next show how to calculate S_Σ explicitly.

Recall that group and algebra elements may be identified, in J, and that we have $e^{\boldsymbol{x}} = \boldsymbol{x}$, and $(\boldsymbol{x}, \boldsymbol{y}) = [\boldsymbol{x}, \boldsymbol{y}]$. Hence

$$(e^{\boldsymbol{x}}, e^{\boldsymbol{y}}) = (\boldsymbol{x}, \boldsymbol{y}) = [\boldsymbol{x}, \boldsymbol{y}] = e^{[\boldsymbol{x}, \boldsymbol{y}]}, \quad \boldsymbol{x}, \boldsymbol{y} \in J. \tag{86}$$

Since $\boldsymbol{e}_3 = [\boldsymbol{e}_1, \boldsymbol{e}_2]$, we have

$$(e^{\boldsymbol{e}_1}, e^{\boldsymbol{e}_2}) = e^{\boldsymbol{e}_3}, \text{ so } e^{\boldsymbol{e}_2} e^{\boldsymbol{e}_1} = e^{\boldsymbol{e}_1} e^{\boldsymbol{e}_2} (e^{\boldsymbol{e}_2}, e^{\boldsymbol{e}_1}) = e^{\boldsymbol{e}_1} e^{\boldsymbol{e}_2} e^{-\boldsymbol{e}_3}. \tag{87}$$

It follows that any element of D can be written in the form

$$\boldsymbol{d} \equiv e^{n_1 \boldsymbol{e}_1} e^{n_2 \boldsymbol{e}_2} e^{n_3 \boldsymbol{e}_3}, \quad n_1, n_2, n_3 \in \mathbb{Z}. \tag{88}$$

Thus any element of D can be represented as the product of three group elements, each of which has the form $e^{n\boldsymbol{e}_i}, i = 1, 2, 3$. Then calculations of Cermelli and Parry [6] gives that $\boldsymbol{d} = d_i \boldsymbol{e}_i$ where

$$(d_i) = (n_1, n_2, n_3 + \tfrac{1}{2} n_1 n_2), \quad n_1, n_2, n_3 \in \mathbb{Z}. \tag{89}$$

From (85), we need to calculate $\pi(\boldsymbol{d}\boldsymbol{g}_0)$,where $\boldsymbol{g}_0 = \boldsymbol{\sigma}(\boldsymbol{x}_0)$. So put $\boldsymbol{x}_0 = (x_1^0, x_2^0)$, $\boldsymbol{\sigma}(\boldsymbol{x}_0) = (0, x_1^0, x_2^0)$ to find after some manipulation that

$$\begin{aligned}\boldsymbol{\pi}(\boldsymbol{d}\boldsymbol{\sigma}(\boldsymbol{x}_0)) &= \boldsymbol{\pi}(\boldsymbol{\psi}((n_1, n_2, n_3 + \tfrac{1}{2} n_1 n_2), (0, x_1^0, x_2^0)) \\ &= (x_1^0, x_2^0) + n_1(0, x_1^0) + (n_2, n_3 + n_1 n_2), \quad n_1, n_2, n_3 \in \mathbb{Z}.\end{aligned} \tag{90}$$

Let $K \equiv \{n_1(0, x_1^0); n_1 \in \mathbb{Z}\}$ be the set of integer multiples of $(0, x_1^0)$, and note that for fixed n_1,

$$\{(n_2, n_3 + n_1 n_2); n_2, n_3 \in \mathbb{Z}\} = \mathbb{Z}^2. \tag{91}$$

Then

$$\begin{aligned} S_\Sigma &= \pi(D\sigma(\boldsymbol{x}_0)) \\ &= \boldsymbol{x}_0 + K + \mathbb{Z}^2. \end{aligned} \tag{92}$$

Now we adopt what seems to a reasonable extension of the crystallographic restriction, in this case—we assume that S_Σ is discrete. It transpires that this rules out the case where x_1^0 is irrational [42], but if $x_1^0 = p/q$, with $p, q \in \mathbb{Z}$ relatively prime, then there exist $k, l \in \mathbb{Z}$ such that $kp + lq = 1$, so $\frac{1}{q} = k\left(\frac{p}{q}\right) + l$, and $\left\{n_1 x_1^0 + k; n_1, k \in \mathbb{Z}\right\}$ is the set of all integer multiples of $\frac{1}{q}$. **Hence S_Σ is a simple lattice**, in the sense of traditional crystallography, containing $\boldsymbol{x}_0$, with basis $(1, 0)$, $(0, 1/q)$. So once again, in judiciously chosen coordinates, the elements of the discrete structure coincide with those of a perfect lattice, even though the underlying continuous structure involves non-commuting vector fields.

7 Geometrical Setting

In this section we allow that M has any finite dimension, say n, although in practice and in our examples we only consider $n \leq 3$. To begin with we recast some of the earlier material in more modern geometrical language (see for example [18, 19, 23, 24, 50]), to fix ideas, then we use some of the associated concepts and results to extend the earlier material. This allows us to discuss notions of connection, torsion, and curvature in a manner which seems to fit naturally with some of the earlier assumptions.

As postulated earlier, the kinematic state of the continuous solid crystal body M is given by a continuous lattice defined by the frame field (a smooth section) $\mathbf{l} : M \to L(M)$ where $L(M)$ denotes the bundle of linear frames of M. In general, a differentiable manifold may not admit a global section of its frame bundle $L(M)$. However, our approach is local, so there is no loss of generality in assuming that the continuous lattice $\mathbf{l}$ is globally defined. Alternatively, the reader may think about the body manifold M as an open neighbourhood of $\mathbb{R}^n$.

We assume that the linearly independent smooth vector fields $l_i : M \to TM$, $i = 1, \cdots, n$, defining the lattice $\mathbf{l}$, where TM denotes the tangent space of M, form an m-dimensional Lie subalgebra $\mathfrak{l}$ of the algebra $\chi(M)$ of all smooth vector fields on M.[2] A particular case of this assumption was made in Sect. 4.4. In general, we shall call the algebra $\mathfrak{l}$ the **lattice algebra** of the corresponding crystal state, where $n \leq m < \infty$, and postulate that it is complete, i.e., it contains only complete vector fields.[3] Hence, there exists a Lie group, say G, acting smoothly on the body manifold M and such that its Lie algebra is isomorphic to the lattice algebra $\mathfrak{l}$. More precisely, we have the following result.

Theorem 4 *Consider a continuous lattice defined by n linearly independent smooth vector fields $l_i : M \to TM$, $i = 1, \cdots, n$. Let $\mathfrak{l} \subset \chi(M)$ be the subalgebra of vector fields defined by the fields l_i, $i = 1, \cdots, n$. Assume that $\mathfrak{l}$ is finite dimensional and complete. Then, there exists a connected Lie group G contained (as an abstract subgroup) in the group of all diffeomorphisms of M, Diff(M), such that the natural left action $\Lambda : G \times M \to M$ of the group G on the body manifold M is smooth and the lattice algebra $\mathfrak{l}$ is isomorphic to the Lie algebra, say $\mathfrak{g}$, of the group G.*

That is, let $\phi : G \to \mathrm{Diff}(M)$ be the homomorphism from the group G into the group of all diffeomorphisms of M. Define the smooth action of G on M by

[2] For the motivation of this assumption see [13].

[3] A vector field on a manifold M is considered complete if the corresponding flow is globally defined.

$$\Lambda(g, x) = \phi(g)(x), \ g \in G, x \in M. \tag{93}$$

We assume that the action Λ, which is effective, is also transitive. This means that given the **orbit map** $\Lambda_x : G \to M$ such that $\Lambda_x(g) = \Lambda(g, x)$, $g \in G$, $x \in M$, the orbit of every point $x \in M$ is identical to the manifold M, that is, $\Lambda_x(G) = M$. Moreover, for any $x \in M$, the mapping Λ_x is a morphism of the left action of the group G on itself into the action Λ of G on M but it is not necessarily an isomorphism unless the action Λ is free. Indeed, given a point $x \in M$, let G_x be the isotropy group of the action Λ at x. That is, (c.f., the corresponding definition in Sect. 4.5) let

$$G_x = \{g \in G : \Lambda(g, x) = x\}, \ \ x \in M. \tag{94}$$

Note that the mapping Λ_x is a bijection onto the body manifold M if and only if the action Λ is transitive and the isotropy group G_x is trivial. Note also that due to the transitivity of the action Λ the isotropy groups at different points, say $x, y \in M$, $x \neq y$, are conjugate subgroups of G. Namely, $G_y = hG_xh^{-1}$ where $h \in G$ is such that $\Lambda(h, x) = y$.

Given the orbit map $\Lambda_x : G \to M$, consider its tangent map $d\Lambda_x : TG \to TM$ such that $d_g\Lambda_x : T_gG \to T_{\Lambda_x(g)}M$, $g \in G$, $x \in M$. It can be shown that it establishes an isomorphism between the algebra of right-invariant vectors fields on the group G (the right Lie algebra of G) and the lattice algebra $\mathfrak{l}$. In particular, when evaluated at the identity e of the group G, the mapping $d_e\Lambda_x : T_eG \to T_xM$ identifies the tangent space T_eG with the lattice algebra $\mathfrak{l}$. This, in turn, implies [29]:

Theorem 5 *Given a continuous lattice* $\mathbf{l} : M \to L(M)$ *defining a finite-dimensional lattice algebra* $\mathfrak{l}$*, consider the induced smooth action* $\Lambda : G \times M \to M$ *of the connected group* G *whose Lie algebra* $\mathfrak{g}$ *is isomorphic to the algebra* $\mathfrak{l}$*. Let* G_x *be the isotropy group of the action* Λ *at* $x \in M$*. Then, assuming that the action* Λ *is transitive, the quotient space* G/G_x *is a homogeneous manifold*[4] *which can be diffeomorphically identified with the underlying body manifold* M*.*

Indeed, as the action Λ is transitive and the isotropy group G_x is a closed subgroup (a submanifold) of the group G, the rank of the orbit map Λ_x is constant allowing one to identify the body manifold M with the quotient space G/G_x. That is, selecting $x \in M$ and the corresponding isotropy group G_x, define the **realization mapping** $\widehat{\Lambda(x)} : G/G_x \to M$ by

[4]Given a Lie group G and its closed subgroup G_x, the quotient space G/G_x is called a homogeneous manifold if it admits a structure of smooth manifold.

$$\widehat{\Lambda(x)}(hG_x) = \Lambda_x(h) = \Lambda(h, x), \ h \in G, \tag{95}$$

where hG_x denotes the left coset of G_x generated by the element $h \in G$. It can be shown that any realization $\widehat{\Lambda(x)}$ is a diffeomorphism commuting with the natural left action of the group G on the quotient G/G_x. In other words, the action Λ and the choice of the isotropy group G_x induce a principal bundle structure on the group G with the projection $\pi : G \to M$ such that $\pi(g) = \Lambda(g, x)$, $g \in G$, and the isotropy group G_x as its structure group. Indeed, the group G_x acts freely on the right on the total space (the group) G and the isotropy group G_x is the kernel of the projection π as $\pi(gh) = \Lambda(gh, x) = \Lambda(g, \Lambda(h, x)) = \Lambda(g, x) = \pi(g)$ for any $g \in G$ and $h \in G_x$. In addition, the projection π is obviously differentiable. As the tangent map $d_e\pi : T_eG \to T_xM$, where $d_e\pi = d_e\Lambda_x$, is surjective, it identifies the tangent space to M at x with the quotient algebra $\mathfrak{g}/\mathfrak{g}_0$ where $\mathfrak{g}_0 = \ker d_e\pi$ denotes the Lie algebra of the isotropy group G_x and where we identified the tangent space T_eG with the Lie algebra of all left (right) invariant vector fields on the group G.

Example 1 To illustrate the basic elements of our model, we consider here a three-dimensional continuous lattice on $M = \mathbb{R}^3$ defined by the vector fields

$$l_1 = (1, z, 0), \ l_2 = (0, 0, 1), \ l_3 = (z, \frac{1}{2}(z^2 - x^2), -x). \tag{96}$$

The Lie algebra $\mathfrak{l}$ generated by the given vector fields is spanned by $l_1 = (1, z, 0)$, $l_2 = (0, 0, 1)$, $l_3 = (z, \frac{1}{2}(z^2 - x^2), -x)$, and $l_4 = (0, 0, 1)$ as $[l_1, l_2] = 0$, $[l_1, l_3] = l_4$, $[l_2, l_3] = 0$ while $[l_2, l_4] = 0$ and $[l_3, l_4] = l_1$. Viewing the vector fields l_i, $1 = 1, 2, 3, 4$, as the infinitesimal generators of one-parameter groups acting on $\mathbb{R}^3$, it can be shown that the lattice $\mathbf{l}$ (96) induces the action of the four-parameter group $G = H \rtimes SO(2)$ on $\mathbb{R}^3$, being the semi-direct product of the three-dimensional Heisenberg group H [49], and the special orthogonal group $SO(2)$.[5] Identifying the Heisenberg group with $\mathbb{R}^3$, $\mathbb{R}^3$ with the Cartesian product $\mathbb{C} \times \mathbb{R}$ and the special orthogonal group $SO(2)$ with a unit circle in $\mathbb{C}$, the action of the group G on $\mathbb{R}^3$ can be represented by the mapping $\Lambda : G \times \mathbb{R}^3 \to \mathbb{R}^3$ such that

$$\Lambda(((w, t), e^{i\theta}), (a, b)) = (w + ae^{i\theta}, t + b + \frac{1}{2}\mathrm{Im}(\overline{w}ae^{-i\theta})), \tag{97}$$

where the point $(a, b) \in \mathbb{C} \times \mathbb{R}$ is given by $(a, b) = (x + iz, y - \frac{1}{2}xz)$ and where $(w, t) = (\alpha + i\gamma, \beta - \frac{1}{2}\alpha\gamma)$ for any $(x, y, z) \in \mathbb{R}^3$ and $(\alpha, \beta, \gamma) \in H$. As the group operation in $H \rtimes SO(2)$ is given by

[5]The process of composing the actions of one-parameter subgroups to obtain the whole group G acting on the corresponding base space is only valid if the group G is connected [35].

$$((w_1,t_1),e^{i\theta_1})((w_2,t_2),e^{i\theta_2})=((w_1+w_2e^{-i\theta_1},t_1+t_2+\frac{1}{2}\mathrm{Im}(\overline{w}_1w_2e^{-i\theta_1})),e^{i(\theta_1+\theta_2)}) \tag{98}$$

one can show that Λ defines indeed the left action of G on $\mathbb{R}^3$ [49].

Given an arbitrary point $(x, y, z) \in \mathbb{R}^3$, consider now the orbit map $\Lambda_{(x,y,z)} : G \to \mathbb{R}^3$ of the group action Λ. Its tangent map $d_g\Lambda_{(x,y,z)} : T_gG \to T_{\Lambda(g,(x,y,z))}\mathbb{R}^3$, where $g = (\alpha, \beta, \gamma, \theta)$, is represented in the standard coordinate systems on G and $\mathbb{R}^3$ by the matrix

$$\begin{pmatrix} 1 & 0\;0 & -x\sin\theta + z\cos\theta \\ z\cos\theta - x\sin\theta & 1\;0 & -\alpha(z\sin\theta + x\cos\theta) - xz\sin 2\theta + \frac{1}{2}(z^2 - x^2)\cos 2\theta \\ 0 & 0\;1 & -x\cos\theta - z\sin\theta \end{pmatrix} \tag{99}$$

inducing at the identity of the group, i.e., $e = (0, 0, 0, 0)$, our lattice algebra $\mathfrak{l}$. On the other hand, as the (left) Lie algebra $\mathfrak{g}$ of the group $H \rtimes SO(2)$ is spanned by the vector fields

$$v_1 = (\cos\theta, -\alpha\sin\theta, -\sin\theta, 0), \;\; v_2 = (0, 1, 0, 0), \tag{100}$$

$$v_3 = (\sin\theta, \alpha\cos\theta, \cos\theta, 0), \;\; v_4 = (0, 0, 0, 1). \tag{101}$$

it is easy to show that the lattice algebra $\mathfrak{l}$ and the Lie algebra $\mathfrak{g}$ are isomorphic. Finally, selecting a point in $\mathbb{R}^3$, e.g., $(x_0, 0, 0)$, one can show that the isotropy group of the action Λ at $(x_0, 0, 0)$ is

$$G_0 = \{((x_0(1 - e^{-i\theta}), \frac{1}{2}x_0^2\sin\theta), e^{i\theta}) : \theta \in \mathbb{R}\} \tag{102}$$

and that its one-dimensional Lie algebra $\mathfrak{g}_0$ is spanned by $(0, x, \frac{1}{2}x^2, \theta)$.

7.1 Canonical Lattice Connection

We shall now consider the group G corresponding to the lattice $\mathbf{l}$ as the total space of a principal bundle on the body manifold M. Our main objective is to identify some of its geometric characteristics such as curvature and torsion.

To this end, let us fix a specific point $x_0 \in M$ and consider the corresponding isotropy group $G_0 = G_{x_0}$. Let $\mathfrak{g}$ denote the (left) Lie algebra of G. As $\mathfrak{g}_0$ is a subalgebra of the algebra $\mathfrak{g}$, there exists a (not necessarily unique) complementary vector space of left-invariant vector fields on TG, say $\mathfrak{D}$, such that $\mathfrak{g} = \mathfrak{g}_0 \oplus \mathfrak{D}$. Note that although $\mathfrak{g}$ and $\mathfrak{g}_0$ are Lie algebras, the vector space $\mathfrak{D}$ may not be an algebra in general. We define the vector space $\mathfrak{D}$ by lifting the continuous lattice (frame) $\mathbf{l}$ to the tangent space T_eG and use the left action of the group G on itself to generate a family of left-invariant linearly independent vector fields on TG. Specifically, given the vector fields $l_i : M \to TM$, $l_i \in \mathbf{l}$, $i = 1, \cdots, n$, we lift them to the tangent space T_eG by requiring that the lifted vectors $\mathfrak{l}_i \in T_eG$, $i = 1, \cdots, n$, are such that

$$d_e\pi(\mathfrak{l}_i) = d_e\Lambda_x(\mathfrak{l}_i) = l_i(x_0),\ i = 1, \cdots, n, \tag{103}$$

where, as before, we identified the Lie algebra $\mathfrak{g}$ with the tangent space to the group G at the identity e and where $\pi : G \to M$ is the bundle projection. As the vector fields l_i, $i = 1, \cdots, n$ are linearly independent and as the projection π is of maximum rank, the induced action of the group G on the lifted frame $\mathfrak{l}$ spans the left-invariant distribution $\mathfrak{D}$ on the manifold (group) G.

The vector space (of left-invariant vector fields) $\mathfrak{D}$ forms, by definition, a horizontal distribution on the principal bundle $\pi : G \to M$. Indeed, it depends smoothly on $g \in G$ and, as the subalgebra $\mathfrak{g}_0$ is the kernel of the projection $d\pi$ the restriction of $d\pi$ to the subspace $\mathfrak{D}$, i.e., $d\pi|_{\mathfrak{D}}$, is an isomorphism onto the tangent space TM. However, although the distribution $\mathfrak{D}$ is horizontal and left-invariant under the induced action of G on its tangent space, it does not, in general, define a principal bundle connection on G. The reason for that is that $\mathfrak{D}$ is not necessarily right-invariant under the action of the isotropy group G_0; the structure group of the bundle $\pi : G \to M$.

To circumvent this deficiency we shall restrict our analysis to continuous lattice structures such that the corresponding homogeneous manifold G/G_0 is **reductive**, that is, there exists a vector space $\mathfrak{D}$ of left-invariant vector fields on TG such that the algebra $\mathfrak{g}$ is the direct sum of the isotropy Lie algebra $\mathfrak{g}_0$ and the vector space $\mathfrak{D}$ and that the space $\mathfrak{D}$ is invariant under the infinitesimal action of the subalgebra $\mathfrak{g}_0$, i.e., $[\mathfrak{g}_0, \mathfrak{D}] \subset \mathfrak{D}$.[6] As the distribution $\mathfrak{D}$ is left-invariant under the induced action of the whole group G, and as the group G is connected, the reductivity requirement of G/G_0 implies that the distribution $\mathfrak{D}$ is right-invariant under the subgroup G_0. Hence, such $\mathfrak{D}$ is a horizontal distribution corresponding to a left-invariant principal bundle connection on $\pi : G \to M$.

Our last step in identifying the torsion and the curvature associated with the given lattice frame $\mathbf{l}$ is to 'reconstruct' the left-invariant principal bundle connection $\mathfrak{D}$ on the bundle of linear frames of the body manifold M, thus associating a specific linear connection with the defective solid continuum. To do this, let us first introduce the concept of the linear isotropy representation of the subgroup G_0. That is, given $g \in G$, consider the mapping $\Lambda_g : M \to M$ such that $\Lambda_g(x) = \Lambda(g, x)$, $x \in M$.

[6] The reductivity of a homogeneous space G/G_0 is usually defined by requiring that there exists a vector space $\mathfrak{D} \subset \mathfrak{g}$ such that the algebra $\mathfrak{g} = \mathfrak{g}_0 \oplus \mathfrak{D}$ and such that it is invariant under the adjoint action of the subgroup G_0. The condition $[\mathfrak{g}_0, \mathfrak{D}] \subset \mathfrak{D}$ implies the invariance of the distribution $\mathfrak{D}$ under the adjoint action of the group G_0, but not vice versa. However, when the group G is a connected Lie group, both conditions are equivalent.

Note that not all homogeneous spaces are reductive [44]. Note also that it is not necessarily easy to determine if a given homogeneous space is indeed reductive as it requires finding a vector space complement to $\mathfrak{g}_0$ in $\mathfrak{g}$ (among all possible complements) which satisfies the said condition of invariance.

In particular, $\Lambda_h(x_0) = x_0$ if and only if $h \in G_0$. Moreover, given $h \in G_0$, the tangent map $d_{x_0}\Lambda_h : T_{x_0}M \to T_{x_0}M$ is a linear isomorphism inducing a $GL(n, \mathbb{R})$-representation of the isotropy group G_0, where $GL(n, \mathbb{R})$ denotes the general linear group of $\mathbb{R}^n$. Namely, let $u_0 : \mathbb{R}^n \to T_{x_0}$ be a linear isomorphism (a **linear frame**) at $x_0 \in M$ assigning to an n-tuple $(\xi_1, \cdots, \xi_n) \in \mathbb{R}^n$ a vector in $T_{x_0}M$ having $\xi_1, \cdots, \xi_n$ coordinates in the specific basis. The **linear isotropy representation** of G_0 is the homomorphism $\lambda : G_0 \to GL(n, \mathbb{R})$ such that

$$\lambda(h) = u_0^{-1} \circ d_{x_0}\Lambda_h \circ u_0, \quad h \in G_0, \tag{104}$$

where the image $\tilde{G}_0 = \lambda(G_0)$ is called the **linear isotropy group** at $x_0 \in M$.

Choosing the specific frame u_0 and having the linear isotropy representation available, we can now 'replicate' the bundle $\pi : G \to M$ as a subbundle of the bundle of linear frames of M. Indeed, given the frame u_0 and selecting a point $x \in M$, the mapping $d_{x_0}\Lambda_g \circ u_0 : \mathbb{R}^n \to T_xM$, such that $\Lambda(g, x_0) = x$, represents a linear frame at x; all due to the transitivity of the action Λ. In addition, acting on such a frame on the right by the linear isotropy group one obtains a selection of linear frames at x. Namely, given $h \in G_0$

$$d_{x_0}\Lambda_g \circ u_0 \circ \left(u_0^{-1} \circ d_{x_0}\Lambda_h \circ u_0\right) = d_{x_0}\Lambda_{gh} \circ u_0 \tag{105}$$

is again a linear frame at x. Collecting all linear frames by varying $g \in G$ and $h \in G_0$ one obtains a reduction of the bundle of linear frames $L(M)$ to the linear isotropy group, that is, a subbundle of $L(M)$ with $\tilde{G}_0 \subset GL(n, \mathbb{R})$ as its structure group. Let $L(M, \tilde{G}_0)$ denote the new subbundle of linear frames with the projection $\tilde{\pi} : L(M, \tilde{G}_0) \to M$ assigning to a frame $d_{x_0}\Lambda_g \circ u_0$ the point $\Lambda(g, x_0)$. Hence, the mapping $g \mapsto d_{x_0}\Lambda_g \circ u_0$ from G to $L(M, \tilde{G}_0)$ defines the bundle isomorphism between $\pi : G \to M$ and $\tilde{\pi} : L(M, \tilde{G}_0) \to M$ over the identity map on M. Moreover, as $\pi : G \to M$ is left-invariant under the action of the group G so is $L(M, \tilde{G}_0)$. That is, given a frame $u \in L(M, \tilde{G}_0)$ the action of the group G is defined by assigning to the pair g, u the frame $gu = d_x\Lambda_g \circ u$ where $g \in G$ and $\tilde{\pi}(u) = x$.

The left-invariant horizontal distribution $\mathfrak{D}$ on the group G can now be replicated on the bundle of frames $L(M, \tilde{G}_0)$ using the **natural lift** of $X \in \mathfrak{g}$ construction[7] thus inducing a left-invariant horizontal distribution $\tilde{\mathfrak{D}}$ on $L(M, \tilde{G}_0)$ corresponding to a linear connection on M. To this end, let $\mathfrak{gl}(n, \mathbb{R})$ denote the Lie algebra of the general Lie group $GL(n, \mathbb{R})$ and let Π be an equivariant linear mapping from the Lie algebra $\mathfrak{g}$ to $\mathfrak{gl}(n, \mathbb{R})$ (a $\mathfrak{gl}(n, \mathbb{R})$-valued one-form on T_eG) such that

$$\Pi(X) = \begin{cases} d\lambda(X), & X \in \mathfrak{g}_0, \\ 0, & X \in \mathfrak{D}, \end{cases} \tag{106}$$

[7] See for example [24].

where $d\lambda$ is the Lie algebra homomorphism from $\mathfrak{g}_0$ into $\mathfrak{gl}(n, \mathbb{R})$ induced by the linear isotropy representation λ.[8] The (lattice) **canonical connection** corresponding to the decomposition $\mathfrak{g} = \mathfrak{g}_0 \oplus \mathfrak{D}$ of the reductive homogeneous space G/G_0 is a left-invariant linear connection on the reduced frame bundle $L(M, \tilde{G}_0)$ defined by the $\mathfrak{gl}(n, \mathbb{R})$-valued one-form (a connection form) ω such that

$$\omega(\tilde{X}) = \Pi(X), \quad X \in \mathfrak{g}, \tag{107}$$

where $\tilde{X}$ is the natural lift of X to the bundle $L(M, \tilde{G}_0)$. Note that it can be shown that the natural lift of any $X \in \mathfrak{g}_0$ is vertical in $L(M, \tilde{G}_0)$ and that the distribution $\tilde{\mathfrak{D}}$ containing the natural lifts of $X \in \mathfrak{D}$ is a horizontal distribution in $L(M, \tilde{G}_0)$. Then we have:

Theorem 6 *Let* $\mathbf{l} : M \to L(M)$ *be a continuous lattice defined on the body manifold* M. *Select the point* $x_0 \in M$ *and the frame* $u_0 : \mathbb{R}^n \to T_{x_0}M$. *Suppose that the homogeneous space* G/G_0 *associated with the lattice* $\mathbf{l}$ *is reductive relative to the decomposition* $\mathfrak{g} = \mathfrak{g}_0 \oplus \mathfrak{D}$ *of the Lie algebra of the Lie group* G *and let* $L(M, \tilde{G}_0)$ *be the corresponding reduction of the bundle of linear frames of* M *to the linear isotropy group* $\tilde{G}_0$. *Finally, identify the vector space* $\mathfrak{D}$ *(viewed as a subspace of* T_eG*) with* $\mathbb{R}^n$ *by means of the isomorphism* $u_0^{-1} \circ d_e\Lambda_{x_0}|_{\mathfrak{D}} : \mathfrak{D} \to \mathbb{R}^n$. *Then, subject to the choice of the frame* u_0, *the torsion and curvature of the left-invariant lattice canonical connection* ω *are given at* x_0 *by*

1. $T(X, Y) = -[X, Y]_{\mathfrak{D}}$,
2. $R(X, Y)Z = -[[X, Y]_{\mathfrak{g}_0}, Z]$

for any $X, Y, Z \in \mathfrak{D}$ *where* $[\cdot, \cdot]_{\mathfrak{D}}$ *and* $[\cdot, \cdot]_{\mathfrak{g}_0}$ *denote the* $\mathfrak{D}$ *and* $\mathfrak{g}_0$ *components of the Lie algebra bracket in* $\mathfrak{g}$, *respectively. In addition, both tensors are left-invariant on* M, *thus, covariantly constant.*

As the torsion and the curvature of the lattice canonical connection are expressed by the Lie bracket of the Lie algebra $\mathfrak{g}$ it is evident that the properties of the connection ω are, de facto, determined by the properties of the group G. Two specific cases are particularly worth mentioning. First, suppose that the reductive decomposition $\mathfrak{g} = \mathfrak{g}_0 \oplus \mathfrak{D}$ is such that the vector space $\mathfrak{D}$ is also a Lie algebra. Then, the canonical connection ω is curvature free (flat) as the Lie bracket $[\mathfrak{D}, \mathfrak{D}]$ has no $\mathfrak{g}_0$ component. This is true, in particular, when the group G is a semi-direct product of the isotropy group G_0 and another Lie group, say D, the Lie algebra of which is isomorphic to $\mathfrak{D}$. If, in addition, the group D is abelian the canonical connection ω is

[8] The equivariance of the one-form Π is the direct consequence of the assumption that the homogeneous manifold G/G_0 is reductive as shown in [13].

trivial as its torsion vanishes as well. Second, suppose that the continuous lattice $\mathbf{l}$ is uniformly defective (i.e., the dislocation density tensor is constant). Consequently, its body manifold M can be viewed as a Lie group acting on itself and the isotropy group G_0 is trivial. Thus, the curvature of the corresponding lattice canonical connection ω vanishes. Moreover, its torsion is given by the Lie algebra constants of the algebra $\mathfrak{g}$. In other words, when the continuous lattice is uniformly defective the long-distance parallelism induced on M by the lattice frame $\mathbf{l}$ is identical to the long-distance parallelism of the corresponding canonical connection ω. In general, however, these two connections are somewhat independent, and complementing, as illustrated by the examples in the next section.

We should also point out that when the lattice canonical connection ω is flat, the left-invariant distribution $\widetilde{\mathfrak{D}}$ is involutive, i.e., there exists a frame field l_i^*, $i = 1, \cdots, n$, on M given by the tangent of the orbit map Λ_x;

$$l_i^*(x) = d_e \Lambda_x(l_i), \; x \in M, \tag{108}$$

defining the long-distance parallelism corresponding to ω.

Example 2 Here we continue to develop Example 1, where the continuous lattice $\mathbf{l}$ is defined by the frame $l_1 = (1, z, 0)$, $l_2 = (0, 0, 1)$, $l_3 = (z, \frac{1}{2}(z^2 - x^2), -x)$ inducing the left action of the semi-direct product of the Heisenberg group and the special orthogonal group $G = H \rtimes SO(2)$ on $\mathbb{R}^3$ and the isotropy group G_0 at $(x_0, 0, 0) \in \mathbb{R}^3$.

As the group G is a semi-direct product of two subgroups its Lie algebra $\mathfrak{g}$ is reductive via the decomposition $\mathfrak{g} = \mathfrak{g}_0 \oplus \mathfrak{h}$ where $\mathfrak{h}$ denotes the Lie algebra of the Heisenberg group while $\mathfrak{g}_0$ is the Lie algebra of the isotropy group isomorphic to $so(2)$. Indeed, define the vector space $\mathfrak{D} = \text{span}\{v_1, v_2, v_3\}$ to realize that it forms a Lie subalgebra (the Lie algebra of H) of $\mathfrak{g}$. Note also that $\mathfrak{D}$ is left-invariant under the action of the whole group G and that it is invariant under the infinitesimal action of the algebra $\mathfrak{g}_0$ as $[\mathfrak{D}, \mathfrak{g}_0] = [\mathfrak{D}, v_4] \subset \mathfrak{D}$. Moreover, as the action $\Lambda : G \times \mathbb{R}^3 \to \mathbb{R}^3$ induces the mapping $\Lambda_{(w,t,\theta)} : \mathbb{R}^3 \to \mathbb{R}^3$ such that $\Lambda_{(w,t,\theta)}(a, b) = \Lambda((w, t, \theta), (a, b))$, where $(w, t, \theta) \in G = H \rtimes SO(2)$, its tangent map $d_{(a,b)}\Lambda_{(w,t,\theta)}$ can be represented by

$$\begin{pmatrix} \cos\theta & 0 & \sin\theta \\ -\alpha\sin\theta - z\sin^2\theta - \frac{1}{2}x\sin 2\theta & 1 & \alpha\cos\theta - x\sin^2\theta + \frac{1}{2}\sin 2\theta \\ -\sin\theta & 0 & \cos\theta \end{pmatrix}. \tag{109}$$

Thus, the linear isotropy representation of the isotropy group $G_0 \cong \mathfrak{so}(2)$ is given as

$$d_{(x_0,0,0)}\Lambda_{(0,0,\theta)} = \begin{pmatrix} \cos\theta & 0 & \sin\theta \\ 0 & 1 & 0 \\ -\sin\theta & 0 & \cos\theta \end{pmatrix}. \tag{110}$$

Finally, as the vector space $\mathfrak{D}$ is a subalgebra of $\mathfrak{g}$, it is involutive and the corresponding lattice canonical connection is flat (curvature free) but it has a non-vanishing torsion as not all Lie brackets $[v_i, v_j]$, $i, j = 1, 2, 3$, vanish. Consequently, the long-distance parallelism of ω is defined by the frame

$$\mathbf{l}^*(x, y, z) = d_e \Lambda_{(x,y,z)}(v_1, v_2, v_3) = \begin{pmatrix} 1 & 0 & 0 \\ z & 1 & 0 \\ 0 & 0 & 1 \end{pmatrix}. \tag{111}$$

The corresponding Christoffel's symbols Γ^i_{jk} are defined [15], by

$$\Gamma^i_{j3} = -\begin{pmatrix} 1 & 0 & 0 \\ z & 1 & 0 \\ 0 & 0 & 1 \end{pmatrix}^{-1} \frac{\partial}{\partial z} \begin{pmatrix} 1 & 0 & 0 \\ z & 1 & 0 \\ 0 & 0 & 1 \end{pmatrix}, \; i, j = 1, 2, 3, \tag{112}$$

where all Γ^i_{j1}, Γ^i_{j2} vanish.

7.2 Examples

In this last section we present a summary[9] of two other examples of continuous lattice structures; one when the lattice canonical connection is not only flat, as in our main example but, in fact, is trivial and the other when the lattice canonical connection is non-flat but symmetric.

Let us consider first the two-dimensional lattice defined in the standard coordinate system on $\mathbb{R}^2$ by the frame $l_1 = (1, 0)$ and $l_2 = (0, -x)$. This lattice is non-uniformly defective and generates the three-dimensional algebra (c.f., Sect. 6)

$$\mathfrak{l} = \mathrm{span}\{(1, 0), (0, -x), (0, 1)\} \tag{113}$$

inducing the left action Λ of the connected group G on $\mathbb{R}^2$ such that

$$\Lambda((a, b, c), (x, y)) = (x + a, y - b(x + a) + c) \tag{114}$$

where $(a, b, c) \in G$, $(x, y) \in \mathbb{R}^2$ and where the group multiplication is $g\overline{g} = (a + \overline{a}, b + \overline{b}, c + \overline{c} + \overline{b}a)$, $g, \overline{g} \in G$.

The (left) Lie algebra $\mathfrak{g}$ of the group G is spanned by the vector fields

$$v_1 = (1, 0, 0), \; v_2 = (0, 1, a), \; v_3 = (0, 0, 1) \tag{115}$$

[9]More detailed presentation can be found in [13].

and it is straightforward to show that it is isomorphic to $\mathfrak{l}$. Selecting a point $(x_0, y_0) \in \mathbb{R}^2$, the isotropy group of the action Λ at (x_0, y_0) is

$$G_0 = \{(0, b, bx_0) : b \in \mathbb{R}\} \tag{116}$$

and the corresponding Lie algebra is spanned by $(0, 1, x_0)$.

To determine if the homogeneous space G/G_0 is reductive, let us define the vector space of left-invariant vector fields on G by

$$\mathfrak{D} = \text{span}\{v_1, v_3\}. \tag{117}$$

When supplemented by the vector field $v_0 = (0, 1, x_0 + a)$ it generates a Lie algebra of left-invariant vector fields on G isomorphic to $\mathfrak{g}$ and such that $\mathfrak{g} = \mathfrak{g}_0 \oplus \mathfrak{D}$. In addition, as $[v_1, v_0] = v_3$ and $[v_3, v_0] = 0$, the vector space $\mathfrak{D}$ is invariant under the infinitesimal action of the isotropy group G_0 proving that G/G_0 is reductive under the decomposition $\mathfrak{g} = \mathfrak{g}_0 \oplus \mathfrak{D}$. Moreover, as the bracket $[v_1, v_3]$ vanishes, the vector space $\mathfrak{D}$ is an abelian Lie algebra. Putting all these facts together, we conclude that the lattice canonical connection ω induced by the distribution $\mathfrak{D}$ is flat and has a vanishing torsion (Theorem 6). In fact, the corresponding long-distance parallelism is defined by the standard frame

$$l^*(x, y) = d_e\Lambda_{(x,y)}(v_1, v_3) = \begin{pmatrix} 1 & 0 \\ 0 & 1 \end{pmatrix} \tag{118}$$

implying that the Christoffel's symbols Γ^i_{jk}, $i, j, k = 1, 2$, of the connection ω vanish identically.

In our second example, we consider the lattice frame $\mathbf{l} : \mathbb{R}^2 \to L(\mathbb{R}^2)$ given by the vector fields

$$l_1 = (y, -x),\ l_2 = (\frac{1}{2}(1 + x^2 - y^2), xy). \tag{119}$$

Calculating the corresponding Lie algebra $\mathfrak{l}$ we have $[l_1, l_2] = (xy, \frac{1}{2}(1 + y^2 - x^2)) = l_3$ while $[l_2, l_3] = l_1$ and $[l_1, l_3] = -l_2$. Thus, the lattice algebra $\mathfrak{l}$ is a three-dimensional Lie algebra isomorphic to the Lie algebra $\mathfrak{so}(3)$ of the special orthogonal group $SO(3)$.[10] It turns out that the algebra $\mathfrak{so}(3)$ is isomorphic to the Lie algebra $\mathfrak{su}(2)$ of the special unitary group $SU(2)$. Moreover, as the group $SU(2)$

[10] To confirm this isomorphism select

$$\begin{pmatrix} 0 & 0 & 0 \\ 0 & 0 & -1 \\ 0 & 1 & 0 \end{pmatrix}, \begin{pmatrix} 0 & -1 & 0 \\ 1 & 0 & 0 \\ 0 & 0 & 0 \end{pmatrix}, \begin{pmatrix} 0 & 0 & -1 \\ 0 & 0 & 0 \\ 1 & 0 & 0 \end{pmatrix}$$

as a basis of $\mathfrak{so}(3)$ [2].

covers $SO(3)$ via the isomorphism $p : SU(2)/\{I, -I\} \to SO(3)$, we replace the action of $SO(3)$ on $\mathbb{R}^2$ by the analogous action of $SU(2)$ on the complex plane $\mathbb{C}$ viewed as $\mathbb{R}^2$. That is, we consider the action $\Lambda : SU(2) \times \mathbb{C} \to \mathbb{C}$ such that

$$\Lambda\left(\begin{pmatrix} a & -\bar{b} \\ b & \bar{a} \end{pmatrix}, z\right) = \frac{b + \bar{a}z}{a - \bar{b}z}, \tag{120}$$

where $\begin{pmatrix} a & -\bar{b} \\ b & \bar{a} \end{pmatrix} \in SU(2), a, b \in \mathbb{C}$, subject to the constraint $a\bar{a} + b\bar{b} = 1$. Such action Λ is transitive with the isotropy group at, say $z_0 = 1$, given by

$$G_0 = \left\{\begin{pmatrix} \alpha & \beta i \\ \beta i & \alpha \end{pmatrix} : \alpha^2 + \beta^2 = 1; \alpha, \beta \in \mathbb{R}\right\}. \tag{121}$$

Selecting

$$E = \frac{1}{2}\begin{pmatrix} 0 & i \\ i & 0 \end{pmatrix}, \quad F = \frac{1}{2}\begin{pmatrix} i & 0 \\ 0 & -i \end{pmatrix}, \quad H = \frac{1}{2}\begin{pmatrix} 0 & 1 \\ -1 & 0 \end{pmatrix} \tag{122}$$

as the basis of the (left) Lie algebra of $SU(2)$, we see that the Lie algebra $\mathfrak{g}_0$ of the isotropy group is spanned by the matrix E and that the vector space $\mathfrak{D} = \text{span}\{H, F\}$ is such that $[\mathfrak{g}_0, \mathfrak{D}] \subset \mathfrak{D}$. Hence, the homogeneous space $G/G_0 = SU(2)/G_0 \cong SU(2)/SO(2)$ is reductive. However, in the contrast with our previous example, the vector space $\mathfrak{D}$ is not a Lie subalgebra of $\mathfrak{su}(2)$ as $[H, F] = E \in \mathfrak{g}_0$. Consequently, the lattice canonical connection ω corresponding to our reductive decomposition is symmetric but non-flat (Theorem 6). Indeed, as $[H, F]_{\mathfrak{D}} = 0$ the components of the curvature tensor are given by the relations:

$$R(H, F)H = [[H, F]_{\mathfrak{g}_0}, H] = [E, H] = F \tag{123}$$

$$R(H, F)F = [[H, F]_{\mathfrak{g}_0}, F] = [E, F] = -H. \tag{124}$$

8 Conclusion

We have explored the relation between discrete and continuous crystal structures in the context of three dimensional nilpotent Lie groups, mostly, and developed the mathematical apparatus required to discuss these ideas in a systematic way, it seems to us. It is a very particular assumption that, given a crystal state Σ, there exists such a nilpotent group (even more particular is the assumption that the relevant group is commutative, which leads to traditional crystallography ...). More general is the notion that the crystal state is such that there is a corresponding finite-dimensional lattice algebra, and so a higher dimensional Lie group acting on the manifold M. This more general notion can be expressed in terms of relations between the directional derivatives of the ddt, and so in terms of constraints on the classifying manifold. From this point of view, what we have done is explore properties of vector fields relating to that part of the classifying manifold where the corresponding constraints are satisfied, and focus on related discrete structures. We have also asked whether or not the points that compose those discrete (group) structures can be represented (in carefully chosen coordinates) in terms that are familiar in crystallography, that is as lattices or collections of lattices, and in the cases that we have discussed it is indeed so that there are such representations. Some more general cases are discussed in [1, 32, 33], for example.

We do not know of general results relating to the classification of discrete group structures relating to (for example) general submanifolds of CM_{Σ}, or the existence of corresponding representations as (collections of) lattices. (Since the precepts of crystallography were shaken by the discovery of quasicrystals, i.e., discrete sets of points with aperiodic structure [47], one must suppose that such representations do not always exist). The major task that we envisage is, loosely, to provide an 'atlas' of CM_{Σ}, showing the possible corresponding discrete crystal structures in some parts, more general discrete structures (e.g., perhaps, quasicrystals) elsewhere. The quantities that define CM_{Σ} are the scalar invariants, or plastic strain variables—with such an atlas one would have insight into how changes in the plastic strain variables induce changes in crystal structure, and this should enable one to construct a mechanics of defective crystals based entirely on a well defined kinematics.

Of course other tasks need to be addressed, to facilitate this programme in a way that resonates with phenomenological theories of inelastic behaviour and encourages more study of the geometry: for example, 'slip' should find modern geometric expression, curvature of a higher dimensional Lie group lacks physical interpretation in terms analogous to the Burgers vector construction. There are technical issues too, but we refer the reader to source material for details.

Acknowledgements We are very grateful to the EPSRC for support provided via Research Grant EP/M024202/1.

References

1. Auslander, L., Green, L., Hahn, F.: Flows on homogeneous spaces, Annals of Mathematics Studies **53**, Princeton University Press, Princeton NJ (1963)
2. Baker A.: Matrix groups, Springer-Verlag, London (2003)
3. Bianchi,L.: Lezioni sulla teoria dei gruppi continui finiti di trasformazioni. Enrico Spoerri, Editore, Pisa (1918)
4. Bilby, B.A.: Continuous distributions of dislocations, in Progress in Solid Mechanics, vol. 1, ed. I. Sneddon, North-Holland, Amsterdam (1960).
5. Cauchy, A.L.: Sur l'equilibre et le mouvement d'un systeme de points materials sollecites par d'attraction ou de repulsion mutuelle, Ex. de Math., **3**, 227–287, (1828)
6. Cermelli, P., Parry, G.P.: The structure of uniform discrete defective crystals, Continuum Mechanics and Thermodynamics, **18**, 47–61 (2006)

7. Chipot, M., Kinderlehrer, D.: Equilibrium configurations of crystals, Arch. Rat. Mech. Anal., **103**, 237–277, (1988)
8. Crainic, M., Fernandes, R.L.: Integrability of Lie brackets, Annals of Mathematics, **157**, 575–620, (2003), https://doi.org/10.4007/annals.2003.157.575
9. Davini, C.: A proposal for a continuum theory of defective crystals, Arch. Ration. Mech. Anal., **96**, 295–317 (1986), https://doi.org/10.1007/BF00251800
10. Davini, C., Parry, G.P.: On defect-preserving deformations in crystals, Int. J.Plasticity **5**, 337–369 (1989), https://doi.org/10.1016/0749-6419(89)90022-3
11. Davini, C. and Parry, G.P.: A complete list of invariants for defective crystals, Proc. R. Soc. Lond. A **432**, 341–365 (1991), Erratum: Proc. R. Soc. Lond. A **434**, 735 (1991), https://doi.org/10.1098/rspa1991.0125
12. Elżanowski, M., Parry G.P.: Material symmetry in a theory of continuously defective crystals, J. Elasticity, **74**, 215–237, (2004)
13. Elżanowski, M., Parry, G.P.: Connection and curvature in crystals with nonconstant dislocation density, Math. Mech. Solids, 1–12 (2018), https://doi.org/10.1177/1081286518791008
14. Elżanowski, M., Preston, S.: On continuously defective elastic crystals, Miskolc Mathematical Notes, **14**(2), 659–670 (2013)
15. Epstein, M.: The Geometrical Language of Continuum Mechanics, Cambridge University Press, Cambridge (2010)
16. Fernandes, R.L., Struchiner, I.: The classifying Lie algebroid of a geometric structure I: Classes of coframes. Trans. Ams. Math. Soc., **366**, 2419–2462 (2014), https://doi.org/10.1090/S0002-9947-2014-05973-4
17. Fonseca, I., Parry, G.P.: Equilibrium configurations of defective crystals. Arch. Ration. Mech. Anal., **120**, 245–283 (1992), https://doi.org/10.1007/BF00375027
18. Gorbatsevich, V.V., Onishchik, A.L., Vinberg, E.B.: Lie Groups and Lie Algebras I, Encyclopedia of Mathematical Sciences, Vol. **20**, Onishchik, A.L. (ed.), Springer-Verlag, Berlin Heiderberg (1993)
19. Gorbatsevich, V.V., Onishchik, A.L., Vinberg, E.B.: Foundations of Lie theory and Lie transformation groups, Springer-Verlag, Berlin, Heidelberg (1997)
20. Green, A.E., Adkins,J.E.: Large Elastic Deformations, (2nd ed.), Oxford University Press, London, Oxford (1970)
21. Jacobsen, N.: Lie algebras, Dover Publications, New York (1979)
22. Johnson, D.L.: Presentations of groups, London Mathematical Society Student Texts **15**, 2nd edn, Cambridge University Press, Cambridge (1997).
23. Kobayashi, S.: Transformation Groups in Differential Geometry, Springer-Verlag, Berlin Heidelberg (1995)
24. Kobayashi, S., Nomizu, K.: Foundations of Differential Geometry, Vol. I, II, John Wiley & Sons, Inc., New York (1996)
25. Kondo, K.: Non-Riemannian geometry of imperfect crystals from a macroscopic viewpoint. *RAAG memoirs of the unifying study of basic problems in engineering and physical science by means of geometry*, **1**, Gakuyusty Bunken Fukin-Day, Tokyo (1955)
26. Lee, E.H.: Finite-strain elastic-plastic theory with application. J. of Appl.Physics., **38**, 19–27, (1967)
27. Lee, E.H.: Elastic-plastic deformation at finite strains, J. of App. Mechanics, **36**, 1–6, (1969)
28. Love, A.E.H.: A treatise on the mathematical theory of elasticity, Dover Publications, New York (1927)
29. Komrakov, B., Churyamov, A., Doubrov, B.: Two-dimensional homogeneous spaces, Preprint Series 17-Matematisk institutt, Universitetet i Oslo (1993)
30. Magnus, W., Karrass, A., Solitar, D.: Combinatorial Group Theory, Dover, New York (1976)
31. Mal'cev, A.: On a class of homogeneous spaces, Izv. Akad. Nauk SSSR, Ser. Mat., **13**, 9–32, (1949), Am. Math. Soc. Translation, **39**, (1949)
32. Nicks, R., Parry, G.P.: On symmetries of crystals with defects related to a class of solvable groups (S_1), Math. Mech. Solids, **17**, 631–651 (2011), https://doi.org/10.1177/1081286511427485

33. Nicks, R., Parry, G.P.: On symmetries of crystals with defects related to a class of solvable groups (S_2), Mathematical Methods in the Applied Sciences, **35**, 1741–1755, (2012)
34. Olver, P.J.: Applications of Lie Groups to Differential Equations, 2nd edn. Graduate Texts in Mathematics, Vol. 107, Springer-Verlag, Berlin (1993)
35. Olver, P.J.: Equivalence, Invariants, and Symmetry, Cambridge University Press, Cambridge (1995)
36. Palais, R.S.: A global formulation of the Lie theory of transformation groups, Mem. Amer. Math. Soc. **22** (1957)
37. Parry, G.P.: Group properties of defective crystal structures, Math. Mech. Solids, **8**, 515–537, (2003)
38. Parry, G.P.: Rotational Symmetries of Crystals with Defects, J. Elasticity, **94**, 147–166, (2009)
39. Parry, G.P.: Elastic Symmetries of Defective Crystals, J. Elasticity, **101**, 101–120, (2010)
40. Parry, G.P.: Discrete structures in continuum descriptions of defective crystals, Phil.Trans.R.Soc. A, **374**, (2015), https://doi.org/10.1098/rsta.2015.0172
41. Parry, G.P., Šilhavý, M.: Elastic invariants in the theory of defective crystals, Proc. R. Soc. Lond. A **455**, 4333–4346 (1999), https://doi.org/10.1098/rspa.1999.0503
42. Parry, G.P., Zyskin, M.: Geometrical structure of two-dimensional crystals with non-constant dislocation density, J.Elast **127**, 249–268 (2017), https://doi.org/10.1007/s10659-016-9612-3
43. Pitteri, M., Zanzotto, G.: Continuum Models for Phase Transitions and Twinning in Crystals, Chapman and Hall, Boca Raton, London, New York, Washington DC (2003)
44. Poor, W.A.: Differential Geometric Structures, McGraw-Hill, New York (1981)
45. Pontryagin, L.S.: Topological Groups, 2nd edn. Gordon and Breach, New York, London, Paris (1955)
46. Rossmann, W.: Lie Groups, An Introduction Through Linear groups, Oxford Graduate Texts in Mathematics; **5**, Oxford University Press, Oxford (2006)
47. Senechal, M.: Quasicrystals and geometry, Cambridge University Press, Cambridge (1995)
48. Thurston, W.:Three Dimensional Geometry and Topology, vol 1, Princeton University Press, Princeton (1997)
49. Tricerri, F., Vanhecke, L.: Homogeneous Structures on Riemannian Manifolds, Cambridge University Press, Cambridge (1983)
50. Varadarajan, V.S.: Lie groups, Lie algebras, and their representations, Prentice–Hall, Englewood Cliffs (1974)

Limits of Distributed Dislocations in Geometric and Constitutive Paradigms

Marcelo Epstein, Raz Kupferman, and Cy Maor

Contents

Abstract The 1950s foundational literature on rational mechanics exhibits two somewhat distinct paradigms to the representation of continuous distributions of defects in solids. In one paradigm, the fundamental objects are geometric structures on the body manifold, e.g., an affine connection and a Riemannian metric, which represent its internal microstructure. In the other paradigm, the fundamental object is the constitutive relation; if the constitutive relations satisfy a property of material uniformity, then it induces certain geometric structures on the manifold. In this paper, we first review these paradigms, and show that they are equivalent if the constitutive model has a discrete symmetry group (otherwise, they are still consistent; however, the geometric paradigm contains more information). We then consider bodies with continuously distributed edge dislocations, and show, in both paradigms, how they can be obtained as homogenization limits of bodies with finitely many dislocations as the number of dislocations tends to infinity. Homogenization in the geometric paradigm amounts to a convergence of manifolds; in the constitutive paradigm it amounts to a Γ-convergence of energy functionals.

M. Epstein
Department of Mechanical and Manufacturing Engineering, University of Calgary, Calgary, AB, Canada
e-mail: mepstein@ucalgary.ca

R. Kupferman · C. Maor (✉)
The Hebrew University of Jerusalem, Jerusalem, Israel
e-mail: raz@math.huji.ac.il; cy.maor@mail.huji.ac.il

R. Segev, M. Epstein (eds.), *Geometric Continuum Mechanics*, Advances in Mechanics and Mathematics 43, https://doi.org/10.1007/978-3-030-42683-5_8

We show that these two homogenization theories are consistent, and even identical in the case of constitutive relations having discrete symmetries.

1 Introduction

1.1 Geometric and Constitutive Paradigms

Geometric Paradigm: Body Manifolds The 1950s foundational literature on rational mechanics exhibits two somewhat distinct paradigms to the representation of continuous distributions of defects in solids. On the one hand, there is a paradigm promoted by Kondo [24], Nye [31], Bilby [2], and later Kröner (e.g., [26]), in which solid bodies are modeled as geometric objects—manifolds—and their internal microstructure is represented by sections of fiber bundles, such as a metric and an affine connection.

More specifically, in [2, 24, 31], the body manifold is assumed to be a smooth manifold $\mathcal{M}$, endowed with a notion of distant parallelism, which amounts to defining a curvature-free affine connection ∇. The connection is generally non-symmetric, and its torsion tensor is associated with the density of dislocations. This geometric model is motivated by an analysis of Burgers circuits, which in the presence of dislocations exhibit geodesic rectangles whose opposite sides are not of equal lengths—a signature of torsion (see Sect. 3 for a discussion of Burgers circuits and Burgers vectors in this setting).

Note that modulo the choice of a basis at a single point, the definition of a distant parallelism is equivalent to a choice of a basis for the tangent bundle at each point (i.e., a global smooth section of the frame bundle). Intuitively, the frame field at each point corresponds to the crystalline axes one would observe under a microscope. Torsion is a measure for how those local bases twist when moving from one point to another.

The choice of local bases induces a Riemannian metric $\mathfrak{g}$, known as a *reference* or an *intrinsic* metric. The intrinsic metric is the metric with respect to which the bases are orthonormal; although no specific constitutive response is assumed ab initio, it is interpreted as the metric that a small neighborhood would assume if it were cut off from the rest of the body, and allowed to relax its elastic energy.

The reference metric $\mathfrak{g}$ induces also a Riemannian (Levi-Civita) connection, denoted ∇^{LC}, which differs from ∇, unless the torsion vanishes. The Riemannian connection, unlike ∇, is generally non-flat; its curvature, if non-zero, is an obstruction for the existence of a strain-free global reference configuration. Finally, a triple $(\mathcal{M}, \mathfrak{g}, \nabla)$, where ∇ is a flat connection, metrically consistent with $\mathfrak{g}$, is known as a *Weitzenböck space* or a *Weitzenböck manifold* [37] (a notion originating from relativity theory, see, e.g., [1, 17]; for its use in the context of distributed dislocations, see, e.g., [19, 21, 32, 38]).

Constitutive Paradigm The second paradigm, due largely to Noll [30] and Wang [36], takes for elemental object a *constitutive relation*. The underlying manifold $\mathcal{M}$

has for role to set the topology of the body, and be a domain for the constitutive relation. In the case of a hyperelastic body, the constitutive relation takes the form of an *energy density* $W : T^*\mathcal{M} \otimes \mathbb{R}^d \to \mathbb{R}$. A constitutive relation is called *uniform* if the energy density at every point $p \in \mathcal{M}$ is determined by an "archetypical" function $\mathcal{W} : \mathbb{R}^d \times \mathbb{R}^d \to \mathbb{R}$, along with a local frame field $E : \mathcal{M} \times \mathbb{R}^d \to T\mathcal{M}$, which specifies how $\mathcal{W}$ is "implanted" into $\mathcal{M}$. Once a uniform constitutive relation has been defined, its pointwise symmetries and its dependence on position may define a so-called *material connection* ∇ along with an *intrinsic Riemannian metric* $\mathfrak{g}$ (described in detail in Sect. 2).

At this point, it is interesting to note Wang's own reflections comparing the geometric approach (in our language) to his [36]:

> It is not possible to make any precise comparison, however, since the physical literature on dislocation theory rarely if ever introduces definite constitutive equations, resting content with heuristic discussions of the body manifolds and seldom taking up the response of bodies to deformation and loading, which is the foundation stone of modern continuum mechanics.

Indeed, in the geometric paradigm, the constitutive relation typically does not appear explicitly. However, the geometric and the constitutive paradigms are consistent with each other. On the one hand, as shown by Wang, a constitutive relation subject to a uniformity property defines an intrinsic metric and a material connection (as will be shown below, the material connection is unique only if $\mathcal{W}$ has a discrete symmetry group). On the other hand, a body manifold endowed with a notion of distant parallelism defines a uniform constitutive relation for every choice of archetypal function $\mathcal{W}$ and implant map at a single point—once $\mathcal{W}$ has been implanted at some $p \in \mathcal{M}$, the whole constitutive relation is determined by parallel transporting this implant to any other point in $\mathcal{M}$ according to ∇; by construction, ∇ is a material connection of that constitutive relation.

This is the viewpoint that we take in this paper, and the one through which we show how homogenization processes in both paradigms are also equivalent with each other (see below). However, Wang's comment above is not unfounded: first, in the case of an archetype with a continuous symmetry group (say, isotropic), there is more than one material connection associated with the constitutive relation, hence from the constitutive point of view it does not make sense to talk about a single parallelism (or Weitzenböck manifold) that represents the body. Second, in certain cases in which the geometric viewpoint assumes a posteriori a constitutive response, the parallelism, or the torsion tensor associated with it, are eventually considered as variables in the constitutive relation [27], resulting in so-called *coupled stresses* [25]. This approach, in which the underlying geometric structure can change, e.g., due to loading, is beyond the scope of the constitutive paradigm (or at least, its time-independent version), and such models will not be considered in this work.

Finally, let us note that there are other approaches to dislocations not covered by the above discussion, which are beyond the scope of this paper. In particular, we will not consider the line of works emanating from Davini [10], and other more recent approaches such as [6, 18], although some of the consequences of the discussions here (e.g., continuous vs. discrete symmetries) may also apply to them.

1.2 Description of the Main Results

The physical notion of dislocations is rooted in discrete structures, such as defective crystal lattices. Thus, when considering distributed dislocations, it is natural to consider a homogenization process, in which a continuous distribution of dislocations (according to a chosen paradigm) is obtained as a limit of finitely many dislocations, as those are getting denser in some appropriate sense. A priori, each of the two paradigms could have its own homogenization theory:

1. Geometric paradigm: Consider body manifolds representing solids with finitely many (singular) dislocations, and study their limit as the number of dislocations tends to infinity.
2. Constitutive paradigm: Consider constitutive relations modeling solids with finitely many (singular) dislocations, and study their limit as the number of dislocations tends to infinity.

The first task belongs to the realm of geometric analysis, and has been addressed in [19, 21], where it was shown that any two-dimensional Weitzenböck manifold can be obtained as a limit of bodies with finitely many dislocations (see Sect. 3 for a precise statement). The second task belongs, for hyperelastic bodies, to the realm of the calculus of variations, and has been addressed in [20] for the special case of isotropic materials.

In this paper, we review the main results of these papers and extend the analysis of [20] to the non-isotropic case. More importantly, we show that the homogenization theories resulting from the geometric and the constitutive paradigms are consistent, and even identical in the case of constitutive relations having discrete symmetries. In particular, both predict the emergence of (the same) torsion as a limit of distributed dislocations.

Our main result in this chapter can be summarized as follows:

Theorem 1 (Equivalence of Homogenization Processes, Informal)

1. *For a body manifold* $(\mathcal{M}, \mathfrak{g}, \nabla)$ *with finitely many dislocations, there is a natural way to define a constitutive relation* $(\mathcal{M}, W)$ *based on a given archetype* $\mathcal{W}$*, for which* ∇ *is a material connection and* $\mathfrak{g}$ *is an intrinsic metric (Proposition 4).*
2. *If the archetype* $\mathcal{W}$ *has a discrete symmetry group, then this relation is bijective; i.e., a constitutive relation* $(\mathcal{M}, W)$ *defines a unique material connection* ∇ *and a unique intrinsic metric* $\mathfrak{g}$ *(Proposition 5).*
3. *If a sequence of body manifolds with n dislocations* $(\mathcal{M}_n, \mathfrak{g}_n, \nabla_n^{LC})$ *converges (in the sense of Theorem 2) to a Weitzenböck manifold* $(\mathcal{M}, \mathfrak{g}, \nabla)$*, then the corresponding constitutive models* $(\mathcal{M}, W_n)$ Γ*-converge to a constitutive model* $(\mathcal{M}, W)$*, for which* ∇ *is a material connection and* $\mathfrak{g}$ *is an intrinsic metric (Theorem 3).*

A sketch of Theorem 1 is shown in Fig. 1.

In addition to Theorem 1, this paper reviews the fundamental notions of the geometric and constitutive paradigms, and their abovementioned equivalence; we

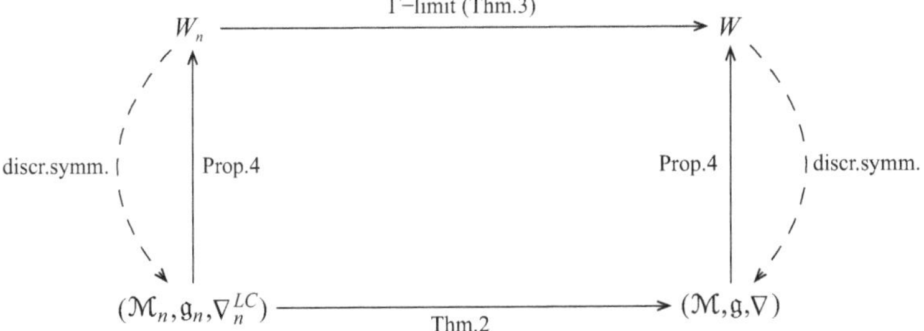

Fig. 1 A sketch of the main result (Theorem 1)

believe that the current presentation is original, and includes several results for which it is difficult (if at all possible) to find in the existing literature precise statements and proofs.

In the rest of this section, we elaborate on our main results. We start by considering a defect-free body: in the geometric paradigm, such a body is modeled as a d-dimensional Riemannian manifold $(\mathcal{M}_0, \mathfrak{g}_0)$, which can be embedded isometrically in Euclidean space $(\mathbb{R}^d, \mathfrak{e})$, where $\mathfrak{e}$ is the standard Euclidean metric. Let ∇_0^{LC} be the Levi-Civita connection of $\mathfrak{g}_0$; since $(\mathcal{M}_0, \mathfrak{g}_0)$ is isometric to a Euclidean domain, the connection ∇_0^{LC} is flat, and parallel transport is path-independent.

To obtain a constitutive relation for that same body, one has to fix an archetype $\mathcal{W}$ and a bijective linear map $(E_0)_p : \mathbb{R}^d \to T_p\mathcal{M}_0$ at some reference point $p \in \mathcal{M}_0$. The two together determine the mechanical response to deformation at p: for $A \in T_p^*\mathcal{M}_0 \otimes \mathbb{R}^d$, the elastic energy density (per unit volume, where the reference volume is the volume form of $(\mathcal{M}_0, \mathfrak{g}_0)$) at p is

$$(W_0)_p(A) = \mathcal{W}(A \circ (E_0)_p).$$

A constitutive relation is obtained by extending $(E_0)_p$ into a ∇_0^{LC}-parallel frame field $E_0 : \mathcal{M}_0 \times \mathbb{R}^d \to T\mathcal{M}_0$ (here is where the path-independence of the parallel transport is required). The elastic energy density is

$$W_0(A) = \mathcal{W}(A \circ E_0), \tag{1}$$

and the elastic energy associated with a map $f : \mathcal{M}_0 \to \mathbb{R}^d$ is

$$I_0(f) = \int_{\mathcal{M}_0} W_0(df)\, d\mathrm{Vol}_{\mathfrak{g}_0}, \tag{2}$$

where $d\mathrm{Vol}_{\mathfrak{g}_0}$ is the Riemannian volume form. As we show in Sect. 2.1, the geometric and the constitutive paradigms are consistent: $\mathfrak{g}_0$ is an *intrinsic metric* for W_0 and ∇_0^{LC} is a *material connection* for W_0; moreover, ∇_0^{LC} is the unique material connection for W_0, provided that $\mathcal{W}$ has a discrete symmetry group.

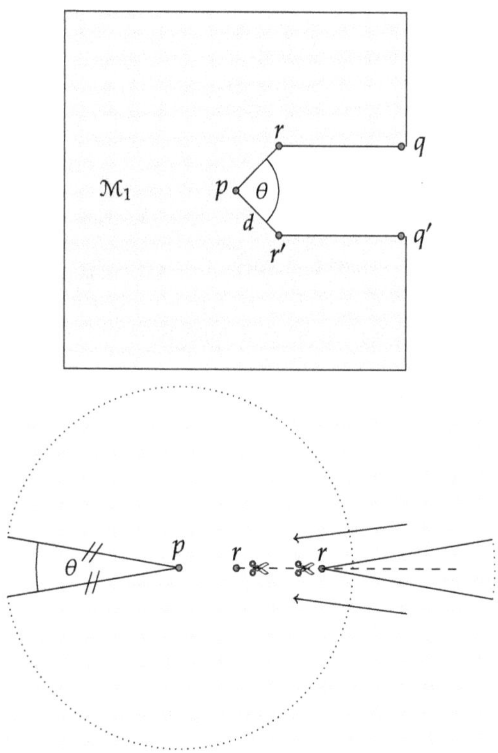

Fig. 2 Two equivalent cut-and-weld constructions generating a body manifold with a single edge dislocation. Top: the segments pr and pr' are identified (i.e., glued) as well as the segments rq and $r'q'$. p and $r \sim r'$ are the only singular points in the manifold (each with conical singularity of the same magnitude and opposite sign). Bottom: a sector whose vertex is denoted by p is removed from the plane and its outer boundaries are glued together, thus forming a cone. The same sector is then inserted into a straight cut along a ray whose endpoint is denoted by r

Consider next a body with a single straight edge-dislocation. Note that straight edge dislocations are in essence two-dimensional; we henceforth restrict ourselves to two dimensions. From the point of view of the geometric paradigm, the body manifold of a body with one edge-dislocation can be described by a Volterra cut-and-weld protocol [35]. There are numerous ways of implementing a Volterra protocol: two ways are depicted in Fig. 2.

The outcome of this cut-and-weld protocol is a topological manifold $\mathcal{M}_1$, which is smooth everywhere except at two points (the points p and $r \sim r'$ in Fig. 2). It is endowed with a metric $\mathfrak{g}_1$, which is locally Euclidean, since locally, every non-singular point has a neighborhood satisfying the abovementioned defining properties of a defect-free body manifold. As there is no *continuous* distribution of dislocations in this picture, the natural connection associated with this body is the Levi-Civita connection ∇_1^{LC} of $(\mathcal{M}_1, \mathfrak{g}_1)$. Moreover, the parallel transport induced by ∇_1^{LC} is path-independent for all simple closed paths that do not encircle only one of the two singular points. This restriction on admissible paths can be replaced by removing from the smooth part a segment connecting the two singular points. Note that the *topological* manifold $\mathcal{M}_1$ is simply connected; however, its smooth component is not. Despite being (almost everywhere) locally Euclidean, it cannot be embedded in the Euclidean plane isometrically.

The procedure for obtaining a constitutive relation within the constitutive paradigm follows the exact same lines as for a defect-free body. One has to fix an archetype $\mathcal{W}$ and a frame at a point $(E_1)_p : \mathbb{R}^2 \to T_p\mathcal{M}_1$; extending $(E_1)_p$ into a parallel frame field $E_1 : \mathcal{M}_1 \otimes \mathbb{R}^2 \to T\mathcal{M}_1$, the elastic energy density $W_1 : T^*\mathcal{M}_1 \otimes \mathbb{R}^2 \to \mathbb{R}$ is given by (1), after changing the subscript 0 to 1. Once again, the two paradigms are consistent, as $\mathfrak{g}_1$ and ∇_1^{LC} are an intrinsic metric and a material connection for the energy density W_1. Note that none of the two pictures makes any explicit mention of torsion.

The generalization of this procedure to a body carrying n singular edge dislocations follows the same lines, performing n Volterra cut-and-weld protocols, thus obtaining a simply connected topological manifold, which is smooth everywhere but at n pairs of singular point. On the geometric side, one obtains a triple $(\mathcal{M}_n, \mathfrak{g}_n, \nabla_n^{LC})$, where the Levi-Civita connection ∇_n^{LC} has trivial holonomy, namely, its parallel transport is path-independent for all simple closed paths that do not encircle only one singular point within a pair. After the choice of an archetype $\mathcal{W}$ and a frame at a point $(E_n)_p : \mathbb{R}^2 \to T_p^*\mathcal{M}_n$, one obtains an energy density W_n, for which $\mathfrak{g}_n$ and ∇_n^{LC} are an intrinsic metric and a material connection.

Next consider the limit of $n \to \infty$. As proved in [20], every two-dimensional body manifold $(\mathcal{M}, \mathfrak{g}, \nabla)$ admitting a global ∇-parallel frame field is a limit of manifolds $(\mathcal{M}_n, \mathfrak{g}_n \nabla_n^{LC})$ with finitely many dislocations. A precise definition of this convergence is stated in Theorem 2; loosely speaking, it means that $\mathcal{M}_n$ can be mapped into $\mathcal{M}$ such that orthonormal ∇_n^{LC}-parallel frame fields E_n are mapped into a frame field asymptotically close to an orthonormal ∇-parallel frame field. Note the *emergence of torsion*, as ∇_n^{LC} is torsion-free for every n, whereas ∇ has non-zero torsion.

We then switch to the constitutive paradigm: as described above, each of the manifolds $(\mathcal{M}_n, \mathfrak{g}_n \nabla_n^{LC})$ defines, upon the choice of an archetype $\mathcal{W}$ and a frame at one point $(E_n)_p$, an energy density W_n and an associated energy I_n. In Theorem 3, we prove that as $n \to \infty$, I_n converges in the sense of Γ-convergence to a limiting functional I, which has an energy density W, where W is the energy density obtained by the same construction using $\mathcal{W}$ and ∇. In particular, W has intrinsic metric $\mathfrak{g}$ and material connection ∇. This "closes the circle," proving that the construction of a uniform energy density from a given body manifold can be extended from finitely many to continuously distributed dislocations.

1.3 Structure of This Paper

In the rest of the paper we formalize the above outline:

- In Sect. 2, we present the main ingredients of the constitutive paradigm, following [36], under the assumption of hyperelasticity. We use a more modern notation and some simplifying assumptions.

Furthermore, we show in Sect. 2 how the constitutive paradigm and the geometric paradigm for describing dislocations are consistent, and equivalent in the case of discrete symmetry group (Propositions 4 and 5), thus establishing the vertical arrows in Fig. 1.

- In Sect. 3, we present in more detail the modeling of dislocations via the geometric paradigm, using the notion of Weitzenböck manifolds. In particular, we explain how Burgers vectors arise in this context, and their relation to the torsion tensor.

  The main part of this section is an overview of recent results [19, 21] concerning the homogenization of dislocations within this paradigm—a convergence of Weitzenböck manifolds (Theorem 2). This establishes the lower horizontal arrow in Fig. 1. For the sake of readability, we omit some of the technical details, and focus on the main ideas of the construction.
- In Sect. 4, we prove the convergence of the elastic energies associated with the converging Weitzenböck manifolds; we show that they Γ-converge to the elastic energy associated with the limiting Weitzenböck manifold (Theorem 3, Corollary 2), thus establishing the upper horizontal arrow in Fig. 1, and concluding the proof of Theorem 1.
- Finally, in Sect. 5 we show explicitly how the torsion tensor appears in the equilibrium equations of elastic bodies with continuously distributed dislocations according to the constitutive paradigm.

2 The Constitutive Paradigm of Noll and Wang

In this section we present some of the basic notions of the Noll-Wang approach. We generally follow [36], although our presentation and some of the proofs are somewhat different. For simplicity, we will assume a hyperelastic model.

Definition 1 (Hyperelastic Body) A hyperelastic body consists of a d-dimensional differentiable manifold, $\mathcal{M}$—the *body manifold*—and an energy-density function (or *constitutive relation*),

$$W : T^*\mathcal{M} \otimes \mathbb{R}^d \to \mathbb{R},$$

which is viewed as a (non-linear) bundle map over $\mathcal{M}$.

For $p \in \mathcal{M}$ and $A \in T_p^*\mathcal{M} \otimes \mathbb{R}^d$, we denote the action of W on A by $W_p(A)$. If ξ is a section of $T^*\mathcal{M} \otimes \mathbb{R}^d$, then $W(\xi)$ is a function on $\mathcal{M}$.

Remark 1 In the terminology of Noll, such a body is called a *simple body* since the constitutive relation at a point depends only on the local deformation (i.e., the first jet of the deformation) at that point.

We will use the following notation: the groups $\mathrm{GL}(d)$, $\mathrm{SO}(d)$ are the standard subgroups of $\mathrm{Hom}(\mathbb{R}^d, \mathbb{R}^d)$; for two oriented inner-product spaces $(V, \mathfrak{g})$, $(W, \mathfrak{h})$ we will denote by $\mathrm{SO}(V, W)$ or $\mathrm{SO}(\mathfrak{g}, \mathfrak{h})$ the set of orientation-preserving isometries $V \to W$, and by $\mathrm{SO}(V)$ the orientation-preserving isometries $V \to V$.

The next definition makes precise the notion of material uniformity, namely, a constitutive relation that is "the same" at every point:

Definition 2 (Material Uniformity) A hyperelastic body is called *uniform* if for every $p \in \mathcal{M}$ there exists a frame, i.e., a linear isomorphism $E_p : \mathbb{R}^d \to T_p\mathcal{M}$ such that,

$$W_p(A) = \mathcal{W}(A \circ E_p) \quad \text{for every } A \in T_p^*\mathcal{M} \otimes \mathbb{R}^d, \tag{3}$$

for some

$$\mathcal{W} : \mathbb{R}^d \otimes \mathbb{R}^d \to \mathbb{R}$$

independent of p.

Remark 2 More precisely, a hyperelastic energy density W is a section of $(T^*\mathcal{M} \otimes \mathbb{R}^d)^* \otimes \wedge^d T^*\mathcal{M}$, i.e., for $A \in T^*\mathcal{M} \otimes \mathbb{R}^d$, $W(A)$ is a d-form. Correspondingly, a body is uniform if there exists an archetype

$$\mathcal{W} : \mathbb{R}^d \otimes \mathbb{R}^d \to \wedge^d \mathbb{R}^d$$

such that $W_p = (E_p)^*\mathcal{W}$. Since, eventually, we will only consider solid bodies with a given Riemannian volume form, it is more convenient to consider W as a scalar density with respect to this volume form, and $\mathcal{W}$ is a scalar density with respect to the canonical volume form in $\mathbb{R}^d$. The given volume form then appears when considering the energy functional and not merely the scalar energy density, as in (2) or Definition 8.

Material uniformity is the weakest sense in which a constitutive relation is independent of position; it is defined independently of any coordinate system. It is a type of what is sometimes called "homogeneity" (though this term has another significance in [36]). The function $\mathcal{W}$ is sometimes called an *archetype*, whereas the frame E_p is sometimes called an *implant map*, because it shows how the archetype $\mathcal{W}$ is implanted into the material. Note that for a given uniform constitutive relation, neither the archetype $\mathcal{W}$ nor the implant map E_p is unique. If $(\mathcal{W}, E_p)$ is an archetype-implant pair at $p \in \mathcal{M}$, then so is $(\mathcal{W}', E_p \circ S)$, where $S \in \mathrm{GL}(d)$, and for every $B \in \mathrm{Hom}(\mathbb{R}^d, \mathbb{R}^d)$,

$$\mathcal{W}'(B) = \mathcal{W}(B \circ S^{-1}).$$

Moreover, the implant map may not be unique even for a fixed $\mathcal{W}$, depending on the symmetries of $\mathcal{W}$ (see below).

Definition 3 (Smooth Body) A uniform hyperelastic body is called *smooth* if there exists an archetype $\mathcal{W}$, a cover of $\mathcal{M}$ with open sets U^α, and implants $E^\alpha = \{E_p^\alpha\}_{p\in U^\alpha}$, such that the sections E^α are smooth.

Example 1 Let $\mathfrak{g}$ be a smooth Riemannian metric on $\mathcal{M}$, and consider the energy density

$$W(A) = \operatorname{dist}^2(A, \mathrm{SO}(\mathfrak{g}, \mathfrak{e})), \tag{4}$$

where $\mathrm{SO}(\mathfrak{g}, \mathfrak{e})$ at $p \in \mathcal{M}$ is the set of orientation-preserving isometries $T_p\mathcal{M} \to \mathbb{R}^d$, and the distance in $T_p^*\mathcal{M}\otimes\mathbb{R}^d$ is induced by the inner product $\mathfrak{g}_p$ on $T\mathcal{M}$ and the Euclidean inner product $\mathfrak{e}$ on $\mathbb{R}^d$. Then, any orthonormal frame $E_p \in \mathrm{SO}(\mathbb{R}^d, T_p\mathcal{M})$ is an implant map, with archetype

$$\mathcal{W}(\cdot) = \operatorname{dist}^2(\cdot, \mathrm{SO}(d)). \tag{5}$$

This body is smooth, as we can choose locally smooth orthonormal frames. Note that the implant map is non-unique, as it may be composed with any smooth section of $\mathrm{SO}(d)$ over $\mathcal{M}$. This example illustrates why we do not require the existence of a global section $\{E_p\}_{p\in\mathcal{M}}$ in the definition of smoothness; such sections may not exist regardless of W, for example, because of topological obstructions on $\mathcal{M}$ (e.g., if $\mathcal{M}$ is a sphere).

Definition 4 (Symmetry Group) Let $\mathcal{M}$ be a uniform hyperelastic body. The *symmetry group* of the body associated with an archetype $\mathcal{W}$ is a group $\mathcal{G} \leq \mathrm{GL}(d)$, defined by

$$\mathcal{W}(B \circ g) = \mathcal{W}(B) \quad \text{for every } B \in \mathbb{R}^d \otimes \mathbb{R}^d \text{ and } g \in \mathcal{G}.$$

The body is called a *solid* if there exists a $\mathcal{W}$ such that $\mathcal{G} \leq \mathrm{SO}(d)$ (or sometimes if $\mathcal{G} \leq O(d)$). In this case, we shall only consider such $\mathcal{W}$ as admissible, and call $\mathcal{W}$ *undistorted*.

It is easy to see that if $\mathcal{W}$ and $\mathcal{W}'$ are archetypes for the same constitutive relation, then their symmetry groups $\mathcal{G}$ and $\mathcal{G}'$ are conjugate, i.e., there exists a $g \in \mathrm{GL}(d)$, such that $\mathcal{G}' = g^{-1}\mathcal{G}g$. Thus, a hyperelastic body is a solid if and only if it has an archetype $\mathcal{W}$, whose symmetry group is conjugate to a subgroup of $\mathrm{SO}(d)$.

The intrinsic right-symmetry of the constitutive relation is determined by W rather than by $\mathcal{W}$. The symmetry group of W at a point $p \in \mathcal{M}$ is a subgroup

$$\mathcal{G}_p \leq \mathrm{GL}(T_p\mathcal{M}).$$

If $(\mathcal{W}, E_p)$ is an archetype-implant pair at p, and $\mathcal{G}$ is the symmetry group of $\mathcal{W}$, then for every $g \in \mathcal{G}$ and $A \in T_p^*\mathcal{M} \otimes \mathbb{R}^d$,

$$W_p(A) = \mathcal{W}(A \circ E_p) = \mathcal{W}(A \circ E_p g) = W_p(A \circ E_p g E_p^{-1}),$$

i.e.,

$$\mathcal{G}_p = E_p \mathcal{G} E_p^{-1}.$$

Consequently, the space of all implant maps that correspond to $\mathcal{W}$ at p is $E_p\mathcal{G}$.

Example 2 In Example 1, the symmetry group of W at $p \in \mathcal{M}$ is $\mathrm{SO}(T_p\mathcal{M})$ (where $T_p\mathcal{M}$ is endowed with the inner product $\mathfrak{g}_p$). $\mathcal{W}$ is undistorted if and only if the implant map E_p at every $p \in \mathcal{M}$ satisfies

$$E_p^{-1}\,\mathrm{SO}(T_p\mathcal{M})E_p = \mathrm{SO}(d).$$

In particular, the archetype (5) is undistorted.

Thus far, we only considered point-symmetries of W in the form of symmetry groups. We next consider symmetries of W associated with pairs of points in the manifold:

Definition 5 (Material Connection) A *material connection* of $(\mathcal{M}, W)$ is an affine connection ∇ on $\mathcal{M}$ whose parallel transport operator Π leaves W invariant. That is, for every $p, q \in \mathcal{M}$, $A \in T_q^*\mathcal{M} \otimes \mathbb{R}^d$, and path γ from p to q,

$$W_p(A \circ \Pi_\gamma) = W_q(A),$$

where $\Pi_\gamma : T_p\mathcal{M} \to T_q\mathcal{M}$ is the parallel transport along γ.

In general, a material connection may fail to exist (there may be topological obstructions), or may not be unique. The following proposition relates the uniqueness of a material connection to the nature of the symmetry group (a less general version of this result appears in [36]):

Proposition 1 *Let $(\mathcal{M}, W)$ be a smooth uniform hyperelastic body with symmetry group $\mathcal{G}$. If $\mathcal{G}$ is discrete, then there exists a unique locally flat material connection.*[1]

Proof Assume two material connections, whose parallel transport operators are Π^1 and Π^2. Let γ be a curve starting at $p \in \mathcal{M}$, and let $A \in T_{\gamma(t)}^*\mathcal{M} \otimes \mathbb{R}^d$ for some $t \geq 0$. Then,

$$W_p(A \circ \Pi^1_{\gamma|_{[0,t]}}) = W_{\gamma(t)}(A) = W_p(A \circ \Pi^2_{\gamma|_{[0,t]}}).$$

[1]Strictly speaking, the intrinsic condition is that $\mathcal{G}_p$ is discrete for some $p \in \mathcal{M}$ (and therefore for every $p \in \mathcal{M}$). By locally flat, we mean that the curvature tensor vanishes; globally flat implies also a trivial holonomy. Note that the term *flat* has a different interpretation in [36], where it describes a curvature- and torsion-free connection.

Setting $A = B \circ (\Pi^1_{\gamma|[0,t]})^{-1}$ for $B \in T^*_p\mathcal{M} \times \mathbb{R}^d$, we obtain that

$$W_p(B) = W_p(B \circ (\Pi^1_{\gamma|[0,t]})^{-1} \circ \Pi^2_{\gamma|[0,t]}),$$

hence

$$(\Pi^1_{\gamma|[0,t]})^{-1}\Pi^2_{\gamma|[0,t]} \in \mathcal{G}_p$$

for every t. Since the left-hand side is continuous in t and $\mathcal{G}_p$ is a discrete group, $(\Pi^1_{\gamma|[0,t]})^{-1}\Pi^2_{\gamma|[0,t]}$ is constant. Since at $t = 0$ it is the identity,

$$\Pi^1_{\gamma|[0,t]} = \Pi^2_{\gamma|[0,t]}$$

for every t. Finally, since γ is arbitrary, $\Pi^1 = \Pi^2$.

We next prove existence of a locally flat material connection. Let $\cup_\alpha U^\alpha = \mathcal{M}$ be a cover of $\mathcal{M}$, and let $\{E^\alpha_p\}_{p\in U^\alpha}$ be implant maps. For a curve $\gamma \subset U^\alpha$ starting at p and ending at q, define

$$\Pi_\gamma = E^\alpha_q \circ (E^\alpha_p)^{-1}. \tag{6}$$

For a general curve $\gamma \subset \mathcal{M}$, partition it into curves $\gamma = \gamma_n * \ldots * \gamma_1$ (where $*$ is the concatenation operator), where each $\gamma_i \subset U^{\alpha_i}$ for some α_i, and use the above definition. In order to show that Π_γ is well defined, we need to show that this definition is independent of the concatenation. To this end, it is enough to show that if $\gamma \subset U^\alpha \cap U^\beta$, then the definition of Π_γ with respect to either U^α or U^β is the same.

Indeed, consider the function of p,

$$(E^\alpha_p)^{-1}E^\beta_p : U^\alpha \cap U^\beta \to \mathrm{GL}(R^d).$$

Since for any $A \in T^*_p\mathcal{M} \otimes \mathbb{R}^d$,

$$\mathcal{W}(A \circ E^\alpha_p) = W(A) = \mathcal{W}(A \circ E^\beta_p),$$

it follows that $(E^\alpha_p)^{-1}E^\beta_p \in \mathcal{G}$ for any $p \in U^\alpha \cap U^\beta$. Since $\mathcal{G}$ is discrete, it follows that this is a constant function of p, that is $(E^\alpha_p)^{-1}E^\beta_p = B \in \mathrm{GL}(\mathbb{R}^d)$ for every p. We therefore have that for $p, q \in U^\alpha \cap U^\beta$,

$$E^\alpha_q(E^\alpha_p)^{-1} = E^\alpha_q BB^{-1}(E^\alpha_p)^{-1} = E^\alpha_q(E^\alpha_q)^{-1}E^\beta_q(E^\beta_p)^{-1}E^\alpha_p(E^\alpha_p)^{-1} = E^\beta_q(E^\beta_p)^{-1},$$

and therefore Π_γ is well defined. Finally, for a closed curve γ, starting and ending at p, and contained in one of the domains U^α, it follows from the definition that

$\Pi_\gamma = \mathrm{Id}_{T_p\mathcal{M}}$, hence the holonomy of Π is locally trivial, which implies that the curvature tensor of the connection associated with Π is zero. Note, however, that the holonomy of Π may be non-trivial in general (for non-simply-connected manifolds). □

Note that if there exists a global continuous implant section $\{E_p\}_{p\in\mathcal{M}}$ (for an archetype $\mathcal{W}$), then the connection defined by (6) (without the α superscript) is well defined regardless of the symmetry group, and moreover, it is not only locally flat, but has a trivial holonomy (that is, a path-independent parallel transport). In fact, the existence of a material connection with a trivial holonomy is equivalent to the existence of a global implant section $\{E_p\}_{p\in\mathcal{M}}$. Indeed, let ∇ be such a connection, and let E_{p_0} be an implant at $p_0 \in \mathcal{M}$, then

$$E_p := \Pi_\gamma E_{p_0} \tag{7}$$

is a global continuous implant section (here γ is an arbitrary curve connecting p_0 and p).

In the case of a solid body, there is an additional intrinsic geometric construct associated with the body.

Definition 6 (Intrinsic Metric) Let $(\mathcal{M}, W)$ be a smooth solid body with an undistorted archetype $\mathcal{W}$ and implant maps $\{E_p\}_{p\in\mathcal{M}}$. The *intrinsic Riemannian metric* of $\mathcal{M}$ associated with $\mathcal{W}$ is defined by

$$\mathfrak{g}_p(X, Y) = \mathfrak{e}(E_p^{-1}(X), E_p^{-1}(Y)), \qquad \text{for every } X, Y \in T_p\mathcal{M}, \tag{8}$$

where $\mathfrak{e}$ is the Euclidean inner product in $\mathbb{R}^d$.

This definition depends on $\mathcal{W}$ (see Example 3 below), but not on the choice of implants E_p. Indeed, if E_p and E'_p are two implants at p, then, since $\mathcal{M}$ is a solid, $g = E_p^{-1}E'_p \in \mathcal{G} \le \mathrm{SO}(d)$, and therefore

$$\mathfrak{e}(E_p^{-1}(X), E_p^{-1}(Y)) = \mathfrak{e}\left(g \circ {E'_p}^{-1}(X), g \circ {E'_p}^{-1}(Y)\right) = \mathfrak{e}({E'_p}^{-1}(X), {E'_p}^{-1}(Y)),$$

where we used in the last step the $\mathrm{SO}(d)$ invariance of the Euclidean metric. Note also that the existence of a Riemannian metric on $\mathcal{M}$ that is invariant under the action of $\mathcal{G}_p$ implies that $\mathcal{M}$ is solid [36, Proposition 11.2].

Proposition 2 *If ∇ is a material connection and $\mathfrak{g}$ is an intrinsic metric of a solid $\mathcal{M}$ with an archetype $\mathcal{W}$, then ∇ is metrically consistent with $\mathfrak{g}$ (equivalently, the induced parallel transport is an isometry).*

Proof Let $p, q \in \mathcal{M}$, and let γ be a curve from p to q. Let Π_γ the parallel transport of ∇ along γ, $X, Y \in T_p\mathcal{M}$, and let E_q be an implant at q. Then

$$\mathfrak{g}_q(\Pi_\gamma X, \Pi_\gamma Y) = \mathfrak{e}(E_q^{-1} \circ \Pi_\gamma X, E_q^{-1} \circ \Pi_\gamma Y) = \mathfrak{g}_p(X, Y),$$

where in the right-most equality we used the fact that $\Pi_\gamma^{-1} \circ E_q$ is an implant at p, for the same archetype $\mathcal{W}$. This equality shows that ∇ is metrically consistent with $\mathfrak{g}$. □

Corollary 1 ([36, Proposition 11.6]) *A solid body* $(\mathcal{M}, W)$ *is equipped with at most one torsion-free material connection, in which case it is the Levi-Civita connection of all intrinsic metrics of* $(\mathcal{M}, W)$.

Proposition 2 states that all material connections are metrically consistent with every intrinsic metric. In isotropic solids, i.e., solids whose symmetry group is $\mathrm{SO}(d)$, the converse is also true: every metrically consistent connection is a material connection (note the strong contrast to the case of a discrete symmetry group, Proposition 1).

Proposition 3 *Let* $(\mathcal{M}, W)$ *be an isotropic solid and let* ∇ *be a connection metrically consistent with some intrinsic metric* $\mathfrak{g}$. *Then* ∇ *is a material connection. In particular, any isotropic solid admits a torsion-free connection—the Levi-Civita connection of any intrinsic metric.*[2]

Proof Let $\mathcal{W}$ be an undistorted archetype and let $E = \{E_p\}_{p\in\mathcal{M}}$ be an implant map (the proof below does not require any smoothness assumptions of E, and thus we can assume the existence of a global implant map without loss of generality). Suppose that $\mathfrak{g}$ is an intrinsic metric for W, and let ∇ be an affine connection metrically consistent with $\mathfrak{g}$. Since $(\mathcal{M}, W)$ is isotropic and $\mathcal{W}$ is undistorted, we have (by definition) that its symmetry group is $\mathrm{SO}(d)$.

Let now $\Pi_\gamma : T_p\mathcal{M} \to T_q\mathcal{M}$ be the parallel transport of ∇ along a curve γ from p to q. Since ∇ is metrically consistent with respect to $\mathfrak{g}$, $\Pi_\gamma \in \mathrm{SO}(\mathfrak{g}_p, \mathfrak{g}_q)$. Using the fact that for any $r \in \mathcal{M}$, $E_r \in \mathrm{SO}(\mathfrak{e}, \mathfrak{g}_r)$ (by the very definition of an intrinsic metric), we have that $E_q^{-1} \circ \Pi_\gamma \circ E_p \in \mathrm{SO}(d)$. Therefore, since $\mathcal{W}$ is $\mathrm{SO}(d)$-invariant, we have that for any $A \in T_q^*\mathcal{M} \otimes \mathbb{R}^d$,

$$W_p(A\circ\Pi_\gamma) = \mathcal{W}(A\circ\Pi_\gamma\circ E_p) = \mathcal{W}(A\circ E_q\circ(E_q^{-1}\circ\Pi_\gamma\circ E_p)) = \mathcal{W}(A\circ E_q) = W_q(A).$$

□

The fact that an isotropic solid always has a torsion-free material connection (or more generally, it has many material connections with different torsions) suggests that the equilibrium equations of such a body are independent of the torsion tensor. Indeed, it can be shown explicitly (see Sect. 5) that W only depends on the metric.

Example 3 Consider once again Example 1. Then $\mathfrak{g}$ is an intrinsic metric, corresponding to the archetype

$$\mathcal{W}(B) = \mathrm{dist}^2(B, \mathrm{SO}(d)),$$

[2]This proposition is a more general version of [36, Proposition 11.8].

and implants $E_p \in \mathrm{SO}(\mathfrak{e}, \mathfrak{g})$. However, $c^2\mathfrak{g}$, $c > 0$, is also an intrinsic metric, corresponding to the archetype

$$\mathcal{W}(B) = \mathrm{dist}^2(cB, \mathrm{SO}(d))$$

and implants $E_p \in c^{-1}\,\mathrm{SO}(\mathfrak{e}, \mathfrak{g})$. It can be shown that there are no other intrinsic metrics in this case. The phenomenon whereby the intrinsic metric is unique up to a multiplicative constant holds for every isotropic solid.

Remark 3 In two dimensions, a solid archetype is either isotropic or it has a discrete symmetry; in three dimensions, a body can also be transversely isotropic (see [36, p. 60]). In this case, the material connection is not unique, but the Levi-Civita connection of an intrinsic metric may not be a material connection. More on transversely isotropic materials can be found in [36, Proposition 11.9] and [13, Proposition 5].

2.1 *Relation Between Geometric and Constitutive Paradigms*

As presented in the introduction, a body with distributed dislocations is modeled in the geometric paradigm as a Weitzenböck manifold $(\mathcal{M}, \mathfrak{g}, \nabla)$, where ∇ is curvature-free and metrically consistent with $\mathfrak{g}$. For simplicity, assume that ∇ also has trivial holonomy (an assumption that often appears implicitly in this paradigm), hence the parallel transport operator of ∇ is path-independent (a property known as *distant parallelism* or *teleparallelism*). We denote the parallel transport from p to q by Π_p^q.

To relate the geometric body manifold to the constitutive paradigm, assume a given undistorted solid archetype $\mathcal{W}$ and an implant E_{p_0}, which is an orthonormal basis (with respect to $\mathfrak{g}_{p_0}$) at some $p_0 \in \mathcal{M}$. The pair $(\mathcal{W}, E_{p_0})$ determines the mechanical response of the body at the point p_0. Parallel transporting E_{p_0} using (7), we obtain a parallel frame field $\{E_p\}_{p\in\mathcal{M}}$, which is orthonormal, since ∇ is metrically consistent with $\mathfrak{g}$.

An implant field $E = \{E_p\}_{p\in\mathcal{M}}$ and an archetype $\mathcal{W}$ define a unique energy density using (3). Note that this is the only energy density W with a material connection ∇ for which $\mathcal{W}$ is an archetype with an implant E_{p_0} at $p_0 \in \mathcal{M}$.

We have thus proved the following:

Proposition 4 *Fix a solid (undistorted) archetype* $\mathcal{W} \in C(\mathbb{R}^d \times \mathbb{R}^d)$.

1. *Given a Weitzenböck manifold* $(\mathcal{M}, \mathfrak{g}, \nabla)$ *with trivial holonomy and an orthonormal basis* $E_p \in \mathrm{SO}(\mathfrak{e}, \mathfrak{g}_p)$ *at some* $p \in \mathcal{M}$, *there exists a unique energy density* W, *such that* $\mathcal{M}$ *is uniform with archetype* $\mathcal{W}$, *and implant map* E_p *at* p, *and such that* $\mathfrak{g}$ *is an intrinsic metric and* ∇ *is a material connection.*

2. *Moreover, all energy densities W having an archetype $\mathcal{W}$, an intrinsic metric $\mathfrak{g}$, and a material connection ∇ can be constructed this way. In particular, W is unique up to a global rotation—the choice of a basis at one point.*

A somewhat more intrinsic version of this proposition would be that a Weitzenböck manifold $(\mathcal{M}, \mathfrak{g}, \nabla)$ and a response function W_p at a single point defines a unique energy density consistent with $\mathfrak{g}$ and ∇ (without the need to define E_p and $\mathcal{W}$). However, since the same archetype $\mathcal{W}$ can be implanted into different bodies (thus making sense of different bodies having "the same" response function), and since we are eventually interested in this paper in sequences of elastic bodies, it is useful to take $\mathcal{W}$ as a basic building block, as done in Proposition 4.

In the case of a discrete symmetry group, the constitutive model $(\mathcal{M}, W)$ induces a unique geometric model $(\mathcal{M}, \mathfrak{g}, \nabla)$; this follows readily from Proposition 1, and the discussion following Definition 6:

Proposition 5 *Let $(\mathcal{M}, W)$ be a uniform solid material with an undistorted archetype $\mathcal{W}$ having a discrete symmetry group. Then, the material connection ∇ and the intrinsic metric $\mathfrak{g}$ associated with $\mathcal{W}$ are unique (if the symmetry group is not discrete, $\mathfrak{g}$ is still uniquely determined however not ∇).*

Another way of describing the relation between the geometric and constitutive paradigms is the following:

(a) The triple $(\mathcal{M}, \mathcal{W}, E)$, where $\mathcal{W} : \mathbb{R}^d \otimes \mathbb{R}^d \to [0, \infty)$ and E is a frame field, determines a uniform body $(\mathcal{M}, W)$ uniquely by (3).
(b) On the other hand, by declaring E to be a parallel orthonormal field, we obtain a Weitzenböck manifold $(\mathcal{M}, \mathfrak{g}, \nabla)$.

In fact, $(\mathcal{M}, \mathcal{W}, E)$ contains slightly more information than both $(\mathcal{M}, W)$ and $(\mathcal{M}, \mathfrak{g}, \nabla)$: given $\mathcal{W}$, $(\mathcal{M}, \mathcal{W}, E)$ can be derived from $(\mathcal{M}, \mathfrak{g}, \nabla)$ uniquely, up to a global rotation (choice of E_p at one point), and in the case of a discrete symmetry group, the same holds for deriving $(\mathcal{M}, \mathcal{W}, E)$ from $(\mathcal{M}, W)$.

3 Homogenization of Dislocations: Geometric Paradigm

In this section we describe the results of [19, 21], showing how a smooth Weitzenböck manifold $(\mathcal{M}, \mathfrak{g}, \nabla)$, representing a body with continuously distributed dislocations (the torsion tensor of ∇ representing their density), can be obtained as a limit of bodies with finitely many dislocations. These results are for two-dimensional bodies, hence we are only considering edge dislocations.

Bodies with Finitely Many Edge Dislocations To set the scene for the geometric homogenization of elastic bodies, we start by defining a two-dimensional body with finitely many (edge) dislocations. As illustrated in Fig. 2, we view each dislocation as a pair of disclinations of opposite sign (a curvature dipole).

Definition 7 A *body with finitely many singular edge dislocations* is a compact two-dimensional manifold with boundary $\mathcal{M}$, endowed with Riemannian metric $\mathfrak{g}$, which is almost-everywhere smooth and locally flat. The singularities are concentrated on a finite, even number of points, such that

1. The metric $\mathfrak{g}$, restricted to a small enough neighborhood around a singular point, is a metric of a cone.
2. One can partition the singular points into pairs (*curvature dipoles*), such that the geodesics connecting each pair (*dislocation cores*) do not intersect.
3. The Levi-Civita connection ∇^{LC}, defined on the complement of those segments is path-independent.

A body with finitely many dislocations is a Weitzenböck manifold $(\mathcal{M}, \mathfrak{g}, \nabla^{LC})$, and whenever we refer to a smooth field over $\mathcal{M}$ (say a frame field), it is understood as being smooth on complement of the dislocation cores.

The assumption on the Levi-Civita connection being path-independent implies that the two cone defects in each pair (that is, the difference between 2π and the total angle around the cone) are of the same magnitude but of different signs. That is, they are curvature dipoles. In particular, the construction in Fig. 2 yields a body with a single dislocation according to this definition.

Another approach for modeling bodies with finitely many dislocation was presented in [14, 15]; instead of assuming a frame field describing lattice directions, one assumes a co-frame, that is, a family of 1-forms (called *layering forms*). This slightly different viewpoint enables the use of distributional 1-forms—*de-Rham currents*—for describing the singular dislocations. This viewpoint is quite close to the one presented here, although in some sense it requires less structure. Recently, a homogenization result in this context has been proved [23], which is similar conceptually to the one presented here. However, the notion of convergence used in [23] is very weak compared to Theorem 2, and therefore much more difficult to relate to the convergence of associated energy functionals, which is the main result of this paper.

Burgers Circuits and Vectors We now present in more detail how Burgers vectors appear in the context of Weitzenböck manifolds. Let $\mathcal{M}$ be a manifold, endowed with a connection ∇. A *Burgers circuit* is a closed curve $\gamma : [0, 1] \to \mathcal{M}$, and its associated *Burgers vector* is defined by

$$\mathbf{b}_\gamma = \int_0^1 \Pi_{\gamma(t)}^{\gamma(0)} \dot{\gamma}(t)\, dt \in T_{\gamma(0)}\mathcal{M},$$

where $\Pi_{\gamma(t)}^{\gamma(0)} : T_{\gamma(t)}\mathcal{M} \to T_{\gamma(0)}\mathcal{M}$ is the parallel transport of ∇ along γ (see, e.g., [2, Sec. 4] or [36, Sec. 10]). Thus, as in the classical material science context, the Burgers vector is the sum of the tangents to the curve; in order to make sense of this on manifolds one has first to parallel transport all the tangent vectors to the same tangent space.

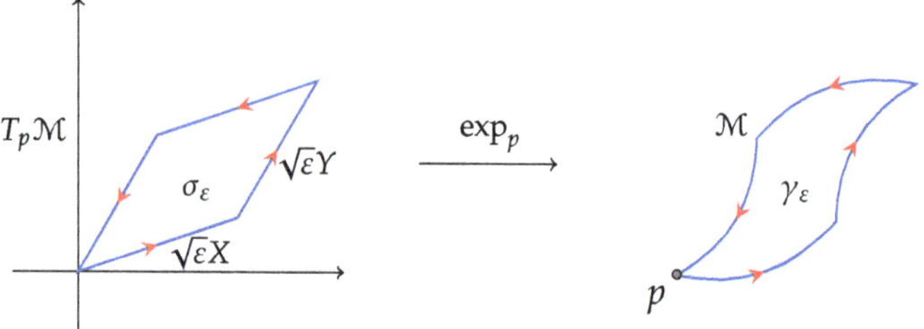

Fig. 3 The Burgers vector associated a loop γ_ε in $\mathcal{M}$, where γ_ε is the image under the exponential map of a parallelogram σ_ε in $T_p\mathcal{M}$ (whose edges are $\sqrt{\varepsilon}X$ and $\sqrt{\varepsilon}Y$), tends asymptotically to $\varepsilon\, T(X, Y)$

Burgers vectors are closely related to the *torsion tensor*,

$$T(X, Y) = \nabla_X Y - \nabla_Y X - [X, Y].$$

The torsion T is an infinitesimal Burgers vector in the following sense: Let $p \in \mathcal{M}$ and let $\exp_p : T_p\mathcal{M} \to \mathcal{M}$ be the exponential map of ∇ at p.[3] Let $\sigma_\varepsilon : [0, 1] \to T_p\mathcal{M}$ be the parallelogram from the origin built from the vectors $\sqrt{\varepsilon}X$, $\sqrt{\varepsilon}Y$, and let $\gamma_\varepsilon = \exp_p(\sigma_\varepsilon)$ (see Fig. 3). Then

$$\left.\frac{d}{d\varepsilon}\right|_{\varepsilon=0} \mathbf{b}_{\gamma_\varepsilon} = T(X, Y).$$

This result is due to Cartan; see [33, Chapter III, Section 2] for a proof.

In the case of a body with finitely many dislocations $(\mathcal{M}, \mathfrak{g}, \nabla^{LC})$ (according to Definition 7), the Burgers vector for any curve that does not encircle one of the dislocation cores is zero. This follows from the fact that every simply connected submanifold of $\mathcal{M}$ which does not contain dislocations is isometrically embeddable into Euclidean plane, and that the Burgers vector of any closed curve in the plane is zero. To quantify the Burgers vector associated with a curve encircling a single dislocation, consider the manifold depicted in Fig. 2. One can then see that the magnitude of the Burgers vector is

$$\mathrm{b} = 2d \sin(\theta/2), \tag{9}$$

where d is the length of the dislocation core (the distance between the two singular points forming the curvature dipole), and θ is the magnitude of the cone defect (see Fig. 4). For a general Burgers circuit, the Burgers vector is the sum of the

[3]Actually, any map $\phi : T_p\mathcal{M} \to \mathcal{M}$ with $\phi(0) = p$ whose differential at the origin is the identity will do.

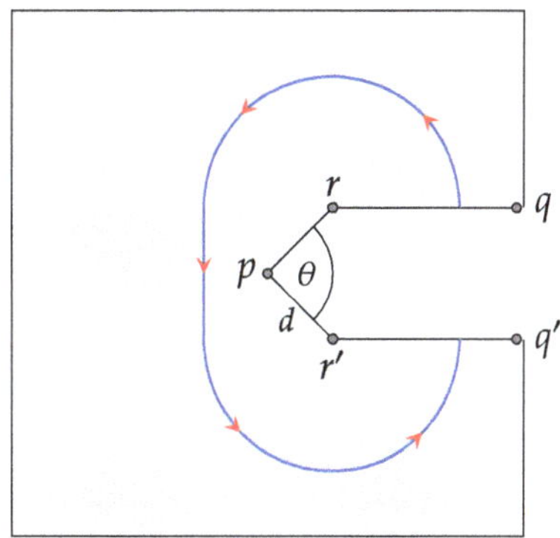

Fig. 4 A Burgers circuit yielding a Burgers vector whose magnitude is $2d\sin(\theta/2)$, where θ is the disclination angle and d is the distance between two disclinations forming the edge-dislocation. The vector points downwards from a chosen base point of the circuit

contributions of the dislocation cores it encircles (after parallel transporting each contribution to the base point).

It follows that by changing d and θ in Fig. 2, while keeping $\mathrm{b} = 2d\sin(\theta/2)$ fixed, we can obtain "the same" dislocation in different ways, in the sense that a Burgers circuit around the dislocation core will not be able to distinguish between the two. Nevertheless, the choice of d and θ will be important from the viewpoint of convergence of bodies with dislocations, as depicted in the sketch of the proof below.

Main Result: Convergence in the Geometric Paradigm We now describe a version of the main theorem of [21], stating that in the geometric paradigm, every two-dimensional body with distributed dislocations is a limit of bodies with finitely many dislocations.

Theorem 2 (Homogenization of Dislocations, Geometric Paradigm) *For every compact two-dimensional Weitzenböck manifold* $(\mathcal{M}, \mathfrak{g}, \nabla)$ *and parallel orthonormal frame* E*, there exists a sequence of bodies with finitely many dislocations* $(\mathcal{M}_n, \mathfrak{g}_n, \nabla_n^{LC})$ *and parallel orthonormal frames* E_n*, such that there exist homeomorphisms* $F_n : \mathcal{M}_n \to \mathcal{M}$*, whose restrictions to the smooth part of* $\mathcal{M}_n$ *are smooth embeddings, satisfying*

$$\|dF_n \circ E_n - E\|_{L^\infty} \to 0. \tag{10}$$

Note that an orthonormal parallel frame E contains all the geometric information of the Weitzenböck manifold: since $E : \mathbb{R}^d \to T\mathcal{M}$ is orthonormal, it induces $\mathfrak{g}$ by pushing forward the Euclidean metric on $\mathbb{R}^d$ (as in (8)), and since it is parallel, it induced the parallel transport of ∇ (see (6)). Therefore, the notion of convergence in Theorem 2, which is defined through the convergence of orthonormal parallel frames, induces the convergence of the entire structure $(\mathcal{M}_n, \mathfrak{g}_n, \nabla_n^{LC}) \to (\mathcal{M}, \mathfrak{g}, \nabla)$ of the Weitzenböck manifolds.

We can also view Theorem 2 as a theorem about the convergence of manifolds endowed with frame fields $(\mathcal{M}_n, E_n) \to (\mathcal{M}, E)$, where each of the manifolds

$(\mathcal{M}_n, E_n)$ induces the structure of a body with edge dislocations as in Definition 7. This viewpoint, while maybe somewhat less natural from a geometric perspective, will be useful in the next section (convergence in the constitutive paradigm), when we associate these manifolds with a fixed archetype and consider E_n and E as implant maps.

3.1 Sketch of Proof of Theorem 2

Theorem 2 is an approximation result: given a manifold $(\mathcal{M}, \mathfrak{g}, \nabla)$, we approximate it with a sequence of manifolds of a specific type (Definition 7).

Approximation by Disclinations Before we describe the main idea of this approximation, it is illustrative to present a similar one, which is somewhat more intuitive—the approximation of a Riemannian surface by locally flat surfaces with disclinations. Given a surface $(\mathcal{M}, \mathfrak{g})$, we approximate it as follows:

1. First, assume that $\mathcal{M}$ does not have a boundary. Take a geodesic triangulation of the manifold—a set of points in $\mathcal{M}$, connected by minimizing geodesics that do not intersect, such that the resulting partition $\mathcal{M}$ consists of geodesic triangles (such triangulations exist; see, for example, [3, Note 3.4.5.3]). If $\mathcal{M}$ has a boundary, triangulate a subdomain $\mathcal{M}' \subset \mathcal{M}$, such that the distance between $\partial\mathcal{M}'$ and $\partial\mathcal{M}$ is small (of the order of the distance between the vertices).
2. Construct a manifold by replacing each triangle with a Euclidean triangle with the same edge lengths. Since $\mathcal{M}$ is (generally) not flat, the angles of the original geodesic triangles differ from the angles of their Euclidean counterparts (by the Gauss–Bonnet theorem, the angles of each geodesic triangle generally do not sum up to π).

This way we obtain a topological manifold which is smooth and flat everywhere but at the vertices, which are cone singularities (disclinations)—the angles around each vertex do not generally sum up to 2π, since they differ from the angles of the original geodesic triangulation. This approximation of the surface is similar to the approximation of a sphere by a football (soccer ball), using triangles rather than pentagons and hexagons.

By choosing finer and finer triangulations, say, triangulations in which the edge lengths are of order $1/n$ for $n \gg 1$, it is clear (intuitively) that one obtains better and better approximations of the original manifold; they converge as metric spaces to the original manifold (see [12] for an explicit estimate) while the distribution-valued curvatures converge to the smooth curvature of $\mathfrak{g}$ (see [7]).

The Approximating Sequence for Theorem 2 The idea behind the proof of Theorem 2 is very similar: Construct a fine geodesic triangulation of the Weitzenböck manifold $(\mathcal{M}, \mathfrak{g}, \nabla)$, and then replace each triangle with a locally flat one to obtain a body with finitely many dislocations. The difference between the two constructions is in the triangulation and in the type of locally flat replacements.

1. Take a triangulation of $(\mathcal{M}, \mathfrak{g}, \nabla)$ in which the edges are ∇-geodesics; those differ generally from the Levi-Civita geodesics and are not even locally length-minimizing. At the nth stage, we choose the triangulation such that the length of each edge is between (say) $1/n$ and $3/2n$, and all the angles are bounded between δ and $\pi - \delta$ for some $\delta > 0$ independent of n (to ensure that all the triangles are uniformly non-degenerate as $n \to \infty$). The existence of a geodesic triangulation, based on a non-Levi-Civita connection, is not trivial; it is proved in [21, Proposition 3.1]. Denote the skeleton of this triangulation (the union of all the edges) by X_n.
2. Since the Gauss–Bonnet theorem holds for a metrically consistent connection (see [21, Theorem B.1]), and since ∇ is metrically consistent and has zero curvature, the angles of each geodesic triangle sum up to π. In other words, if a geodesic triangle has edge lengths a, b, c and angles α, β, γ, then $\alpha+\beta+\gamma = \pi$; the angles are however "wrong" in the sense that generally $\alpha \neq \alpha_0$, $\beta \neq \beta_0$, and $\gamma \neq \gamma_0$, where $\alpha_0, \beta_0, \gamma_0$ are the angles of the Euclidean triangle having edge lengths a, b, c. Since the geodesic triangles are uniformly regular, the angles do not deviate much from the angle of the Euclidean triangle,
$$|\alpha - \alpha_0|, |\beta - \beta_0|, |\gamma - \gamma_0| = O(1/n). \tag{11}$$
See [21, Corollary 2.7].[4]
3. As stated above, the Euclidean triangle having side lengths a, b, c does not have angles α, β, γ; however, if Condition (11) holds and $\alpha + \beta + \gamma = \pi$, then there exists a manifold containing a single dislocation (according to Definition 7), whose boundary is a triangle whose edge lengths and angles are a, b, c and α, β, γ [21, Proposition 3.3] (see Fig. 5). The only additional parameter entering in this construction is the Burgers vector associated with the perimeter of the triangle, and whose magnitude is of order $O(1/n^2)$. The precise location of the dislocation core inside the "triangle" is arbitrary (as long as it does not intersect the boundary), as is the choice of the parameters θ and d (see (9)).
4. The approximation of $(\mathcal{M}, \mathfrak{g}, \nabla)$ is obtained by replacing each triangle in the triangulation with a "dislocated" triangle having the same edge lengths and angles. Denote the resulting manifold by $(\mathcal{M}_n, \mathfrak{g}_n, \nabla_n^{LC})$, and the skeleton of the triangulation on $\mathcal{M}_n$ by Y_n. Since the angles in each triangle in Y_n are the same as in the corresponding triangle in X_n, it follows that the angles around each vertex in Y_n sum up to 2π. In other words, there are no cone defects (disclinations) at the vertices of the triangulation; the only singularities in $\mathcal{M}_n$ are the dislocation cores within each triangle. Hence, $(\mathcal{M}_n, \mathfrak{g}_n, \nabla_n^{LC})$ is a body with finitely many dislocations according to Definition 7 (see Fig. 6)

[4]The estimate (11) does not appear in this corollary explicitly; it follows from its fourth part, using the fact a small triangle on $\mathcal{M}$ with edges that are Levi-Civita geodesics is, to leading order, Euclidean (this follows from standard triangle comparison results).

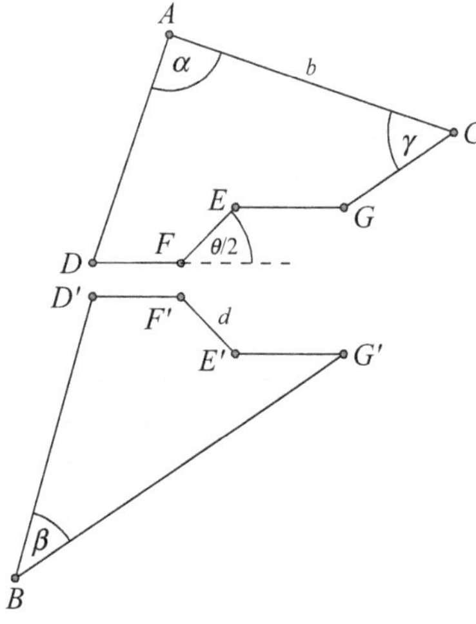

Fig. 5 A triangle containing a single edge dislocation. Given angles α, β, γ adding up to π and edge lengths a, b, c, we construct a defective triangle by identifying the edges DF and $D'F'$, FE and $F'E'$, and EG and $E'G'$, such that $CG + G'B = a$, $AC = b$, and $AD + D'B = c$

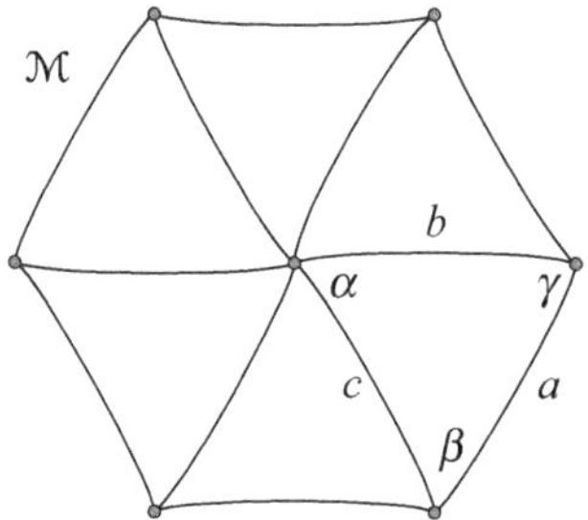

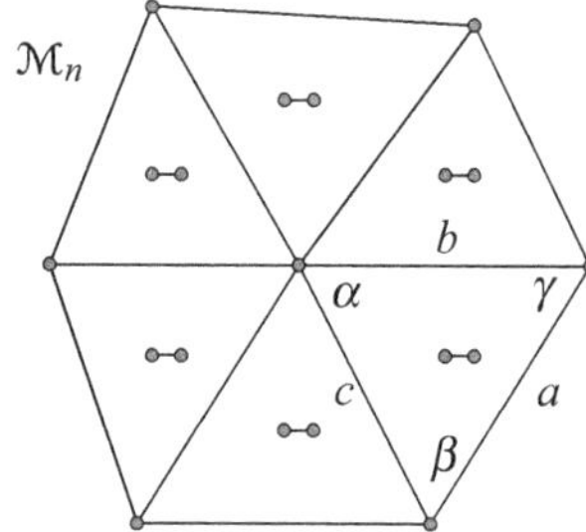

Fig. 6 Approximating the smooth Weitzenböck manifold $(\mathcal{M}, \mathfrak{g}, \nabla)$ by manifolds $(\mathcal{M}_n, \mathfrak{g}_n, \nabla_n^{LC})$ with singularities. Each ∇-geodesic triangle in $(\mathcal{M}, \mathfrak{g}, \nabla)$ is replaced by locally Euclidean triangle, having the same angles and side lengths, and containing a single dislocation (the core of each dislocation is sketched here as a segment inside the triangle)

Convergence of the Approximating Sequence The next step is to show that $(\mathcal{M}_n, \mathfrak{g}_n, \nabla_n^{LC})$ converges to $(\mathcal{M}, \mathfrak{g}, \nabla)$ in the sense of Theorem 2. That is, show that given a ∇-parallel orthonormal frame E on $\mathcal{M}$, there exist ∇_n^{LC}-parallel orthonormal frames E_n on $\mathcal{M}_n$ and maps $F_n : \mathcal{M}_n \to \mathcal{M}$ such that (10) holds.

Given E, the construction of E_n is very natural: let $\{e_1, e_2\}$ be the standard basis of $\mathbb{R}^2$. Let $p_n \in \mathcal{M}$ be a vertex in X_n, the nth triangulation of $\mathcal{M}$, and let q_n be its corresponding vertex in Y_n. Each of the vectors $E_{p_n}(e_1), E_{p_n}(e_2)$ is a $\mathfrak{g}$-unit vector in $T_{p_n}\mathcal{M}$, which is uniquely defined by its angles with the ∇-geodesics in X_n emanating from p_n. Define $(E_n)_{q_n}(e_i)$ to be the $\mathfrak{g}_n$-unit vector in $T_{q_n}\mathcal{M}_n$ which forms the same angles with the corresponding geodesics emanating from q_n. This defines E_n everywhere by ∇_n^{LC}-parallel transport. Note that this relation between $E_{p_n}(e_i)$ and $E_{q_n}(e_i)$ actually holds for *any* vertex $p_n \in X_n \subset \mathcal{M}$ and corresponding vertex $q_n \in Y_n \subset \mathcal{M}_n$. This follows from the construction, since X_n consists of

∇-geodesics and Y_n consists of ∇_n-geodesics, and the angles in the corresponding triangles match.

The construction of F_n is more subtle. Since X_n and Y_n have the same graph structure, and the lengths of its corresponding edges are the same, there is a natural map between these skeletons (the isometry of their graph metric); it is natural to define the restriction of F_n to Y_n to be this map. Next, note that at every corresponding pair of vertices $p_n \in \mathcal{M}$, $q_n \in \mathcal{M}_n$, the frame fields induce an isometry $A := E_{p_n} \circ ((E_n)_{q_n})^{-1} : T_{q_n}\mathcal{M}_n \to T_{p_n}\mathcal{M}$. Define F_n in a neighborhood of q_n by

$$F_n(q) := \exp^{\nabla}_{p_n}\left(A \circ (\exp^{\nabla_n}_{q_n})^{-1}(q)\right).$$

By construction, this map respects the mapping of Y_n to X_n, and moreover, $d_{q_n}F_n$ maps $(E_n)_{q_n}$ to E_{p_n}, and hence $|dF_n \circ E_n - E|$ is small near p_n. In [21, Section 4], it is proved that F_n can be extended in this way to a map that satisfies $|dF_n \circ E_n - E| = O(1/n)$ uniformly everywhere outside a small neighborhood, of diameter $o(1/n)$, of the dislocation core. Note that [21] aims at a slightly different notion of convergence (compared to Theorem 2), hence this statement is not explicit in [21]; however, the proof of Proposition 4.3 in [21] yields this result.

It remains to analyze the vicinity of a dislocation core. Recall that in the construction of $\mathcal{M}_n$, only the Burgers vector inside each triangle was taken into account. For understanding the behavior of F_n near the dislocation core, and only there, the exact construction of the dislocation plays a role: in [21], a dislocation of magnitude $O(1/n^2)$ is built using an arbitrary, but fixed, dislocation angle $\theta \approx 1$, whereas the size of the dislocation core is $d = O(1/n^2)$. In this case, extensions of F_n to the dislocation core only satisfy that $|dF_n \circ E_n - E|$ is bounded near the core (an explicit construction can be seen in [19, Section 3.2]). This only yields L^p convergence in (10), for any $p < \infty$, but not L^∞, which is enough for the version of Theorem 2 that appear in [21], but not to Theorem 2 as stated here (which is needed for the next section). If however one takes $\theta = o(1)$ and $d = o(1/n)$ (such that the dislocation magnitude (9) is as prescribed), F_n can be extended to the dislocation core such that (10) holds.[5]

[5]In [19, Section 3.2], choosing $\theta = o(1)$, $d = o(1/n)$ implies, in the notation of [19], $n^{-1} \ll D \ll 1$, which then implies L^∞ convergence (see the proof of [19, Proposition 2]). The general case is very similar, since we are only considering minuscule pieces of the manifolds, in which the only geometry that plays a role is the structure of the singular points (everything else is uniformly close to the trivial Euclidean plane). See also [20, Section 2.3.2, Example 2].

4 Homogenization of Dislocations: Constitutive Paradigm

Our aim in this section is to prove a homogenization theorem for dislocations within the constitutive paradigm, thus proving the third and final part of Theorem 1. To this end, some assumptions about the archetype $\mathcal{W} : \mathbb{R}^d \times \mathbb{R}^d \to [0, \infty)$ are required:

1. Growth conditions:
$$\alpha(-1 + |A|^p) \leq \mathcal{W}(A) \leq \beta(1 + |A|^p), \tag{12}$$
for some $p \in (1, \infty)$ and $\alpha, \beta > 0$.
2. Quasiconvexity[6]:
$$\mathcal{W}(A) \leq \int_{(0,1)^d} \mathcal{W}(A + d\varphi(x))\, dx \qquad \text{for every } \varphi \in C_c^\infty((0, 1)^d, \mathbb{R}^d).$$
3. Solid symmetry group: $\mathcal{G}(\mathcal{W}) \leq \mathrm{SO}(d)$.

Remark 4 It is usual to assume that $\mathcal{W}$ is frame-indifferent and that $\mathcal{W}(A) = 0$ iff $A \in \mathrm{SO}(d)$, but both assumptions are not required for the theorem. Moreover, quasiconvexity and (12) imply that $\mathcal{W}$ satisfies the p-Lipschitz property [8, Proposition 2.32]:

$$|\mathcal{W}(A) - \mathcal{W}(B)| \leq C(1 + |A|^{p-1} + |B|^{p-1})|A - B|, \tag{13}$$

for some $C > 0$ (and in particular $\mathcal{W}$ is continuous).

Example 4 We describe now two simple examples of archetypes $\mathcal{W}$ satisfying the above hypotheses—one isotropic and one having a discrete symmetry group:

1. The isotropic archetype $\mathcal{W}_{\mathrm{iso}}(A) = \mathrm{dist}^p(A, \mathrm{SO}(d))$ (as in Example 1) satisfies all the hypotheses but for quasiconvexity. This can be rectified by replacing $\mathcal{W}_{\mathrm{iso}}$ with its quasiconvex envelope $Q\mathcal{W}_{\mathrm{iso}}$, which is an isotropic archetype satisfying all of the hypotheses. In two dimensions, we can write $Q\mathcal{W}_{\mathrm{iso}}$ explicitly for every $p \geq 2$ [11, 34]:
$$Q\mathcal{W}_{\mathrm{iso}}(A) = \begin{cases} \mathrm{dist}^p(A, \mathrm{SO}(d)) & \mu_1 + \mu_2 \geq 1 \\ (1 - 2\det A)^{p/2} & \mu_1 + \mu_2 \leq 1, \end{cases}$$
where $\mu_1 \geq |\mu_2| \geq 0$ are the signed singular values of A (i.e., if $\sigma_1 \geq \sigma_2 \geq 0$ are the singular values, $\mu_1 = \sigma_1$ and $\mu_2 = (\mathrm{sgn} \det A)\sigma_2$). In higher dimensions, $Q\mathcal{W}_{\mathrm{iso}}$ is not known explicitly; however, it is known that

[6]The quasiconvexity assumption is natural from a variational point of view, as it guarantees the existence of an energy minimizer of the functional; see also Remark 6.

$$c\mathcal{W}_{\text{iso}} \le Q\mathcal{W}_{\text{iso}} \le \mathcal{W}_{\text{iso}}$$

for some constant $c > 0$ (see [22, Proposition 10]).

2. An example of an archetype having a discrete symmetry group is

$$\mathcal{W}_{\text{cubic}}(A) = \sum_{i=1}^{d} \beta_i \left(|Ae_i| - 1\right)^2,$$

where $\beta_i > 0$ are parameters and $\{e_i\}$ is the standard basis of $\mathbb{R}^d$. This energy density penalizes stretching along each of the lattice directions e_i. Once again, this function is not quasiconvex, and its quasiconvex envelope is given by [29, Lemma 4.1]

$$Q\mathcal{W}_{\text{cubic}}(A) = \sum_{i=1}^{d} \beta_i \left(|Ae_i| - 1\right)_+^2,$$

where for $f \in \mathbb{R}$, f_+ denotes the maximum between f and zero. While $Q\mathcal{W}_{\text{cubic}}$ satisfies all the assumptions, it is somewhat non-physical. For example, it does not penalize for compression (this is due to the fact that $\mathcal{W}_{\text{cubic}}$ is invariant under orientation reversal). By adding to $\mathcal{W}$ penalization for volume change (as in [22]) or simply by considering $Q\mathcal{W}_{\text{cubic}} + Q\mathcal{W}_{\text{iso}}$ one obtains an archetype satisfying all the hypotheses and having a discrete symmetry group.

Remark 5 The assumption $\mathcal{W} < \infty$ excludes physically relevant archetypes in which $\mathcal{W}(A)$ diverges as A becomes singular (see, e.g., [5, Theorem 4.10-2]). The requirement $\mathcal{W} < \infty$ is due to purely technical reasons that commonly appear in Γ-convergence results in elasticity when the elastic energy is $O(1)$.

In the rest of this section, it is easier to consider that $\mathcal{M}$ is endowed with an orthonormal parallel frame field E rather than a flat connection ∇; as stated above, this is completely equivalent modulo a global rotation of E.

Definition 8 Let $\mathcal{W}$ be an archetype satisfying the above conditions. Let $(\mathcal{M}, \mathfrak{g}, E)$ be a Riemannian manifold with an orthonormal frame field E. The elastic energy associated with $(\mathcal{M}, \mathfrak{g}, E)$ and $\mathcal{W}$ is

$$I(f) = \int_{\mathcal{M}} \mathcal{W}(df \circ E)\, \text{dVol}_{\mathfrak{g}} \qquad f \in W^{1,p}(\mathcal{M}; \mathbb{R}^d).$$

Note that $\mathfrak{g}$ is an intrinsic metric for this energy, and that the connection ∇, defined by declaring E parallel, is a material connection.

As standard in these type of problems, we extend I to $L^p(\mathcal{M}; \mathbb{R}^d)$ by

$$\tilde{I}(f) = \begin{cases} \int_{\mathcal{M}} \mathcal{W}(df \circ E)\, \text{dVol}_{\mathfrak{g}} & f \in W^{1,p}(\mathcal{M}; \mathbb{R}^d) \\ +\infty & f \in L^p(\mathcal{M}; \mathbb{R}^d) \setminus W^{1,p}(\mathcal{M}; \mathbb{R}^d). \end{cases}$$

In order to define convergence of the energy functionals, each defined on a different manifold $\mathcal{M}_n$, we need a notion of convergence of maps $f_n : \mathcal{M}_n \to \mathbb{R}^d$:

Definition 9 $(\mathcal{M}, \mathfrak{g})$ be a Riemannian manifold, and let $\mathcal{M}_n$ be topological manifolds. Let $F_n : \mathcal{M}_n \to \mathcal{M}$ be homeomorphisms. We say that a sequence of maps $f_n : \mathcal{M}_n \to \mathbb{R}^d$ converges to a map $f : \mathcal{M} \to \mathbb{R}^d$ in L^p if

$$\|f_n \circ F_n^{-1} - f\|_{L^p(\mathcal{M};\mathbb{R}^d)} \to 0.$$

Theorem 3 (Γ-Convergence of Elastic Energies) *Let $\mathcal{W}$ be an archetype satisfying the above assumptions. Let $(\mathcal{M}, \mathfrak{g}, E)$, $(\mathcal{M}_n, \mathfrak{g}_n, E_n)$ be Riemannian manifolds with orthonormal frames. Let $\tilde{I}$, $\tilde{I}_n$ be their associated elastic energies according to Definition 8. If there exists Lipschitz homeomorphisms $F_n : \mathcal{M}_n \to \mathcal{M}$ such that*

$$\|dF_n \circ E_n - E\|_{L^\infty} \to 0, \tag{14}$$

then $\tilde{I}_n \to \tilde{I}$ in the sense of Γ-convergence, relative to the convergence induced by F_n, as defined in Definition 9 (note that for Lipschitz maps, $dF_n \in L^\infty(T\mathcal{M}_n, F_n^*T\mathcal{M})$, hence the convergence is well defined).

Remark 6 If $\mathcal{W}$ is not quasiconvex (but (13) holds), then it follows from slight changes in the proof below that $\tilde{I}_n$ converges to the functional associated with $(\mathcal{M}, \mathfrak{g}, E)$ and the archetype $Q\mathcal{W}$, which is the quasiconvex envelope of $\mathcal{W}$. Note that it is still true that $\mathfrak{g}$ is an intrinsic metric and that ∇ is a material connection, hence Fig. 1 still holds.

Combining Theorems 2 and 3, we conclude the proof of Theorem 1:

Corollary 2 *Every two-dimensional body with a continuous distribution of dislocations $(\mathcal{M}, \mathfrak{g}, E)$ is a limit of bodies with finitely many dislocations $(\mathcal{M}_n, \mathfrak{g}_n, E_n)$ in the sense of Theorem 2 (equivalently (14)). Given an archetype $\mathcal{W}$, the elastic energies associated with $(\mathcal{M}_n, \mathfrak{g}_n, E_n)$ according to Definition 8 Γ-converge to the elastic energy associated with $(\mathcal{M}, \mathfrak{g}, E)$.*

Remark 7 Note that we do not rescale the elastic energies of the bodies with dislocations, that is, we are considering energies that are of order 1. This fits the typical heuristics for energies of dislocations: that a dislocation with a Burgers vector of magnitude ε will have a self-energy (or core energy) of order $\varepsilon^2 \log|\varepsilon|$, and that the interaction energy between two such dislocations will be of order ε^2 (see, e.g., [4, 16], which treats this in a linear case where ε^2 is factored out). Indeed, in our case $(\mathcal{M}_n, \mathfrak{g}_n, E_n)$ contains an order of n^2 dislocations of order $\varepsilon \approx n^{-2}$, so the self-energy is of order $n^2 \cdot \varepsilon^2 \log|\varepsilon| \to 0$, while the interaction energy is of order $n^4 \cdot \varepsilon^2 \approx 1$. To the best of our knowledge, this is the first rigorous framework in which an order 1 energy limit of bodies of dislocations is obtained in non-linear settings.

Note also that for coercive archetypes, that is, archetypes that satisfy $\mathcal{W}_{\text{iso}}(A) \geq c\,\text{dist}^p(A, \text{SO}(d))$ for some $c > 0$, the limiting energy associated with $(\mathcal{M}, \mathfrak{g}, E)$ is bounded away from zero if $\mathfrak{g}$ is non-flat, that is, there are no stress-free configurations.

4.1 Proof of Theorem 3

Let $\tilde{I}_\infty$ be the Γ-limit of a (not-relabeled) subsequence of $\tilde{I}_n$. Such a subsequence always exists by the general compactness theorem of Γ-convergence (see Theorem 8.5 in [9] for the classical result, or Theorem 4.7 in [28] for the case where each functional is defined on a different space). It is enough to prove that $\tilde{I}_\infty = \tilde{I}$. Indeed, since by the compactness theorem, every sequence has a Γ-converging subsequence, the Urysohn property of Γ-convergence (see Proposition 8.3 in [9]) implies that if all converging subsequences converge to the same limit, then the entire sequence converges to that limit.

From (14) it follows that

1. dF_n and dF_n^{-1} are uniformly bounded.
2. $(F_n)_\star \mathfrak{g}_n \to \mathfrak{g}$ in L^∞, and in particular, $(F_n)_\star \text{dVol}_{\mathfrak{g}_n} \to \text{dVol}_{\mathfrak{g}}$ in L^∞.

Lemma 1 (Infinity Case) *Let $f \in L^p(\mathcal{M}; \mathbb{R}^d) \setminus W^{1,p}(\mathcal{M}; \mathbb{R}^d)$. Then,*

$$\tilde{I}_\infty(f) = \infty = \tilde{I}(f).$$

Proof Suppose, by contradiction, that $\tilde{I}_\infty(f) < \infty$. Let $f_n \to f$ be a recovery sequence, namely,

$$\lim_{n\to\infty} \tilde{I}_n(f_n) = I_\infty(f) < \infty.$$

Without loss of generality we may assume that $\tilde{I}_n(f_n) < \infty$ for all n, and in particular, $f_n \in W^{1,p}(\mathcal{M}_n, \mathbb{R}^d)$. The coercivity of W_n implies that

$$\sup_n \int_{\mathcal{M}_n} |df_n|^p_{\mathfrak{g}_n,\mathfrak{e}}\, \text{dVol}_{\mathfrak{g}_n} < \infty.$$

Thus, f_n is uniformly bounded in $W^{1,p}$, and since dF_n^{-1} are uniformly bounded, $f_n \circ F_n^{-1}$ is also uniformly bounded in $W^{1,p}(\mathcal{M}; \mathbb{R}^d)$, hence weakly converges (modulo a subsequence). By the uniqueness of the limit, this limit is f, hence $f \in W^{1,p}(\mathcal{M}; \mathbb{R}^d)$, which is a contradiction. □

Lemma 2 (Upper Bound) *For every $f \in W^{1,p}(\mathcal{M}; \mathbb{R}^d)$,*

$$\tilde{I}_\infty(f) \leq \tilde{I}(f).$$

Proof Let $f \in W^{1,p}(\mathcal{M};\mathbb{R}^d)$. Define $f_n = f \circ F_n \in W^{1,p}(\mathcal{M}_n;\mathbb{R}^d)$. Trivially, $f_n \to f$ in L^p according to Definition 9 and by the definition of the Γ-limit,

$$\tilde{I}_\infty(f) \le \liminf_n \tilde{I}_n(f_n).$$

It follows from the uniform convergence $dF_n \circ E_n \to E$ and $(F_n)_\star \mathrm{dVol}_{\mathfrak{g}_n} \to \mathrm{dVol}_{\mathfrak{g}}$, using the p-Lipschitz property (13), that

$$\lim_n \tilde{I}_n(f_n) = \tilde{I}(f),$$

that is

$$\lim_n \int_{\mathcal{M}_n} \mathcal{W}(df \circ dF_n \circ E_n)\, \mathrm{dVol}_{\mathfrak{g}_n} = \int_{\mathcal{M}} \mathcal{W}(df \circ E)\, \mathrm{dVol}_{\mathfrak{g}}. \tag{15}$$

□

Lemma 3 (Lower Bound) *For every* $f \in W^{1,p}(\mathcal{M};\mathbb{R}^d)$,

$$\tilde{I}_\infty(f) \ge \tilde{I}(f).$$

Proof Let $f \in W^{1,p}(\mathcal{M};\mathbb{R}^d)$, and let $f_n \in L^p(\mathcal{M};\mathbb{R}^d)$ be a recovery sequence for f, that is $f_n \circ F_n^{-1} \to f$ in L^p and $\tilde{I}_n(f_n) \to \tilde{I}_\infty(f)$. In particular, it follows that we can assume without loss of generality that $f_n \in W^{1,p}$, and that f_n are uniformly bounded in $W^{1,p}$. Therefore, $f_n \circ F_n^{-1} \rightharpoonup f$ in $W^{1,p}(\mathcal{M};\mathbb{R}^d)$. We need to show that

$$\lim_n \tilde{I}_n(f_n) \ge \tilde{I}(f). \tag{16}$$

Note that since $f \in W^{1,p}(\mathcal{M};\mathbb{R}^d)$ and $f_n \in W^{1,p}(\mathcal{M}_n;\mathbb{R}^d)$, $\tilde{I}(f) = I(f)$ and $\tilde{I}_n(f_n) = I_n(f_n)$. Since $dF_n \circ E_n \to E$ and $(F_n)_\star \mathrm{dVol}_{\mathfrak{g}_n} \to \mathrm{dVol}_{\mathfrak{g}}$ uniformly, and $df_n \circ dF_n^{-1}$ are uniformly bounded in L^p, the p-Lipschitz property (13) implies that

$$\begin{aligned} \lim_n I_n(f_n) &= \lim_n \int_{\mathcal{M}_n} \mathcal{W}(df_n \circ E_n)\, \mathrm{dVol}_{\mathfrak{g}_n} \\ &= \lim_n \int_{\mathcal{M}_n} \mathcal{W}(df_n \circ dF_n^{-1} \circ E)\, \mathrm{dVol}_{\mathfrak{g}} = \lim_n I(f_n \circ F_n^{-1}). \end{aligned} \tag{17}$$

Since $\mathcal{W}$ is quasiconvex and satisfies (12), $I(\cdot)$ is lower semicontinuous with respect to the weak topology of $W^{1,p}(\mathcal{M};\mathbb{R}^d)$ [8, Theorem 8.11]. Since $f_n \circ F_n^{-1}$ converges weakly to f in $W^{1,p}(\mathcal{M};\mathbb{R}^d)$,

$$\lim_n I(f_n \circ F_n^{-1}) \ge I(f),$$

which together with (17) implies (16). □

5 The Role of Torsion in the Equilibrium Equations

In this section we analyze explicitly the equilibrium equations for a hyperelastic solid body having a continuous distribution of dislocations, and in particular, we address the role of torsion. We will explain why torsion does not enter explicitly in the equilibrium of an isotropic body. Similar equations are derived in [36, Section 12] (without the hyperelasticity assumption). Throughout this section we use the Einstein summation convention.

Let $\mathcal{W} \in C^2(\mathbb{R}^d \times \mathbb{R}^d)$ be a solid undistorted archetype, and let $(\mathcal{M}, W)$ be a uniform solid material having $\mathcal{W}$ as an archetype with respect to an implant map $E = \{E_p\}_{p\in\mathcal{M}}$. We denote the (matrix) argument of $\mathcal{W}$ by $B = (B_1 \mid \ldots \mid B_d)$, and by $\partial\mathcal{W}/\partial B_i : \mathbb{R}^d \times \mathbb{R}^d \to \mathbb{R}^d$ the derivative of $\mathcal{W}$ with respect to the column B_i (this is a vector).

The implant map E is a parallel frame of a flat material connection ∇ (defined by (6)) and it defines a metric $\mathfrak{g}$ via (8). E is a d-tuple of vector fields which we denote by $E_1, \ldots, E_d$. Its co-frame $E^1, \ldots, E^d$ is the d-tuple of one-forms defined by $E^i(E_j) = \delta^i_j$. The torsion tensor of ∇ is given by

$$T(E_i, E_j) = -[E_i, E_j] =: T^k_{ij} E_k,$$

as follows from the definition of the torsion tensor $T(X, Y) = \nabla_X Y - \nabla_Y X - [X, Y]$, since E_i are parallel, which means $\nabla E_i = 0$.

The elastic energy functional corresponding to this elastic body is

$$I(f) = \int_{\mathcal{M}} W(df)\,\mathrm{dVol}_{\mathfrak{g}} = \int_{\mathcal{M}} \mathcal{W}(df \circ E)\, E^1 \wedge \ldots \wedge E^d,$$

defined on functions $f : \mathcal{M} \to \mathbb{R}^d$. The Euler–Lagrange equations corresponding to this functional are, in a weak formulation,

$$\int_{\mathcal{M}} \frac{\partial\mathcal{W}}{\partial B_i}(df \circ E) \cdot E_i(h)\,\mathrm{dVol}_{\mathfrak{g}} = 0 \qquad \forall h \in C^\infty_c(\mathcal{M}; \mathbb{R}^d),$$

where $E_i(h) = dh(E_i) : \mathcal{M} \to \mathbb{R}^d$, and $\cdot$ is the standard inner product in $\mathbb{R}^d$. The strong formulation of the Euler–Lagrange equations is

$$E_i\left(\frac{\partial\mathcal{W}}{\partial B_i}(df \circ E)\right) + \frac{\partial\mathcal{W}}{\partial B_i}(df \circ E)\;\mathrm{div}\, E_i = 0,$$

or more explicitly,

$$\frac{\partial^2\mathcal{W}}{\partial B_i \partial B_j}(df \circ E)\, E_i E_j(f) + \frac{\partial\mathcal{W}}{\partial B_i}(df \circ E)\;\mathrm{div}\, E_i = 0,$$

where div E_i is defined by the relation

$$d(\iota_{E_i}\mathrm{dVol}_{\mathfrak{g}}) = \mathrm{div}\, E_i\, \mathrm{dVol}_{\mathfrak{g}},$$

where ι is the contraction operator. Using the fact that $\mathrm{dVol}_{\mathfrak{g}} = E^1 \wedge \ldots \wedge E^d$,

$$\iota_{E_i}\mathrm{dVol}_{\mathfrak{g}} = (-1)^{i+1} E^1 \wedge \ldots \wedge E^{i-1} \wedge E^{i+1} \wedge \ldots \wedge E^d,$$

hence

$$d(\iota_{E_i}\mathrm{dVol}_{\mathfrak{g}}) = (-1)^{i+1} \Big(dE^1 \wedge \ldots \wedge E^{i-1} \wedge E^{i+1} \wedge \ldots \wedge E^d + \ldots$$
$$\ldots + (-1)^{d-1} E^1 \wedge \ldots \wedge E^{i-1} \wedge E^{i+1} \wedge \ldots \wedge dE^d \Big).$$

By the definition of the exterior derivative, and the fact that $E^k(E_i) = \delta_i^k$,

$$dE^k(E_i, E_j) = E_i(E^k(E_j)) - E_j(E^k(E_i)) - E^k([E_i, E_j]) = T_{ij}^l E^k(E_l) = T_{ij}^k$$

and therefore $dE^k = T_{ij}^k E^i \wedge E^j$, so $d(\iota_{E_i}\mathrm{dVol}_{\mathfrak{g}})$ simplifies to

$$d(\iota_{E_i}\mathrm{dVol}_{\mathfrak{g}}) = -T_{ji}^j\, \mathrm{dVol}_{\mathfrak{g}},$$

hence $\mathrm{div}\, E_i = -T_{ji}^j$. It follows that the Euler–Lagrange equations are

$$\frac{\partial^2 \mathcal{W}}{\partial B_i \partial B_j}(df \circ E)\, E_i E_j(f) - T_{ji}^j \frac{\partial \mathcal{W}}{\partial A_i}(df \circ E) = 0.$$

The trace of the torsion appears explicitly in the equations; however, the torsion also appears, more implicitly, as the antisymmetric part $E_i E_j - E_j E_i = T_{ij}^k E_k$ of the first addend.

If the solid is isotropic, then the equilibrium equations are independent of the torsion. Isotropy means that

$$\mathcal{W}B \circ R = \mathcal{W}B \qquad \text{for any } R \in \mathrm{SO}(d).$$

Using polar decomposition, this implies that there exists a function $\widetilde{\mathcal{W}} : \mathrm{Sym}_+(d) \to \mathbb{R}$, where $\mathrm{Sym}_+(d)$ is the set of positive-semidefinite $d \times d$ symmetric matrices, such that

$$\mathcal{W}(B) = \widetilde{\mathcal{W}}(BB^T)$$

[5, Theorem 3.4-1] (if one allows B to be orientation reversing, then $\widetilde{\mathcal{W}}$ also depends on the orientation of B, but this does not affect the argument below and therefore we ignore this subtlety). It follows that

$$I(f) = \int_{\mathcal{M}} W(df)\,\mathrm{dVol}_{\mathfrak{g}} = \int_{\mathcal{M}} \mathcal{W}(df \circ E)\,\mathrm{dVol}_{\mathfrak{g}} = \int_{\mathcal{M}} \widetilde{\mathcal{W}}((df \circ E)(df \circ E)^T)\,\mathrm{dVol}_{\mathfrak{g}}.$$

Choosing coordinates on $\mathcal{M}$, we can think of df and E as matrices. In this case, since E is an orthonormal frame for $\mathfrak{g}$, $EE^T = \mathfrak{g}^*$, the $\mathfrak{g}$-metric on $T^*\mathcal{M}$ (whose coordinates are $\mathfrak{g}^{ij}$). Therefore, in coordinates,

$$I(f) = \int_{\mathcal{M}} \widetilde{\mathcal{W}}(df_x \circ \mathfrak{g}_x^* \circ df_x^T)\,\sqrt{|\mathfrak{g}|}(x)\,dx.$$

In a more abstract language,

$$I(f) = \int_{\mathcal{M}} \tilde{\mathcal{W}}(f_\star \mathfrak{g}^*)\,\mathrm{dVol}_{\mathfrak{g}},$$

where $f_\star \mathfrak{g}^*$ is the push-forward by f of the metric $\mathfrak{g}^*$ from $T^*\mathcal{M}$ to $\mathbb{R}^d$. Either way, it is clearly seen that the energy (and therefore the equilibrium equations) only depend on $\mathfrak{g}$ and not on the frame E, and therefore not on the connection ∇ and its torsion which are derived from E.

Acknowledgements This project was initiated in the Oberwolfach meeting “Material Theories” in July 2018. RK was partially funded by the Israel Science Foundation (Grant No. 1035/17), and by a grant from the Ministry of Science, Technology and Space, Israel and the Russian Foundation for Basic Research.

References

1. H.I. Arcos, and J.G. Pereira, *Torsion Gravity: a Reappraisal*, International Journal of Modern Physics D **13** (2004), no. 10, 2193–2240.
2. B.A. Bilby, R. Bullough, and E. Smith, *Continuous distributions of dislocations: A new application of the methods of Non-Riemannian geometry*, Proc. Roy. Soc. A **231** (1955), 263–273.
3. M. Berger, *A panoramic view of Riemannian geometry*, Springer, 2002.
4. P. Cermelli and G. Leoni, *Renormalized energy and forces on dislocations*, SIAM journal on mathematical analysis **37** (2005), no. 4, 1131–1160.
5. P.G. Ciarlet, *Mathematical elasticity, volume 1: Three-dimensional elasticity*, Elsevier, 1988.
6. D. Christodoulou and I. Kaelin, *On the mechanics of crystalline solids with a continuous distribution of dislocations*, Advances in Theoretical and Mathematical Physics **17** (2013), no. 2, 399–477.
7. J. Cheeger, W. Müller, and R. Schrader, *On the curvature of piecewise flat spaces*, Commun. Math. Phys. **92** (1984), 405–454.
8. B. Dacorogna, *Direct methods in the calculus of variations*, 2nd ed., Springer, 2008.
9. G. dal Maso, *An introduction to Γ-convergence*, Birkhäuser, 1993.
10. C. Davini, *A proposal for a continuum theory of defective crystals*, Arch. Rat. Mech. Anal. **96** (1986), 295–317.
11. G. Dolzmann, *Regularity of minimizers in nonlinear elasticity – the case of a one-well problem in nonlinear elasticity*, Technische Mechanik **32** (2012), 189–194.

12. R. Dyer, G. Vegter, and M. Wintraecken, *Riemannian simplices and triangulations*, Geometriae Dedicata **179** (2015), 91–138.
13. M. Elżanowski, M. Epstein, and J. Śniatycki, *G-structures and material homogeneity*, Journal of Elasticity **23** (1990), no. 2, 167–180.
14. M. Epstein and R. Segev, *Geometric aspects of singular dislocations*, Mathematics and Mechanics of Solids **19** (2014), no. 4, 337–349.
15. ———, *Geometric theory of smooth and singular defects*, International Journal of Non-Linear Mechanics **66** (2014), 105–110.
16. A. Garroni, G. Leoni, and M. Ponsiglione, *Gradient theory for plasticity via homogenization of discrete dislocations*, J. Eur. Math. Soc. **12** (2010), 1231–1266.
17. K. Hayashi and T. Shirafuji, *New general relativity*, Physica D **19** (1979), 3524–3553.
18. M. O. Katanaev, *Geometric theory of defects*, UFN **175** (2005), no. 7, 705–733.
19. R. Kupferman and C. Maor, *The emergence of torsion in the continuum limit of distributed dislocations*, J. Geom. Mech. **7** (2015), 361–387.
20. ———, *Limits of elastic models of converging Riemannian manifolds*, Calc. Variations and PDEs **55** (2016), 40.
21. ———, *Riemannian surfaces with torsion as homogenization limits of locally-Euclidean surfaces with dislocation-type singularities*, Proc. Roy. Soc. Edinburgh **146A** (2016), no. 4, 741–768.
22. ———, *Variational convergence of discrete geometrically-incompatible elastic models*, Calc. Var. PDEs **57** (2018), no. 2, 39.
23. R. Kupferman and E. Olami, Homogenization of edge-dislocations as a weak limit of de-Rham currents, *Geometric continuum mechanics. Advances in mechanics and mathematics* (R. Segev and M. Epstein, eds.), Springer, New York, 2020. https://doi.org/10.1007/978-3-030-42683-5_6.
24. K. Kondo, *Geometry of elastic deformation and incompatibility*, Memoirs of the Unifying Study of the Basic Problems in Engineering Science by Means of Geometry (K. Kondo, ed.), vol. 1, 1955, pp. 5–17.
25. E. Kröner, *The dislocation as a fundamental new concept in continuum mechanics*, Materials Science Research (H. H. Stadelmaier and W. W. Austin, eds.), Springer US, Boston, MA, 1963, pp. 281–290.
26. ———, *The physics of defects*, Les Houches Summer School Proceedings (Amsterdam) (R. Balian, M. Kleman, and J.-P. Poirier, eds.), North-Holland, 1981.
27. ———, *Dislocation theory as a physical field theory*, Meccanica **31** (1996), 577–587.
28. K. Kuwae and T. Shioya, *Variational convergence over metric spaces*, Trans. Amer. Math. Soc. **360** (2008), no. 1, 35–75.
29. M. Lewicka and P. Ochoa, *On the variational limits of lattice energies on prestrained elastic bodies*, Differential Geometry and Continuum Mechanics (G.-Q. G. Chen, M. Grinfeld, and R. J. Knops, eds.), Springer, 2015, pp. 279–305.
30. W. Noll, *A mathematical theory of the mechanical behavior of continuous media*, Arch. Rat. Mech. Anal. **2** (1958), 197–226.
31. J.F. Nye, *Some geometrical relations in dislocated crystals*, Acta Met. **1** (1953), 153–162.
32. A. Ozakin and A. Yavari, *Affine development of closed curves in Weitzenböck manifolds and the burgers vector of dislocation mechanics*, Math. Mech. Solids **19** (2014), 299–307.
33. J.A. Schouten, *Ricci-calculus*, Springer-Verlag Berlin Heidelberg, 1954.
34. M. Šilhavý, *Rank 1 convex hulls of isotropic functions in dimension 2 by 2*, Math. Bohem. **126** (2001), 521–529.
35. V. Volterra, *Sur l'équilibre des corps élastiques multiplement connexes*, Ann. Sci. Ecole Norm. Sup. Paris 1907 **24** (1907), 401–518.
36. C.-C. Wang, *On the geometric structures of simple bodies, a mathematical foundation for the theory of continuous distributions of dislocations*, Arch. Rat. Mech. Anal. **27** (1967), 33–93.
37. R. Weitzenböck, *Invariantentheorie*, ch. XIII, Sec 7, Nordhoff, Groningen, 1923.
38. A. Yavari and A. Goriely, *Weyl geometry and the nonlinear mechanics of distributed point defects*, Proc. Roy. Soc. A **468** (2012), 3902–3922.

On the Homogeneity of Non-uniform Material Bodies

V. M. Jiménez, M. de León, and M. Epstein

Contents

Abstract To any simple body with a given smooth constitutive equation, a groupoid, known as the material groupoid, can be associated naturally. When the body is non-uniform, however, the material groupoid is generally not differentiable. In such cases, the analysis can be based on a new differential geometric construct called the material distribution, to which we can associate other physically meaningful objects, such as a material foliation, with the help of which we have the possibility to study and rigorously classify non-uniform bodies. Thus, the material distribution and its associated singular foliation result in a rigorous and unique subdivision of the material body into strictly smoothly uniform sub-bodies,

V. M. Jiménez (✉)
Instituto de Ciencias Matemáticas (CSIC-UAM-UC3M-UCM), Madrid, Spain
e-mail: victor.jimenez@icmat.es

M. de León
Instituto de Ciencias Matemáticas (CSIC-UAM-UC3M-UCM), Madrid, Spain

Real Academia de Ciencias Exactas, Físicas y Naturales, Madrid, Spain
e-mail: mdeleon@icmat.es

M. Epstein
Department of Mechanical and Manufacturing Engineering, University of Calgary, Calgary, AB, Canada
e-mail: mepstein@ucalgary.ca

R. Segev, M. Epstein (eds.), *Geometric Continuum Mechanics*, Advances in Mechanics and Mathematics 43, https://doi.org/10.1007/978-3-030-42683-5_9

laminates, filaments and isolated points. Furthermore, the material distribution permits us to define a measure of uniformity of a simple body as well as more general definitions of homogeneity for non-uniform bodies.

1 Introduction

The multifaceted relation between continuum mechanics and differential geometry has a rich history, particularly apparent since the turn of the last century. Various theories of elastic beams and shells had already required the use of the classical results of differential geometry of curves and surfaces embedded in a Euclidean space. The astonishing work of the Cosserat brothers anticipated certain aspects of modern differential geometry in the style of Cartan. But the more intimate kinship between the two disciplines can be said to have started in earnest after the Second World War, when continuum mechanics found its own mathematical bearings and, thanks to the work of various groups of like-minded scholars, it established its foundations on a rigorous basis.

The most obvious link is provided by the very use of the word 'continuum', which, just as in general relativity, is a physicist's way to allude to a differentiable manifold. In general expository texts, a material body $\mathcal{B}$ (the material continuum) is identified with a connected 3-dimensional manifold (without boundary) that can be covered with a single coordinate chart. Moreover, the setting (the space-time continuum) is implicitly identified with an affine bundle over the real line, with $\mathbb{R}^3$ as its typical fibre and the orthogonal group as its structure group. A *configuration* is regarded as an embedding of $\mathcal{B}$ in $\mathbb{R}^3$. One of the achievements of the modern formulation [16] consists in showing that most of these limitations can be removed without substantially affecting the physical content of the theory. In [16], moreover, a consistent theory of stress is shown to emerge from a setting that, in some aspects, mimics the differential geometric approach to Classical Mechanics, the configuration space being identified with the infinite-dimensional manifold of configurations.

In this chapter, however, we would like to highlight a different facet of the relation between differential geometry and continuum mechanics. While as early as 1907 Volterra, working on multiple connected elastic bodies, had already provided a characterization of certain defects, which we now call dislocations and disclinations, two generations passed before theories of continuous distributions of such defects were attempted. Some of the pioneers in this field were Kondo, Kroener, Frank, Bilby, Nye and Eshelby, to name but a few. All of these people were well acquainted with differential geometry and also had a keen awareness of the experimental results pertaining to the identification of dislocations and of their motion. The macroscopic phenomenon of metal plasticity was, in fact, correctly attributed to these causes. Various differential geometric constructs were proposed to represent different types of defects with great success.

A different approach, based on the epistemological stance that in a genuinely continuum mechanical theory all the information about the material response should be encoded in the constitutive laws and nothing else, was proposed by Noll [15] and extended by Wang [20] and Bloom [1]. Although many of the results obtained by this school of thought were the same as those of its predecessors, there were significant differences. One of these is the role played in Noll's theory by the material symmetry group, which may be a continuous group. Indeed, if all the information about the material constitution is gathered from macroscopic response functionals, the discrete symmetry groups of the underlying crystal lattices may be enough to guarantee larger groups for the macroscopic material response, such as full isotropy or transverse isotropy.

The main idea behind Noll's approach is contained in the definition of *material isomorphism*. If W denotes a constitutive variable (such as the *Cauchy stress*, or the elastic energy per unit mass) and if the material is *simple* (in the sense that only the first gradient F of the deformation is the relevant independent variable for the material response), the *constitutive equation* at a point $X \in \mathcal{B}$ is of the form $W = W(F, X)$. The particular form of this function depends on the *reference configuration* chosen to express the deformation and its gradient. This freedom of choice manifests itself in the fact that a change of reference configuration results, at most, in the right action of the general linear group $Gl(3, \mathbb{R})$ on the independent variable F. We may, therefore, say that a material response can be identified with a whole orbit (in some space of functions) generated by this right action. Given a reference configuration and two points, X and Y, in $\mathcal{B}$, we say that they are *materially isomorphic* if their constitutive functions belong to one and the same orbit. In other words, X and Y are materially isomorphic if there exists a linear map $P_{XY} : T_X\mathcal{B} \to T_Y\mathcal{B}$ such that

$$W(F, Y) = W(F P_{XY}, X), \tag{1}$$

for all deformation gradients F. A body is *materially uniform* if all its points are mutually materially isomorphic.

The material isomorphism P_{XY} is, in general, not unique. Indeed, if G_X is a material symmetry at X, and if P_{XY} is a material isomorphism, so is $P_{XY}\, G_X$. Moreover, the material symmetry groups at X and Y are mutually conjugate, the conjugation being obtained precisely by any material isomorphism P_{XY}. On the basis of these facts, assuming the body to be smoothly uniform, Noll introduced the notion of *material parallelisms* and their associated curvature-free *material connections*, a concept further generalized by Wang in [20]. The presence of defects, in the case of a discrete symmetry group, is measured by the non-vanishing of the torsion of the unique material connection, a result identical to that obtained by earlier formulations. In Noll's terminology, the absence of defects coincides with the notion of *local homogeneity* of the body. The case of continuous symmetry groups requires the identification of other obstacles to integrability. The work of Elżanowski et al. [4] presents a formulation of these ideas in the terminology of G-structures.

In later developments, for general (not necessarily uniform) bodies, groupoids and smooth distributions [9, 11, 13] were shown to be useful tools to express in geometrical terms the structure implied by the constitutive law. The *material groupoid* $\Omega(\mathcal{B})$ over $\mathcal{B}$ of a simple material consists of all material isomorphisms P for all pairs of body points X, Y. It follows from this definition that $\mathcal{B}$ is (smoothly) uniform if, and only if, $\Omega(\mathcal{B})$ is a transitive (Lie) subgroupoid of $\Pi^1(\mathcal{B}, \mathcal{B})$, where $\Pi^1(\mathcal{B}, \mathcal{B})$, called the *1-jets groupoid on* $\mathcal{B}$, is the Lie groupoid over $\mathcal{B}$ of all linear isomorphisms between the tangent spaces $T_X\mathcal{B}$ and $T_Y\mathcal{B}$, for $X, Y \in \mathcal{B}$.

The main results of this chapter can be summarized as follows: In the general case of non-uniformity, $\Omega(\mathcal{B})$ is not a Lie subgroupoid of $\Pi^1(\mathcal{B}, \mathcal{B})$, and, in particular, an associated Lie algebroid is not available. Instead, we have introduced the *material distribution* $A\Omega^T(\mathcal{B})$ (see [11] or [13]). $A\Omega^T(\mathcal{B})$ is generated by the (local) left-invariant vector fields on $\Pi^1(\mathcal{B}, \mathcal{B})$ which are in the kernel of the tangent map TW of W. Due to the groupoid structure, we can still associate two new objects with $A\Omega^T(\mathcal{B})$, denoted by $A\Omega(\mathcal{B})$ and $A\Omega^\sharp(\mathcal{B})$, as defined by the following diagram:

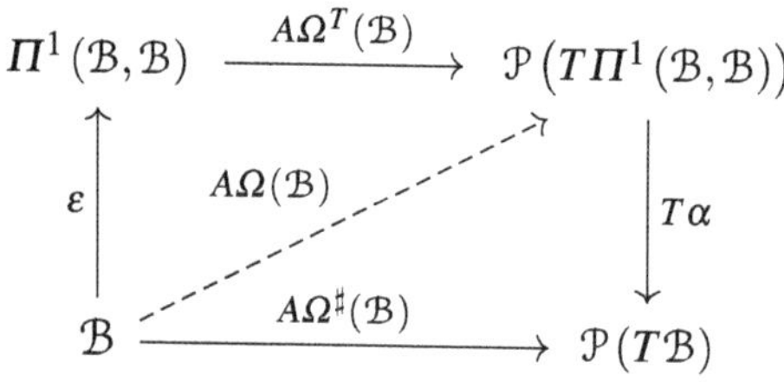

Here $\mathcal{P}(E)$ defines the power set of E, $\epsilon(X)$ is the identity map of $T_X\mathcal{B}$ and $\alpha : \Pi^1(\mathcal{B}, \mathcal{B}) \to \mathcal{B}$ denotes the source map of the groupoid.

By construction, the distributions $A\Omega^T(\mathcal{B})$ and $A\Omega^\sharp(\mathcal{B})$ are integrable (in the sense of Stefan [17] and Sussmann [18]), and they provide two foliations, $\overline{\mathcal{F}}$ on $\Pi^1(\mathcal{B}, \mathcal{B})$ and $\mathcal{F}$ on $\mathcal{B}$, such that $\Omega(\mathcal{B})$ is union of leaves of $\overline{\mathcal{F}}$. As a consequence, we have that $\mathcal{B}$ can be covered by a foliation of some kind of smoothly uniform 'sub-bodies', called *material submanifolds*.[1] The material distribution is also a tool apt to provide a general classification of smoothly non-uniform bodies and Ms the possibility to distinguish various degrees of uniformity. In addition, homogeneity may be generalized in such a way that any simple body can be tested to be homogeneous. A first step in this direction was taken in [10] where the authors give a homogeneity condition for bundles and laminated bodies.

Next, we consider a more general situation. We study the problem from a purely mathematical framework, since we are convinced that this analysis should be relevant not only for its applications to continuum mechanics, but also for the general theory of groupoids.

So, let $\overline{\Gamma} \subseteq \Gamma$ be a subgroupoid of a Lie groupoid $\Gamma \rightrightarrows M$; notice that we are not assuming, in principle, any differentiable structure on $\overline{\Gamma}$. Even in that case, we can construct a generalized distribution $A\overline{\Gamma}^T$ over Γ generated by the (local)

[1]These submanifolds are not sub-bodies in the usual sense of continuum mechanics [21], because their dimensions are not necessarily equal to the dimension of $\mathcal{B}$.

left-invariant vector fields on Γ whose flow at the identities is totally contained in $\overline{\Gamma}$. This distribution $A\overline{\Gamma}^T$ will be called the *characteristic distribution of* $\overline{\Gamma}$. Again, due to the groupoid structure, we can still associate two new objects with $A\overline{\Gamma}^T$, denoted by $A\overline{\Gamma}$ and $A\overline{\Gamma}^\sharp$ analogously to the above diagram.

The paper is structured as follows: Sects. 2 and 3 are devoted to a brief introduction to groupoids and algebroids, respectively. In particular we show, as an example, the groupoid of 1-jets of local automorphisms on a manifold M. In the next section we study the characteristic distribution, a general smooth distribution associated with any subgroupoid of a Lie groupoid. Here we present new interesting results which could be applied to different fields. There are also new results describing deeper the characteristic distribution. Thus, in Sect. 5 we apply this construction to the theory of simple bodies, generating in this way the so-called material groupoid and material distribution. By using these two mathematical objects we introduce the concept of *graded uniformity* of a simple body. Section 6 proposes a new definition of homogeneity for non-uniform bodies which generalizes the known definition for smoothly uniform bodies. Some characterizations are given related to the integrability of the material groupoid and the material distribution. Finally, we study an example of non-uniform body in which the homogeneity is checked.

2 Groupoids

In this section, we shall give a brief introduction to *Lie groupoids*. The standard reference on groupoids is [14].[2]

Definition 1 Let M be a set. A *groupoid* over M is given by a set Γ provided with the following maps: $\alpha, \beta : \Gamma \to M$ (*source* and *target maps*, respectively), $\epsilon : M \to \Gamma$ (*identities map*), $i : \Gamma \to \Gamma$ (*inversion map*) and $\cdot : \Gamma_{(2)} \to \Gamma$ (*composition law*) where for each $k \in \mathbb{N}$,

$$\Gamma_{(k)} := \{(g_1, \ldots, g_k) \in \Gamma^k \ : \ \alpha(g_i) = \beta(g_{i+1}), \ i = 1, \ldots, k-1\},$$

satisfying the following properties:

(1) α and β are surjective and, for each $(g, h) \in \Gamma_{(2)}$, we have

$$\alpha(g \cdot h) = \alpha(h), \quad \beta(g \cdot h) = \beta(g).$$

(2) Associativity of the composition law, i.e.,

$$g \cdot (h \cdot k) = (g \cdot h) \cdot k, \ \forall (g, h, k) \in \Gamma_{(3)}.$$

[2]For a short introduction, see [19] (written in Spanish). Other recommended references are [6] and [22].

(3) For all $g \in \Gamma$,

$$g \cdot \epsilon\left(\alpha\left(g\right)\right) = g = \epsilon\left(\beta\left(g\right)\right) \cdot g.$$

In particular,

$$\alpha \circ \epsilon \circ \alpha = \alpha, \quad \beta \circ \epsilon \circ \beta = \beta.$$

Since α and β are surjective we get

$$\alpha \circ \epsilon = Id_{\Gamma}, \quad \beta \circ \epsilon = Id_{\Gamma}.$$

(4) For each $g \in \Gamma$,

$$i\left(g\right) \cdot g = \epsilon\left(\alpha\left(g\right)\right), \quad g \cdot i\left(g\right) = \epsilon\left(\beta\left(g\right)\right).$$

Then,

$$\alpha \circ i = \beta, \quad \beta \circ i = \alpha.$$

These maps $\alpha, \beta, i, \epsilon$ will be called *structure maps*. In what follows, we will denote this groupoid by $\Gamma \rightrightarrows M$.

If Γ is a groupoid over M, then M is also denoted by $\Gamma_{(0)}$ and it is often identified with the set $\epsilon(M)$ of identity elements of Γ. Γ is also denoted by $\Gamma_{(1)}$. The map $(\alpha, \beta) : \Gamma \to M \times M$ is called the *anchor* of the groupoid.

Remark 1 For pictorial purposes, it is always useful to think of a groupoid as a set Γ of *arrows*. Each arrow $g \in \Gamma$ has a *source* (or tail) and a *target* (or tip), both of which belong to a set M of *objects*. Two *projection maps*, designated by α and β, assign to each arrow g its source $\alpha(g) \in M$ and its target $\beta(g) \in M$, respectively. The *composition* operation (also called product) can be applied only to those pairs whose arrows are joined in a tip-to-tail fashion. That is, if $g, h \in \Gamma$, and if $\beta(h) = \alpha(g)$, and only in this case, there is a well-defined composition $g \cdot h$, as shown in Fig. 1. For each object $X \in M$, there is a (unique) *identity* of *unit* arrow at X, denoted by $\epsilon(X)$, satisfying $\alpha(\epsilon(X)) = \beta(\epsilon(X))$. Pictorially, every unit is a loop-shaped arrow. Whenever these unit elements can be composed with other elements, on the left or on the right, they do not affect the result. Finally, for each arrow g, there exists an *inverse* arrow g^{-1} such that $\alpha(g) = \beta(g^{-1})$ and $\beta(g) = \alpha(g^{-1})$, and such that $g \cdot g^{-1} = \epsilon(\beta(g))$ and $g^{-1} \cdot g = \epsilon(\alpha(g))$.

Now, we define the morphisms in the category of groupoids.

Definition 2 If $\Gamma_1 \rightrightarrows M_1$ and $\Gamma_2 \rightrightarrows M_2$ are two groupoids, then a morphism from $\Gamma_1 \rightrightarrows M_1$ to $\Gamma_2 \rightrightarrows M_2$ consists of two maps $\Phi : \Gamma_1 \to \Gamma_2$ and $\phi : M_1 \to M_2$ such that for any $g_1 \in \Gamma_1$

$$\alpha_2\left(\Phi\left(g_1\right)\right) = \phi\left(\alpha_1\left(g_1\right)\right), \qquad \beta_2\left(\Phi\left(g_1\right)\right) = \phi\left(\beta_1\left(g_1\right)\right), \tag{2}$$

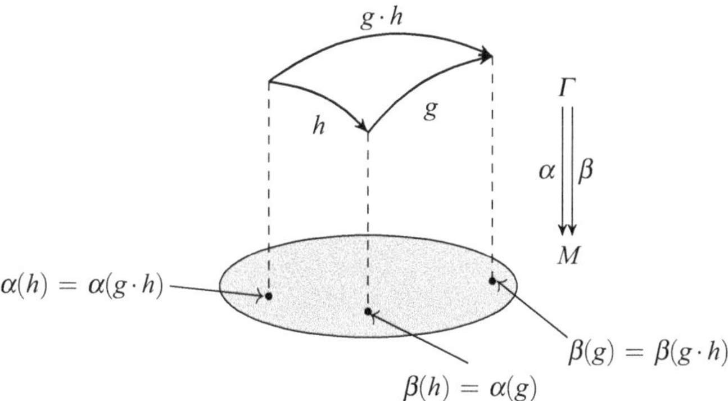

Fig. 1 Schematic representation of a groupoid

where α_i and β_i are the source and the target map of $\Gamma_i \rightrightarrows M_i$, respectively, for $i = 1, 2$, and such that Φ preserves the composition, i.e.,

$$\Phi\left(g_1 \cdot h_1\right) = \Phi\left(g_1\right) \cdot \Phi\left(h_1\right), \ \forall \left(g_1, h_1\right) \in \left(\Gamma_1\right)_{(2)}.$$

We will denote this morphism by (Φ, ϕ) or by Φ (because, using Eq. (2), ϕ is completely determined by Φ).

Observe that, as a consequence, Φ preserves the identities, i.e., denoting by ϵ_i the section of identities of $\Gamma_i \rightrightarrows M_i$ for $i = 1, 2$, we have

$$\Phi \circ \epsilon_1 = \epsilon_2 \circ \phi.$$

Using this definition we define a *subgroupoid* of a groupoid $\Gamma \rightrightarrows M$ as a groupoid $\Gamma' \rightrightarrows M'$ such that $M' \subseteq M$, $\Gamma' \subseteq \Gamma$ and the corresponding inclusion map is a morphism of groupoids.

Remark 2 There is a more abstract way of defining a groupoid. A groupoid is a "small" category (the class of objects and the class of morphisms are sets) in which each morphism is invertible.
If $\Gamma \rightrightarrows M$ is the groupoid, then M is the set of objects and Γ is the set of morphisms. In this sense, we can think about a groupoid as a set M of objects and a set Γ of invertible maps between objects of M. Then, for each map $g \in \Gamma$, $\alpha(g)$ is the domain of g, $\beta(g)$ is the codomain g and $i(g)$ is the inverse of g. For all $X \in M$, $\epsilon(X)$ is the identity map at X and, finally, the operation $\cdot$ can be thought as the composition of maps.
A groupoid morphism is a functor between these categories, which is a more natural definition.

Now, we present the most basic examples of groupoids.

Example 1 A group is a groupoid over a point. In fact, let G be a group and e the identity element of G. Then, $G \rightrightarrows \{e\}$ is a groupoid, where the operation of the groupoid, $\cdot$, is just the operation in G.

Example 2 For any set M, we can consider the product space $M \times M$. Then $M \times M$ has a groupoid structure over M such that

$$(X, Y) \cdot (Z, X) = (Z, Y),$$

for all $X, Y, Z \in M$. The groupoid $M \times M \rightrightarrows M$ is said to be the *pair groupoid of M*.
Note that, if $\Gamma \rightrightarrows M$ is an arbitrary groupoid over M, then the anchor $(\alpha, \beta) : \Gamma \to M \times M$ is a morphism from $\Gamma \rightrightarrows M$ to the pair groupoid of M.

Next, we introduce the notion of orbits and isotropy group.

Definition 3 Let $\Gamma \rightrightarrows M$ be a groupoid with α and β the source map and target map, respectively. For each $X \in M$, we denote

$$\Gamma_X^X = \beta^{-1}(X) \cap \alpha^{-1}(X),$$

which is called the *isotropy group of* Γ at X. The set

$$\mathcal{O}(X) = \beta\left(\alpha^{-1}(X)\right) = \alpha\left(\beta^{-1}(X)\right),$$

is called the *orbit* of X, or *the orbit* of Γ through X.

If $\mathcal{O}(X) = M$ for all $X \in M$, or equivalently if $(\alpha, \beta) : \Gamma \to M \times M$ is a surjective map, then the groupoid $\Gamma \rightrightarrows M$ is called *transitive*.

Furthermore, the preimages of the source map α of a groupoid are called *α-fibres*. Those of the target map β are called *β-fibres*. We will usually denote the α-fibre (resp. β-fibre) at a point X by Γ_X (resp. Γ^X).

Definition 4 Let $\Gamma \rightrightarrows M$ be a groupoid with α and β the source and target map, respectively. We may define the left translation by $g \in \Gamma$ as the map $L_g : \Gamma^{\alpha(g)} \to \Gamma^{\beta(g)}$, given by

$$h \mapsto g \cdot h.$$

Similarly, we may define the right translation on g, $R_g : \Gamma_{\beta(g)} \to \Gamma_{\alpha(g)}$.

Note that,

$$Id_{\Gamma^X} = L_{\epsilon(X)}. \tag{3}$$

So, for all $g \in \Gamma$, the left (resp. right) translation on g, L_g (resp. R_g), is a bijective map with inverse $L_{i(g)}$ (resp. $R_{i(g)}$), where $i : \Gamma \to \Gamma$ is the inverse map.

Different structures (topological and geometrical) can be imposed on groupoids, depending on the context we are dealing with. We are interested in a particular case, the so-called Lie groupoids.

Definition 5 A *Lie groupoid* is a groupoid $\Gamma \rightrightarrows M$ such that Γ and M are smooth manifolds, and all the structure maps are smooth. Furthermore, the source and the target maps are submersions.
A *Lie groupoid morphism* is a groupoid morphism which is differentiable.

Definition 6 Let $\Gamma \rightrightarrows M$ be a Lie groupoid. A *Lie subgroupoid* of $\Gamma \rightrightarrows M$ is a Lie groupoid $\Gamma' \rightrightarrows M'$ such that Γ' and M' are submanifolds of Γ and M, respectively; and the pair given by the inclusion maps $j_{\Gamma'} : \Gamma' \hookrightarrow \Gamma$ $j_{M'} : M' \hookrightarrow M$ becomes a morphism of Lie groupoids.

Observe that, taking into account that $\alpha \circ \epsilon = Id_M = \beta \circ \epsilon$, we conclude that ϵ is an injective immersion.

On the other hand, in the case of a Lie groupoid, L_g (resp. R_g) is clearly a diffeomorphism for every $g \in \Gamma$.

Example 3 A Lie group is a Lie groupoid over a point.

Example 4 Let M be a manifold. The pair groupoid $M \times M \rightrightarrows M$ is a Lie groupoid.

Next, we will introduce an example which will be fundamental for our treatment.

Example 5 Let M be a manifold, and denote by $\Pi^1(M, M)$ the set of all vector space isomorphisms $L_{X,Y} : T_X M \to T_Y M$ for $X, Y \in M$ or, equivalently, the space of the 1-jets of local diffeomorphisms on M. An element of $\Pi^1(M, M)$ will by denoted by $j^1_{X,Y}\psi$, where ψ is a local diffeomorphism from M into M such that $\psi(X) = Y$.

$\Pi^1(M, M)$ can be seen as a transitive groupoid over M where, for all $X, Y \in M$ and $j^1_{X,Y}\psi, j^1_{Y,Z}\varphi \in \Pi^1(M, M)$, we have

(i) $\alpha\left(j^1_{X,Y}\psi\right) = X$
(ii) $\beta\left(j^1_{X,Y}\psi\right) = Y$
(iii) $j^1_{Y,Z}\varphi \cdot j^1_{X,Y}\psi = j^1_{X,Z}(\varphi \circ \psi)$

This groupoid is called the *1-jets groupoid on M*. In fact, let $\left(x^i\right)$ and $\left(y^j\right)$ be local coordinate systems on M sets $U, V \subseteq M$. Then, we can consider a local coordinate system on $\Pi^1(M, M)$ given by

$$\Pi^1(U, V) : \left(x^i, y^j, y^j_i\right), \tag{4}$$

where, for each $j^1_{X,Y}\psi \in \Pi^1(U, V)$

- $x^i\left(j^1_{X,Y}\psi\right) = x^i(X)$.
- $y^j\left(j^1_{X,Y}\psi\right) = y^j(Y)$.
- $y^j_i\left(j^1_{X,Y}\psi\right) = \left.\dfrac{\partial\left(y^j \circ \psi\right)}{\partial x^i}\right|_X$.

Using these coordinates one can easily check that $\Pi^1(M, M)$ is a transitive Lie groupoid.

3 Lie Algebroids

3.1 Introduction

As an algebraic structure, a groupoid can be colourfully described as a group 'on steroids'. Indeed, a group can be regarded as the particular case of a groupoid whose set of objects M is a singleton. All the elements of a group can be mutually composed (multiplied) and there is a single, uniquely defined, unit element e. In a general groupoid $\Gamma \rightrightarrows M$, however, M is an arbitrary set, but, except for the totally intransitive case, Γ is not just the union of its individual vertex groups. The elements of Γ are 'arrows' that may have different sources and targets. Arrows can be composed only if they satisfy the extra condition of being in tandem, tip-to-tail fashion. Moreover, rather than a single unit element, each element X of the set of objects M carries its own (unique) unit $\epsilon(X)$, an arrow in the form of a loop, so to speak.

Recall that a *Lie algebra* is a vector space endowed with an antisymmetric binary operation called a *Lie bracket*. Lie algebras are defined independently from groups, but the fundamental work of Sophus Lie (1842–1899) demonstrated the intimate connection that exists between Lie algebras and Lie groups, that is, groups that are also manifolds in which the operations of multiplication and inversion are smooth.

The Lie algebra of a Lie group represents an infinitesimal version of the latter in a precise sense. Its underlying vector space can be identified with the tangent space of the Lie group at the unit element. The vehicle to this identification is provided by the notion of left- (or right-) invariant vector fields on the Lie group. Similarly, the concept of a *Lie algebroid* can be introduced independently and eventually related to the notion of Lie groupoid. As an infinitesimal version of the latter, however, it involves certain tangent spaces to the groupoid Γ at each of its unit elements. Again, these notions are intermediated via left- (or right-) invariant vector fields on the groupoid. As everything else pertaining to groupoids, these notions acquire a further degree of sophistication as compared with their group counterparts. Although certainly premature for this introduction, we take the liberty of depicting, in Fig. 2, a schematic drawing that may serve as an intuitive basis for a mental representation of the concepts that will be advanced below in a more precise fashion.

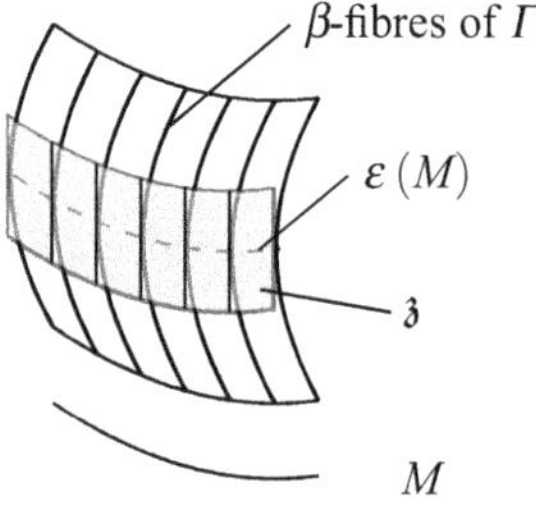

Fig. 2 A preliminary mental picture of the Lie algebroid $\mathfrak{z}$ of a Lie groupoid Γ (right) as compared with the Lie algebra $\mathfrak{g}$ of a Lie group $\mathcal{G}$ (left). The β-fibre at $X \in M$ is the collection of all the arrows arriving at X

3.2 Definition

Let $\pi : A \to M$ denote a *vector bundle* over a *base manifold* M, and let $\Gamma(A)$ denote the space of its smooth global sections $\sigma : M \to A$. A *Lie algebroid* structure on this vector bundle is obtained by specifying a bilinear *bracket* operation $[\cdot, \cdot] : \Gamma(A) \times \Gamma(A) \to \Gamma(A)$ and a vector-bundle morphism $\sharp : A \to TM$, called the *anchor map*. These maps must satisfy the following properties:

1. Skew-symmetry:

$$[\rho, \sigma] = -[\sigma, \rho] \qquad \forall \rho, \sigma \in \Gamma(A). \tag{5}$$

2. Jacobi identity:

$$[\rho, [\sigma, \tau]] + [\tau, [\rho, \sigma]] + [\sigma, [\tau, \rho]] = \mathbf{0} \qquad \forall \rho, \sigma, \tau \in \Gamma(A). \tag{6}$$

3. Consistency:

$$[\rho, \sigma]^{\sharp} = \left[\rho^{\sharp}, \sigma^{\sharp}\right] \qquad \forall \rho, \sigma \in \Gamma(A). \tag{7}$$

4. Leibniz rule:

$$[\rho, f\sigma] = f[\rho, \sigma] + \rho^{\sharp}(f)\,\sigma \qquad \forall \rho, \sigma \in \Gamma(A), f \in \mathcal{C}^{\infty}(M). \tag{8}$$

Remark 3 The first two properties are self-explanatory. The third property can be shown to be a consequence of the other ones. For compactness of notation, we have indicated by $\rho^{\sharp}$ the image $\sharp(\rho) \in \Gamma(TM)$ of the section $\rho \in \Gamma(A)$. Moreover, the bracket appearing on the right-hand side of (7) is the ordinary *Lie bracket* of vector fields in TM. The fourth property requires some further clarification, as it displays the reason behind the need for an anchor map. In an arbitrary vector bundle, there is in principle no canonical action of the vectors in the bundle on a smooth real-valued function $f \in \mathcal{C}^{\infty}(M)$ defined on the base manifold M. It is only in the *tangent bundle* TM that such an action exists, providing us with the *directional derivative* $\mathbf{v}(f)$ of f in the direction of $\mathbf{v} \in TM$.

A Lie algebroid is *transitive* if the anchor map $\sharp$ is surjective. It is *totally intransitive* if the anchor is the zero map (assigning to each vector in A the zero tangent vector at the corresponding point of the base manifold). The reason for this terminology will become apparent later.

3.3 The Lie Algebroid of a Lie Groupoid

3.3.1 The β-Bundle

Consider the disjoint union Γ^M of all the β-fibres of a Lie groupoid Γ, that is,

$$\Gamma^M = \bigcup_{X \in M} \Gamma^X. \tag{9}$$

This set, which we call the β-bundle, can be regarded as a fibre bundle over the base manifold M with projection β. In terms of arrows, Γ^M looks like a spider colony, each fibre Γ^X being a spider with legs arriving at X and issuing from some point $Y \in M$, as shown schematically in Fig. 3 [10]. Notice that the total set of this fibre bundle is the same as the total set of the original transitive groupoid Γ. They both consist of the set of all arrows.

3.3.2 Left-Invariant Vector Fields on a Lie Groupoid

Remember that, in any groupoid $\Gamma \rightrightarrows M$ we can define the concept of left translation. In fact, for each $g, h \in \Gamma$ such that $\beta(h) = \alpha(g)$, the left translation of h by g is given by

$$L_g(h) = gh. \tag{10}$$

A vector field $\Theta : \Gamma \to T\Gamma$ on Γ is *left-invariant* if

$$TL_g(\Theta(h)) = \Theta\left(L_g(h)\right) \qquad \forall g, h \in \Gamma. \tag{11}$$

Of necessity, a left-invariant vector field must be β-*vertical*, that is, it must dwell on the tangent spaces of the β-fibres of $\Gamma \rightrightarrows M$.

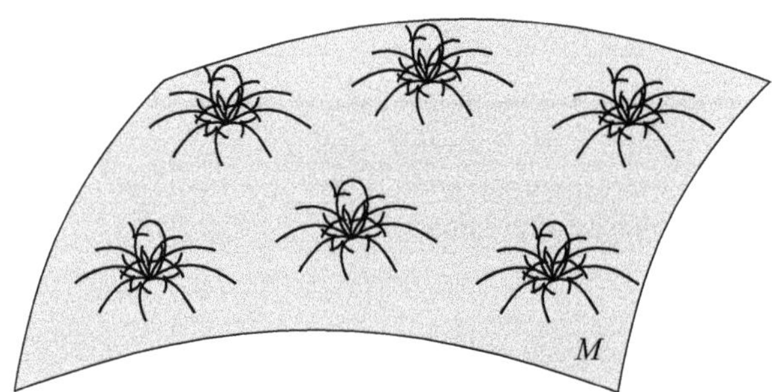

Fig. 3 The β-bundle Γ^M as a spider colony

A left-invariant vector field on a Lie groupoid is completely determined by its values at the unit elements $\epsilon(X)$, for all $X \in M$. Indeed, Eq. (11), we obtain

$$TL_g(\Theta(\epsilon \circ \alpha(g))) = \Theta\left(L_g(\epsilon \circ \alpha(g))\right) = \Theta(g(\epsilon \circ \alpha(g))) = \Theta(g). \quad (12)$$

Recall that, given any smooth vector field Θ on a manifold M, the fundamental theorem of the theory of ODEs guarantees the existence and uniqueness of maximal smooth *integral curves* defined at each point of M. If $\gamma_x = \gamma_x(t)$ is the integral curve containing the point $x \in M$, the curve parameter can be adjusted by a mere translation such that $\gamma_x(0) = x$. By definition of integral curve, we have

$$\Theta(x) = \left.\frac{d\gamma_x(t)}{dt}\right|_{t=0}. \quad (13)$$

Moreover, every smooth vector field acts as the *infinitesimal generator* of a *local flow* φ_t^Θ. For each t in a certain interval of $\mathbb{R}$ containing the origin, φ_t^Θ is a diffeomorphism of M defined by the prescription

$$\varphi_t^\Theta(x) = \gamma_x(t). \quad (14)$$

Clearly, $\varphi_0^\Theta = Id_M$ and $\varphi_{-t}^\Theta = \left(\varphi_t^\Theta\right)^{-1}$. Applying these concepts to Eq. (11) we obtain the result of the following lemma:

Lemma 1 *A (β-vertical) vector field Θ on a Lie groupoid is left-invariant if, and only if, its local flow commutes with left translations, that is,*

$$L_g \circ \varphi^\Theta(h) = \varphi_t^\Theta \circ L_g(h), \quad (15)$$

for all g, h such that $\alpha(g) = \beta(h)$.

3.3.3 The Associated Lie Algebroid

After the foregoing properties of left-invariant vector fields on a Lie groupoid $\Gamma \rightrightarrows M$ have been established, we introduce the vector bundle $\pi : A\Gamma \to M$ whose fibre at each $X \in M$ is the tangent space to the β-fibre of Γ at the identity $\epsilon(X)$. It is clear that a section of this vector bundle corresponds exactly to a left-invariant vector field on Γ. Since each map L_g is a diffeomorphism between two β-fibres, and since the left-invariant vector fields are tangent to these fibres, it follows from Lemma 1 that the ordinary Lie bracket between two left-invariant vector fields is again left-invariant. Therefore, given two sections of $A\Gamma$, we can define a Lie algebroid bracket operation by considering the Lie bracket of the corresponding vector fields in Γ and then considering its value at the unit section.

To complete the determination of the Lie algebroid associated with the Lie groupoid Γ, we declare the anchor map to be given by the restriction $A\alpha$ of

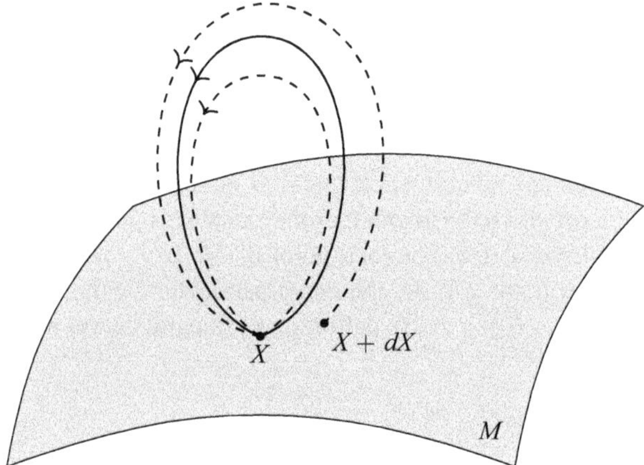

Fig. 4 Intuitive view of the Lie algebroid of a Lie groupoid. The identity $\epsilon\,(X)$ is drawn as a solid arrow, while elements in its vicinity are drawn as dashed arrows. The anchor map assigns to each dashed arrow the Ming between its tail and X

$T\alpha : T\Gamma \to TM$ to $A\Gamma$. An intuitive idea of the anchor map (and of the meaning of the Lie algebroid) can be gathered from Fig. 4. Starting from the identity loop-like arrow at a point $X \in M$, we explore its vicinity in Γ by keeping the tip of the arrow fixed at X, so as to stay always in the same β-fibre Γ^X. If we keep the tail of the arrow also at X (that is, if we explore just the loop-like neighbours), we are clearly moving within the vertex group at X. As a result, we recover the Lie algebra of this vertex group. In the case of the material groupoid $\Omega\,(\mathcal{B})$ introduced above, we obtain the Lie algebra of the *material symmetry group* at X.

Let us further explore the vicinity of the unit $\epsilon\,(X)$ by considering an arrow h with its tip at X, but with its tail elsewhere, at say $X+dX$. The differential projection $\alpha\,(h) - \alpha\,(\epsilon\,(X))$ is precisely dX. Thus, intuitively enough, we see how the map $A\alpha$ acts as the anchor of the algebroid. We see, moreover, that the Lie algebra of the vertex group at X is precisely the kernel of the anchor at X. Finally, if $A\alpha$ is a surjective map, there are arrows between X and every point in an M neighbourhood of X in M. This picture perfectly justifies the terminology introduced above for transitive and totally intransitive algebroids. In the case of the material groupoid $\Omega\,(\mathcal{B})$ we conclude that a smoothly uniform body has a transitive *material Lie algebroid.*

Example 6 A Lie algebra is a Lie algebroid.

Example 7 Let M be a manifold. The tangent $TM \rightrightarrows M$ is a Lie algebroid.

4 Characteristic Distribution

This section is devoted to construct the so-called *characteristic distribution*. This object arises from the need of working with a groupoid which does not have a structure of Lie groupoid. In fact, this object endows the groupoid with a kind of 'differentiable' structure. For a detailed study of the characteristic distribution, see [13].

Let $\Gamma \rightrightarrows M$ be a Lie groupoid and $\overline{\Gamma}$ be a subgroupoid of Γ (not necessarily a Lie subgroupoid of Γ) over the same manifold M. We will denote by $\overline{\alpha}$, $\overline{\beta}$, $\overline{\epsilon}$ and $\overline{i}$ the restrictions of the structure maps of Γ to $\overline{\Gamma}$ (see the diagram below),

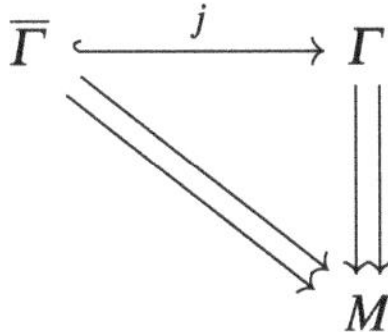

where j is the inclusion map. Now, we can construct a distribution $A\overline{\Gamma}^T$ over the manifold Γ in the following way:

$$g \in \Gamma \mapsto A\overline{\Gamma}^T_g \leq T_g\Gamma,$$

such that $A\overline{\Gamma}^T_g$ is generated by the (local) left-invariant vector fields $\Theta \in \mathfrak{X}_{loc}(\Gamma)$ whose flow at the identities is totally contained in $\overline{\Gamma}$, i.e.,

(i) Θ is tangent to the β-fibres,

$$\Theta(g) \in T_g\Gamma^{\beta(g)},$$

for all g in the domain of Θ.

(ii) Θ is invariant by left translations,

$$\Theta(g) = T_{\epsilon(\alpha(g))}L_g\left(\Theta\left(\epsilon\left(\alpha\left(g\right)\right)\right)\right),$$

for all g in the domain of Θ.

(iii) The (local) flow φ_t^{Θ} of Θ satisfies

$$\varphi_t^{\Theta}\left(\epsilon\left(X\right)\right) \in \overline{\Gamma},$$

for all $X \in M$.

Notice that, the set of local vector fields on Γ satisfying (i), (ii) and (iii) is not empty. In fact, the zero vector field fulfils these conditions. It is remarkable that condition (iii) is equivalent to the following:

(iii)′ The (local) flow φ_t^{Θ} of Θ at $\overline{g}$ is totally contained in $\overline{\Gamma}$, for all $\overline{g} \in \overline{\Gamma}$.

Then, roughly speaking, $A\overline{\Gamma}^T$ is generated by the left-invariant vector fields whose flows cannot cross $\overline{\Gamma}$. The distribution $A\overline{\Gamma}^T$ is called the *characteristic distribution of* $\overline{\Gamma}$.

For the sake of simplicity, we will denote the family of the vector fields which satisfy conditions (i), (ii) and (iii) by $\mathcal{C}$. The elements of $\mathcal{C}$ will be called *admissible vector fields*.

By using the structure of groupoid of Γ and $\overline{\Gamma}$, we can construct a smooth distribution $A\overline{\Gamma}^{\sharp}$ on M and a generalized vector bundle $A\overline{\Gamma}$ such that for each $X \in M$, the fibres are defined by

$$A\overline{\Gamma}_X = A\overline{\Gamma}^T_{\epsilon(X)}$$

$$A\overline{\Gamma}^{\sharp}_X = T_{\epsilon(X)}\alpha\left(A\overline{\Gamma}_X\right).$$

Therefore, the following diagram is commutative:

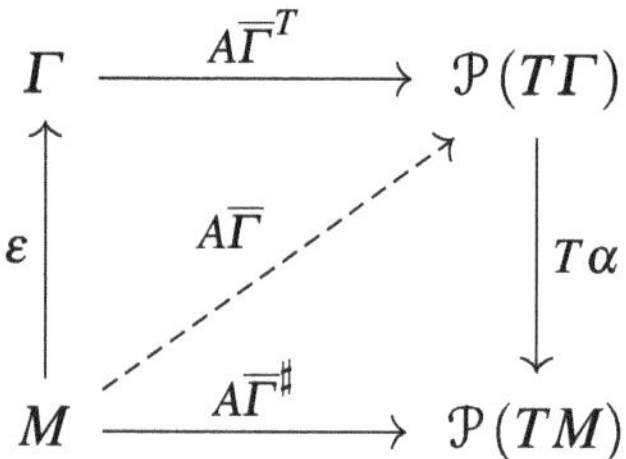

where $\mathcal{P}(E)$ defines the power set of E.

The distribution $A\overline{\Gamma}^{\sharp}$ is called *base-characteristic distribution of* $\overline{\Gamma}$. It is remarkable that both distributions are characterized by $A\overline{\Gamma}$ in the following way:

$$A\overline{\Gamma}^T_g = T_{\epsilon(\alpha(g))}L_g\left(A\overline{\Gamma}_{\alpha(g)}\right), \ \forall g \in \Gamma.$$

Summarizing, associated with $\overline{\Gamma}$, we have three mathematical objects $A\overline{\Gamma}$, $A\overline{\Gamma}^T$ and $A\overline{\Gamma}^{\sharp}$. Next, let us describe the importance of these objects.

Consider a left-invariant vector field Θ on Γ whose (local) flow φ_t^{Θ} at the identities is contained in $\overline{\Gamma}$. Then, the characteristic distribution $A\overline{\Gamma}^T$ is invariant by the flow φ_t^{Θ}, i.e., for all $g \in \Gamma$ and t in the domain of φ_g^{Θ} we have

$$T_g\varphi_t^{\Theta}\left(A\overline{\Gamma}^T_g\right) = A\overline{\Gamma}^T_{\varphi_t^{\Theta}(g)}. \tag{16}$$

Consider an arbitrary $v_g = \Upsilon(g) \in A\overline{\Gamma}^T_g$ with $\Upsilon \in \mathcal{C}$. Then,

$$T_g\varphi_t^{\Theta}\left(v_g\right) = T_g\varphi_t^{\Theta}\left(\Upsilon\left(g\right)\right) = \frac{\partial}{\partial s_{|0}}\left(\varphi_t^{\Theta} \circ \varphi_s^{\Upsilon}\left(g\right)\right),$$

where φ_s^{Υ} is the flow of Υ.
Let us consider the (local) vector field on Γ given by

$$\Xi\left(h\right) = \{\left(\varphi_t^{\Theta}\right)^* \Upsilon\}\left(h\right) = T_{\varphi_{-t}^{\Theta}(h)}\varphi_t^{\Theta}\left(\Upsilon\left(\varphi_{-t}^{\Theta}\left(h\right)\right)\right).$$

Obviously, $\Xi \in \mathfrak{C}$ (the flow of Ξ is given by $\varphi_t^{\Theta} \circ \varphi_s^{\Upsilon} \circ \varphi_{-t}^{\Theta}$). Furthermore,

$$T_g\varphi_t^{\Theta}\left(v_g\right) = \Xi\left(\varphi_{-t}^{\Theta}\left(g\right)\right).$$

So, $T_g\varphi_t^{\Theta}\left(A\overline{\Gamma}^T_g\right) \subseteq A\overline{\Gamma}^T_{\varphi_t^{\Theta}(g)}$. The other content is proved in an analogous way.

Thus, $A\overline{\Gamma}^T$ is invariant by the generating family of vector fields $\mathfrak{C}$. Now, let us recall a classical theorem, due to Stefan [17] and Sussmann [18], which characterizes the integrability of singular distributions.

Theorem 1 (Stefan–Sussmann) *Let D be a smooth singular distribution on a smooth manifold M. Then the following three conditions are equivalent:*

(a) D is integrable.
(b) D is generated by a family C of smooth vector fields, and is invariant with respect to C.
(c) D is the tangent distribution $D^{\mathfrak{F}}$ of a smooth singular foliation $\mathfrak{F}$.

Hence, there exists a foliation $\overline{\mathfrak{F}}$ on Γ which integrates $A\overline{\Gamma}^T$, i.e., at each point $g \in \Gamma$ the leaf $\overline{\mathfrak{F}}\left(g\right)$ at g satisfies that

$$T_g\overline{\mathfrak{F}}\left(g\right) = A\overline{\Gamma}^T_g.$$

The set of the leaves of $\overline{\mathfrak{F}}$ at points in $\overline{\Gamma}$ is called the *characteristic foliation of* $\overline{\Gamma}$. Note that the characteristic foliation of $\overline{\Gamma}$ does not define a foliation on $\overline{\Gamma}$ because $\overline{\Gamma}$ is not necessarily a manifold.
We already have the following result:

Theorem 2 *Let $\Gamma \rightrightarrows M$ be a Lie groupoid and $\overline{\Gamma}$ be a subgroupoid of Γ (not necessarily a Lie groupoid) over M. Then, there exists a foliation $\overline{\mathfrak{F}}$ of Γ such that $\overline{\Gamma}$ is a union of leaves of $\overline{\mathfrak{F}}$.*

In this way, $\overline{\Gamma}$, which is not a manifold, has some kind of 'differentiable' structure via the foliation $\overline{\mathfrak{F}}$.
Let us highlight the following assertions:

(i) For each $\overline{g} \in \overline{\Gamma}$, then

$$\overline{\mathfrak{F}}\left(g\right) \subseteq \overline{\Gamma}^{\beta(g)}.$$

(ii) For each $g, h \in \Gamma$ such that $\alpha(g) = \beta(h)$, we have

$$\overline{\mathcal{F}}(g \cdot h) = g \cdot \overline{\mathcal{F}}(h).$$

(iii) Let $\Theta \in \mathfrak{X}_{loc}(\Gamma)$ be a left-invariant vector field on Γ. Then, $\Theta \in \mathcal{C}$ if, and only if,

$$\Theta_{|\overline{\mathcal{F}}(g)} \in \mathfrak{X}\left(\overline{\mathcal{F}}(g)\right), \tag{17}$$

for all g in the domain of Θ

The construction of the characteristic distribution imposes some condition of maximality.

Corollary 1 *Let $\overline{\mathcal{G}}$ be a left-invariant foliation of Γ such that $\overline{\Gamma}$ is a union of leaves of $\overline{\mathcal{G}}$. Then, the characteristic foliation $\overline{\mathcal{F}}$ is coarser than $\overline{\mathcal{G}}$, i.e.,*

$$\overline{\mathcal{G}}(g) \subseteq \overline{\mathcal{F}}(g), \ \forall g \in \Gamma.$$

Proof *Let $\mathcal{D}$ be the family of (local) vector fields tangent to the foliation $\overline{\mathcal{G}}$. Then, restricting to the identities and generating by left invariance we obtain a new family of (local) vector fields generating the tangent distribution to the foliation $\overline{\mathcal{G}}$. In fact, this family is obviously a subset of $\mathcal{C}$ (the family of admissible vector fields).*

Particularly, $\overline{\Gamma}^X$ *is a submanifold of Γ for all $X \in M$ if, and only if, $\overline{\Gamma}^X = \overline{\mathcal{F}}(\epsilon(X))$ for all $X \in M$.*

Analogously, the base-characteristic distribution $A\overline{\Gamma}^\sharp$ is integrable. Its associated foliation $\mathcal{F}$ of M will be called the *base-characteristic foliation of $\overline{\Gamma}$.*

Let us apply these results to a particular example. Let M be a manifold and $M \times M$ the pair groupoid (Example 2). Then, any transitive subgroupoid of M is the pair groupoid $N \times N$ of a subset $N \subseteq M$. Then, using Theorem 2 we have the following result:

Theorem 3 *Let M be a manifold and N be a subset of M. Then, there exists a maximal foliation $\mathcal{F}$ of M such that N is union of leaves.*

Proof Let $N \times N \rightrightarrows N$ be the transitive pair groupoid of N. We may consider a (generally intransitive) subgroupoid $[(M - N) \times (M - N)] \sqcup [N \times N] \rightrightarrows M$ of $M \times M \rightrightarrows M$, where $M - N$ is the subset of M consisting of the points at M outside N. So, we may consider $\overline{\mathcal{F}}$ and $\mathcal{F}$ its characteristic foliation and base-characteristic foliation, respectively.

Then, for each $X \in N$ we have that

$$\overline{\mathcal{F}}(X, X) \subseteq N \times \{X\}.$$

In fact, it satisfies that

$$\overline{\mathcal{F}}(X, X) = \mathcal{F}(X) \times \{X\}. \tag{18}$$

Hence, N is the union of the leaves of the base-characteristic foliation at points of N and we already have our foliation. Let us now study the condition of maximality of the foliation.
Let $\mathcal{G}$ be another foliation of M such that N is union of leaves. Then, for each $(X, Y) \in M \times M$ we may define

$$\overline{\mathcal{G}}(X, Y) = \mathcal{G}(X) \times \{Y\}.$$

Then, the family $\overline{\mathcal{G}} = \{\overline{\mathcal{G}}(X, Y)\}_{(X,Y) \in M \times M}$ defines a left-invariant foliation of $M \times M$ such that $[(M - N) \times (M - N)]$ is union of leaves. Thus, the maximality condition of the characteristic foliation (Corollary 1) ends the proof.

Notice that the maximal foliation given in Theorem 3 permits us to endow N with differential structure which generalizes the structure of manifold. Indeed, N is a submanifold of M if, and only if, N consists of just one leaf of the foliation.
Let Θ be an admissible vector field of the subgroupoid $[(M - N) \times (M - N)] \sqcup [N \times N] \rightrightarrows M$. Then, the projection

$$\theta = T\alpha \circ \Theta \circ \epsilon$$

on M is a vector field on M such that its flow at point of N is confined in N. Conversely, any vector fields θ whose flows at point of N is inside N may be lifted to an admissible vector field Θ by imposing that

$$\Theta(X, Y) = (\theta(X), 0) \in T_X M \times T_Y M, \ \forall X, Y \in M. \tag{19}$$

Thus, the foliation given in the Theorem 3 can be described by the vector fields on M whose flow at points of N is contained in N.

Example 8 Let $\sim$ be an equivalence relation on a manifold M, i.e., a binary relation that is reflexive, symmetric and transitive. Then, define the subset $\mathcal{O}$ of $M \times M$ given by

$$\mathcal{O} := \{(X, Y) \ : \ X \sim Y\}. \tag{20}$$

Then, $\mathcal{O}$ is a subgroupoid of $M \times M$ over M. In fact, this fact is equivalent to the properties of being reflexive, symmetric and transitive. For each $X \in M$, we denote by $\mathcal{O}_X$ to the orbit around X,

$$\mathcal{O}_X := \{Y \ : \ X \sim Y\}.$$

Notice that the orbits divide M into a disjoint union of subsets. However, these are not (necessarily) submanifolds.
On the other hand, the base-characteristic foliation gives us a foliation $\mathcal{F}$ of M such that

$$\mathcal{F}(X) \subseteq \mathcal{O}_X, \ \forall X \in M.$$

This foliation is maximal in the sense that there is no any other coarser foliation of M whose leaves are contained in the orbits (see Theorem 4 and Corollary 4).

Another example give rise to the so-called *material distributions*. This example will be presented in the next section.
Next, the groupoid structure is used to endow the leaves of $\mathcal{F}(X)$ with a Lie groupoid structure. First, by using the foliated atlas associated with $\mathcal{F}$ and $\overline{\mathcal{F}}$, we can prove the following result:

Proposition 1 *Let $\Gamma \rightrightarrows M$ be a Lie groupoid and $\overline{\Gamma}$ be a subgroupoid of Γ with $\overline{\mathcal{F}}$ and $\mathcal{F}$ the characteristic foliation and the base-characteristic foliation of $\overline{\Gamma}$, respectively. Then, for all $X \in M$, the mapping*

$$\alpha_{|\overline{\mathcal{F}}(\epsilon(X))} : \overline{\mathcal{F}}(\epsilon(X)) \to \mathcal{F}(X),$$

is a surjective submersion.

As an interesting consequence we have the next corollary.

Corollary 2 *Let $\Gamma \rightrightarrows M$ be a Lie groupoid and $\overline{\Gamma}$ be a subgroupoid of Γ. Then, the manifolds $\overline{\mathcal{F}}(\epsilon(X)) \cap \Gamma_X$ are Lie subgroups of Γ_X^X for all $X \in M$.*

Another interesting consequence is that we can improve Corollary 1

Corollary 3 *Let $\overline{\mathcal{G}}$ be a foliation of Γ such that $\overline{\Gamma}$ is a union of leaves of $\overline{\mathcal{G}}$ and*

$$\overline{\mathcal{G}}(g) \subseteq \Gamma^{\beta(g)}, \ \forall g \in \Gamma.$$

Then, the characteristic foliation $\overline{\mathcal{F}}$ is coarser than $\overline{\mathcal{G}}$, i.e.,

$$\overline{\mathcal{G}}(g) \subseteq \overline{\mathcal{F}}(g), \ \forall g \in \Gamma.$$

Proof Let us consider $\mathcal{D}$ as the family of (local) vector fields tangent to the foliation $\overline{\mathcal{G}}$. Fix $g \in \Gamma$ and $v_g \in T_g\overline{\mathcal{G}}(g)$. We may assume that there exists $\Theta \in \mathcal{D}$ such that

$$\Theta(g) = v_g. \tag{21}$$

By using Proposition 1, we may have a local section σ_g of $\alpha_{|g\cdot\overline{\mathcal{F}}(\epsilon(\alpha(g)))} : g \cdot \overline{\mathcal{F}}(\epsilon(\alpha(g))) \to \mathcal{F}(\alpha(g))$ with $\sigma_g(\alpha(g)) = g$. So, we will define the following (local) left-invariant vector field Υ^{σ_g} on $g \cdot \overline{\mathcal{F}}(\epsilon(\alpha(g)))$ characterized by

$$\Upsilon^{\sigma_g}(\epsilon(Y)) = T_{\sigma_g(Y)}L_{\sigma_g(Y)^{-1}}\left(\Theta\left(\sigma_g(Y)\right)\right). \tag{22}$$

Thus, the flow of Υ^{σ_g} is given by

$$\varphi_t^{\Upsilon^{\sigma_g}}(h) = h \cdot \left(\sigma_g(\alpha(h))^{-1}\right) \cdot \varphi_t^{\Theta}\left(\sigma_g(\alpha(h))\right).$$

Hence, Υ^{σ_g} generates an admissible vector field. Furthermore,

$$\Upsilon^{\sigma_g}(g) = \Theta(g) = v_g,$$

i.e., $v_g \in A\overline{\Gamma}^T_g$.

Notice that, taking into account this result, we may "relax" conditions of the family of admissible vector fields. In fact, the characteristic distribution is generated by the (local) vector fields $\Theta \in \mathfrak{X}_{loc}(\Gamma)$ such that

(i) Θ is tangent to the β-fibres,

$$\Theta(g) \in T_g\Gamma^{\beta(g)},$$

for all g in the domain of Θ.

(ii) The (local) flow φ_t^{Θ} of Θ satisfies

$$\varphi_t^{\Theta}(\overline{g}) \in \overline{\Gamma},$$

for all $\overline{g} \in \overline{\Gamma}$.

Let us now construct an algebraic structure of a groupoid over the leaves of $\mathcal{F}$. We will consider the groupoid $\overline{\Gamma}(\mathcal{F}(X))$ generated by $\overline{\mathcal{F}}(\epsilon(X))$ by imposing that for all $\overline{g}, \overline{h} \in \overline{\mathcal{F}}(\epsilon(X))$,

$$\overline{g}, \overline{g}^{-1}, \overline{h}^{-1} \cdot \overline{g} \in \overline{\Gamma}(\mathcal{F}(X)).$$

Notice that,

$$\overline{\mathcal{F}}(\epsilon(X)) = \overline{\mathcal{F}}\left(\overline{h}\right) = \overline{h} \cdot \overline{\mathcal{F}}\left(\epsilon\left(\alpha\left(\overline{h}\right)\right)\right).$$

Therefore,

$$\overline{\mathcal{F}}\left(\overline{h}^{-1}\right) = \overline{h}^{-1} \cdot \overline{\mathcal{F}}(\epsilon(X)) = \overline{\mathcal{F}}\left(\epsilon\left(\alpha\left(\overline{h}\right)\right)\right).$$

On the other hand, let be $\overline{t} \in \overline{\mathcal{F}}\left(\epsilon\left(\alpha\left(\overline{h}\right)\right)\right)$. Then,

$$\overline{\mathcal{F}}\left(\overline{h} \cdot \overline{t}\right) = \overline{h} \cdot \overline{\mathcal{F}}\left(\overline{t}\right) = \overline{h} \cdot \overline{\mathcal{F}}\left(\epsilon\left(\alpha\left(\overline{h}\right)\right)\right) = \overline{\mathcal{F}}(\epsilon(X)).$$

i.e., $\overline{h} \cdot \overline{t} \in \overline{\mathcal{F}}(\epsilon(X))$ and, hence, $\overline{t}$ can be written as $\overline{h}^{-1} \cdot \overline{g}$ with $\overline{g} \in \overline{\mathcal{F}}(\epsilon(X))$. Thus, we have proved that

$$\overline{\mathcal{F}}\left(\epsilon\left(\alpha\left(\overline{h}\right)\right)\right) \subset \overline{\Gamma}(\mathcal{F}(X)),$$

for all $\overline{h} \in \overline{\mathcal{F}}(\epsilon(X))$. In fact, by following the same argument we have that

$$\overline{\Gamma}(\mathcal{F}(X)) = \sqcup_{\overline{g} \in \overline{\mathcal{F}}(\epsilon(X))} \overline{\mathcal{F}}(\epsilon(\alpha(\overline{g}))), \tag{23}$$

i.e., $\overline{\Gamma}(\mathcal{F}(X))$ can be depicted as a disjoint union of fibres at the identities.

Let us now show that $\overline{\Gamma}(\mathcal{F}(X))$ is, in fact, a subgroupoid of $\overline{\Gamma}$. Consider two arbitrary elements $\overline{g}, \overline{h} \in \overline{\Gamma}(\mathcal{F}(X))$ with $\alpha\left(\overline{h}\right) = \beta(\overline{g})$. Then, we may assume that we are in one of the following options:

(i) $\overline{g}, \overline{h} \in \overline{\mathcal{F}}(\epsilon(X))$. Then,

$$\overline{\mathcal{F}}\left(\overline{h} \cdot \overline{g}\right) = \overline{h} \cdot \overline{\mathcal{F}}(\overline{g}) = \overline{h} \cdot \overline{\mathcal{F}}(\epsilon(X)) = \overline{\mathcal{F}}\left(\overline{h}\right) = \overline{\mathcal{F}}(\epsilon(X)).$$

i.e., $\overline{h} \cdot \overline{g} \in \overline{\mathcal{F}}(\epsilon(X)) \subset \overline{\Gamma}(\mathcal{F}(X))$.

(ii) $\overline{g}^{-1}, \overline{h} \in \overline{\mathcal{F}}(\epsilon(X))$. Then,

$$\overline{\mathcal{F}}\left(\overline{h} \cdot \overline{g}\right) = \overline{h} \cdot \overline{\mathcal{F}}(\overline{g}) = \overline{h} \cdot \overline{\mathcal{F}}(\epsilon(\beta(\overline{g}))) = \overline{\mathcal{F}}\left(\overline{h}\right) = \overline{\mathcal{F}}(\epsilon(X)).$$

So, $\overline{h} \cdot \overline{g} \in \overline{\mathcal{F}}(\epsilon(X)) \subset \overline{\Gamma}(\mathcal{F}(X))$.

(iii) $\overline{g}, \overline{h}^{-1} \in \overline{\mathcal{F}}(\epsilon(X))$,

$$\overline{\mathcal{F}}\left(\overline{h} \cdot \overline{g}\right) = \overline{h} \cdot \overline{\mathcal{F}}(\overline{g}) = \overline{h} \cdot \overline{\mathcal{F}}(\epsilon(X)) = \overline{\mathcal{F}}\left(\overline{h}\right) = \overline{\mathcal{F}}\left(\epsilon\left(\beta\left(\overline{h}\right)\right)\right).$$

Hence, $\overline{h} \cdot \overline{g} \in \overline{\mathcal{F}}\left(\epsilon\left(\beta\left(\overline{h}\right)\right)\right) \subset \overline{\Gamma}(\mathcal{F}(X))$ (see Eq. (23)).

It is important to note that $\overline{\Gamma}(\mathcal{F}(X))$ may be equivalently defined as the smallest transitive subgroupoid of $\overline{\Gamma}$ which contains $\overline{\mathcal{F}}(\epsilon(X))$. Observe that the β-fibre of this groupoid at a point $Y \in \mathcal{F}(X)$ is given by $\overline{\mathcal{F}}(\epsilon(Y))$. Hence, the α-fibre at Y is

$$\overline{\mathcal{F}}^{-1}(\epsilon(Y)) = i \circ \overline{\mathcal{F}}(\epsilon(Y)).$$

Furthermore, the Lie groups $\overline{\mathcal{F}}(\epsilon(Y)) \cap \Gamma_Y$ are exactly the isotropy groups of $\overline{\Gamma}(\mathcal{F}(X))$. All these results imply the following one ([13]):

Theorem 4 *For each $X \in M$ there exists a transitive Lie subgroupoid $\overline{\Gamma}(\mathcal{F}(X))$ of Γ with base $\mathcal{F}(X)$.*

Proof Let be $\overline{g} \in \overline{\Gamma}(\mathcal{F}(X))$. Then, by Proposition 1, the restriction

$$\beta_{|\overline{\mathcal{F}}^{-1}(\overline{g})} : \overline{\mathcal{F}}^{-1}\left(\overline{g}^{-1}\right) \to \mathcal{F}(X), \tag{24}$$

is a surjective submersion, where $\overline{\mathcal{F}}^{-1}\left(\overline{g}^{-1}\right) = i \circ \overline{\mathcal{F}}\left(\overline{g}^{-1}\right)$. Using this fact, we will endow with a differentiable structure to $\overline{\Gamma}(\mathcal{F}(X))$. Let be $\overline{g} \in \overline{\Gamma}(\mathcal{F}(X))$. Consider $\sigma_{\overline{g}} : U \to \overline{\mathcal{F}}^{-1}\left(\overline{g}^{-1}\right)$ a (local) section of $\beta_{|\overline{\mathcal{F}}^{-1}\left(\overline{g}^{-1}\right)}$ such that $\sigma_{\overline{g}}(\beta(\overline{g})) = \overline{g}$.

On the other hand, let $\{\Theta_i\}_{i=1}^r$ be a finite collection of vector fields in $\mathcal{C}$ such that $\{\Theta^i(\epsilon(\alpha(\overline{g})))\}_{i=1}^r$ is a basis of $A\overline{\Gamma}^T_{\epsilon(\alpha(\overline{g}))}$. Then, a local chart over $\overline{g}$ can be given by

$$\begin{array}{lll}\varphi^{\Theta}: & W \times U & \to \overline{\Gamma} \subseteq \Gamma \\ & (t_1, \ldots, t_r, Z) \mapsto \sigma_{\overline{g}}(Z) \cdot [\varphi_{t_r}^{\Theta^r} \circ \cdots \circ \varphi_{t_1}^{\Theta^1}(\epsilon(\alpha(\overline{g})))]\end{array},$$

where $\varphi_t^{\Theta^i}$ is the flow of Θ^i. These local charts are enough to prove that the restrictions of the source and target maps are submersions.

So, we have divided the manifold M into leaves $\mathcal{F}(X)$ which have a maximal structure of transitive Lie subgroupoids of Γ.

Corollary 4 *Let $\mathcal{G}$ be a foliation of M such that for each $X \in M$ there exists a transitive Lie subgroupoid $\Gamma(X)$ of Γ over the leaf $\mathcal{G}(X)$ contained in $\overline{\Gamma}$. Then, the base-characteristic foliation $\mathcal{F}$ is coarser than $\mathcal{G}$, i.e.,*

$$\mathcal{F}(X) \subseteq \mathcal{G}(X), \ \forall X \in M.$$

Furthermore it satisfies that

$$\Gamma(X) \subseteq \overline{\Gamma}(\mathcal{F}(X)).$$

Proof Let $\mathcal{G}$ be a foliation of M in the condition of the corollary. Then, we consider the family of manifolds given by the β-fibres $\Gamma(X)^X$. Then, by left translations we generate a foliation of Γ into submanifolds. By using Corollary 3 we have finished.

As a particular consequence we have that: *$\overline{\Gamma}$ is a transitive Lie subgroupoid of Γ if, and only if, $M = \mathcal{F}(X)$ and $\overline{\Gamma} = \overline{\Gamma}(\mathcal{F}(X))$ for some $X \in M$.*

Remark 4 This construction of the characteristic distribution associated with a subgroupoid $\overline{\Gamma}$ of a Lie groupoid Γ generalizes the known correspondence between Lie groupoids and Lie algebroids. Indeed, $A\overline{\Gamma}$ is the associated Lie algebroid with $\overline{\Gamma}$ if $\overline{\Gamma}$ is a Lie subgroupoid of Γ.

5 Material Groupoid and Material Distribution

In this section we will apply the results of Sect. 4 to the case of continuum mechanics. First, let us fix the fundamental notions.

A *body* $\mathcal{B}$ is a 3-dimensional differentiable manifold which can be covered with just one chart. An embedding $\phi : \mathcal{B} \to \mathbb{R}^3$ is called a *configuration of* $\mathcal{B}$ and its 1-jet $j^1_{X,\phi(X)}\phi$ at $X \in \mathcal{B}$ is called an *infinitesimal configuration at* X. We usually identify the body with any one of its configurations, say ϕ_0, called *reference configuration*. Given any arbitrary configuration ϕ, the change of configurations $\kappa = \phi \circ \phi_0^{-1}$ is

called a *deformation*, and its 1-jet $j^1_{\phi_0(X),\phi(X)}\kappa$ is called an *infinitesimal deformation at* $\phi_0(X)$.
In the case of simple bodies, the mechanical response of the material is characterized by one function W which depends, at each point $X \in \mathcal{B}$, on the gradient of the deformations evaluated at the point. Thus, W is defined as a differentiable map

$$W : \Pi^1(\mathcal{B}, \mathcal{B}) \to V,$$

which does not depend on the final point with respect to the reference configuration, i.e., for all $X, Y, Z \in \mathcal{B}$

$$W\left(j^1_{X,Y}\phi\right) = W\left(j^1_{X,Z}\left(\phi_0^{-1} \circ \tau_{Z-Y} \circ \phi_0 \circ \phi\right)\right), \; \forall j^1_{X,Y}\phi \in \Pi^1(\mathcal{B}, \mathcal{B}), \qquad (25)$$

where V is a real vector space and τ_v is the translation map on $\mathbb{R}^3$ by the vector v. This map is called *mechanical response*. There are other equivalent definitions ([4, 5, 9] or [12]) of this function . We will use this definition for convenience.

Now, consider a situation in which an M neighbourhood of a point $Y \in \mathcal{B}$ is diffeomorphic to an M neighbourhood of another point $X \in \mathcal{B}$ such that the diffeomorphism cannot be detected by a mechanical experiment. Then, roughly speaking, we will say that Y and X are made of the same material. In the case when this property is satisfied for any point in $\mathcal{B}$ we will say that the body is *uniform*.

Definition 7 A body $\mathcal{B}$ is said to be *uniform* if for each two points $X, Y \in \mathcal{B}$ there exists a local diffeomorphism ψ from an M neighbourhood $U \subseteq \mathcal{B}$ of X to an M neighbourhood $V \subseteq \mathcal{B}$ of Y such that $\psi(X) = Y$ and

$$W\left(j^1_{Y,\kappa(Y)}\kappa \cdot j^1_{X,Y}\psi\right) = W\left(j^1_{Y,\kappa(Y)}\kappa\right), \qquad (26)$$

for all infinitesimal deformation $j^1_{Y,\kappa(Y)}\kappa$. The 1-jet $j^1_{X,Y}\psi$ is called a *material isomorphism*.

Let us now consider the family of all material isomorphisms denoted by $\Omega(\mathcal{B})$. It is a straightforward exercise to prove that $\Omega(\mathcal{B})$ has a natural groupoid structure by using the composition of 1-jets as the composition law of the groupoid. A material isomorphism from X to X is said to be a *material symmetry*. We will denote the structure maps of the material groupoid $\Omega(\mathcal{B})$ by $\overline{\alpha}$, $\overline{\beta}$, $\overline{\epsilon}$ and $\overline{i}$, which are, indeed, the restrictions of the corresponding ones on $\Pi^1(\mathcal{B}, \mathcal{B})$.
$\Omega(\mathcal{B})$ is a subgroupoid of the Lie groupoid of the 1-jets $\Pi^1(\mathcal{B}, \mathcal{B})$. However, $\Omega(\mathcal{B})$ is not necessarily a Lie subgroupoid of $\Pi^1(\mathcal{B}, \mathcal{B})$ (see the examples below) and, hence, we are in the conditions of Sect. 4.

Taking into account the continuity of the mechanical response W, we have that for any $X \in \mathcal{B}$ the group of material symmetries $\Omega(\mathcal{B})^X_X$ is a closed subgroup of $\Pi^1(\mathcal{B}, \mathcal{B})^X_X$. So, it follows that:

Proposition 2 *Let $\mathcal{B}$ be a simple body. Then, for all $X \in \mathcal{B}$ the set of all material symmetries $\Omega(\mathcal{B})_X^X$ is a Lie subgroup of $\Pi^1(\mathcal{B},\mathcal{B})_X^X$.*

Notice that this result does not imply that $\Omega(\mathcal{B})$ is a Lie subgroupoid of $\Pi^1(\mathcal{B},\mathcal{B})$. This is a consequence of the fact that β-fibres of $\Omega(\mathcal{B})$ could have different dimensions.

Now, let us express the uniformity as a known property of Lie groupoids.

Proposition 3 *Let $\mathcal{B}$ be a body. $\mathcal{B}$ is uniform if, and only if, $\Omega(\mathcal{B})$ is a transitive subgroupoid of $\Pi^1(\mathcal{B},\mathcal{B})$.*

Next, we will consider another (slightly more restrictive) notion of uniformity.

Definition 8 A body $\mathcal{B}$ is said to be *smoothly uniform* if for each point $X \in \mathcal{B}$ there is a neighbourhood U around X such that for all $Y \in U$ and $j^1_{Y,X}\phi \in \Omega(\mathcal{B})$ there exists a smooth field of material isomorphisms P at X from $\epsilon(X)$ to $j^1_{Y,X}\phi$.

Observe that a smooth field of material isomorphisms P at X is just a (local) differentiable section of the restriction of $\overline{\alpha}$ to $\Omega^X(\mathcal{B})$

$$\overline{\alpha}^X : \Omega^X(\mathcal{B}) \to \mathcal{B}.$$

The existence of these smooth fields of material isomorphism can be equivalently expressed by saying that $\overline{\alpha}^X$ is a surjective submersion. Immediately, we can prove that smooth uniformity implies uniformity.
It is obvious that $\mathcal{B}$ is smoothly uniform if, and only if, for each two points $X, Y \in \mathcal{B}$ there are two M neighbourhoods $U, V \subseteq \mathcal{B}$ of X and Y, respectively, and $P : U \times V \to \Omega(\mathcal{B}) \subseteq \Pi^1(\mathcal{B},\mathcal{B})$, a smooth section of the anchor map $(\overline{\alpha}, \overline{\beta})$. When $X = Y$ we can assume that $U = V$ and P is a morphism of groupoids over the identity map, i.e.,

$$P(Z,T) = P(R,T)\,P(Z,R), \ \forall\, T, R, Z \in U.$$

So, we have the following corollary of Proposition 2:

Corollary 5 *Let $\mathcal{B}$ be a body. $\mathcal{B}$ is smoothly uniform if, and only if, $\Omega(\mathcal{B})$ is a transitive Lie subgroupoid of $\Pi^1(\mathcal{B},\mathcal{B})$.*

Remark 5 Let Θ be an admissible left-invariant vector field on $\Pi^1(\mathcal{B},\mathcal{B})$ (see Sect. 4), i.e., $\varphi_t^\Theta(\epsilon(X)) \in \Omega(\mathcal{B})$ for all $X \in \mathcal{B}$. Then, for all $g \in \Pi^1(\mathcal{B},\mathcal{B})$, we have that

$$\begin{aligned} TW(\Theta(g)) &= \frac{\partial}{\partial t}\left(W\left(\varphi_t^\Theta(g)\right)\right)\Big|_0 \\ &= \frac{\partial}{\partial t}\left(W\left(g\cdot\varphi_t^\Theta(\epsilon(\alpha(g)))\right)\right)\Big|_0 \\ &= \frac{\partial}{\partial t}\left(W(g)\right)\Big|_0 = 0. \end{aligned}$$

Therefore,

$$TW\left(\Theta\right)=0. \tag{27}$$

The converse is proved in the same way.
So, the characteristic distribution $A\Omega\left(\mathcal{B}\right)^{T}$ of the material groupoid is generated by the left-invariant vector fields on $\Pi^{1}\left(\mathcal{B},\mathcal{B}\right)$ which are in the kernel of TW. This characteristic distribution will be called *material distribution*. The base-characteristic distribution $A\Omega\left(\mathcal{B}\right)^{\sharp}$ will be called *body-material distribution*. Let us recall that the left-invariant vector fields on $\Pi^{1}\left(\mathcal{B},\mathcal{B}\right)$ which satisfy Eq. (27) are called admissible vector fields and the family of these vector fields is denoted by $\mathcal{C}$.

Denote by $\overline{\mathcal{F}}\left(\epsilon\left(X\right)\right)$ and $\mathcal{F}\left(X\right)$ the foliations associated with the material distribution and the body-material distribution, respectively. For each $X\in\mathcal{B}$, we will denote the Lie groupoid $\Omega\left(\mathcal{B}\right)\left(\mathcal{F}\left(X\right)\right)$ by $\Omega\left(\mathcal{F}\left(X\right)\right)$.

Notice that, strictly speaking, in continuum mechanics a *sub-body* of a body $\mathcal{B}$ is an M submanifold of $\mathcal{B}$ but, here, the foliation $\mathcal{F}$ gives us submanifolds of different dimensions. So, we will consider a more general definition so that, a *material submanifold (or generalized sub-body) of* $\mathcal{B}$ is just a submanifold of $\mathcal{B}$. A generalized sub-body $\mathcal{P}$ inherits certain material structure from $\mathcal{B}$. In fact, we will measure the material response of a material submanifold $\mathcal{P}$ by restricting W to the 1-jets of local diffeomorphisms ϕ on $\mathcal{B}$ from $\mathcal{P}$ to $\mathcal{P}$. However, it easy to observe that a material submanifold of a body is not exactly a body. See [11] for a discussion on this subject.

Then, as a corollary of Theorem 4 and Corollary 1, we have the following result:

Theorem 5 *For all* $X\in\mathcal{B}$, $\Omega\left(\mathcal{F}\left(X\right)\right)$ *is a transitive Lie subgroupoid of* $\Pi^{1}\left(\mathcal{B},\mathcal{B}\right)$. *Thus, any body* $\mathcal{B}$ *can be covered by a maximal foliation of smoothly uniform material submanifolds.*

Notice that, in this case "maximal" means that any other foliation $\mathcal{G}$ by smoothly uniform material submanifolds is thinner than $\mathcal{F}$, i.e.,

$$\mathcal{G}\left(X\right)\subseteq\mathcal{F}\left(X\right),\ \forall X\in\mathcal{B}.$$

Remark 6 Imagine that there is, at least, a 1-jet $g\in\Omega^{X}\left(\mathcal{B}\right)$ for some $X\in\mathcal{B}$ such that

$$g\notin\overline{\mathcal{F}}\left(\epsilon\left(X\right)\right).$$

Then, we are not including g inside any of the transitive Lie subgroupoids $\Omega\left(\mathcal{F}\left(X\right)\right)$. Thus, these material isomorphisms are being discarded.
Nevertheless

$$\overline{\mathcal{F}}\left(g\right)=g\cdot\overline{\mathcal{F}}\left(\epsilon\left(\alpha\left(g\right)\right)\right), \tag{28}$$

and, indeed, $\overline{\mathcal{F}}(\epsilon(\alpha(g)))$ is contained in $\Omega(\mathcal{F}(\alpha(g)))$, i.e., using Eq. (28), we can reconstruct $\overline{\mathcal{F}}(g)$.

Finally, using the body-material distribution, we will be able to define a more general notion of smooth uniformity. This notion was introduced in [10]. We will end up using the foliation by uniform sub-bodies to interpret it over the material groupoid.

Definition 9 Let be a body $\mathcal{B}$ and a body point $X \in \mathcal{B}$. Then, $\mathcal{B}$ is said to be *uniform of grade p at X* if $A\Omega(\mathcal{B})^{\sharp}_{X}$ has dimension p. $\mathcal{B}$ is *uniform of grade p* if it is uniform of grade p at all the points.

Note that, smooth uniformity is a particular case of graded uniformity. In fact, $\mathcal{B}$ is smoothly uniform if, and only if, $\mathcal{B}$ is uniform of grade n. Equivalently, $\mathcal{B}$ is uniform of grade 3 if, and only if, $A\Omega(\mathcal{B})^{\sharp}_{X}$ has dimension 3 for all $X \in \mathcal{B}$, i.e., there exists just one leaf of the material foliation equal to $\mathcal{B}$. Hence, the material groupoid $\Omega(\mathcal{B})$ is a Lie subgroupoid of $\Pi^1(\mathcal{B}, \mathcal{B})$ whose $\overline{\beta}$-fibres integrate the material distribution.

Corollary 6 *Let be a body $\mathcal{B}$ and let $X \in \mathcal{B}$ be a body point. $\mathcal{B}$ is uniform of grade p at X if, and only if, the uniform leaf $\mathcal{F}(X)$ at X has dimension p.*

Corollary 7 *Let $\mathcal{B}$ be a body. $\mathcal{B}$ is uniform of grade p if, and only if, the body-material foliation is regular of rank p.*

It is important to highlight that the body-material foliation has certain condition of maximality. In fact, suppose that there exists another foliation $\mathcal{G}$ of $\mathcal{B}$ by smoothly uniform material submanifolds. Then, for all $X \in \mathcal{B}$ we have that

$$\mathcal{G}(X) \subseteq \mathcal{F}(X), \ \forall X \in \mathcal{B}.$$

So, we have the following results:

Corollary 8 *Let be a body $\mathcal{B}$ and let $X \in \mathcal{B}$. $\mathcal{B}$ is uniform of grade greater or equal to p at X if, and only if, there exists a foliation $\mathcal{G}$ of $\mathcal{B}$ by smoothly uniform submanifolds such that the leaf $\mathcal{G}(X)$ at X has dimension greater or equal to p.*

Corollary 9 *Let $\mathcal{B}$ be a body. $\mathcal{B}$ is uniform of grade p if, and only if, the body can be foliated by smoothly uniform material submanifolds of dimension p.*

6 Homogeneity

This section is devoted to deal with the definition of homogeneity. As we already know, a body is uniform if the function W satisfies Equation (26). In addition, a body is said to be *homogeneous* if we can choose a global section of the material groupoid which is constant on the body. More precisely:

Definition 10 A body $\mathcal{B}$ is said to be *homogeneous* if it admits a global configuration ψ which induces a global section of (α, β) in $\Omega(\mathcal{B})$, P, i.e., for each $X, Y \in \mathcal{B}$

$$P(X, Y) = j^1_{X,Y}\left(\psi^{-1} \circ \tau_{\psi(Y)-\psi(X)} \circ \psi\right),$$

where $\tau_{\psi(Y)-\psi(X)} : \mathbb{R}^3 \to \mathbb{R}^3$ denotes the translation on $\mathbb{R}^3$ by the vector $\psi(Y) - \psi(X)$. $\mathcal{B}$ is said to be *locally homogeneous* if there exists a covering of $\mathcal{B}$ by homogeneous M sets. $\mathcal{B}$ is said to be *(locally) inhomogeneous* if it is not (locally) homogeneous.

Notice that local homogeneity is clearly more restrictive than smooth uniformity. In fact, in this case, the smooth fields of material isomorphisms (see Definition 8) are induced by particular (local) configurations. However, in a purely intuitive picture, homogeneity can be interpreted as the absence of defects. So, it makes sense to develop a concept of some kind of homogeneity for non-uniform materials which measures the absence of defects and generalizes the known one. In the literature we can already find some partial answer of this question ([2, 8] for FGM's and [7, 10] for laminated and bundle materials).

Recall that the material distributions are characterized by the commutativity of the following diagram:

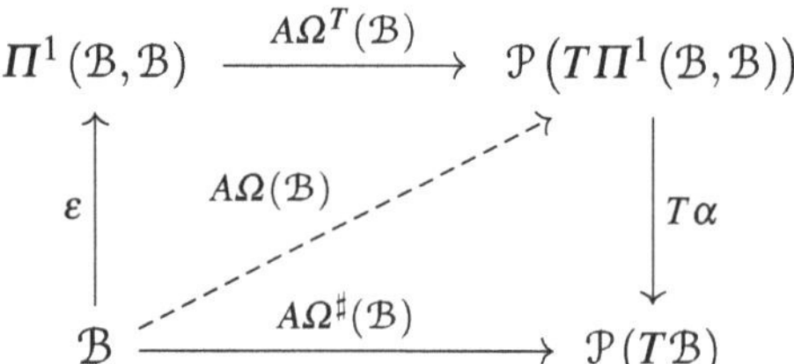

As we have proved in the previous section, the body-material foliation $\mathcal{F}$ divides the body into smoothly uniform components.

Let us now provide the intuition behind the definition of homogeneity of a non-uniform body. A non-uniform body will be *(locally) homogeneous* when each smoothly uniform material submanifold $\mathcal{F}(X)$ is (locally) homogeneous and all the uniform material submanifolds can be straightened at the same time.

Thus, we need to clarify what we understand by homogeneity of submanifolds of $\mathcal{B}$.

Definition 11 Let $\mathcal{B}$ be a simple body and $\mathcal{N}$ be a submanifold of $\mathcal{B}$. $\mathcal{N}$ is said to be *homogeneous* if, and only if, for all point $X \in \mathcal{N}$ there exists a local configuration ψ of $\mathcal{B}$ on an M subset $U \subseteq \mathcal{B}$, with $\mathcal{N} \subseteq U$, which satisfies that

$$j^1_{Y,Z}\left(\psi^{-1} \circ \tau_{\psi(Z)-\psi(Y)} \circ \psi\right)$$

is a material isomorphism for all $Y, Z \in U \cap \mathcal{N}$. We will say that $\mathcal{N}$ is *locally homogeneous* if there exists a covering of $\mathcal{N}$ by M subsets U_a of $\mathcal{B}$ such that $U_a \cap \mathcal{N}$ are homogeneous submanifolds of $\mathcal{B}$. $\mathcal{N}$ is said to be *(locally) inhomogeneous* if it is not (locally) homogeneous.

Notice that, the definitions of homogeneity and local homogeneity for smoothly uniform materials (Definition 10) are generalized by this one whether $\mathcal{N} = \mathcal{B}$ or $\mathcal{N}$ is just an M subset of $\mathcal{B}$.

Now, taking into account that $\mathcal{F} = \{\mathcal{F}(X)\}_{X \in \mathcal{B}}$ is a foliation, there is a kind of compatible atlas which is called a *foliated atlas*. In fact, $\{((X_a^I), U_a)\}_a$ is a foliated atlas of $\mathcal{B}$ associated with $\mathcal{F}$ whenever for each $X \in U_a \subseteq \mathcal{B}$ we have that $U_a := \{-\epsilon < X_a^1 < \epsilon, \ldots, -\epsilon < X_a^3 < \epsilon\}$ for some $\epsilon > 0$, such that the k-dimensional disk $\{X_a^{k+1} = \ldots = X_a^3 = 0\}$ coincides with the path-connected component of the intersection of $\mathcal{F}(X)$ with U_a which contains X, and each k-dimensional disk $\{X_a^{k+1} = c_{k+1}, \ldots, X_a^3 = c_3\}$, where $c_{k+1}, \ldots, c_3$ are constants, is wholly contained in some leaf of $\mathcal{F}$. Intuitively, this atlas straightens (locally) the partition $\mathcal{F}$ of $\mathcal{B}$.

The existence of these kinds of atlases and the maximality condition over the smoothly uniform material submanifolds $\mathcal{F}(X)$ induces us to give the following definition:

Definition 12 Let $\mathcal{B}$ be a simple body. $\mathcal{B}$ is said to be *locally homogeneous* if, and only if, for all point $X \in \mathcal{B}$ there exists a local configuration ψ of $\mathcal{B}$, with $X \in U$, which is a foliated chart and it satisfies that

$$j^1_{Y,Z}\left(\psi^{-1} \circ \tau_{\psi(Z)-\psi(Y)} \circ \psi\right)$$

is a material isomorphism for all $Z \in U \cap \mathcal{F}(Y)$. We will say that $\mathcal{B}$ is homogeneous if $U = \mathcal{B}$. The body $\mathcal{B}$ is said to be *(locally) inhomogeneous* if it is not (locally) homogeneous.

It is remarkable that, as we have said above, all the uniform leaves $\mathcal{F}(X)$ of a homogeneous body are homogeneous. Therefore, the definition of homogeneity for a smoothly uniform body coincides with Definition 10. Notice also that, the condition that all the leaves $\mathcal{F}(X)$ are homogeneous is not enough in order to have the homogeneity of the body $\mathcal{B}$ because there is also a condition of compatibility with the foliation structure of $\mathcal{F}$.

Let us recall a result given in [4] (see also [5] or [20]) which characterizes the homogeneity by using G-structures.

Denote by $F\mathcal{B}$ the frame bundle of $\mathcal{B}$. An element of $F\mathcal{B}$ is called a linear frame at a point $X \in \mathcal{B}$; it is a 1-jet of a local diffeomorphism $f : \mathbb{R}^3 \to \mathcal{B}$ at 0 with $f(0) = X$. Then, the structure group of $F\mathcal{B}$ is the group of 3×3-regular matrices in $\mathbb{R}$, $Gl(3, \mathbb{R})$.

A *G-structure over* $\mathcal{B}$, denoted by $\omega_G(\mathcal{B})$, is a reduced subbundle of $F\mathcal{B}$ with structure group G, *which is* a Lie subgroup of $Gl(3, \mathbb{R})$ (a good reference about frame bundles is [3]).
So, fix $\overline{g}_0$ be a frame at $Z \in \mathcal{B}$. Then, assuming that $\mathcal{B}$ is smoothly uniform, the set

$$\Omega(\mathcal{B})_Z \cdot \overline{g}_0 := \{\overline{g} \cdot \overline{g}_0 \ : \ \overline{g} \in \Omega(\mathcal{B})_Z\},$$

where $\cdot$ defines the composition of 1-jets, is a $\Omega(\mathcal{B})_Z^Z$-structure over $\mathcal{B}$.

Proposition 4 *Let be a frame $\overline{g}_0 \in F\mathcal{B}$. If $\mathcal{B}$ is homogeneous, then the G-structure given by $\Omega(\mathcal{B}) \cdot \overline{g}_0$ is integrable. Conversely, $\Omega(\mathcal{B}) \cdot \overline{g}_0$ is integrable implies that $\mathcal{B}$ is locally homogeneous.*

Thus, the next step will be to give a similar result for this generalized homogeneity. Because of the lack of uniformity we have to use groupoids instead of G-structures.

Let $\mathbb{S} := \{\mathbb{S}(x) \ : \ x \in \mathbb{R}^n\}$ be a canonical foliation of $\mathbb{R}^n$, i.e., for all $x = (x^1, \ldots, x^n) \in \mathbb{R}^n$ the leaf $\mathbb{S}(x)$ at x

$$\mathbb{S}(x) := \{\left(y^1, \ldots, y^p, x^{p+1}, \ldots, x^n\right) \ : \ y^i \in \mathbb{R}, \ i = 1, \ldots, p\},$$

for some $1 \leq p \leq n$.
Notice that for any foliation $\mathcal{G}$ on a manifold Q there exists a map

$$p_{\mathcal{G}} : Q \to \{0, \ldots, dim(Q)\},$$

such that for all $x \in Q$

$$p_{\mathcal{G}}(x) = dim(\mathcal{G}(x)).$$

$p_{\mathcal{G}}$ will be called *grade of* $\mathcal{G}$. $\mathcal{G}$ is a *regular foliation* if, and only if, the grade of $\mathcal{G}$ is constant.
It is important to remark that in the case of $\mathbb{S}$ the grade $p_{\mathbb{S}}$ characterizes the foliation $\mathbb{S}$. Thus, with abuse of notation, we could say that the map $p_{\mathbb{S}}$ is the foliation.

Let $\mathbb{S}$ be a canonical foliation of $\mathbb{R}^n$ with grade $p = p_{\mathbb{S}}$. Thus, as a generalization of the frame bundle of $\mathbb{R}^n$, we define the *p-graded frame groupoid* as the following subgroupoid of $\Pi^1(\mathbb{R}^n, \mathbb{R}^n)$,

$$\Pi^1_p\left(\mathbb{R}^n, \mathbb{R}^n\right) = \{j^1_{x,y}\psi \in \Pi^1\left(\mathbb{R}^n, \mathbb{R}^n\right) \ : \ y \in \mathbb{S}(x)\}.$$

Notice that the restriction of $\Pi^1_p(\mathbb{R}^n, \mathbb{R}^n)$ to any leaf $\mathbb{S}(x)$ is a transitive Lie subgroupoid of $\Pi^1(\mathbb{R}^n, \mathbb{R}^n)$ with all the isotropy groups isomorphic to $Gl(n, \mathbb{R})$. However, the groupoid $\Pi^1_p(\mathbb{R}^n, \mathbb{R}^n)$ is not necessarily a Lie subgroupoid of $\Pi^1(\mathbb{R}^n, \mathbb{R}^n)$. In fact, $\Pi^1_p(\mathbb{R}^n, \mathbb{R}^n)$ is a Lie subgroupoid of $\Pi^1(\mathbb{R}^n, \mathbb{R}^n)$ if, and only if, $\mathbb{S}$ is regular foliation.

A *standard flat G-reduction of grade p* is a subgroupoid $\Pi^1_{G,p}(\mathbb{R}^n, \mathbb{R}^n)$ of $\Pi^1_p(\mathbb{R}^n, \mathbb{R}^n)$ such that the restrictions $\Pi^1_{G,p}(\mathbb{S}(x), \mathbb{S}(x))$ to the leaves $\mathbb{S}(x)$ are transitive Lie subgroupoids of $\Pi^1(\mathbb{R}^n, \mathbb{R}^n)$ on the leaf $\mathbb{S}(x)$. It is remarkable that in this case all the isotropy groups of $\Pi^1_{G,p}(\mathbb{S}(x), \mathbb{S}(x))$ are conjugated.
Clearly, all the structures introduced in this section can be restricted to any M subset of $\mathbb{R}^n$.
Let $\psi : U \to \overline{U}$ be a (local) configuration on $U \subseteq \mathcal{B}$. Then, ψ induces a Lie groupoid isomorphism,

$$\begin{aligned} \Pi\psi : \Pi^1(U,U) &\to \Pi^1\left(\overline{U},\overline{U}\right) \\ j^1_{X,Y}\phi &\mapsto j^1_{\psi(X),\psi(Y)}\left(\psi\circ\phi\circ\psi^{-1}\right). \end{aligned}$$

Proposition 5 *Let $\mathcal{B}$ be a simple body. If $\mathcal{B}$ is homogeneous the material groupoid is isomorphic (via a global configuration) to a standard flat G-reduction. Conversely, if the material groupoid is isomorphic (via a local configuration) to a standard flat G-reduction, then $\mathcal{B}$ is locally homogeneous.*

Notice that, in the context of principal bundles, a G-structure is integrable if, and only if, there exists a local configuration which induces an isomorphism from the G-structure to a standard flat G-structure.

Finally, we will use the material distribution to give another characterization of homogeneity.
Let $\mathcal{B}$ be a homogeneous body with $\psi = \left(X^I\right)$ as a (local) homogeneous configuration. Then, by using the fact that ψ is a foliated chart, we have that the partial derivatives are tangent to $A\Omega(\mathcal{B})^\sharp$, i.e., for each $X \in U$

$$\frac{\partial}{\partial X^L}_{|X} \in A\Omega(\mathcal{B})^\sharp_X,$$

for all $1 \leq L \leq dim(\mathcal{F}(X)) = K$. Thus, there are local functions $\Lambda^J_{I,L}$ such that for each $L \leq K$ the (local) left-invariant vector field on $\Pi^1(\mathcal{B}, \mathcal{B})$ given by

$$\frac{\partial}{\partial X^L} + \Lambda^J_{I,L}\frac{\partial}{\partial Y^J_I}$$

is tangent to $A\Omega(\mathcal{B})^T$, where $\left(X^I, Y^J, Y^J_I\right)$ are the induced coordinates of $\left(X^I\right)$ in $\Pi^1(\mathcal{B}, \mathcal{B})$. Equivalently, the local functions $\Lambda^J_{I,L}$ satisfy that

$$\frac{\partial W}{\partial X^L} + \Lambda^J_{I,L}\frac{\partial W}{\partial Y^J_I} = 0,$$

for all $1 \leq L \leq K$. Next, since for each two points $X, Y \in U$ the 1-jet given by $j^1_{X,Y}\left(\psi^{-1}\circ\tau_{\psi(Y)-\psi(X)}\circ\psi\right)$ is a material isomorphism, we can choose $\Lambda^J_{I,L} = 0$.

Proposition 6 *Let $\mathcal{B}$ be a simple body. $\mathcal{B}$ is homogeneous if, and only if, for each $X \in \mathcal{B}$ there exists a local chart $\left(X^I\right)$ on $\mathcal{B}$ at X such that,*

$$\frac{\partial W}{\partial X^L} = 0, \tag{29}$$

for all $L \leq dim\,(\mathcal{F}(X))$.

Notice that Eq. (29) implies that the partial derivatives of the coordinates $\left(X^I\right)$ up to $dim\,(\mathcal{F}(X))$ are tangent to the material distribution and, therefore, the coordinates are foliated. So, Eq. (29) gives us an apparently more straightforward way to express this general homogeneity.

7 Example

We will devote this section to study the notion of homogeneity given in Definition 12 for non-uniform bodies. In particular, we will present an example of a homogeneous non-uniform material body.

Let $\mathcal{B}$ be a simple material body for which there exists a reference configuration ψ_0 from $\mathcal{B}$ to the 3-dimensional M cube $\mathcal{B}_0 = (-1, 1)^3$ in $\mathbb{R}^3$ that, in terms of the Cauchy stress $\mathbf{t}$, induces the following mechanical response:

$$\begin{aligned} \mathbf{t} : \Pi^1\left(\mathcal{B}_0, \mathcal{B}_0\right) &\rightarrow \mathfrak{gl}\left(3, \mathbb{R}\right) \\ j^1_{X,Y}\phi &\mapsto f\left(X^1\right)\left(F \cdot F^T - I\right), \end{aligned}$$

such that

$$f\left(X^1\right) = \begin{cases} 1 & if\ X^1 \leq 0 \\ 1 + e^{-\frac{1}{X^1}} & if\ X^1 > 0 \end{cases}$$

where $\mathfrak{gl}\left(3, \mathbb{R}\right)$ is the algebra of matrices, F is the Jacobian matrix of ϕ at X with respect to the canonical basis of $\mathbb{R}^3$ and I is the identity matrix. Here, the (global) canonical coordinates of $\mathbb{R}^3$ are denoted by $\left(X^I\right)$ and $X = \left(X^1, X^2, X^3\right)$ with respect to these coordinates. In these coordinates, we allow the summation convention to be in force regardless of the placement of the indices. We also identify the coordinate system in the spatial configuration with that of the reference configuration.

Notice that f is constant up to 0 and strictly increasing thereafter. For this reason, one can immediately conclude that $\mathcal{B}_0$ is not uniform. In fact, there are no material isomorphisms joining any two points $\left(X^1, X^2, X^3\right)$ and $\left(Y^1, Y^2, Y^3\right)$ such that

$$f\left(X^1\right) \neq f\left(Y^1\right).$$

So, let us study the derivatives of $\mathbf{t}$ in order to find the grades of uniformity of the points of the body $\mathcal{B}_0$.[3] We obtain

$$\frac{\partial t^{ij}}{\partial F^k_M} = f\left(X^1\right)\left[\delta^i_k F^j_M + \delta^j_k F^i_M\right]$$

$$\frac{\partial t^{ij}}{\partial X^1} = \frac{\partial f}{\partial X^1}\left(F^i_K F^j_K - \delta^{ij}\right)$$

$$\frac{\partial t^{ij}}{\partial X^I} = 0, \quad \text{for } I \geq 2.$$

We are looking for left-invariant (local) vector fields Θ on $\Pi^1(\mathcal{B}_0, \mathcal{B}_0)$ satisfying

$$\Theta\left(t^{ij}\right) = 0. \tag{30}$$

Let $\left(X^I, Y^I, F^i_J\right)$ be the induced coordinates of $\left(X^I\right)$ on $\Pi^1(\mathcal{B}_0, \mathcal{B}_0)$. A left-invariant vector field Θ can be expressed as follows:

$$\Theta\left(X^I, Y^J, F^j_I\right) = \left(\left(X^I, Y^J, F^j_I\right), \delta X^I, 0, F^j_L \delta P^L_I\right).$$

Hence, Θ satisfies Eq. (30) if, and only if,

$$\Theta\left(t^{ij}\right) = f\left(X^1\right)\left(F^i_L F^j_M + F^j_L F^i_M\right)\delta P^L_M + \delta X^1 \frac{\partial f}{\partial X^1}\left(F^i_K F^j_K - \delta^{ij}\right) = 0. \tag{31}$$

Let us focus first on the open set given by the restriction $X^1 < 0$. Then, Eq. (31), turns into the following:

$$\left(F^i_L F^j_M + F^j_L F^i_M\right)\delta P^L_M = 0 \quad \forall i, j = 1, 2, 3 \tag{32}$$

for every Jacobian matrix $F = \left(F^j_L\right)$ of a local diffeomorphism ϕ on $\mathcal{B}_0$. Since the bracketed expression is symmetric in L and M for every i and j, it follows that δP is a skew-symmetric matrix. We remark that this condition does not impose any restriction on the components δX^I of the admissible vector fields on the base vectors $\partial/\partial X^I$. In other words, any family of local functions $\{\delta X^I, \delta P^L_M\}$ on the open restriction $\{X^1 < 0\}$ of the body $\mathcal{B}_0$, such that $\delta P = \left(\delta P^L_M\right)$ is a skew-symmetric matrix, generates a vector field

$$\Theta\left(X^I, Y^J, F^j_I\right) = \left(\left(X^I, Y^J, F^j_I\right), \delta X^I, 0, F^j_L \delta P^L_I\right),$$

[3]The grades of uniformity for this example were first studied in [11], where the components of the Cauchy stress were identified, less precisely, with those of the second Piola–Kirchhoff stress.

which satisfies Eq. (30). It follows that the body characteristic distribution of the sub-body $(-1, 1)^3 \cap \{X^1 < 0\}$ is a regular distribution of dimension 3. Therefore, this sub-body is uniform, as one would expect from the constancy of the function f thereat. Note also that the part lost when projecting the characteristic distribution onto the body, namely the skew-symmetric matrices δP, consists precisely of the Lie algebra of the orthogonal group. This is nothing but the manifestation of the fact that our sub-body is isotropic.

Next we will study the open subset of $\mathcal{B}_0$ such that $X^1 > 0$. For this case, Eq. (30) is satisfied if, and only if,

$$f\left(X^1\right)\left(F^i_L F^j_M + F^j_L F^i_M\right)\delta P^L_M + \delta X^1 \frac{\partial f}{\partial X^1}\left(F^i_K F^j_K\right) = \delta X^1 \frac{\partial f}{\partial X^1}\delta^{ij}. \tag{33}$$

The function on the left-hand side of this equation is homogeneous of degree 2 with respect to the matrix coordinate F, but the function on the right-hand side does not depend on F. Consequently, Eq. (33) can be identically satisfied if, and only if,

$$\delta X^1 \frac{\partial f}{\partial X^1} = 0. \tag{34}$$

Notice that, the map f is strictly monotonic (and, hence, a submersion) at the open subset given by the condition $X^1 > 0$. Then, for any point X in this open subset we have that

$$T_X f^{-1}\left(f\left(X^1\right)\right) = Ker\left(T_X f\right),$$

i.e., the tangent space of the level set $f^{-1}\left(f\left(X^1\right)\right)$, which is the plane $Y^1 = X^1$, consists of vectors $V = \left(V^1, V^2, V^3\right)$ such that

$$V^1 \left.\frac{\partial f}{\partial X^1}\right|_X = 0.$$

In this way, a vector field Θ satisfies Eq. (30) if, and only if, δP is skew-symmetric and the projection $T\alpha \circ \Theta \circ \epsilon$ is tangent to the vertical planes $Y^1 = C$. Therefore, for each point $X = \left(X^1, X^2, X^3\right)$ with $X^1 > 0$, the uniform leaf is given by the plane $Y^1 = X^1$. As a consequence, the uniform leaf at the points satisfying $X^1 = 0$ is, again, the plane $Y^1 = 0$.

We conclude that the body is uniform of grade 3 for all points $X = \left(X^1, X^2, X^3\right) \in \mathcal{B}_0$ such that $X^1 < 0$, and it is uniform of grade 2 otherwise. It should be remarked that the plane $X^1 = 0$ is uniform of grade 2, even though its points are materially isomorphic to those in the subset with $X^1 < 0$.

Finally, the material body $\mathcal{B}_0$ is homogeneous. In fact, let us consider the canonical (global) coordinates $\left(X^i\right)$ of $\mathbb{R}^3$ restricted to $\mathcal{B}_0$. Then,

$$\frac{\partial t^{ij}}{\partial X^2} = \frac{\partial t^{ij}}{\partial X^3} = 0,$$

i.e., by using Proposition 6, $\mathcal{B}_0$ is homogeneous and the coordinates $\left(X^i\right)$ are homogeneous coordinates.

Acknowledgements This work has been partially supported by MINECO Grants MTM2016-76-072-P and the ICMAT Severo Ochoa projects SEV-2011-0087 and SEV-2015-0554. V.M. Jiménez wishes to thank MINECO for a FPI-PhD Position.

References

1. F. Bloom. *Modern differential geometric techniques in the theory of continuous distributions of dislocations*, volume 733 of *Lecture Notes in Mathematics*. Springer, Berlin, 1979.
2. C. M. Cámpos, M. Epstein, and M. de León. Functionally graded media. *International Journal of Geometric Methods in Modern Physics*, 05(03):431–455, 2008.
3. L. A. Cordero, C. T. Dodson, and M. de León. *Differential Geometry of Frame Bundles*. Mathematics and Its Applications. Springer Netherlands, Dordrecht, 1988.
4. M. Elżanowski, M. Epstein, and J. Śniatycki. G-structures and material homogeneity. *J. Elasticity*, 23(2–3):167–180, 1990.
5. M. Elżanowski and S. Prishepionok. Locally homogeneous configurations of uniform elastic bodies. *Rep. Math. Phys.*, 31(3):329–340, 1992.
6. M. Epstein. *The Geometrical Language of Continuum Mechanics*. Cambridge University Press, Cambridge, 2010.
7. M. Epstein. Laminated uniformity and homogeneity. *Mechanics Research Communications*, 2017.
8. M. Epstein and M. de León. Homogeneity without uniformity: towards a mathematical theory of functionally graded materials. *International Journal of Solids and Structures*, 37(51):7577 – 7591, 2000.
9. M. Epstein and M. de León. Unified geometric formulation of material uniformity and evolution. *Math. Mech. Complex Syst.*, 4(1):17–29, 2016.
10. M. Epstein, V. M. Jiménez, and M. de León. Material geometry. *Journal of Elasticity*, pages 1–24, Oct 2018.
11. V. M. Jiménez, M. de León, and M. Epstein. Material distributions. *Mathematics and Mechanics of Solids*, 0(0):1081286517736922, 0.
12. V. M. Jiménez, M. de León, and M. Epstein. Lie groupoids and algebroids applied to the study of uniformity and homogeneity of material bodies. *Journal of Geometric Mechanics*, 11(3):301–324, 2019.
13. V. M. Jiménez, M. de León, and M. Epstein. Characteristic distribution: An application to material bodies. *Journal of Geometry and Physics*, 127:19–31, 2018.
14. K. C. H. Mackenzie. *General theory of Lie groupoids and Lie algebroids*, volume 213 of *London Mathematical Society Lecture Note Series*. Cambridge University Press, Cambridge, 2005.
15. W. Noll. Materially uniform simple bodies with inhomogeneities. *Arch. Rational Mech. Anal.*, 27:1–32, 1967/1968.
16. R. Segev. Forces and the existence of stresses in invariant continuum mechanics. *Journal of Mathematical Physics*, 27:163–170, 1986.
17. P. Stefan. Accessible sets, orbits, and foliations with singularities. *Proc. London Math. Soc. (3)*, 29:699–713, 1974.

18. H. J. Sussmann. Orbits of families of vector fields and integrability of distributions. *Trans. Amer. Math. Soc.*, 180:171–188, 1973.
19. J. N. Valdés, Á. F. T. Villalón, and J. A. V. Alarcón. *Elementos de la teoría de grupoides y algebroides*. Universidad de Cádiz, Servicio de Publicaciones, Cádiz, 2006.
20. C. C. Wang. On the geometric structures of simple bodies. A mathematical foundation for the theory of continuous distributions of dislocations. *Arch. Rational Mech. Anal.*, 27:33–94, 1967/1968.
21. C. C. Wang and C. Truesdell. *Introduction to rational elasticity*. Noordhoff International Publishing, Leyden, 1973.
22. A. Weinstein. Groupoids: unifying internal and external symmetry. A tour through some examples. In *Groupoids in analysis, geometry, and physics (Boulder, CO, 1999)*, volume 282 of *Contemp. Math.*, pages 1–19. Amer. Math. Soc., Providence, RI, 2001.

Correction to: Geometric Continuum Mechanics

Reuven Segev and Marcelo Epstein

Correction to:
R. Segev, M. Epstein (eds.),
Geometric Continuum Mechanics,
Advances in Mechanics and Mathematics 43,
https://doi.org/10.1007/978-3-030-42683-5

This book has been inadvertently published with the volume number 42 which has now been modified as 43.

The updated online version of this book can be found at
https://doi.org/10.1007/978-3-030-42683-5

R. Segev, M. Epstein (eds.), *Geometric Continuum Mechanics*, Advances in Mechanics and Mathematics 43, https://doi.org/10.1007/978-3-030-42683-5_10

GPSR Compliance
The European Union's (EU) General Product Safety Regulation (GPSR) is a set of rules that requires consumer products to be safe and our obligations to ensure this.

If you have any concerns about our products, you can contact us on

ProductSafety@springernature.com

In case Publisher is established outside the EU, the EU authorized representative is:

Springer Nature Customer Service Center GmbH
Europaplatz 3
69115 Heidelberg, Germany

www.ingramcontent.com/pod-product-compliance
Ingram Content Group UK Ltd.
Pitfield, Milton Keynes, MK11 3LW, UK
UKHW020140300726
14059UKWH00007B/27